SEA-DOO/BOMBARDIER

Personal Watercraft
1992-97 REPAIR MANUAL
ALL MODELS

SELOC®

I0803581

Managing Partners Dean F. Morgantini
Barry L. Beck

Executive Editor Kevin M. G. Maher, A.S.E.

Production Managers Melinda Possinger
Ronald Webb

Author Clarence Coles

Manufactured in USA

104 Willowbrook Lane
West Chester, PA 19382
ISBN 13: 978-0-89330-043-2
ISBN 10: 0-89330-043-8
6789012345 9876543210

www.selocmarine.com
1-866-SELOC55

FOREWORD

This is a comprehensive tune-up and repair manual for Bombardier Sea-Do Personal Watercraft produced by Bombardier, Lashine, Quebec, Canada, from 1992 through 1997.

The book has been designed and written to cover stock, factory "out the door" engines and jet pumps. Modifications for higher than manufacturer's rated performance are so extensive and varied, no attempt has been made to include them in this volume. (Actually, for such coverage, a separate comprehensive book would be required. In such cases the publisher's recommendation is to follow the aftermarket instructions with the particular product or service.)

All Seloc manuals, including this title, are directed toward the professional mechanic, the do-it-yourselfer, and the student developing her/his mechanical skills.

ACKNOWLEDGEMENTS

Seloc Publications wish to express their appreciation for the help and suggestions of Bombardier/Sea-Doo during the production of this manual. A special recognition is extended to Sunwest Sports of Corona, California, for providing us with certain units to complete our model coverage. Special recognition is extended to Dave Madsen for his time and expertise during the photographic sessions of the disassembling and assembling work on the Sea-Doo units included in this manual. His expertise and skill certainly earn him the title of Sea-Doo PWC mechanic -- **FIRST CLASS**.

Clarence W. Coles
Seloc Publications

TABLE OF CONTENTS

1
DESCRIPTION & OPERATION

1-1 BRIEF HISTORY

The jet drive system for propelling a craft through the water arrived on the scene in the mid 1960's with the jet drive boat. In those early days, the jet drive system was mated only with high performance powerplants -- engines in the 454 cu. in. class and larger. For this reason, during the "gas crunch" in the 1970's the jet drives were labeled as inefficient and as "gas hogs".

In addition to these two negative terms, they earned the reputation as "bad boy" boats due to their noisy "straight" exhaust, high rpm operation, and their almost unbelievable maneuverability. These combined factors did little to enhance their image and certainly restricted their popularity.

With new and improved technology, personal watercraft arrived on the scene about the mid 1970's. Personal watercraft, as we know them today, were developed using the same principles as the jet boats, and originally powered with a single cylinder two-stroke engine.

In order to meet the demand for more speed and the ability to carry more than just

From harbors, bays, sounds, and estuaries at sea level to high mountain lakes, personal watercraft are giving their owners hours of fun on the water.

one person, watercraft manufacturers were quick to respond. Today, most craft are powered with a two or three-cylinder two-stroke powerplant, coupled to a single stage pump.

Aftermarket shops and manufacturers have come into existence all across the United States and Canada. Their output of specialty products and services permit the owner to gain more speed and win over the competition at racing events wherever enough water is available.

As mentioned in the "Foreword", this book has been designed and written to cover stock factory engines and jet drives. Modifications for higher than manufacturer's rated performance are so extensive and varied, no attempt has been made to include them in this volume. In such cases the publisher's recommendation is to follow the after market instructions with the particular product or service.

Series Covered

The following Bombardier Sea-Doo series produced from 1992 thru 1997 are covered in this manual. Earlier models are covered in Seloc's Personal Watercraft Volume II -- Bombardier Sea-Doo thru 1991.

Model	Approx. Yr. of Production
SP	1992 - 1997
XP	1992 - 1997
GTS	1992 - 1997
GTX	1992 - 1997
SPX	1993 - 1997
SPI	1993 - 1997
GSX	1996 - 1997
GTI	1996 - 1997
HX	1996 - 1997
GS	1997
GSI	1997

Other Seloc Personal Watercraft Manuals

This manual is only one in a series of Personal Watercraft manuals produced by Seloc Publications, as follows:

- Volume I - Kawasaki -- early days thru 1991
- Volume II - Bombardier SeaDoo -- early days
- Volume IIA - Bombardier SeaDoo -- 1992 thru 1997
- Volume III - Yamaha - early days thru 1991
- Volume IIIA - Yamaha -- 1992 thru 1997
- Volume IA - Kawasaki -- 1992 thru 1998
- Volume IV - Polaris -- 1992 thru 1998

1-2 PRINCIPLES OF OPERATION

One of the first lessons to be learned in any elementary physics class is Newton's basic law: "For every force, there is an opposite and equal force". This statement is the basic principle of the jet pump. Water is "sucked" and "scooped" in from under the craft by a powerful pump rotating at incredible speed and then discharged, "blown" out, sternward in the opposite direction. In this manner the watercraft is propelled forward.

The personal watercraft covered in this manual are all equipped with a twin-cylinder water cooled two-stroke engine, matched with a single stage (one impeller) jet pump.

On certain models, a reverse gate is swung down over the pump outlet nozzle forcing the exhausted water back in a forward direction thus moving the craft sternward.

Personal watercraft jet pumps may be classified as "axial flow" or "mixed flow". All Bombardier Sea-Doo Personal Watercraft are equipped with an axial flow jet pump.

Axial Flow

Water in an axial pump moves in a single axis, as depicted in the adjacent illustration -- thus the word "axial" is used. In simple layman's terms -- water is ingested and discharged parallel to the axis of impeller rotation, as shown.

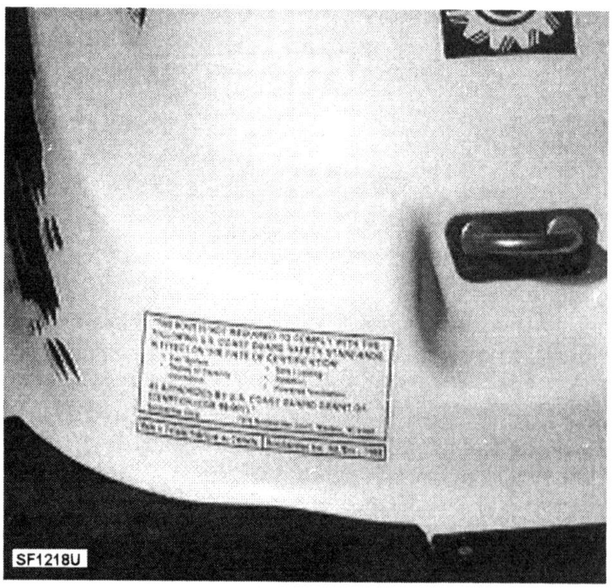

Several decals may be affixed to the stern of the hull to indicate compliance with, or exceptions granted to normal coast guard regulations.

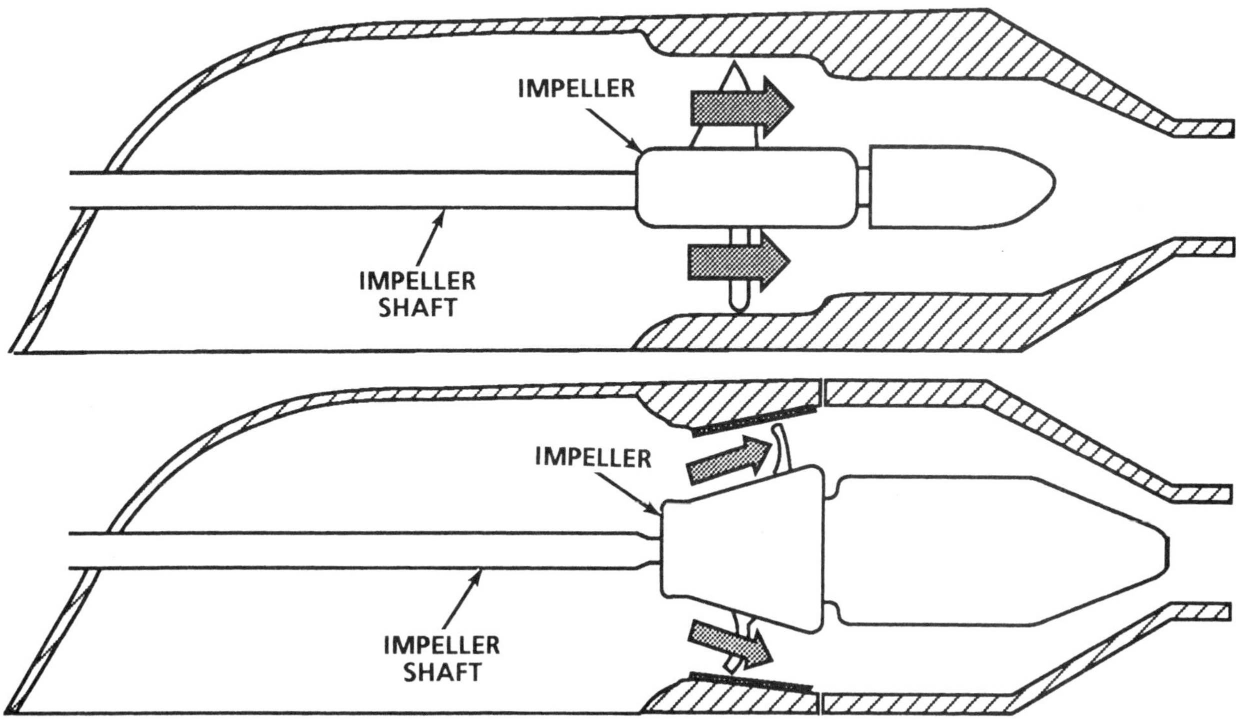

Line drawing to depict water flow through an axial flow pump (top), and a mixed flow pump (bottom). Notice on the mixed flow pump, how the water is directed upward at an able to the axis of impeller rotation. Notice on the axial flow pump how the water passes through parallel to the axis of impeller rotation.

EDUCATIONAL INFORMATION

Even though Bombardier Sea-Doo personal watercraft utilize only an axial flow pump, the following paragraphs are included in this chapter to acquaint the reader with a mixed flow pump -- how it differs from the axial flow pump and why.

Mixed Flow

Many experts consider the mixed flow system to the most efficient for personal watercraft application. The water flow is discharged at an inclined angle to the axis of impeller rotation, as indicated by the arrows in the accompanying illustration. The impeller blades are inclined in such a manner to cause the water to be thrown off at a 45 degree angle to the impeller shaft within the pump case. From the pump case, the water is then forced out through the outlet nozzle and steering nozzle. The vanes in the pump case help straighten the water flow through the nozzle.

The inside of the pump case is inclined at an angle to the water flow. Therefore, this design pump incorporates shim material to space the impeller blades at a specified distance from the pump case. Mixed flow pump efficiency decreases by 30% as blade-to-pump case clearance increases by 0.04" (1mm).

Axial or Mixed?

Axial and mixed flow pumps are intended for different applications. However, one is not necessarily better than the other. The manufacture considers the typical application or operator usage in determining the pump design. The axial pump can be operated at higher engine rpm and is more suitable for racing. A mixed flow pump adds thrust at low engine rpm due to the pump design. Therefore, the mixed flow pump performs more efficiently at low engine rpm. This type pump is more suitable for a craft equipped to carry extra weight.

A jet pump seldom requires service except for the following:

Impeller-to-pump case clearance becomes excessive -- mixed flow pump only.

Impeller is damaged.

The oil seal around the impeller shaft fails -- allowing water to enter the hull.

1-3 ENGINE AND JET DRIVE

The basic principle and arrangement of engine and pump are almost identical for all manufacturers.

The engine crankshaft is coupled either directly to the pump shaft through a coupler containing a rubber "shock absorber" or through a short driveshaft without any gear reduction. An impeller mounted on the pump shaft draws water into an opening in the hull through a suction and intake casting. Once the craft has attained forward motion the intake also serves as a "scoop" adding to the volume of water moved through the pump.

Volume and Velocity

The outlet nozzle is slightly funnel shaped. This design coupled with the capacity of the pump (impeller), causes the volume of water entering the pump to exit the nozzle with increased velocity. This principle is similar to air passing through the venturi of a carburetor.

The amount of water ejected sternward from the nozzle at high velocity is the force propelling the craft through the water. This force can actually be measured and calculated in foot pounds or Newton meters. The greater the velocity of water mass moving through the nozzle, the greater the thrust to move the craft.

Cooling System

A fitting on the jet pump, aft of the impeller, siphons off cooling water. This water is delivered to the exhaust manifold -- the hottest part of the engine -- and the tuned pipe. The water is then channeled around the cylinder walls, cylinder head, and finally discharged from the exhaust outlet. This mixture of water and exhaust gases has the effect of cooling and somewhat quieting the exhaust emmission.

On all models, a bypass hose connects the exhaust pipe water jacket to an outlet on the aft end of the hull. When water is discharged as a "tattle-tale" stream, the operator is assured cooling water is circulating through the engine properly.

Water passes through the exhaust expansion chamber first. A small portion of this water is injected directly into the exhaust gases, quieting the exhaust and increasing performance. The remaining water is directed into the exhaust manifold, then enters the crankcase, is channeled around the cylinder walls, cylinder head then directed overboard via the exhaust pipe and small bleed outlets.

1-4 IMPELLERS

Impellers provide thrust using a combination of water flow and water pressure in a closed environment.

The blades of an impeller overlap, thus water is trapped and forced through the impeller into the pump case and nozzle. The water moving past the impeller resists cavitation because it is under pressure. Blades in the pump case are angled exactly opposite to those of the impeller. These blades redirect the water to establish a concentrated flow through the nozzle.

The manufacturer recommends a radius of about 1/64" (0.3-0.5mm) for the leading edge of the impeller blade. A sharper radius will result in cavitation and a greater radius will reduce pump efficiency.

Cavitation Burns

Cavitation burns are the worst enemy of an impeller. These burns literally "eat away" material from the impeller, leaving holes and weakening the impeller structure. In extreme cases entire blades have been known to "depart" from the impeller

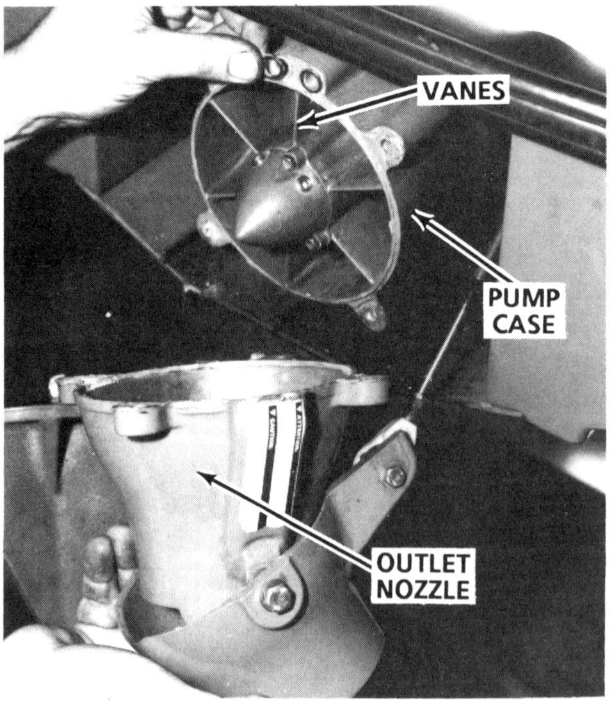

The vanes in the pump case straighten water flow and the conical shape of the outlet nozzle increases flow velocity.

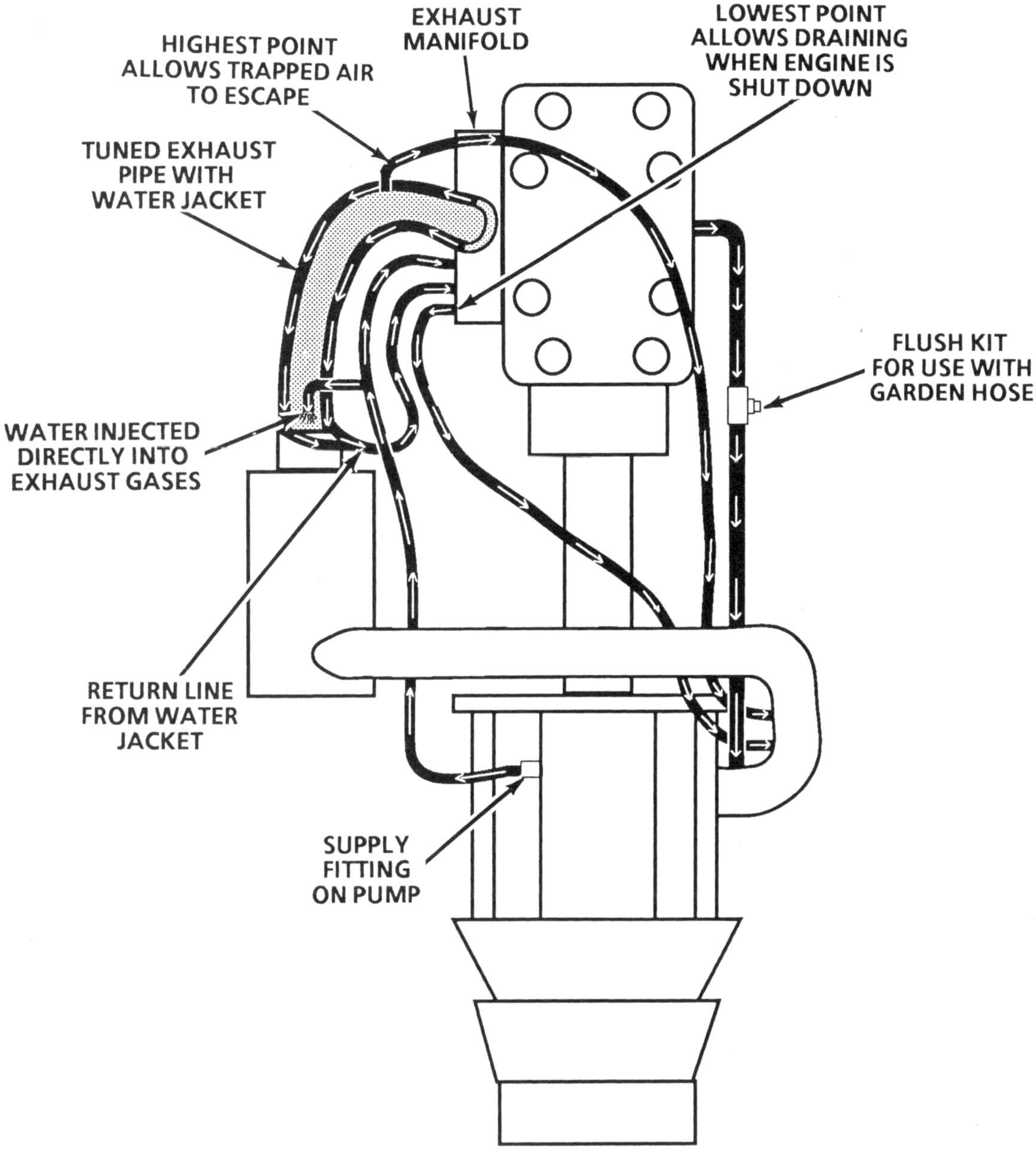

Simple line drawing to illustrate typical water flow from the pump, around the tuned exhaust pipe, through the engine and back to the pump. Other cooling circuits are also shown.

hub because the cavitation burns at the base of the blade were so severe.

Cavitation burns are the result of imperfections (damage) on the impeller blades or air mixed with the water flow.

Cavitation burns may be caused by:

- Wave jumping -- air is sucked into the pump as the craft leaves the surface of the water.
- Bad seal between the pump case and outlet nozzle -- allowing air to enter.
- Leading edge of the impeller becoming damaged.

To explain exactly what causes cavitation burns, first, let us examine the last cause. If the impeller leading edge is damaged and "mushrooms" over the blade, a low pressure area will form under the mushroomed lip.

At atmospheric pressure, water boils at 212 degrees F. Anytime there is an imperfection on the surface of an impeller blade and water passes over that imperfection, the water pressure is lowered. If the pressure is lowered, water will boil at a much lower temperature. Therefore, air bubbles will form in the water boiling under the mushroomed lip.

The impeller of a jet pump is a precisely machined and dynamically balanced aluminum spiral. Significant damage to a blade will affect the balance.

The bubbles will creep down the surface of the blade and accumulate at the impeller hub. A high pressure area is formed at the base of the blade. Here, the air bubbles will collapse and reform back into water with a release of energy. This energy is absorbed by the impeller and results in material being eaten away.

The impeller should be dressed to a slightly rounded edge. It is impossible to achieve a perfectly straight sharp edge, because one side will always be convex and the other side concave. The concave side will form an area of low pressure and encourage cavitation, as described above.

The other two causes of cavitation -- defective seals and wave jumping -- occur when air is sucked in with the water flow. The same principles apply, as just described, in the previous paragraphs. Air bubbles will creep down the surface of the blade and accumulate at the impeller hub. A high pressure area is formed at the base of the blade. Here, the air bubbles will collapse and reform back into water with a release of energy. This energy is absorbed by the impeller and results in material being "eaten away".

Cooling Water and Bilge Hoses

On most pumps, three hoses are attached to the pump case and outlet nozzle. One hose channels some of the water flowing through the pump to the exhaust manifold -- the hottest part of the engine -- to cool the block during engine operation. The other two hoses siphon water out of the bottom of the hull. These hoses are referred to as the bilge system.

The cooling hose has water flowing **FROM** the pump. The bilge hoses carry water **TO** the pump. All three hoses attach to the pump in a similar manner.

What determines the direction of water flow?

The answer to this question is in a simple explanation of high and low pressure areas along the inside surface of the pump.

Close inspection of the area inside the pump at the cooling water hose fitting reveals a smooth rounded shoulder with no obstruction or obstacle to impede the flow of water. A high pressure area develops here and draws the water down the hose attached to the fitting.

Further inspection of the area inside the housing at the bilge hose fitting reveals a small protrusion around the opening. This protrusion causes disturbance to the water flow and creates an area of low pressure. This low pressure area will have the effect of emptying air and water from the hose aft with the impeller water flow -- similar to the action of a vacuum cleaner, as depicted in the accompanying illustration. If the other end of the hose is submerged in water inside the hull, the wa-

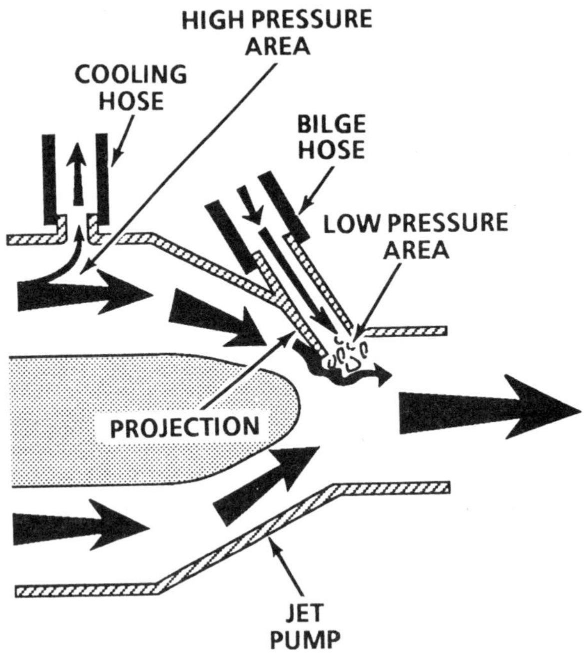

The direction of water flow through the cooling hose and the bilge hose is opposite in this drawing of a "typical" jet pump. The cooling hose routes water from the pump to the engine. Water is "vacuumed" from the bilge through the bilge hose to the pump where it is forced out through the nozzle.

ter will be "vacuumed" out. In this manner, the bilge system keeps the bilge relatively "dry".

Many an owner -- with good intentions -- thinking his action would smooth water flow and increase pump performance -- has filed the protrusion described in the previous paragraph smooth with the surrounding area of the pump. During operation of the watercraft, the bilge system worked in **REVERSE.** Water was pumped from the impeller housing into the bilge area, quickly filling the bilge and possibly sinking the craft before the rider had time to realize what was happening.

Bilge Breather Fitting

A bilge breather fitting is installed at the highest point of the bilge line. This fitting has a small breather hole to prevent siphoning water from the pump into the bilge. If the engine was shut down with the bilge hose filled with water and the jet pump fitting was positioned higher than the end of the bilge line, water would flow back and fill the bilge area. The small breather hole in the fitting prevents this back-siphoning.

1-5 REPLACEABLE PARTS

Any time the engine or jet pump is disassembled, all manufacturers and the authors heartily recommend all parts included in an overhaul kit be installed. An engine kit usually includes all gaskets, seals, O-rings, and other parts required to rebuild the engine. Pistons, rings, bearings, crankshafts and even blocks should be inspected to determine if they are fit for further service. The pump kits usually include all gaskets, seals, O-rings, and other parts required to restore full efficiency and "like new" condition to the jet pump unit.

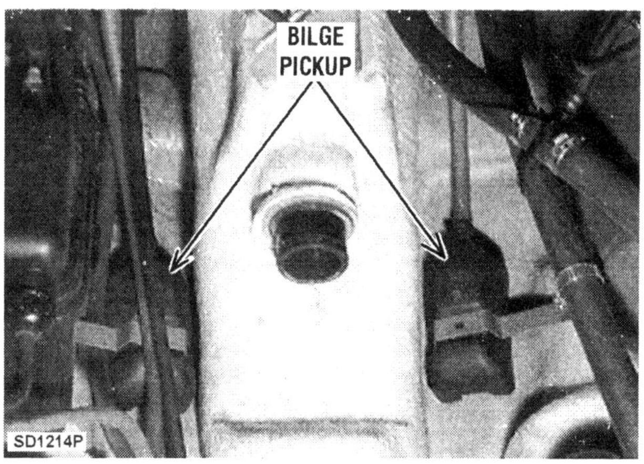

Two bilge pickups actually vacuum the water out of the bilge and discharge overboard through the jet pump. The engine has been removed for photographic clarity.

1-6 DEBRIS REMOVAL AND ENGINE OVERHEATING

As may be expected, a clogged impeller will slow the craft. The volume of water passing through the nozzle and water velocity will be reduced. A clogged impeller may cause restricted cooling water through the engine.

Many times a cooling water supply hose, if not properly secured with a hose clamp, may "blow off". The engine loses its supply of cooling water and internal engine damage quickly occurs.

All Bombardier Sea-Doos are equipped with an overheat warning horn. If the horn sounds, **SHUT DOWN** the engine immediately and remove the restriction or check the cooling hose connection.

Caution must be exercised when moving the craft off a beach to avoid sucking in mud, sand, pebbles, or rocks through the intake. Such ingestion could cause damage to the impeller, the vanes in the impeller duct, or clog engine cooling passages.

The engine cooling passages must be flushed with clear water on a regular basis. This is accomplished by simply connecting a garden hose to the cooling water hose,

1-7 REVERSE CAPABILITY

Certain Models of Bombardier Sea-Doo Watercraft are equipped with a reverse "gate". This gate is installed in such a manner to swing down over the steering nozzle and forces the water from the jet pump nozzle to be directed forward -- thus moving the craft sternward.

With the gate in the fully raised position, all water is directed to the rear of the craft and the craft moves forward.

If the gate is lowered partially, some of the water is directed forward and some is allowed to move sternward -- the craft is "static" in the water and the gate could be considered in a "neutral" position.

When the gate is allowed to move to the fully lowered position, almost all of the water is directed forward and the craft moves sternward -- the gate could be considered in a "reverse" position.

Adjustment of the reverse cable is covered in Chapter 11, Control Adjustments.

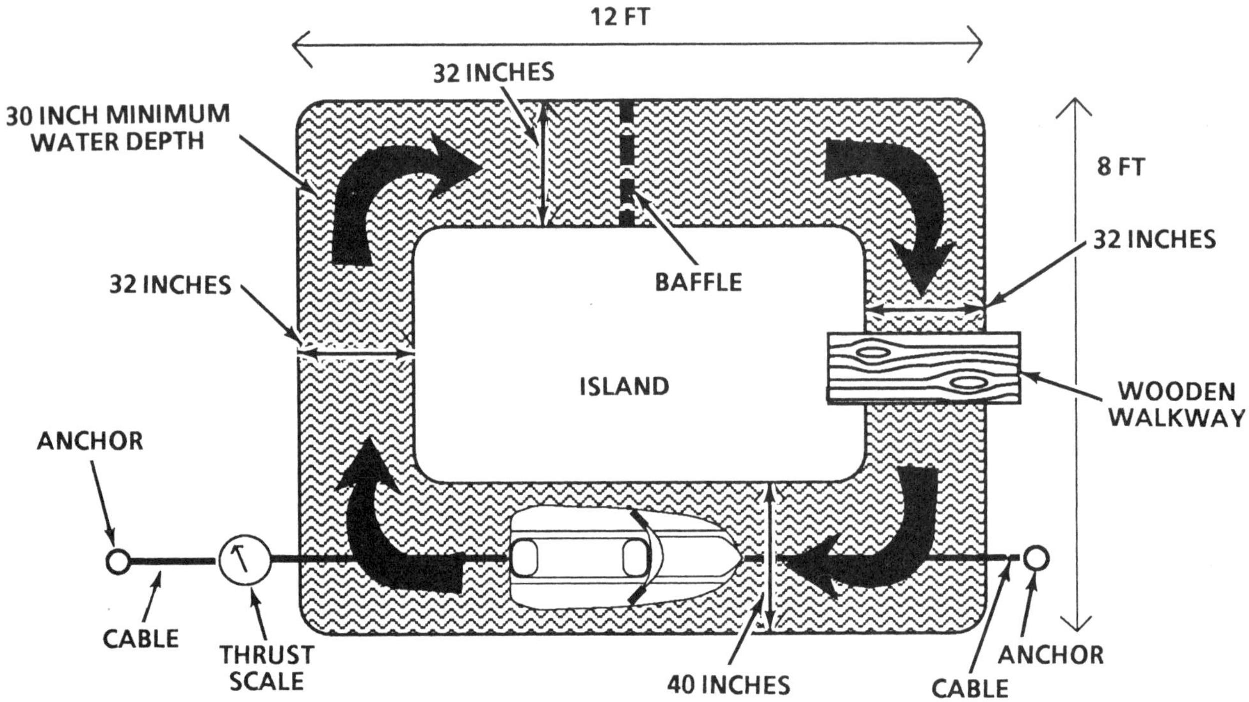

Layout of a typical test tank for testing personal watercraft. Such a tank is most valuable in verifying the completed work following service on the engine or the pump -- especially when making a timing adjustment.

1-8 TEST TANKS

Many times the performance of tune-up tasks, or adjustments are called for while the craft is in the water with the engine running. With an outboard or stern drive, the mechanic always has the boat between him/her and the wet stuff. On a personal watercraft, this is impossible without getting into the water, unless a test tank is used.

The advantages of using a test tank might be listed as:

a- Simulate actual water conditions. The craft may be operated without an exhaust extraction pipe and without using an auxiliary cooling water supply.

b- Test the engine at high rpm, placing a load on the impeller.

c- Reveal water leaks in the engine compartment and hull.

d- Uncover exhaust leaks.

e- Identify fuel leaks.

f- Detect engine noises.

g- Perform carburetor adjustments.

h- Check for bad impellers -- slippage.

i- Spot problems which might occur only after running for say 30 minutes. Therefore, prevent being stranded in the middle of a lake or far from shore if a serious malfunction should occur.

The only area a test tank cannot help diagnose is a craft handling problem.

The accompanying illustration depicts the type of test tank manufacturers use, with **APPROXIMATE** dimensions. This type of tank is partially sunk into the ground. The craft sits in a body of circulating water and the mechanic has access to both sides of the craft without getting wet and has a dry place for tools, etc.

The tank design with an island in the center allows water to circulate around the tank. With no circulation, the water will foam, and the jet pump will cavitate and lose thrust. A series of baffle plates installed on the opposite side to the craft, slow and straighten the water flow. The walls must be strong enough to hold the weight of the water and to contain the thrust of the pump, but coated to seal the walls and not scratch the sides of the craft.

The bow is secured with a stout line anchored to the edge of the tank to prevent the craft from "wandering". The stern is secured to the opposite edge of the tank to prevent the craft from moving forward. A thrust scale is sometimes attached to the stern line to determine the thrust exerted by the pump.

Seek out dealers with a test tank and attempt to use/rent their facilities to test the completed work.

2
SAFETY

2-1 INTRODUCTION

Personal watercrafting is one of the fastest growing leisure activites in the world. Riding a personal watercraft can be an exciting, exhilarating, and safe experience provided the operator exercises prudent behavior and remains in constant control of the vehicle.

Personal watercraft may be likened to a firearm. Weapons do not injure people, but the individual handling them is the true culprit. In a similar manner, if the personal watercraft is operated in an unresponsible manner the driver, passengers, and others using the same waterway may be in danger.

The greatest hazard in operating a personal watercraft stems not from the craft itself, but from the behavior of the operator. The Personal Watercraft Industry Association (PWIA) and the National Marine Manufacturers Association (NMMA) recognize there are a number of problems due to irresponsible craft operators contributing to tarnishment of the sport. Specifically, these problems range from alcohol and substance abuse to excessive noise, speed, and reckless maneuvering. Insensitive operators can disrupt an otherwise enjoyable day for other craft users through misconduct on the water. These individuals often endanger their own lives as well as the lives of others by losing full control of their craft.

Organized events, races, and exhibitions, properly planned and controlled have proven to be a spectactular event appealing to a wide range and number of participants and spectators.

Craft Classification

Any craft equipped with propulsion machinery is classified as a motorboat.

Personal watercraft are classified as "Class A inboard boats" by the Coast Guard and therefore are subject to most of the same laws and requirements as more conventional larger craft. A decal, usually affixed somewhere in the stern area, lists exceptions the Coast Guard has granted for this type of water vehicle.

Information

In 1989, several states recognized the need to impose specific regulations governing operation of personal watercraft. Rules and regulations differ from state to state and are constantly being revised. Therefore, it would be an impossible task to present current legislation on personal watercraft activites. Before the printing ink for this publication was dry, additions, deletions, and revisions would already be in effect.

For current information, the reader may call 1-800-336-BOAT for the name and address of the local Boating Law Adminis-

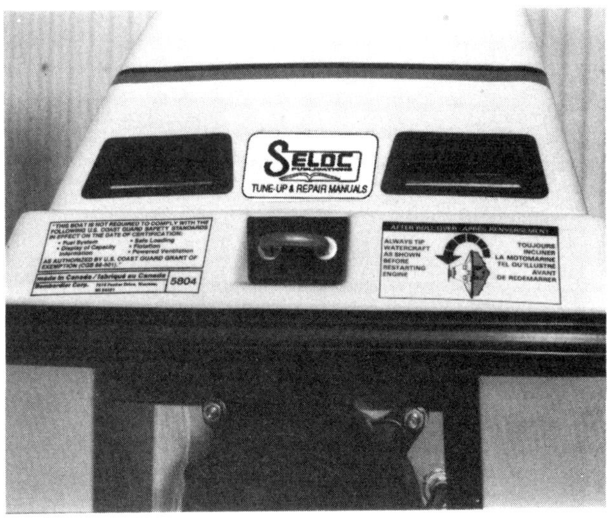

Several decals may be affixed to the stern of the hull to indicate compliance with or exceptions granted to normal Coast Guard regulations.

trator. The reader may also contact the PWIA Government Relations Representative, Washington, D.C.

Regulation Enforcement

The U.S. Coast Guard and state enforcement officers have the authority to stop any craft to check for compliance with federal or state law. They can order a craft to return to the closest dock and remain out of the water until the hazardous condition of the craft is corrected.

Examples are:

Inadequate number of Personal Flotation Devices.

Missing/Inoperative fire extinguisher.

Overloading.

Operation at night without proper lights.

Fuel leakage.

Fuel transportation (other than approved fuel tank).

Failure to meet ventilation requirements.

Failure to meet carburetor backfire flame arrestor requirements.

Excessive leakage or accumulation of water in the bilge.

The U.S. Coast Guard and state enforcement officers have the authority to stop any craft if they determine the craft is being operated in a negligent manner. Negligent operation is defined as "the failure to excercise the degree of care necessary to prevent the endangering of life, limb, or property of any person".

Examples are:

Operating while under the influence of drugs or alcohol.

Excessive speed in a congested area.

Excessive speed in stormy or foggy conditions.

Coming too close to another vessel.

Operating in a swimming area where bathers are present.

Towing water skiers where obstructions exist or where a fall might cause skier to be struck by another vessel.

Operating in the vicinity of a dam.

Cutting through a marine parade or regatta.

2-2 MINIMUM LEGAL REQUIREMENTS FOR EQUIPMENT ONBOARD CRAFT

All personal watercraft owners **MUST** provide the following equipment:

One approved Type I, II, III or IV PFD (personal flotation device) for each person on board or being towed on water skis, etc.

An efficient sound producing device.

A B-I approved hand portable fire extinquisher.

Visual distress signals for nighttime use, if operated on coastal waters.

Personal Flotation Devices (PFDs)

The Coast Guard requires an approved life-saving device be worn by each person on board. Devices approved are identified by a tag indicating Coast Guard approval. Such devices may be life preservers, or buoyant vests. Ring buoys, or buoyant cushions are not acceptable.

Life preservers have been classified by the Coast Guard into five type categories. All PFDs presently acceptable on recreational craft fall into one of these five designations. Only four of the five catagories are stipulated by legal requirements. Type V is not acceptable for use on a personal watercraft.

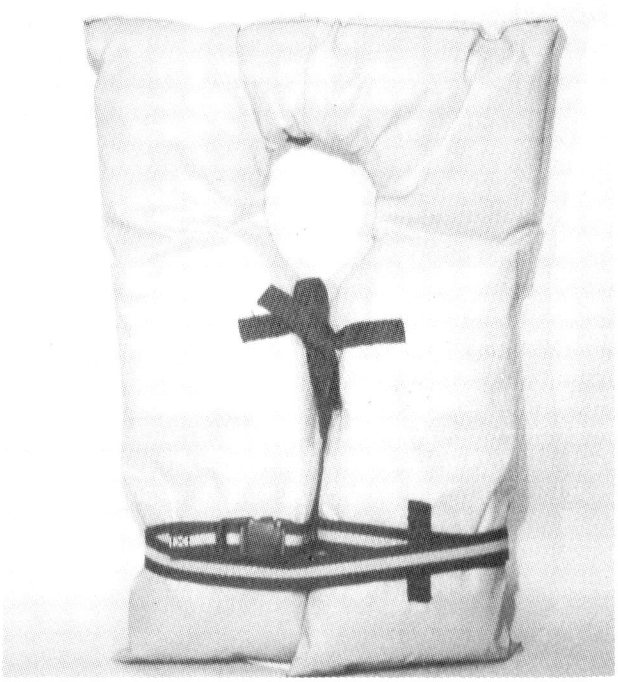

Type I PFD Coast Guard approved life jacket. This type flotation device provides the greatest amount of buoyancy.

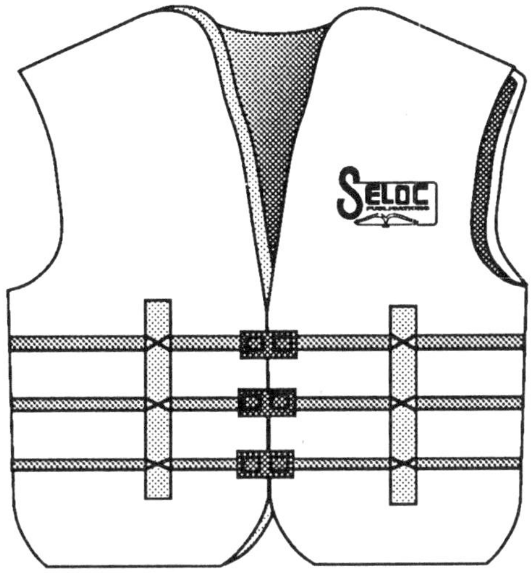

A Type III PFD requires the wearer to be active to remain upright in the water. This type device is comfortable and allows the wearer to actually swim while wearing it in the water.

A PFD **MUST** be U.S. Coast Guard approved, in good and serviceable condition, and of an appropriate size for the person intending to wear it.

It is an accepted fact that most boating people own life preservers, but too few actually wear them. There is little or no excuse for not wearing one because the modern comfortable designs available today do not subtract from an individual's boating pleasure.

PFDs may have a long serviceable life, if proper care is exercised and the device is not abused. Wipe them clean and dry before storing in a well ventilated area during the off-season. **NEVER** use a PFD as a fender. A crushed PFD loses its buoyancy and becomes useless.

Type I PFD has the greatest required buoyancy and is designed to turn most **UNCONSCIOUS** persons in the water from a face down position to a vertical or slightly backward position. The adult size device provides a minimum buoyancy of 22 pounds and the child size provides a minimum buoyancy of 11 pounds. The Type I PFD provides the greatest protection to its wearer and is most effective for all waters and conditions.

Type II PFD is designed to turn its wearer in a vertical or slightly backward position in the water. The turning action is not as pronounced as with a Type I. The device will not turn as many different type persons under the same conditions as the Type I. An adult size device provides a minimum buoyancy of 15½ pounds, the medium child size provides a minimum of 11 pounds, and the infant and small child sizes provide a minimum buoyancy of 7 pounds.

Type III PFD is designed to permit the wearer to place himself (herself) in a vertical or slightly backward position. The Type III device has the same buoyancy as the Type II PFD but it has little or no turning ability. Many of the Type III PFD are designed to be particularly useful when water skiing, sailing, hunting, fishing, or engaging in other water sports. Several of this type will also provide increased hypothermia protection.

Type IV PFD is designed to be thrown to a person in the water and grasped and held by the user until rescued. It is **NOT** designed to be worn. The most common Type IV PFD is a ring buoy or a buoyant cushion.

ONE LAST WORD

Common sense dictates the "cheap and cheerful" approach does not pay when applied to a life preserving device. Select and buy the very best. The life you save may be your own.

NOW, THESE WORDS

Due to the nature and maneuvering characteristics of personal watercraft, and from a "practical" standpoint, the following paragraph on bells, horns, and whistles hardly apply. The information is included to make the manual "complete" and because according to the U.S. Coast Guard they are required. Also, the authors feel the reader may "skipper" or "crew" on a larger craft and therefore, might be required to use the information presented here.

Bells, Horns, and Whistles

All craft must carry some means of producing an efficient sound signal and the sound must be audible for a specific distance and duration, depending on the size of the craft. For personal watercraft, an audible sound must carry for one half mile and be of four to six second duration.

Law enforcement craft are the **ONLY** boats allowed to use sirens.

Personal watercraft operators are required to sound fog and maneuvering signals when underway in fog or rain as a means of communication between craft which are not visible. This system was devised to alert nearby craft of the presence of others in the water and as a means of communicating maneuvering intentions. Signals indicating maneuvering tactics are short one second blasts.

One blast indicates the craft is turning to starboard.
Two blasts indicate the craft is turning to port.

Three blasts indicate the craft is in reverse.
Five or more blasts indicate - stay away - danger.

A prolonged blast lasting 4 to 6 seconds duration every two minutes indicates a straight line course in poor visibility.

These signals should be memorised by every personal watercraft operator, not necessarily because the operator would give them, but in the event of a head on collision with say a large powerboat, the skipper may signal his intention to turn either to port or starboard using this system of communication.

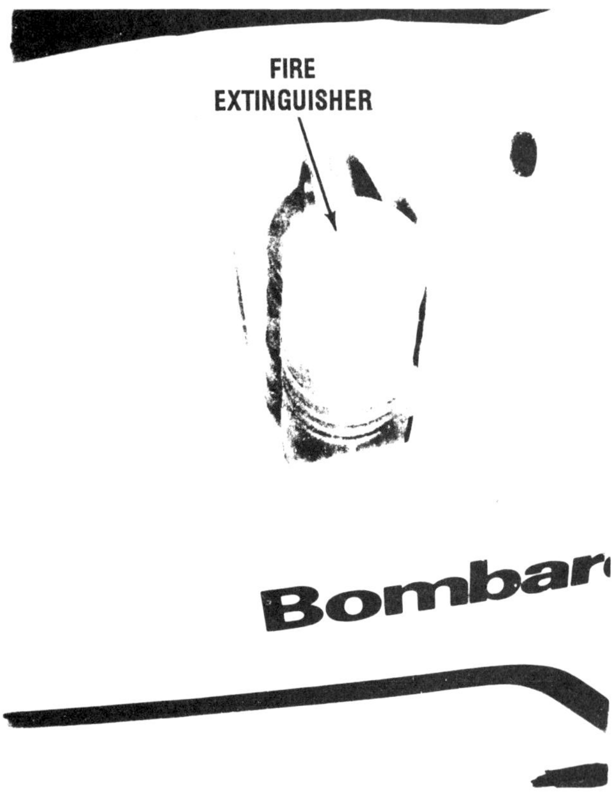

Personal watercraft are required to carry a Type B-1 fire extinguisher aboard. The compartment is in front of the handlebars where it is readily available.

Fire Extinguishers

Personal watercraft are classified as "Class A inboard boats" by the Coast Guard and therefore are required to carry a B-I fire extinguisher.

All fire extinguishers must bear Underwriters Laboratory (UL) "Marine Type" approved labels. With the UL certification, the extinguisher does not have to have a Coast Guard approval number.

B-I contains 1-1/4 gallons foam, or 4 pounds carbon dioxide, or 2 pounds dry chemical agent, or 2-1/2 pounds Halon.

READ labels on fire extinguishers. If the extinguisher is U.L. listed, it is approved for marine use.

Any personal watercraft can be rolled over and held under water for a few seconds to extinguish the fire. Turning the craft over in the water will prevent oxygen from entering the engine compartment. Without oxygen, the fire will soon exhaust the remaining oxygen in the compartment and then go out. **DO NOT** open the engine compartment until the smoke has completely stopped, then open with great caution and have a fire extinguisher at hand.

Visual Distress Signals

Coast Guard Regulations require personal watercraft to carry night visual distress signals when used on coastal waters - even though it is illegal to operate a personal watercraft at night.

Coastal waters include: the Great Lakes, the territorial seas and those waters directly connected to the Great Lakes and the territorial seas, up to a point where the waters are less than two miles wide.

These signals assist rescuers in locating and aiding people in distress. This information is useful to every personal watercraft owner, to know how to respond to others in distress.

For night use: Pyrotechnic visual distress signaling devices (red flares, hand held or aerial) **MUST** be Coast Guard Approved, in serviceable condition and stowed to be readily accessible. If they are marked with a date showing the serviceable life, this date must not have passed.

Coast Guard Approved pyrotechnic devices carry an expiration date. This date can **NOT** exceed 42 months from the date of manufacture and at such time the device can no longer be counted toward the minimum requirements.

The only non-pyrotechnic visual distress signaling device approved by the coast guard for use at night is an electric distress light -- not a flashlight but an approved electric distress light which **MUST** automatically flash the international **SOS** distress signal (. . . - - - - . . .)four to six times each minute.

2-3 MINIMUM LEGAL REQUIREMENTS FOR REGISTRATION OF CRAFT

Federal regulations require all personal watercraft to be issued a registration number and have the number displayed in a specific manner on the craft.

The "Certificate of Number" is issued by the state and must be carried on board at all times. The Hull Identification Number (HIN) is necessary to register the craft with the state.

The registration number must be displayed as follows:

The figures must be read from left to right.

The figures must be displayed on the forward half of each side of the bow of the craft.

The figures must be in bold, block letters of good proportion and must not be less than 3 inches high.

The figures must be of contrasting color to the craft hull, or background.

The figures must be as high above the waterline as practical.

No figures other than those assigned to the craft can be displayed on the forward half of the craft.

Letters must be separated from numbers by spaces or hyphens.

Validation decals (if required by state) must be displayed within 6 inches of the registration number.

Elaborate color schemes and custom detailing are very much a part of the fun of owning a personal watercraft. **HOWEVER**, the paint design must not obstruct the registration numbers or the validation decal. Like an automobile without a license plate, sooner or later a craft with no visible registration numbers will attract the attention of the law.

Hull Identification Number

Federal regulations require all personal watercraft to have a Hull Identification Number (HIN) permanently attached to the craft in two separate locations: on the starboard side of the transom, above the waterline. The HIN must also be displayed in an unexposed location, which is usually left to the discretion of the manufacturer.

2-4 SAFETY PRACTICES

Riding Techniques

Of all watercraft, large and small, the personal watercraft are the most responsive to steering and maneuverability. Therefore, a personal watercraft operator should take it upon herself/himself to stay clear of other boaters and especially swimmers.

A beginner should operate the craft at a low speed until the operator feels comfortable and is aware of its limitations and responses to steering.

While riding the craft, keep feet, hands, hair, and clothing away from the pump intake.

Full Throttle Operation

Prolonged operation at full throttle will raise the cylinder head temperature by about 100° above normal. Some units are equipped with a thermo-switch engine temperature sensor. This sensor is connected directly to the CDI unit and grounded to the engine kill switch. If the engine temperature exceeds a maximum value specified by

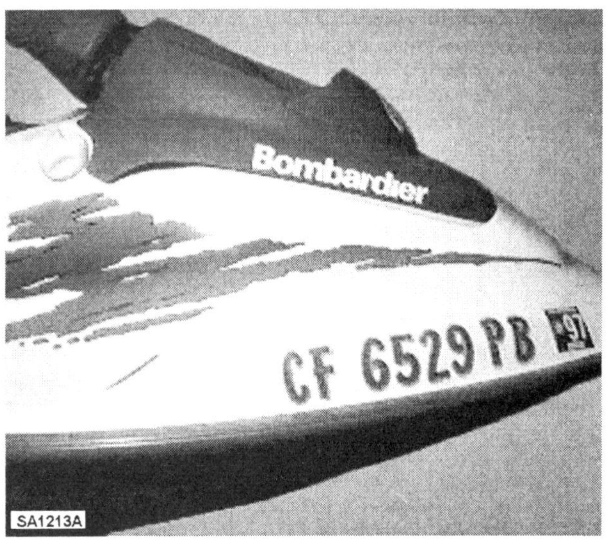

The validation decal, required by most states, and the registration number, required by all states, must be prominently displayed -- port and starboard -- on the forward portion of the hull.

the manufacturer, the ignition will be grounded and the engine shut down to prevent damage through overheating.

Jumping Waves

In the very early days of personal watercraft, jumping could cause extensive internal engine damage. When the craft left the water, unchecked engine rpm rapidly increased to as high as 12,000 rpm. As soon as the craft re-entered the water, engine rpm instantly decreased to about 6,000 rpm. This sudden increase and decrease in engine speed dramatically shortened engine life. In those days, repeated jumps were known to cause bent crankshafts, sheared flywheel keys, and broken connecting rods.

Kawasaki engineers developed an rpm limiter system to prevent high engine rpm and an almost "runaway" condition with the design of a rpm limiter. This limiter restricts engine speed to about 8,000 rpm. Thus, wave jumping, using a ramp, and a host of other almost unbelievable manuevers could be performed without undue stress on the engine. Today, with an rpm limiter system installed, a bent crankshaft or broken connecting rod is almost non-existent.

Alcohol and Substance Use

Operating a personal watercraft while under the influence of either alcohol or drugs is both illegal and life threatening. This fact cannot be stated too strongly. Such substances affect vision and slow reaction times. The danger is quadrupled in a craft on the water compared to on the shore.

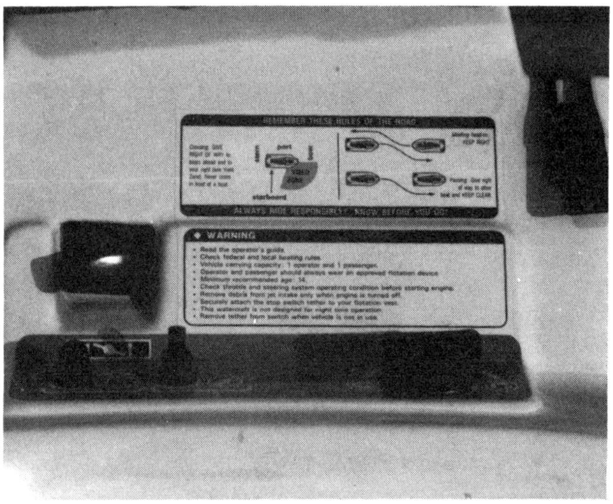

Decals located under the handlebars to remind the operator of the most basic "Rules of the Road".

Age Restrictions

Children under 14 years of age should not operate a watercraft unless accompanied on a two passenger craft by an adult experienced operator. Specific minimum age limits have been imposed by many states.

Speed Restrictions

Where no speed limits are posted, commonsense dictates the craft should be operated so it will pose no danger to other water users or the operator. Many states now have enforced speed limits for certain areas and conditions.

When entering "no wake" or 5 mph zones, such as fishing areas, swimming areas or marinas, the operator must slow to headway speed, the slowest speed at which appropriate steering capability can be maintained.

Individuals responsible for law enforcement on the water are keeping an ever vigilant eye on personal watercraft in many areas. Horseplay such as "wetting down" or "buzzing" other water users may result in a "ticket" for negligent operation.

Operation at Night

Personal watercraft are not equipped with lights and therefore, it is illegal to operate such a craft on the water after sunset.

The law requires any craft operated between sunset and sunrise to display certain colored lights mounted in specific locations, depending on the size and class of the craft. These lights will indicate the position and intended direction of the craft to operators of other watercraft in the area.

Personal watercraft are extremely difficult to see at night because of their small size. Even with lights, because of their almost unbelievable maneuverability, confusing signals would be sent to other traffic.

Shallow Water Operation

Under normal conditions, personal watercraft may be operated in very shallow waters, and it is possible to maneuver the craft directly onto a slopping beach.

Caution must be exercised when moving the craft off a beach to avoid sucking in mud, sand, pebbles, or rocks through the intake. Such ingestion could cause damage to the impeller, the vanes in the pump bowl, or clog engine cooling passages.

WARNING

Expelled sand and small pebbles may cause INJURY to persons in the area.

A grated section set into the hull bottom helps prevent the ingestion of small rocks and pebbles.

Carry about a foot of stiff wire aboard for use in probing into the impeller area to dislodge debris where fingers cannot reach.

Engine Compartment Ventilation

All motorboats (personal watercraft included) built after April 25, 1940 and before August 1, 1980, powered by a gasoline engine or by fuels having a flashpoint of 110° F or less **MUST** have the following, which is quoted from a recent Coast Guard publication:

> At least two ventilation ducts fitted with cowls or their equivalent for the purpose of properly and efficiently ventilating the bilges of every engine and fuel tank compartment. Each duct must be at least two inches in internal diameter.
>
> There shall be at least one exhaust duct installed so as to extend from the lower portion of the bilge to cowls in the open air, and at least one intake duct installed so as to extend to a point at least midway to the bilge or at least below the level of the carburetor air intake.

Flame Arrestors

A flame arrestor, as the name suggests, safeguards against flame caused by engine backfire.

A gasoline engine installed in a motorboat or motor vessel after April 25, 1940, except outboard motors, must have a Coast Guard Approved flame arrestor fitted to the carburetor. This requirement applies to personal watercraft because the engine is enclosed.

Fuel System

All parts of the fuel system should be selected and installed to provide maximum service and protection against leakage.

The capacity of the fuel filter must be large enough to handle the demands of the engine as specified by the engine manufacturer.

All fittings and outlets must come out the top of the tank.

In order to obtain maximum circulation of air around fuel tanks, the tank should not come in contact with the hull except through the necessary supports. The supporting surfaces and hold-downs must fasten the tank firmly and they should be insulated from the tank surfaces. This insulation material should be non-abrasive and non-absorbent material.

Taking On Fuel

The fuel tank should be kept almost full to allow for expansion, but still prevent water from entering the system through condensation caused by temperature changes. Water droplets forming are one of the greatest enemies of the fuel system. By keeping the tank almost full, the air space in the tank is kept to an absolute minimum and there is less room for moisture to form. It is a good practice not to store fuel in the tank over an extended period, say for six months. Today, fuels contain ingredients that change into gums when stored for any length of time. These gums and varnish products will cause carburetor problems and poor spark plug performance. An additive (Sta-Bil) is available and can be used to prevent gums and varnish from forming.

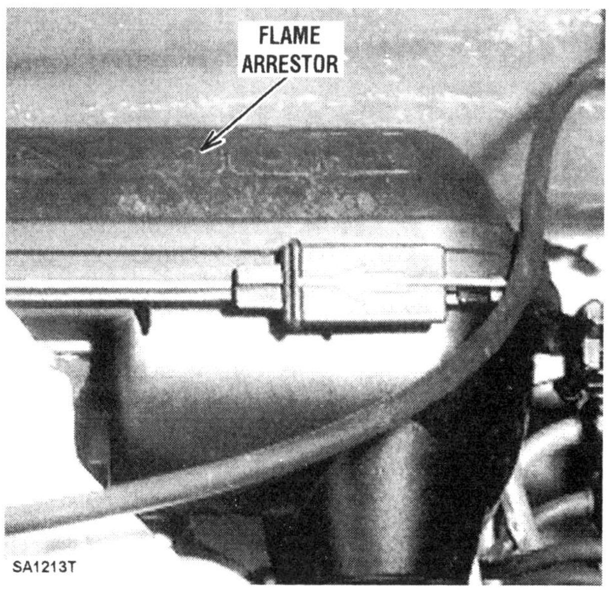

Carburetors on personal watercraft are required to be fitted with a flame arrestor because the engine compartment is an enclosed area. Other safety features include spill-proof battery filler caps -- conventional batteries -- one-way check valves for fuel and oil hoses -- to prevent spillage under varied conditions, and optional maintenance free sealed batteries.

Excessive Noise

Every manufacturer designs a stock exhaust system to keep noise to a minimum. A poorly maintained or modified exhaust system will produce excessive and disturbing noise. This is a form of air polution and is illegal in most states.

Automotive Replacement Parts

When replacing fuel, electrical, and other parts, check to be sure they are marine type and Coast Guard Approved. Automotive parts are not made to the high standards of marine parts. The carburetors must **NOT** leak fuel; electrical equipment **MUST** be able to operate in a gas fume enclosed area without exploding; etc. Automotive parts could cause a fire endangering the operator, passengers, and the craft. The part may look the same and even have a similar number, but if it is not **MARINE** it is not safe to use and will not pass Coast Guard inspection.

Overloading

Each personal watercraft is designed to carry a specific number of persons. Attempting to board more passangers than the craft is designed to carry, will upset weight distribution causing a marked decrease in performance and possible loss of control.

2-5 BOATING ACCIDENT REPORTS

In the United States, new federal and state regulations require an accident report be filed with the nearest State Boating Authority within 48 hours, if a person is lost, disappears, or is injured to the degree of needing medical treatment beyond first aid.

Accidents involving only property or equipment damage **MUST** be reported within 10 days, if the damage is in excess of $500.00 or a total boat loss. A **$1,000.00 PENALTY** may be assessed for failure to submit the report.

WORDS OF ADVICE

Take time to make a copy of the report to keep for your records or for the insurance company. Once the report is filed, the Coast Guard will not give out a copy, even to the person who filed the report.

The report must give details of the accident and include:

1- The date, time, and exact location of the occurrence.

2- The name of each person who died, was lost, or injured.

3- The number and name of the vessel.

4- The names and addresses of the owner and operator.

If the operator cannot file the report for any reason, each person on board **MUST** notify the authorities, or determine that the report has been filed.

In nautical terms, the front of the craft is the **bow;** the rear is the **stern;** the right side, when facing forward, is the **starboard** side; and the left side is the **port** side. One easy way to remember this basic fundamental is to consider the words "port" and "left" both have four letters and go together.

All directional references in this manual use this terminology. Therefore, the direction from which an item is viewed is of no consequence, because **starboard** and **port** **NEVER** change no matter where the individual is located, or standing, even on his or her head.

2-6 SECURITY

As mentioned in the opening paragraphs of this chapter, personal watercrafting is one of the fastest growing leisure activites in the world. Like most personal possessions, personal watercraft have value and valuables are subject to be lost to unscrupulous characters -- thieves. The best advice for a personal watercraft owner is "lock it or lose it".

Do not leave the craft unattended on the beach or dock. **ALWAYS** remove the kill switch tether from the switch and of course, the ignition key, if the unit is equipped with these features. Certain aftermarket security devices are available for installation to prevent theft of personal watercraft. These include: A "secret" ignition grounding switch, which grounds one side of the ignition coil, and a hidden fuel shut-off valve.

Most owners transport their craft on trailers or in the bed of a pickup truck. If loaded on a trailer, chain and lock the craft to the trailer. Invest in a hitch lock so the trailer cannot be detached from the towing vehicle. If the craft is loaded on a pickup truck, chain and lock it to the bed, or to a large and heavy object.

3
TUNING

3-1 INTRODUCTION

The efficiency, reliability, fuel economy and enjoyment available from the engine performance are all directly dependent on having the engine properly tuned. The importance of performing service work in the sequence detailed in this chapter cannot be overemphasized. Before making any adjustments, check the specifications in the Appendix. **NEVER** rely on memory when making critical adjustments.

Before beginning to tune any engine, check to be sure the engine has satisfactory compression. An engine with worn or broken piston rings, burned pistons, or scored cylinder walls, cannot be made to perform properly no matter how much time and expense is spent on the tune-up. Poor compression must be corrected or the tune-up will not give the desired results.

A practical maintenance program that is followed throughout the year, is one of the best methods of ensuring the engine will give satisfactory performance at any time.

The extent of the engine tune-up is usually dependent on the time lapse since the last service. A complete tune-up of the entire engine would entail almost all of the work outlined in this manual. A logical sequence of steps will be presented in general terms. If additional information or detailed service work is required, the chapter containing the instructions will be referenced.

Each year higher compression ratios are built into modern marine engines and the electrical systems become more complex, especially with electronic (capacitor discharge) units. Therefore, the need for reliable, authoritative, and detailed instructions becomes more critical. The information in this chapter and the referenced chapters fulfill that requirement.

3-2 TUNE-UP SEQUENCE

During a major tune-up, a definite sequence of service work should be followed to return the engine to the maximum performance desired. This type work should not be confused with attempting to locate

A clean, properly tuned, and adequately maintained power unit can multiply by many fold the enjoyment derived from owning and operating a personal watercraft.

problem areas of "why" the engine is not performing satisfactorily. Such work is classified as "troubleshooting". In many cases, these two areas will overlap, because many times a minor or major tune-up will correct the malfunction and return the system to normal operation.

The following list is a suggested sequence of tasks to perform during the tune-up service work. The tasks are merely listed here. In most cases, procedures are given in subsequent sections of this chapter. For more detailed instructions, see the referenced chapter, for the unit being serviced.

1- Perform a compression check of each cylinder. See Chapter 7 -- **IGNITION**.

2- Inspect the spark plugs to determine their condition. Test for adequate spark at the plug. See Chapter 7 -- **IGNITION.**

3- Start the engine in a body of water or connect a "Flushing Kit" with a garden hose and check the water flow through the engine. See Chapter 8 -- **ENGINE.**

4- Check the carburetor adjustments and the need for a carburetor overhaul. See Chapter 6 -- **FUEL**

5- Check fuel pump for adequate performance and delivery. See Chapter 6 -- **FUEL.**

6- Make a general inspection of the ignition system. See Chapter. 7 -- **IGNITION.**

7- Test the cranking motor and the solenoid. See Chapter 9 -- **ELECTRICAL**

8- Check the internal wiring.

Performing a compression test on a two cylinder engine. Both spark plug leads have been grounded to prevent placing an unnecessary extra load on the ignition coil.

3-3 COMPRESSION CHECK

A compression check is extremely important, because an engine with low or uneven compression between cylinders **CANNOT** be tuned to operate satisfactorily.

Therefore, it is essential that any compression problem be corrected before proceeding with the tune-up procedure.

If the engine shows any indication of overheating, such as discolored or scorched paint, inspect the piston skirts visually thru the exhaust ports, using a mirror and flashlight, for possible scoring. It is possible for a cylinder with satisfactory compression to be scored slightly.

Checking Compression

Remove the spark plug wires. **ALWAYS** grasp the molded cap and pull it loose with a twisting motion to prevent damage to the connection. Remove both spark plugs and remember from which cylinder they came for evaluation later. Ground both spark plug leads to the engine to render the ignition system inoperative while performing the compression check.

Insert a compression gauge into the forward spark plug opening. Crank the engine with the cranking motor, thru at least four complete revolutions of the crankshaft, with the throttle at the wide-open position, to obtain the highest possible reading. Record the highest reading. Repeat the test and record the compression for the other cylinder. The manufacturer specifies the compression pressure for a new engine should be 142.2 psi (981 kpa). A variation between cylinders is far more important than the actual readings.

A variation of more than 15 psi between the two cylinders indicates the lower compression cylinder is defective and requires attention. The problem may be worn, broken, or sticking piston rings, scored pistons, or worn cylinders.

Use of an engine cleaner will help to free stuck rings and to dissolve accumulated carbon. Follow the directions on the can of cleaner.

3-4 SPARK PLUG INSPECTION

Inspect both spark plugs for badly worn electrodes, glazed, broken, blistered, or lead fouled insulators. Replace both the plugs, if either shows signs of excessive wear.

Make an evaluation of the cylinder performance by comparing the spark plug condition with those shown in Chapter 7 -- **IGNITION**. Check each spark plug to be sure they are both of the same manufacturer and have the same heat range rating.

Inspect the threads in the spark plug opening in the head and clean the threads before installing the plug.

When purchasing new spark plugs, **ALWAYS** ask the dealer if there has been a spark plug change for the engine being serviced. Typically, either NGK BR7ES (Champion RN3C) or NGK BR8ES (Champion RN2C) plugs are recommended for all craft covered, depending on the engine. Set the spark plug gap to .020" (0.5mm) for all series engines covered in this manual.

Crank the engine through several revolutions of the crankshaft to blow out any material which might have become dislodged during cleaning.

Install the spark plugs and tighten them to a torque value of 17 ft lb (24Nm). **ALWAYS** use a new gasket and wipe the seats in the head clean. The gasket must be fully compressed on clean seats to complete the heat transfer process and to provide a gas tight seal in the cylinder. If the torque value is too high, the heat will dissipate too rapidly. Conversely, if the torque value is too low, the dissipation of heat will be too slow.

3-5 ELECTRICAL POWER SUPPLY

BATTERY SERVICE

Remove the battery from the engine compartment for service. Inspect and service the battery, cables and connections. Check for signs of corrosion. Inspect the battery case for cracks or bulges, dirt, acid, and electrolyte leakage.

Check to be sure the breather pipe is attached to the battery and not pinched by any item in the engine compartment.

Clean the top of the battery. The top of a 12-volt battery should be kept especially clean of acid film and dirt, because of the high voltage between the battery terminals and to prevent contact between the acid and the plastic battery case. For best results, first wash the battery with diluted ammonia or baking soda solution to neutralize any acid present. Flush the solution off the battery with clean water. Keep the vent plugs tight to prevent the neutralizing solution or water from entering the cells.

Filling the Battery

Remove each filler cap using a pair of pliers. Fill each cell to the proper level with distilled water. The level of electrolyte should be between the upper and lower level marks during operation. When filling the battery, fill to the upper line and allow the battery to stand for twenty minutes. Check the level again and replenish as necessary.

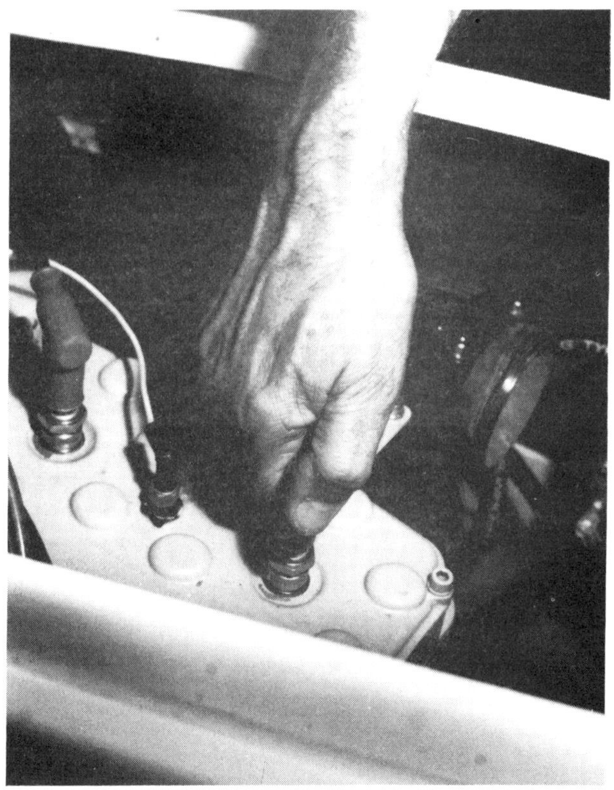

After a compression check or inspection of the spark plugs, install the high tension leads firmly over the spark plugs.

Use a temperature corrected hydrometer to test the specific gravity of the electrolyte. At 20oC (70oF) the hydrometer reading should be 1.26 on the scale. If it is necessary to charge the battery, leave the filler caps lightly resting on the cell openings to allow the gases to escape. The charging current should not exceed 1.9 amps for 10 hours.

After charging, push each filler cap into place and wipe the top of the battery clean before installation into the watercraft.

Clean the battery posts and battery cable ends with a wire brush to ensure good, clean connections. Clean the top surface of the battery and identify the "POS" or ("+") and "NEG" or ("-") embossed symbols. Correctly connect the battery cables to the battery observing polarity. If the cables are connected backwards, the ignition system **WILL BE DESTROYED** the first time the engine is started.

Check to be sure the battery is fastened securely in position. The battery box and cover screws **MUST** be tightened securely to prevent any movement of the battery in the holder under the most violent maneuvers of the watercraft.

*Due to violent maneuvers experienced by a personal watercraft, the battery **MUST** be extremely well-secured. Older models were equipped with a battery cover, as shown. Newer Sea-Doo craft are equipped with convenient restraining straps to secure the battery.*

If the battery posts or cable terminals are corroded, the cables should be cleaned separately with a baking soda solution and a wire brush. Apply a thin coating of Multipurpose Lubricant to the posts and cable clamps before making the connections. The lubricant will help prevent corrosion.

Jumper Cables

If a booster battery is used to start the engine, the jumper cables must be connected correctly and in the proper sequence to prevent damage to either battery, or to the rectifier diodes in the ignition system.

ALWAYS connect a cable from the positive terminal of the "dead" battery to the positive terminal of the good battery **FIRST**. Next, connect one end of the other cable to the negative terminal of the good battery. Finally, connect the other end of the cable to the far side of the **ENGINE** for a good ground. By making the ground connection on the engine, any spark created will not be near the battery. An arc near the battery could result in an explosion, destroying the battery and causing serious personal injury.

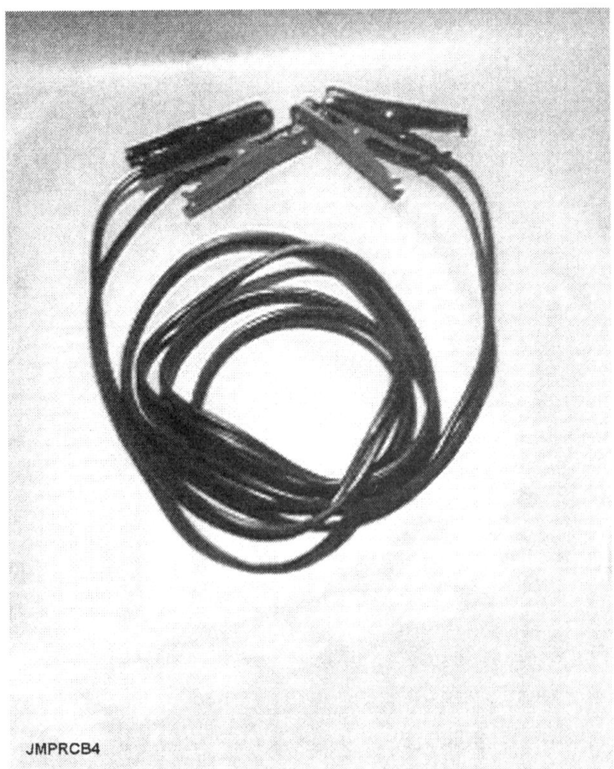

*Common set of jumper cables for use with a second battery to crank and start the engine. **EXTREME** care should be exercised when using a second battery, as explained in the text.*

GOOD WORDS

If the craft must be jump started from a vehicle battery, **DO NOT** perform the procedure while the automobile is operating. Chances are, the car battery will be receiving a charge from the alternator. This extra electrical current will be delivered to the receiver battery. Such a condition would **SURELY** overload the ignition system, causing extensive damage to the components.

If a trickle charger is used on a "dead" battery installed in the craft, one battery lead **MUST** be disconnected prior to connecting the charger, to prevent destroying the diodes in the rectifier/regulator.

NEVER use a trickle charger as a booster to start the engine, because such action will damage the rectifier/regulator module.

3-6 FUEL SYSTEM

Fuel and Fuel Tanks

Take time to check the fuel tank and all fuel lines, fittings, couplings, valves, flexible tank fill, and vent. Turn on the fuel supply valve. If gas was not drained at the end of the previous season, make a careful inspection for gum formation. When gasoline is allowed to stand for long periods of time, particularly in the presence of copper, gummy deposits form. This gum can clog the filters, lines, and passageways in the carburetor.

Commercial additives, such as Sta-Bil, may be used to keep the fuel in the tank fresh. Under favorable conditions, such additives will prevent the fuel from "souring" for up to twelve months.

If the condition of the fuel is in doubt, drain, clean, and fill the tank with fresh fuel, with the proper fuel/oil mixture recommended by the manufacturer.

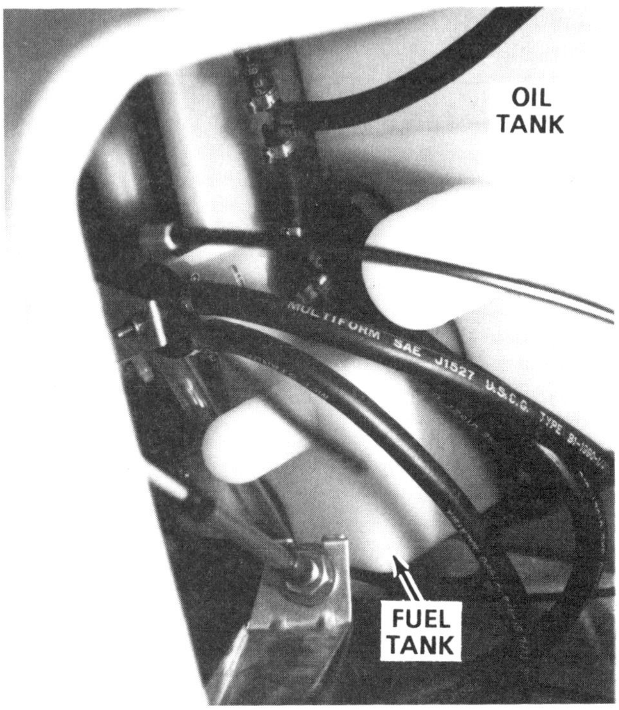

The fuel and oil tank should be drained, every season or two, and inspected for any gum deposits.

CARBURETOR ADJUSTMENTS

Four fuel adjustments may be made on the Mikuni Super BN carburetor, for both single and twin carburetor installations, as follows:

a- Throttle Cable Adjustment -- **ALWAYS** followed by oil injection synchronization.
b- Low Speed Mixture Adjustment.
c- Idle Speed Adjustment -- **ALWAYS** followed by oil pump synchronization.
d- Choke Cable Adjustment.

Each of the listed adjustments are explained in detail in the following paragraphs.

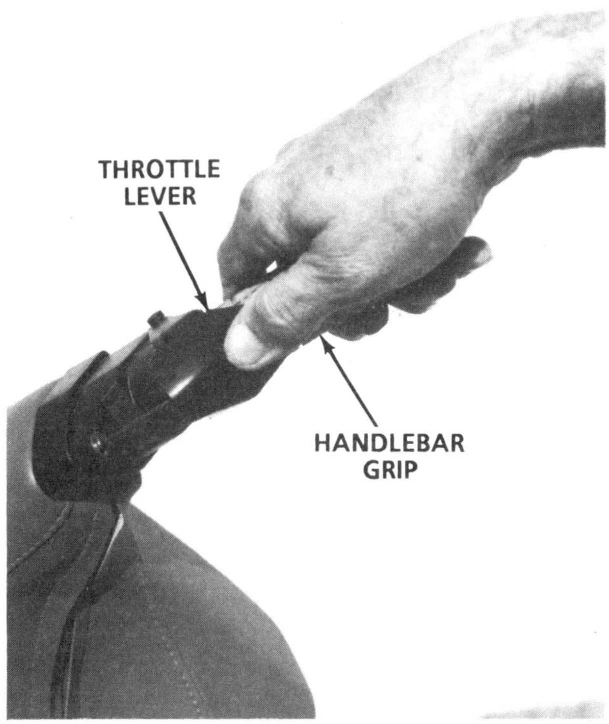

The throttle valve is wide open when the throttle lever is fully applied -- actually making contact with the handlebar grip.

Throttle Cable Adjustment

When no throttle is applied, the idle speed screw must contact the stopper on the carburetor. When the throttle lever is fully applied -- actually making contact with the handlebar grip -- the stop lever on the throttle linkage must contact the other side of the stopper on the carburetor. If these two conditions are not met, an adjustment on the throttle cable is necessary at the cable support bracket. Adjust the throttle cable length by loosening or tightening the two nuts -- one on each side of the support bracket.

After a throttle cable adjustment is made, **ALWAYS** perform a check to ensure the oil injection pump is synchronized with the carburetor. If the oil injection pump is not properly synchronized with the carburetor, serious damage may be caused to the engine due to lack of adequate lubrication.

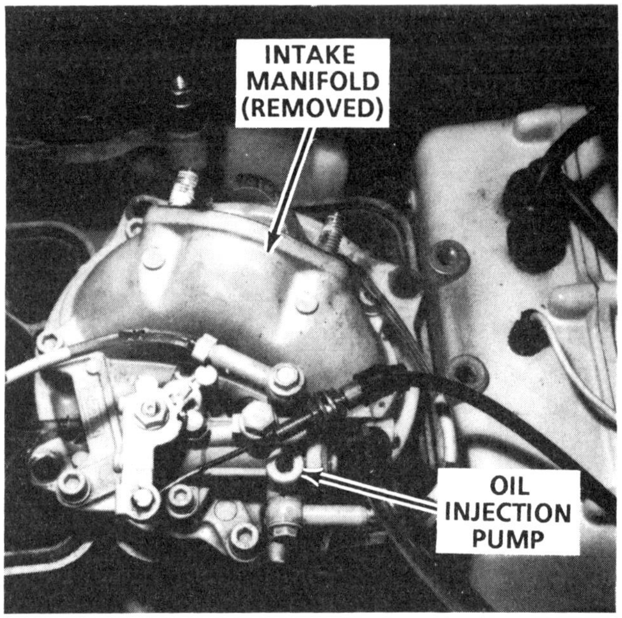

*After a throttle cable adjustment has been made, the oil pump **MUST** be synchronized with the carburetor.*

Low Speed Mixture Screw

The low speed mixture screw must **NEVER** be used to alter engine rpm. Such action could result in severe engine damage. This screw is used to smooth engine operation at a set idle speed. The idle speed is changed by adjustment of the idle speed screw, covered in the following paragraph. If working on a twin carburetor installation, perform the required work on each carburetor in turn.

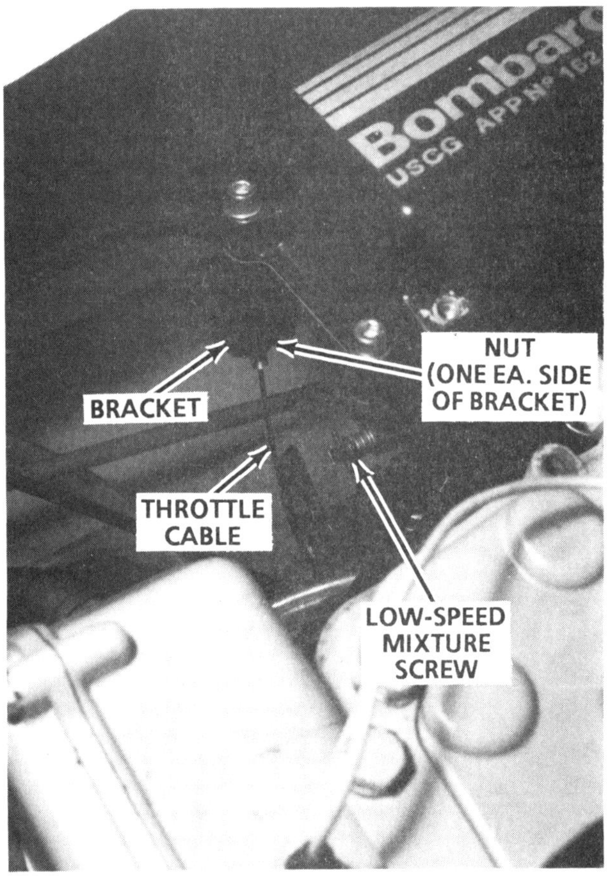

Location of the low-speed mixture screw and throttle cable adjustment nuts on some upright engines. Newer models are equipped with linkage running horizontal to the hull. The 787 engine is canted 30 degrees to port -- the linkage will vary slightly from the illustration.

Now, with the engine shut down, tighten the low speed mixture screw, identified in the accompanying illustration, until the screw is **LIGHTLY** seated, and then back it out 1-1/2 to 2 full turns as a preliminary setting.

Start the engine and allow it to warm to operating temperature.

CAUTION

The following procedure MUST be followed whenever the engine is started with a hose attached for cooling the engine. Start the egine, and THEN turn on the water. The engine MUST be started first to prevent water from backing up into the tuned exhaust pipe and into the engine. Water must circulate anytime the engine is operating to prevent overheating -- which could cause damage to moving parts and possibly cause engine seizure.

NEVER allow the engine to opeate without cooling water for more than 15 seconds. NEVER operate the engine with a flushing attachment for more than five miniutes. The driveline seal is not provided with cooling water

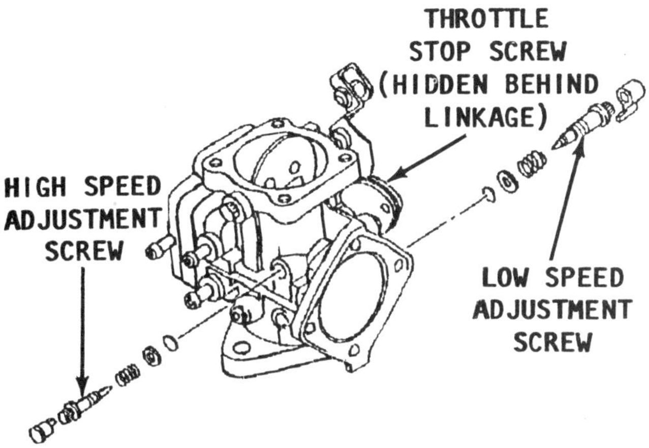

*Exploded line drawing of the Mikuni BN carburetor showing arrangement of parts for the low speed and high speed adjustment screws. Bench adjustment for these two screws is made by seating the screws **LIGHTLY**, and then backing out the number of turns listed in the Appendix for the model and production year. The forward and aft carburetors may have different settings.*

if it is not immersed in water. To shut the engine down, first shut off the water supply, and then shut down the engine. If this sequence of tasks is not followed as stated, the engine could be severely damaged.

REMEMBER
Start the engine.
Turn the water on.
Turn off the water.
Shut down the engine.

To lean the mixture -- rotate the low speed mixture screw **CLOCKWISE.**

To enrichen the mixture -- rotate the low speed mixture screw **COUNTERCLOCKWISE.**

Rotate the low speed mixture screw in the direction which will result in the smoothest engine operation.

Idle Speed Adjustment

The idle speed adjustment screw is located on the throttle shaft linkage. If working on a twin carburetor installation, only the aft carburetor is equipped with an idle speed screw. A link rod connects the two throttle shafts together. Therefore, when an adjustment is made at the idle speed screw on the aft carburetor, the adjustment holds good for both carburetors.

When no throttle is applied, the tip of the screw rests against a stopper cast into the carburetor. To set the idle speed, obtain an induction type tachometer.

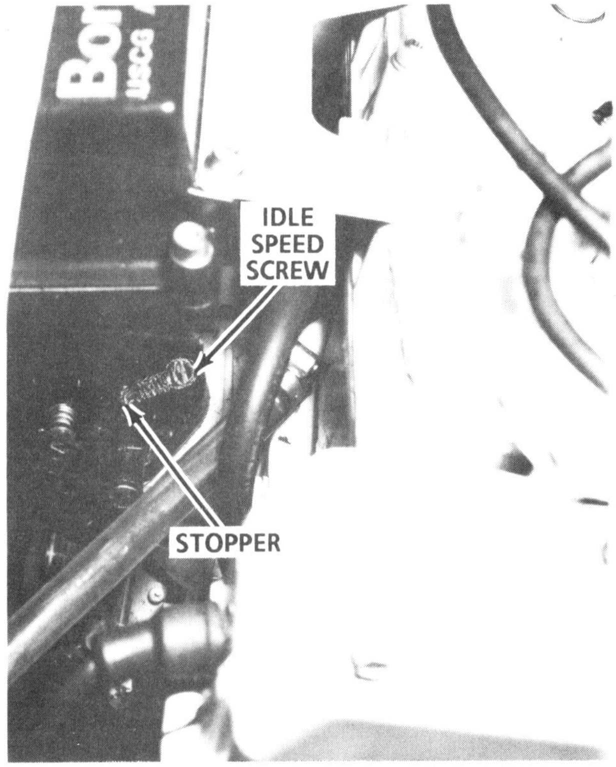

Location of the idle speed screw used to make the idle speed adjustment on a Mikuni carburetor, on a "straight-up" engine.

To increase engine rpm -- rotate the idle speed screw **CLOCKWISE.**

To decrease engine rpm -- rotate the idle speed screw **COUNTERCLOCKWISE.**

Start the engine and allow it to warm to operating temperature.

CAUTION

The following procedure MUST be followed whenever the engine is started with a hose attached for cooling the engine. Start the egine, and THEN turn on the water. The engine MUST be started first to prevent water from backing up into the tuned exhaust pipe and into the engine. Water must circulate anytime the engine is operating to prevent overheating -- which could cause damage to moving parts and possibly cause engine seizure.

NEVER allow the engine to opeate without cooling water for more than 15 seconds. NEVER operate the engine with a flushing attachment for more than five miniutes. The driveline seal is not provided with cooling water if it is not immersed in water. To shut the engine down, first shut off the water supply, and then shut down the engine. If this sequence of tasks is not followed as stated, the engine could be severely damaged.

Rotate the idle speed screw until the engine idles at 1500 rpm -- if using a flush attachment, or 1200 rpm, if the craft is in the water.

After a throttle cable adjustment is made, **ALWAYS** perform a check to ensure the oil injection pump is synchronized with the carburetor. If the oil injection pump is not properly synchronized with the carburetor, serious damage may be caused to the engine due to lack of adequate lubrication. See page 6-42 for oil pump synchronization.

Choke Cable Adjustment

Push the choke knob inward. Observe the choke pivot arm on the side of the carburetor. The arm should be in the position shown in the accompanying illustration, with minimum slack in the choke cable. If the choke needs adjustment: Loosen the small screw securing the choke cable end to the barrel on the choke lever. Push the choke knob inward until it rests against the control panel. Set the choke pivot arm in the fully open position, and then tighten the screw securing the cable to the barrel.

Choke Cable -- Excessive Slack

If there is too much slack in the cable, loosen the top nut around the cable at the cable support bracket and tighten the bottom nut until only a small amount of slack remains.

If the cable is tight -- with no slack -- loosen the bottom nut around the cable at the cable support bracket and tighten the top nut until there is just a small amount of slack in the cable.

If the cable has stretched, or has become kinked or frayed, or tends to stick, the entire cable should be replaced.

Main Jet Words

The main jet is located internally on Mikuni Super BN carburetors. Therefore, the carburetor must be removed to service or replace a main jet. Most Super BN carburetors are equipped with a size 95 or 100 main jet.

Operating the engine on an excessively lean mixture may cause overheating, pre-ignition, and a decrease in performance. If the main jet size is altered, keep a close eye on the condition of the spark plug tip. If the tip begins to show signs of overheating, change to a cooler heat range. Check with the local dealer for the correct number spark plug. Some spark plug manufacturers use a higher number for colder plugs, others use a lower number for colder plugs.

Operating the engine on an excessively lean mixture may cause overheating, pre-ignition, and a decrease in performance. Any time the main jet size is altered, keep a close eye on the condition of the spark plug tip. If the tip begins to show signs of overheating, change to a colder heat range. Check with the local dealer for the correct number spark plug. Some spark plug manufacturers use a higher number for colder plugs, and others use a lower number for colder plugs.

If working on a twin carburetor installation, both jets **MUST** be the same size.

3-7 FUEL PUMPS

Many times, a defective fuel pump diaphragm is mistakenly diagnosed as a problem in the ignition system. The most common problem is a tiny pin hole in the diaphragm. Such a small hole will permit gas to enter the crankcase and wet foul both spark plugs at idle speed. During high speed operation, gas quantity is limited, the plugs are not fouled and will, therefore, fire in a satisfactory manner.

If the fuel pump fails to perform properly, an insufficient fuel supply will be delivered to the carburetor. This lack of fuel will cause the engine to run lean, lose rpm, or cause piston scoring.

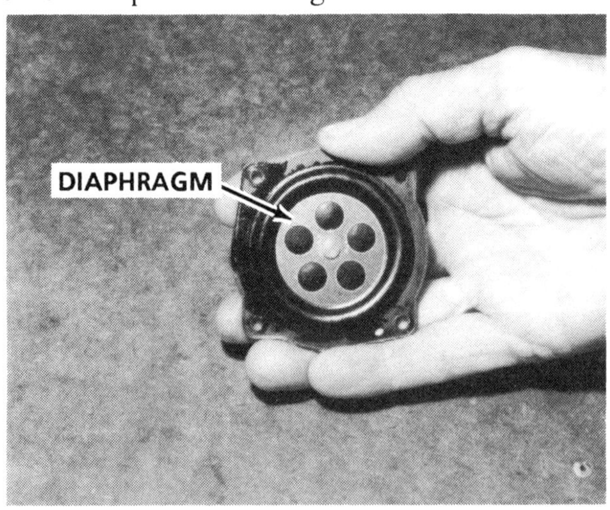

Close view of the fuel pump diaphragm from a carburetor with integral fuel pump. The diaphragm must be carefully inspected during a carburetor overhaul for the tiniest hole or tear. Such damage will "wet foul" the spark plug at idle speed.

NEVER use liquid Neoprene on fuel line fittings. Use Permatex when making fuel line connections. Permatex is available at almost all marine and hardware stores.

To service the fuel pump, see Chapter 6 -- **FUEL.**

3-8 CRANKING MOTOR AND SOLENOID

Cranking Motor Test

Check to be sure the battery is fully charged. Would you believe, many cranking motors are needlessly disassembled, when the battery is actually the culprit.

Lubricate the pinion gear and screw shaft with No. 10 oil.

Connect one lead of a voltmeter to the positive terminal of the cranking motor. Connect the other meter lead to a good ground on the engine. Check the battery voltage under load by depressing the ignition switch and observing the voltmeter reading.

If the reading is 9-1/2 volts or greater, and the cranking motor fails to operate, repair or replace the unit. See Chapter 9 -- **ELECTRICAL.**

Preliminary Tasks Before Testing Cranking Motor Solenoid

Remove the battery cover.

Disconnect the three Red leads from the upper left terminal of the cranking motor solenoid.

Disconnect the single Red large lead from the upper right terminal.

Disconnect the Yellow/Red lead from the lower left terminal.

Disconnect the two Brown leads from the lower right terminal.

Cranking Solenoid Testing

1- Select the Rx1 ohm scale on the ohmmeter. Make contact with the two meter leads across the two upper solenoid terminals. The meter should read an infinite resistance. If the meter shows resistance of less than infinity, the solenoid is defective and **MUST** be replaced.

2- Keep the ohmmeter on the same scale. Obtain a 12-volt battery and a pair of jumper leads. Using the set of jumper leads, connect the battery across the two lower terminals. Make contact with the

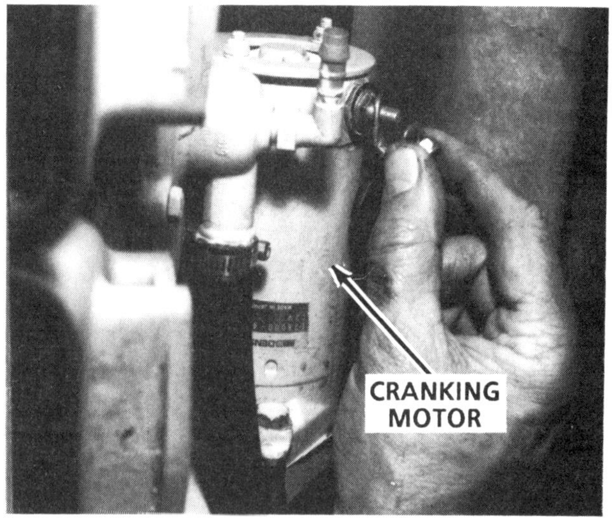

Cranking motor electrical leads being disconnected prior to making tests on the motor. See Chapter 9 -- ELECTRICAL, to service the cranking motor.

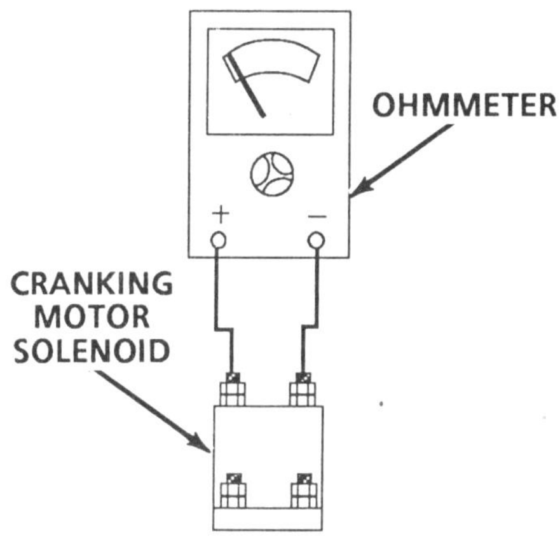

(1)

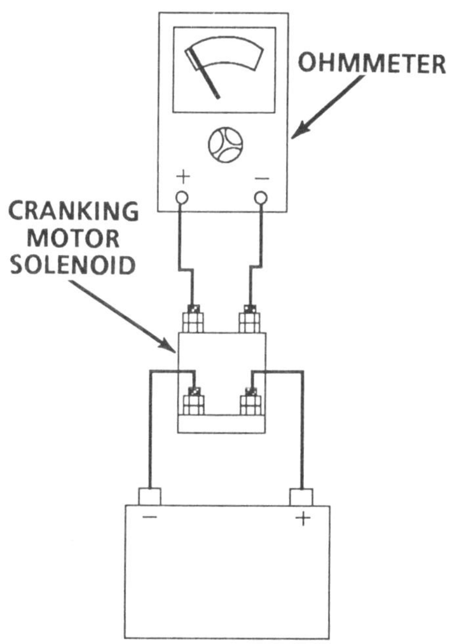

(2)

two meter leads across the two upper solenoid terminals. If the solenoid emits a "click" and the ohmmeter registers zero ohms, the solenoid is good. If no "click" is heard, or if the meter registers a high or infinite resistance, the solenoid is defective and **MUST** be replaced.

If the solenoid failed the test, it must be replaced. Two bolts secure the solenoid to the battery cover. When installing the replacement solenoid, apply Loctite 242 to the threads of the bolts and install the new solenoid in place. Tighten both bolts securely.

Connect the two Brown leads to the lower right terminal.

Connect the Yellow/Red lead to the lower left ternimal.

Connect the three smaller Red leads to the upper left terminal.

Place the cover over the battery. Install and tighten the four securing screws.

3-9 JET PUMP

At press time, only the Bombardier Sea-Doo GT and a very few other personal watercraft were equipped with a reverse gate. If the unit being serviced has a reverse gate, the gate **MUST** be properly adjusted to obtain maximum performance from the craft. When properly adjusted, the gate will permit the pump to deliver its full potential of thrust with minimum drag.

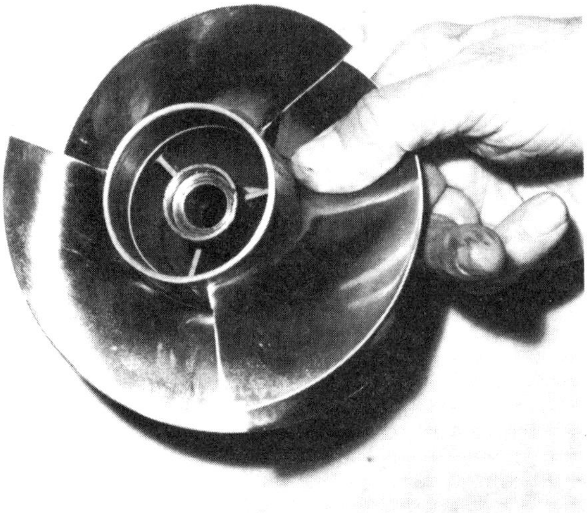

The "bottom line" on a jet pump is the condition and efficiency of the impeller. Adequate care, maintenance and timely replacement will result in maximum fun and safety for the owner and guests using the craft.

Gate Position

The shift rod lever adjustment should be checked from time to time to ensure the gate is firmly against the pump housing, when the unit is in the **FORWARD** position (wide open).

With the gate against the housing, any rattle noises will be avoided as the craft moves through the water. Proper positioning of the gate in forward gear will prevent wave action from accidentally shifting the gate into reverse as the craft is operated through violent maneuvers.

A simple "tuning" task with the jet pump is to make a thorough check of the linkage and gate movement. Such a check will ensure maximum thrust and use of the horsepower developed by the engine.

Impeller

Excessively rounded jet impeller edges will reduce the efficiency of the jet drive.

The term cavitation "burn" is a common expression used throughout the world among people working with pumps, impeller blades, and forceful water movement.

"Burns" on the impeller blades are caused by cavitation air bubbles exploding with considerable force against the impeller blades. The edges of the blades may develop small "dime size" areas resembling a porous sponge, as the aluminum material is actually "eaten" by the condition just described.

3-10 ROTARY VALVE DESCRIPTION AND OPERATION

The most unique feature of the engines used in all Bombardier Sea-Doo watercraft is the rotary valve. Rotary valve two-stroke engines are widely used in GP road racing motorcycle engines.

The purpose of the rotary valve is to mechanically time the oil/air/fuel mixture to be injected at a precise moment into the crankcase. The timing on this engine is so critical, a special section is included in Chapter 8 -- **ENGINE**. Incorrect timing of the rotary valve could lead to difficult starting and backfiring during engine operation.

The oil/air/fuel mixture timing must not be confused with ignition/spark timing. Ignition/spark timing is electronically controlled by the ignition module and cannot be adjusted.

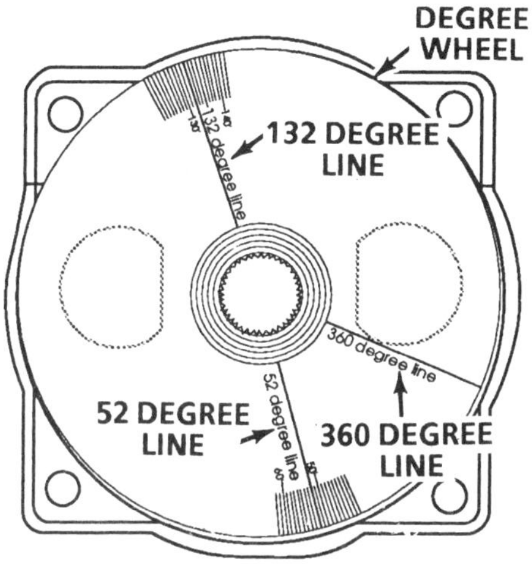

Rotary valve timing using a degree wheel, as explained in the text. The markings are for a specific model and engine.

The intake ports are machined into the starboard side of the lower crankcase; one port per cylinder. A rotary valve is driven off the crankshaft in the center of the two intake ports. The rotary valve is actually a thin metal disc, set against the intake ports. A hole in the center of the valve disc is cut to match the splines on the rotary valve shaft. Therefore, the disc is driven by the shaft.

A large window is cut out of the disc -- a "pie shaped wedge" of varying size, depending on the engine. For example, consider a pie shaped wedge of 132°. As this window rotates past the two intake ports -- one after the other -- the oil/air/fuel mixture is injected into the crankcase. As soon as the trailing edge sweeps past the port, the port is closed off. In this example, each port is opened at 132° BTDC and is closed at 52° ATDC.

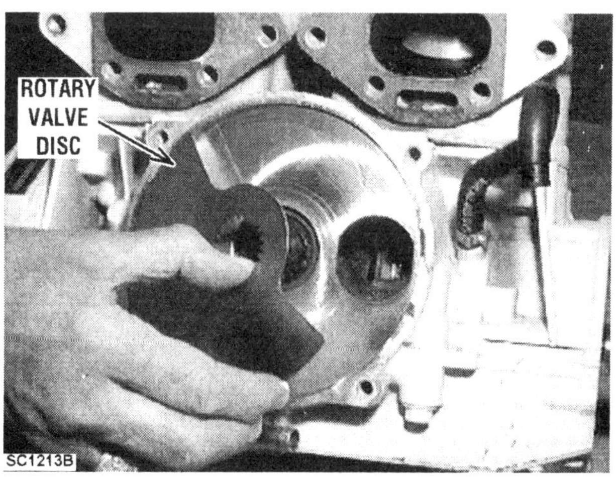

*The Bombardier Sea-Doo engines all have a unique feature -- a rotary valve. A detailed explanation of how this valve functions, is given in the text. If removing the rotary valve disc, be **ABSOLUTELY** sure to scribe a mark as explained in Chapter 8 -- ENGINE.*

If the rotary valve is not installed correctly, the mixture will be injected into the crankcase when the piston is in the wrong position -- at the wrong part of the stroke. The resulting ignition of the mixture will not reach maximum effectiveness and will produce a backfire.

Correct positioning of the rotary valve is quite simple. One edge of the window **MUST** align with a notch cast into the rotary valve housing. It is extremely unlikely the rotary valve -- once installed properly -- will shift during engine operation. The rotary valve housing is an integral part of the upper crankcase. Unfortunately, if a replacement of the upper crankcase is to be made, the rotary valve housing will not have a notch. In this case, the rotary valve **MUST** be aligned with the rotary valve shaft using a degree wheel on the replacement crankcase. This alignment is relatively easy and is covered in Chapter 8 -- **ENGINE.** A cutout paper degree wheel is also provided on a separate page in the Appendix and may be cutout for use.

Rotary Valve Disc Modification

The stock factory "out-the-door" valve disc is machined to provide maximum power for a broad range of rpm operation. The window is a set size to control port timing. Therefore, under most circumstances, the disc should **NOT** be altered.

Some experienced performance seekers, have increased the valve closing delay in very small increments, until the desired results are obtained. Indiscriminate snipping at the valve disc, especially on the port closing side of the window, will cause an appreciable delay in closing the port after the piston has started to move downward from TDC. Such action would cause low speed blow back.

Retarding the intake port closing, will move the engine power peak higher and at the same time, will reduce the power at the lower end of the rpm range.

The intake port opening point has some influence on engine power output, but it is not significant compared to port closing timing. Early intake port opening provides a minimal increase in power. Polishing the intake ports and smoothing out any rough edges on the rotary disc will help engine performance.

RAVE Valve Adjustment

The RAVE valve system requires no tuning, other than periodically checking the Red adjustment knob, located on the top of the black housing cap. Set at the factory, this knob should rest flush with the Black cap. Periodic inspection and cleaning of the sliding valve is outlined in the next Chapter -- **MAINTENANCE.**

3-11 MODIFYING FOR HIGHER PERFORMANCE

Normally Seloc manuals only cover stock engine and jet pump components. However, personal watercraft enthusiasts are notorious for modifying their equipment to gain higher performance. Therefore, a large aftermarket industry has mushroomed in just a few short years to accommodate this need for speed. A few well chosen words on modifying for performance seem to be in order.

The following general guidelines will provide improved two stroke performance. The tasks listed below, in order of effectiveness, or possibly a combination of modifications and components, may give the desired results.

a- To increase bottom end power:
Increase cylinder compression.
Advance the timing.
Use a long chamber exhaust.
Use a low pitch impeller.
Use a larger nozzle bore.

b- To increase top end power:
Install a larger diameter carburetor.
Use a short exhaust chamber.
Go to higher porting specifications.
Use a higher pitched impeller.

c- To increase acceleration:
Increase cylinder compression.
Use a lighter weight driveshaft.
Install an aluminum flywheel.
Check for optimum air flow.

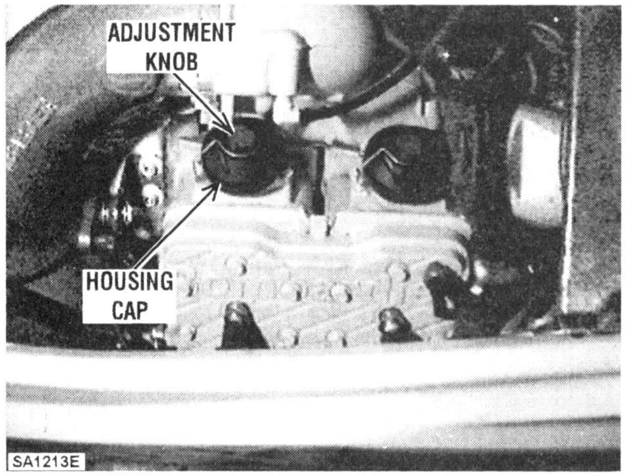

Periodically, check to be sure the red adjustment knob is flush with the top of the rotary valve cap. This is the same calibration adjustment pre-set by the manufacturer prior to shipping the craft.

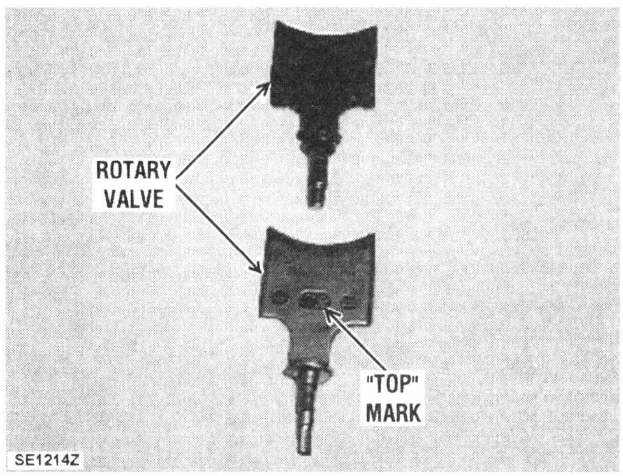

*The manufacturer recommends cleaning the RAVE valve -- Model 787 engines only -- every 100 hours of operation. Check to be sure the "**TOP**" mark is facing **UP** during installation. If the RAVE valve is installed incorrectly, it would surely catch a piston, causing severe damage to the engine.*

4
MAINTENANCE

4-1 INTRODUCTION

The watercraft being prepared for service probably represents a sizeable investment. In order to protect this investment and for the owner to receive the maximum amount of enjoyment from the craft it must be cared for properly while being used and when it is out of the water. Always store the watercraft with the bow slightly higher than the stern and be sure to remove the inner hull drain plug/s. If any type of cover is used to protect the craft -- plastic, canvas, whatever -- be sure to allow for some movement of air through the hull. Proper ventilation will assure evaporation of any condensation formed due to changes in temperature and humidity.

GOOD WORDS

The authors estimate 75% of engine repair work can be directly or indirectly attributed to lack of proper care for the engine. This is especially true of care during the off-season period. There is no way on this green earth for a mechanical engine, particularly a marine engine, to be left sitting idle for an extended period of time, say for six months, and then be ready for instant satisfactory service.

Imagine, if you will, leaving your automobile for six months, and then expecting to turn the key, have it roar to life, and be able to drive off in the same manner as a daily occurrence.

It is critical for a marine engine to be run at least once a month, preferably, in the water, but if this is not possible, then a flush attachment **MUST** be connected whether directly to the engine block or in the hose line from the jet pump to the engine.

Move the craft to a body of water or connect a garden hose to the engine cooling water supply fitting or flush fitting on the cylinder head.

CAUTION

The following procedure MUST be followed whenever the engine is started with a hose attached for cooling the engine. Start the engine, and THEN turn on the water. The engine MUST be started first to prevent water from backing up into the tuned exhaust pipe and into the engine. Water must circulate anytime the engine is operating to prevent overheating -- which could cause damage to moving parts and possibly cause engine seizure.

NEVER allow the engine to operate without cooling water for more than 15 seconds. NEVER operate the engine with a flushing attachment for more than five minutes. The driveline seal is not provided with cooling water if it is not immersed in water. To shut the engine down, first shut off the water supply, and then shut down the engine. If this sequence of tasks is not followed as stated, the engine could be severely damaged.

There is no substitute and few valid excuses for not developing good habits leading to regular maintenance practices. The results of a good program are evident in top performance, maximum enjoyment, and "pride of ownership", all at minimum cost.

During the storage period, and especially after a prolonged period of storage, the steering should be operated from hard-over to hard-over to verify the system is in proper working order.

and then shut down the engine. If this sequence of tasks is not followed as stated, the engine could be severely damaged.

REMEMBER

Start the engine.
Turn the water on.
Turn off the water.
Shut down the engine.

NEVER, AGAIN NEVER, operate the engine at high speed with a flush device attached. The engine, operating at high speed with such a device attached, might very likely **RUNAWAY** from lack of load on the impeller, causing extensive damage.

After the engine has been operated, the shift mechanism, if the unit is so equipped,

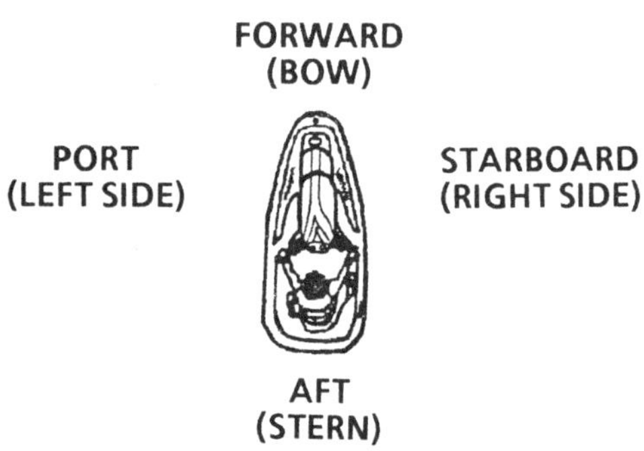

Common terminology used throughout the world and extensively in this manual for reference designation on water craft of ALL sizes. "Port", "Starboard", "Forward", and "Aft", never change, even if standing on your head.

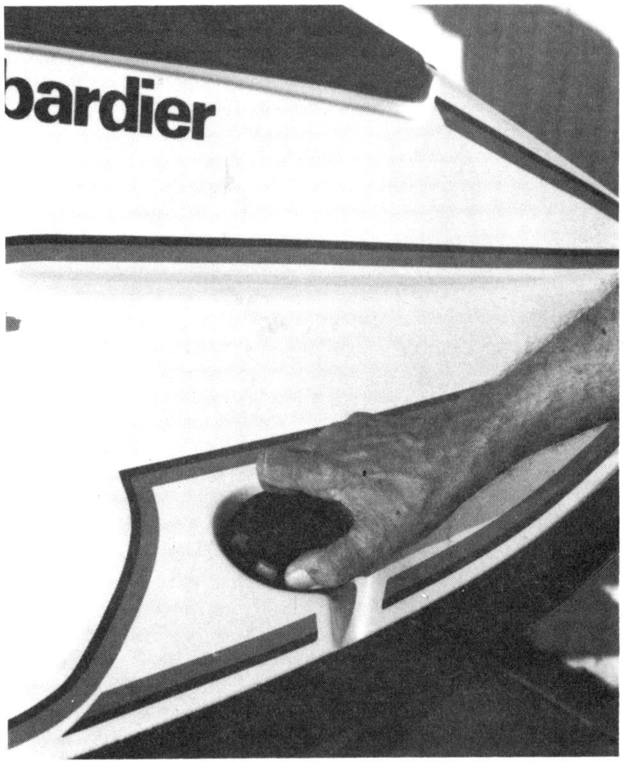

*After a fun day on the water and before moving the craft out of the water, **SLOWLY** open the fuel tank cap to release built-up pressure in the tank.*

should be operated through the full range several times. The steering should also be operated from hard-over to hard-over several times.

Only through a regular maintenance program can the owner expect to receive long life and satisfactory performance at minimum cost.

The material presented in this chapter is divided into four general areas.

1- General information every owner should know.

2- Maintenance tasks to be performed periodically to keep the craft operating at minimum cost.

3- Care necessary to maintain the appearance of the craft and give the owner a "Pride of Ownership" look.

4- Winter storage practices to minimize deterioration during the off-season when the craft is not in use.

In nautical terms, the front of the craft is the **BOW**; the rear is the **STERN**; the right side, when facing forward, is the **STARBOARD** side; and the left side is the **PORT** side. All directional references in this manual use this terminology. Therefore, the direction from which an item is viewed is of no consequence, because **STARBOARD** and **PORT** never change no matter where the individual is located or his/her position -- even if standing on one's head.

AFTER USE TASKS

At the end of a day's use, make a habit of opening the engine compartment to allow fumes to escape and sponging out all water in the bottom of the hull before trailering the craft. Leave the engine cover off when the craft is stored overnight to allow any moisture in the compartment to evaporate. Slowly open the fuel tank cap to release built up pressure. Pressure in the tank will increase considerably during hot weather. Releasing the pressure reduces the chance of an explosion and insures smooth fuel flow when the engine is next started.

If the craft will remain unused for a week, spray storage oil inside the engine and a light film of lubricant, such as WD-40, Super Lube, or Justice Bros. 80, on the engine exterior. Keeping the engine exterior lightly oiled at all times will result in water sliding off the engine and draining away rather than accumulating in casting webs -- resulting in corrosion.

Cooling

The engine cooling passages must be flushed with clear water on a regular basis. This accomplished by simply connecting a "Flushing Kit" per the manufacturer's instructions **OR** by disconnecting the cooling hose from the pump to the engine and connecting a garden hose to the disconnected hose. Such an easy maintenance practice should become a habit and will prevent accumulation of debris which could restrict engine coolant water flow and cause overheating.

Controlling Corrosion

Since man first started out on the water, corrosion has been his enemy. The first form was merely rot in the wood and then it was rust, followed by other forms of destructive corrosion in the more modern materials.

The higher the concentration of salt in the water, the worse the corrosion which will take place.

One defense against corrosion is to use similar metals in the manufacture of the parts exposed to the water.

A second defense against corrosion is to insulate dissimilar metals. This can be accomplished by using an exterior coating of Sea Skin or by insulating them with plastic or rubber gaskets.

Pump Impeller

The pump impeller is a precisely machined and dynamically balanced aluminum spiral. Observe the drilled recesses at exact locations to achieve the delicate balancing.

Excessive vibration of the jet pump may be attributed to an out-of-balance condition caused by the impeller striking rocks, gravel, or by cavitation "burn".

The term cavitation "burn" is a common expression used throughout the world among people working with pumps, impeller blades, and forceful water movement.

"Burns" on the impeller blades are caused by cavitation air bubbles exploding with considerable force against the impeller blades. The edges of the blades may develop small "dime size" areas resem-

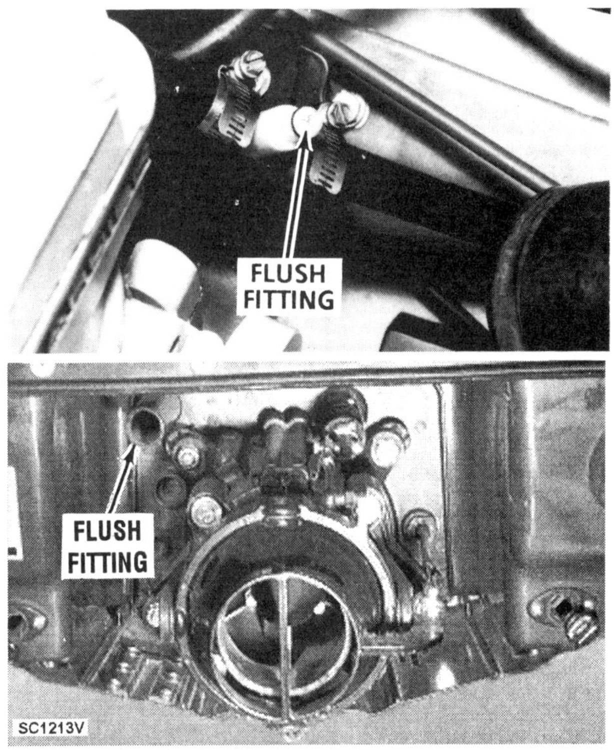

Location of the flush fitting on many of the Bombardier Sea-Doo "upright" engines -- top illustration. A garden hose connected to this fitting is used to flush the engine -- a very important task, especially following operation in salt water. Other models are equipped with a flush fitting located next to the jet pump -- bottom.

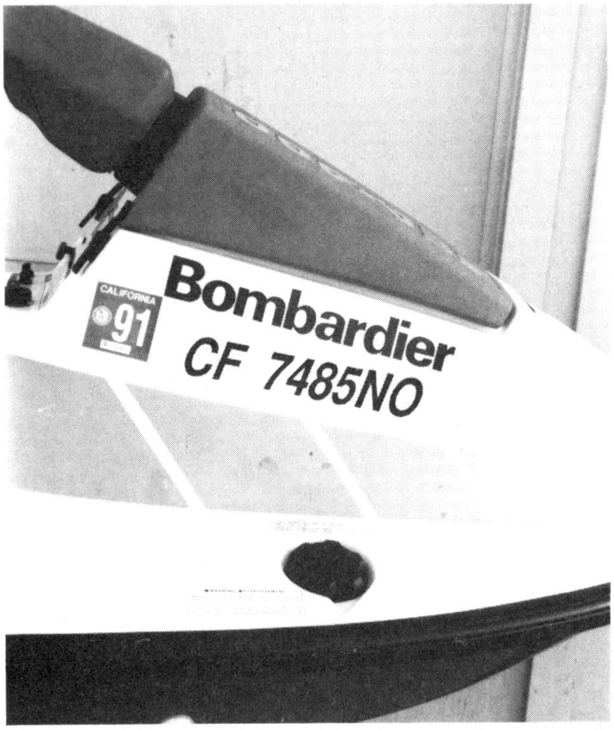

The state license identification number must be prominently displayed on the forward portion of the hull -- port and starboard.

bling a porous sponge, as the aluminum is actually "eaten" by the condition just described.

Excessive rounding of the impeller edges will reduce efficiency and performance. Therefore, the impeller should be inspected at regular intervals. If rounding is detected, the impeller should be removed from the housing, placed on a work bench, and the edges rounded as close to a 1/32" radius as possible, using a file.

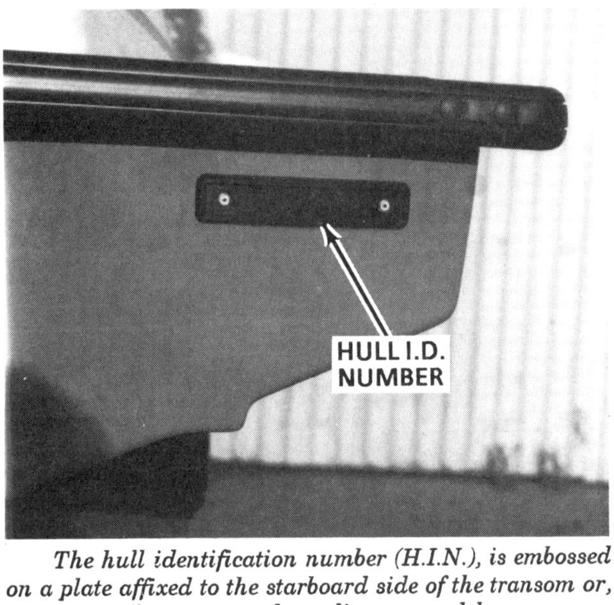

The hull identification number (H.I.N.), is embossed on a plate affixed to the starboard side of the transom or, top center of transom -- depending on model.

Refer to Chapter 10 -- **Jet Pump** for detailed procedures on removal and installation of the impeller.

4-2 SERIAL NUMBERS

Most manufacturers use three different identification numbers stamped, embossed, or decaled somewhere on the craft.

The engine serial number is the manufacturer's key to engine changes. This number identifies the year of manufacture, the qualified horsepower rating, and the parts book identification. If any correspondence or parts are required, the engine serial number **MUST** be used or proper identification is not possible.

U. S. Federal regulations require all personal watercraft to have a Hull Identification Number (HIN) permanently attached to the craft in two separate locations: one on the transom on the starboard side, above the waterline, and the other in an unexposed location, which is usually left to the discretion of the manufacturer. This "HIN" number can be most helpful when attempting to retrieve a stolen craft. The "HIN" number is required in most states in order to properly register the craft and ensure the owner pays all taxes due.

Some manufacturers also use a Primary I. D. Number stamped on a plate attached to the inside of the engine compartment.

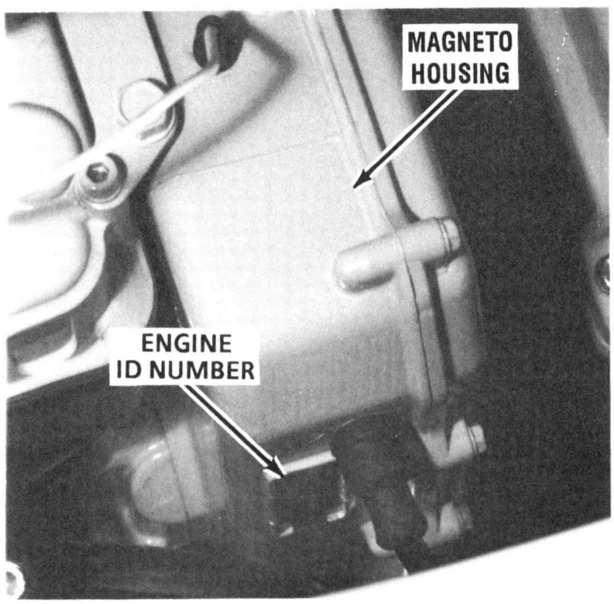

The engine ID number is located on a plate affixed to the starboard side of the magneto housing on 587 through 717 model engines. The ID number is affixed to the aft crankcase, just forward of the aft flywheel.

ONE MORE WORD

As a theft prevention measure, all metal plates with any type of identification number are especially made "wafer" thin. Any attempt to remove these plates will result in cracks across the serial number.

4-3 LUBRICATION -- COMPLETE UNIT

With every type of mechanical invention with moving parts, lubrication plays a prominent role in operation, enjoyment, and longevity of the unit at minimum cost.

If a personal watercraft is operated in salt water, the frequency of applying lubricant to fittings is usually cut in half as compared with the unit being used in fresh water. The few minutes involved in moving around the craft applying lubricant and at the same time making a visual inspection of its general condition will pay in rich rewards with years of continued service.

The first personal watercraft arrived on the scene in the United States in the early 1970's. It is not uncommon to see outboard units well over 20 years of age moving a boat through the water as if the unit had recently been purchased from the current line of models. An inquiry with the proud owner will undoubtedly reveal his main credit for its performance to be regular periodic maintenance. There is no reason this same longevity of performance cannot be true for personal watercraft.

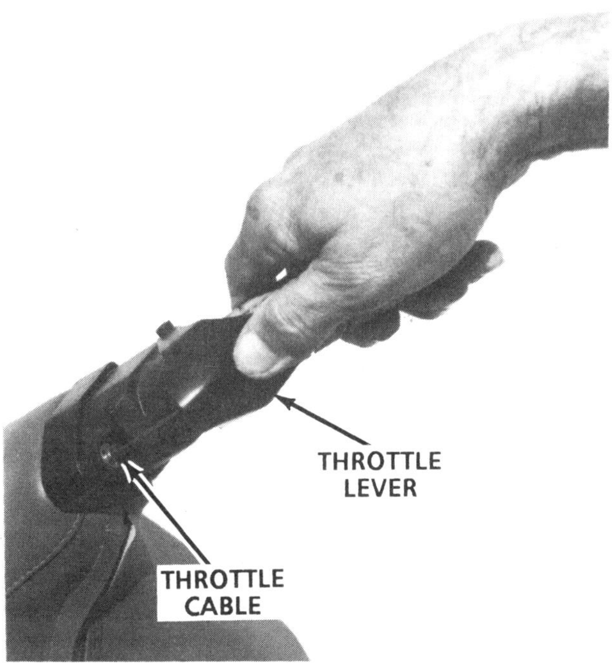

Lubricate the throttle cable at the handlebar every 100 hours of engine operation.

The accompanying chart can be used as a guide to the frequency of maintenance while the craft is being used during the season. Unless otherwise stated, use a good name multi-purpose water resistant lubricant. Detailed lubrication instructions are given below for locations requiring special attention.

In addition to the normal lubrication listed in the lubrication chart, the prudent owner will inspect and make checks on a regular basis, as listed in the accompanying chart.

Throttle Cable Lubrication

Hold the throttle lever in the full open position against the grip. Use a small screwdriver and pry out the small seal from the groove of the throttle stop and a little way along the cable. Spray rust inhibitor along the exposed length of cable. Fill the groove with multi-purpose water resistant lubricant, and then pull the seal down along the cable and back into place inside the groove. Wipe away any excess lubricant.

Steering Cable Lubrication

Pry the barrel of the steering cable free of the ball joint on the steering post arm. Coat the ball joint with lubricant, and then snap the barrel back onto the ball joint.

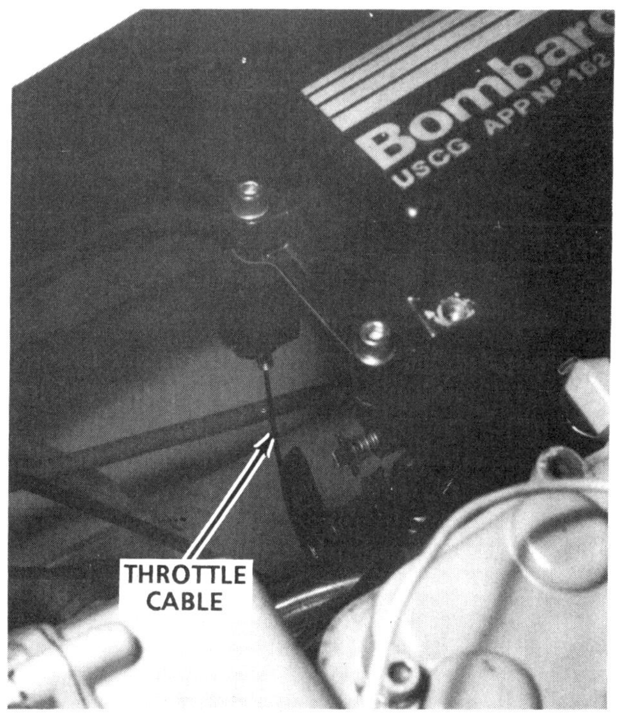

Lubricate the carburetor end of the throttle cable at the same time the handlebar end is lubricated.

Lubrication Points:	Initailly every 10hrs	Initially every 50hrs	Initially every 100hrs	Then every 10hrs	Then every 100 hrs	Then every 200 hrs
Carburetor throttle shaft			❖		❖	
Throttle cable - engine end			❖		❖	
Throttle cable - handlebar end			❖		❖	
Choke cable - engine end			❖		❖	
Choke knob shaft			❖		❖	
Steering nozzle pivot point			❖		❖	
Steering post arm ball joint			❖		❖	
Reverse gate pivot pins			❖		❖	
Jet pump end cap reservoir			❖		❖	
Aft flywheel /driveshaft	❖			❖		
Seat latch			❖		❖	

Maintenance Tasks:

Maintenance Tasks	Initailly every 10hrs	Initially every 50hrs	Initially every 100hrs	Then every 10hrs	Then every 100 hrs	Then every 200 hrs
Flame arrestor element	❖	❖	❖		❖	
Clean and gap spark plugs	❖	❖	❖		❖	
Adjust carburetor idle	❖		❖		❖	
Adjust carburetor mixture	❖		❖		❖	
Drain and inspect fuel tank						❖
Clean and inspect fuel filter	❖	❖	❖		❖	
Inspect fuel line			❖		❖	
Inspect check valve			❖		❖	
Adjust throttle control			❖		❖	
Adjust choke system				❖		
Adjust steering system	❖			❖		
Inspect impeller	❖			❖		
Check impeller to pump case clearance			❖		❖	
Inspect rubber coupling						❖
Inspect electrical connections	❖	❖	❖		❖	
Check battery electrolyte level	❖	❖	❖		❖	
Clean bilge strainer			❖		❖	
Inspect rubber water seal	❖		❖		❖	
Retorque head bolts		❖	❖		❖	
Inspect driveshaft boot	❖	❖	❖		❖	

Fuel/Oil Mixture

A new or rebuilt engine must be properly "broken-in" if it is to develop maximum performance. An engine which is incorrectly broken-in will never be able to achieve the maximum rated horsepower. During the initial breaking-in period, the fuel/oil ratio should be "rich" with oil. Extra two-stroke oil should be added to the fuel tank at a rate of 1 oz. per gallon, for the first 3-5 hours of operation after the unit is brought out of storage. This added oil will **ENSURE** adequate lubrication of moving parts which have been drained of oil during the storage period. Operation of an engine without the proper amount of oil will quickly lead to a significant loss of horsepower -- efficiency of the engine will be compromised.

After the first few hours of operation with extra lubricant, the oil injection system will deliver an adequate amount of oil for proper lubrication.

Experienced two-stroke mechanics run their own units on a higher mixture of oil on the premise it is cheaper to replace a fouled spark plug than to have a complete engine overhaul with rebored cylinders, oversize pistons and rings. These same mechanics who own oil injected units add an extra ounce of oil per each gallon of fuel. The addition of this extra oil may prevent engine seizure in the event of an oil pump failure or clog in an oil delivery line.

Use only outboard **MARINE** oil in the mixture, never automotive oils. Four-stroke automotive engine oil is not formulated to burn completely -- only to lubri-

The steering cable ball joint at the outlet nozzle requires periodic lubrication.

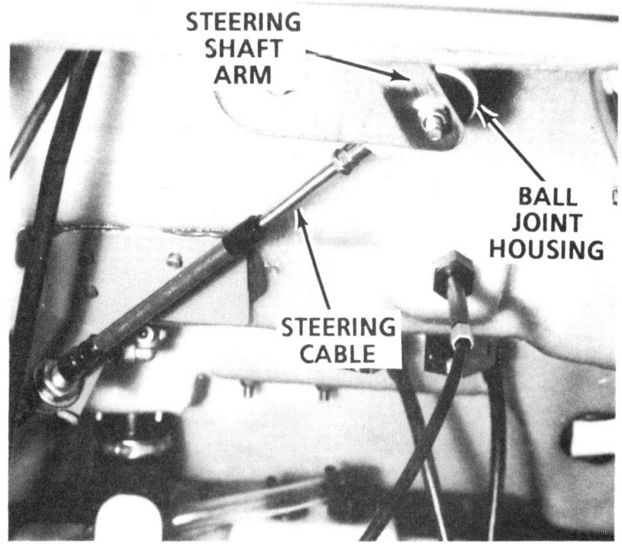

The steering cable ball joint at the steering shaft arm also requires periodic lubrication.

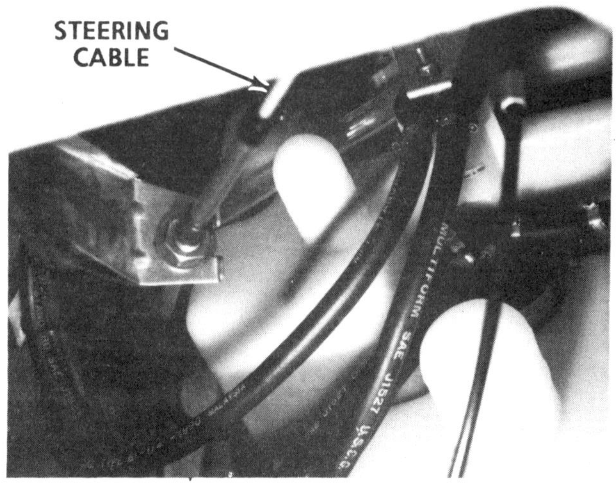

All exposed cable should be lubricated on a regular basis.

Checking the oil level in the oil tank.

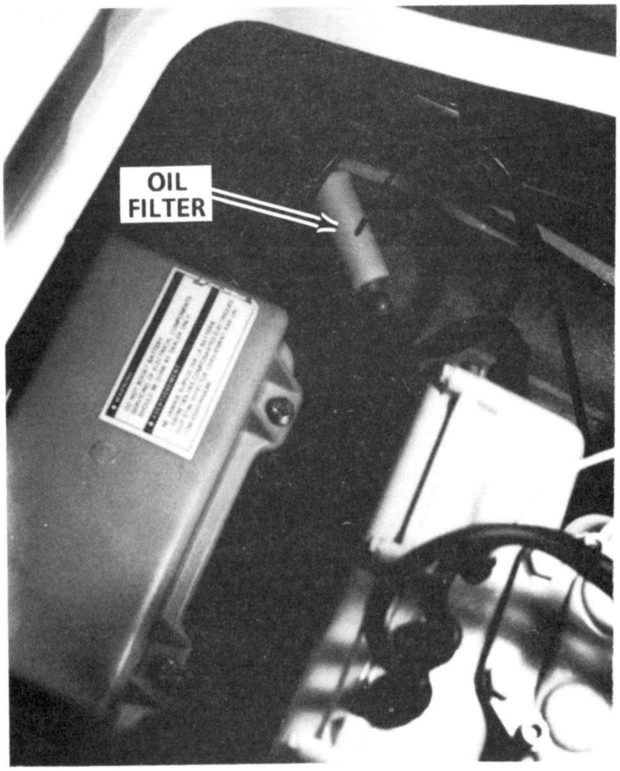

The in-line oil filter is located between the oil tank and the oil injection pump.

cate. Therefore, automotive oil, if used in a two-stroke engine will leave an undesirable residue.

Engine Oil Tank and Filter

A remote oil injection tank is mounted next to the fuel tank. The tank has an oil capacity of 1.0 - 1.5 gallons (4.0 - 6.0 L) -- 787 engine only; 0.7 - 1.5 gallons (3.0 - 6.0 L) -- all other engines.

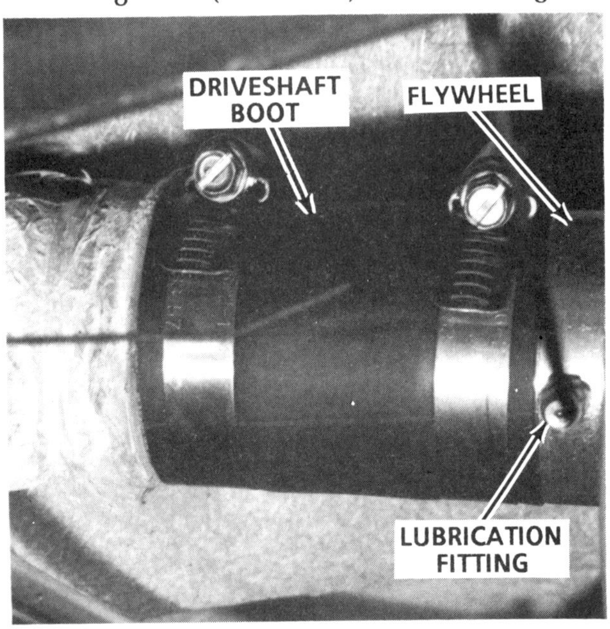

Driveshaft boot equipped on all models prior to 1995. The flywheel lubrication fitting is identified.

Oil level in the tank may be checked by using an externally mounted dipstick -- shown in the accompanying illustration on the previous page.

An in-line oil filter is installed in the line between the oil tank and the oil injection pump. The manufacturer recommends the filter be inspected on a monthly basis during the operating season. The level of oil in the oil tank should be checked each day the craft is operated.

Jet Pump Lubrication

The forward end of the pump shaft is supported by the flywheel. A lubrication fitting on the flywheel hub serves to lubricate the forward end of the driveshaft. Use a standard pump and force bearing lubricant into the fitting just until the boot begins to expand.

The aft end of the pump shaft is supported by a bearing in the impeller housing. This bearing is lubricated through two holes in the casting from a reservoir in the end cap at the aft end of the shaft. The lubricant level in the reservoir should also be checked every three months during the season of operation.

SPECIAL WORDS

The pump may remain in place on the craft, but the nozzle assembly must be removed to gain access to the end cap reservoir. See Chapter 10 -- **JET PUMP** for detailed procedures to remove and install the nozzle.

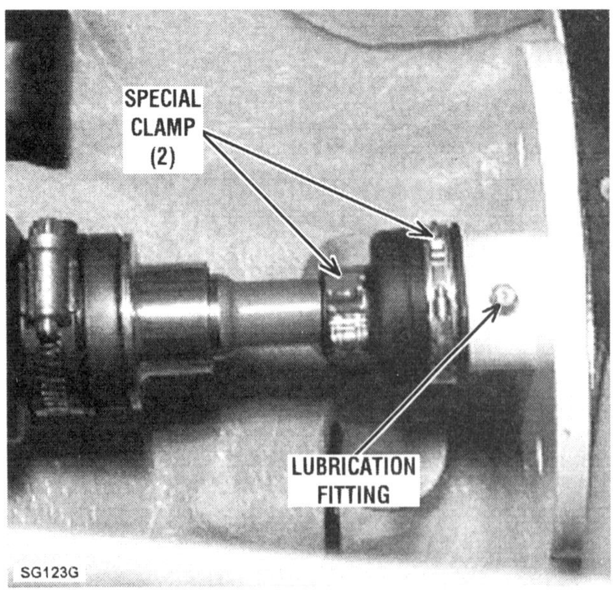

Late type driveshaft boot from 1995 on, with flywheel lubrication fitting identified.

To fill the reservoir, after the nozzle is clear, remove the filler plug. Use only Sea-Doo Jet Pump Synthetic Oil, P/N 293-600-011. **NEVER** mix this synthetic oil with a mineral oil. Pour the oil into the filler opening, until the level reaches the bottom of the opening.

NOTE
Allow the oil to drain into the reservoir for at least five minutes, then add additional oil to bring the level up to the bottom of the opening.

Apply a light coating of Loctite 515 to the threads of the filler plug. Install and tighten the plug securely.

4-4 INSPECTION AND SERVICE

Fuel Tank Check Valve and Oil Tank Check Valve

Before disturbing any hose be sure to have a new hose tie wrap or a small hose clamp to replace the one removed. Original equipment tie wraps must be cut through to remove them from a hose.

Disconnect the ventilation hose from the fitting on the fuel tank pickup. Remove the two tie wraps, one on each end of the check valve.

Blow through the fitting on one face of the valve. Now, blow through the fitting on the other face of the valve. The air should pass through the valve **ONLY** in one direction. A directional arrow should be embossed on the valve indicating the direction of fuel flow -- and the direction air passes. If there is no directional arrow embossed on the valve, use a magic marker, or other device, and mark an arrow to indicate the direction of air and fuel flow. If the check valve will not allow air to pass in one direction, the fuel will remain in the tank and not be drawn out by the fuel pump.

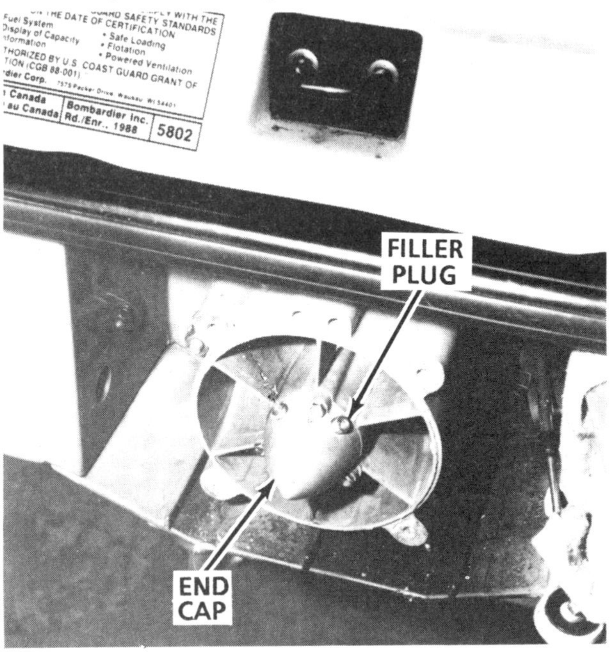

The impeller shaft is lubricated from a reservoir in the end cap at the aft end of the shaft.

Only Sea-Doo Synthetic Oil should be used to lubricate the jet pump.

If the valve allows air to pass in the wrong direction, it will also allow fuel to flow out of the tank through the valve and spill into the hull when the craft is subjected to violent maneuvers. Therefore, replace the valve if it fails to function properly -- allows air to pass in only one direction.

Perform the same check for the oil tank check valve. Both the oil and fuel valves are identical.

A faulty fuel tank check valve will starve the engine of fuel. If the condition becomes really serious, the engine will faulter and die.

A faulty oil check valve will reduce critical lubrication to the engine. Such a condition will quickly cause extensive and expensive damage.

Fuel Filler Cap

Remove the fuel filler cap and inspect the condition of the rubber seal. Replace the seal if it is worn or cracked.

Check the tightness of the large hose clamp around the fuel tube to the tank.

Drain Plug

Remove the drain plug from the transom. Inspect the condition of the O-ring. Replace if necessary. Clean any dirt or

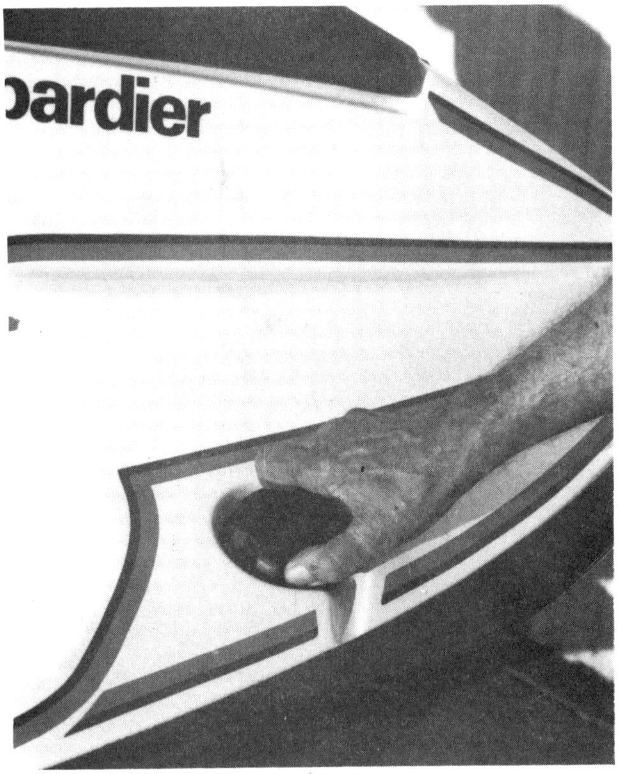

Removing the fuel filler cap to inspect the condition of the rubber seal.

sand from the plug threads. Install the plug and tighten it securely.

4-5 IMPELLER-TO-IMPELLER HOUSING WEAR RING CLEARANCE

SPECIAL WORDS

The following procedure may be performed with the watercraft elevated, as on

*When preparing to inspect the rock grate, observe the warning decal on the stern regarding which direction the craft must be tilted -- **ALWAYS** to port.*

a set of saw horses, or with the craft resting on its side. Either of these positions will provide access to the rock grate and the impeller. The accompanying illustration shows the craft tilted up for clearance.

If the impeller-to-wear ring clearance is to be determined with the water craft raised on saw horses, the battery may be left secured in the craft. **HOWEVER**, if the watercraft is to be positioned on its side, first disconnect the electrical leads at both battery terminals, and then remove the battery.

Once the craft is secured in a favorable position, remove the four bolts securing the rock grate to the jet pump. If the seal between the hull and the pump housing is broken, or the foam packing is damaged, reseal the housing and intake area with silicone sealant to prevent impeller cavitation.

Insert a feeler gauge in the center between each impeller blade and the wear ring. Determine the clearance.

If the clearance is less than 0.040" (1.00mm) for each blade, the condition is considered satisfactory. No action is necessary.

If the clearance is more than the service limit, inspect the condition of the wear

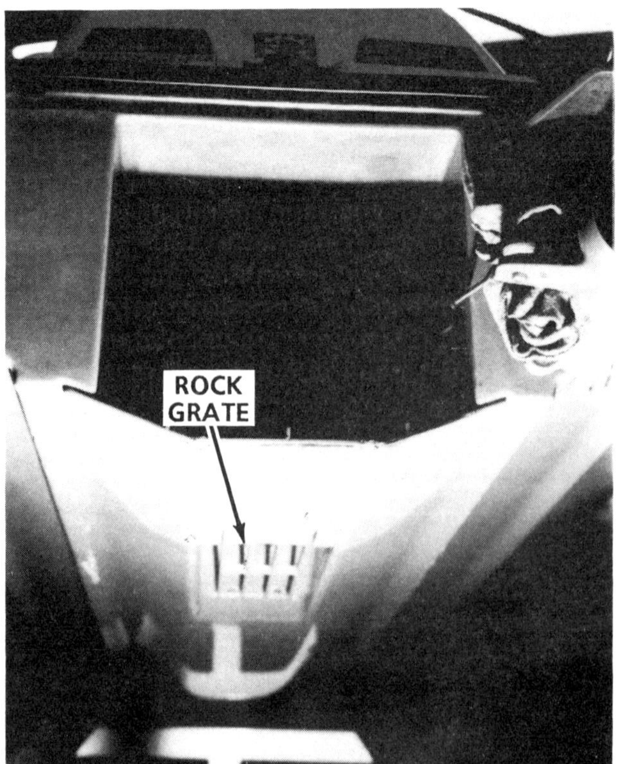

This craft is set on a stand with the bow raised to gain access to the rock grate.

ring. If the wear ring has deep scratches, replace the wear ring. If the wear ring appears in good condition, the problem must be with the impeller. Visually inspect the impeller for nicks, scratches, pitting or a "mushroomed" edge.

If the clearance is not kept to a minimum, the jet pump will lose efficiency by allowing excess water to pass between the blades and the housing. This water has no thrust and will actually slow the stream of water under pressure from the impeller.

The jet pump must be removed from the craft to replace the wear ring or to service the impeller. See Chapter 10 -- **JET PUMP** for detailed illustrated instructions.

Install the rock grate and tighten the attaching bolts securely.

Engine Alignment

The pump shaft is indexed directly into the splined hub on the flywheel. Bombardier Sea-Doo watercraft do not use a rubber disc type coupler between the flywheel and the pump shaft, as found on many other watercraft. Engine alignment is set at the factory and should **NEVER** require adjustment. The rubber engine mounts and shim material pieces should not be disturbed.

If an engine mount should ever need replacing, due to deterioration, **TAKE CARE** to save and use the original pieces of shim material over the mounts to maintain engine alignment.

4-6 PRE-SEASON PREPARATION

Satisfactory performance and maximum enjoyment can be realized if a little time is spent in preparing the unit for service at the beginning of the season. Assuming the unit has been properly stored, as outlined in Section 4-10 of this chapter, a minimum amount of work is required to prepare the unit for action on the water.

The following steps outline an adequate and logical sequence of tasks to be performed before using the craft the first time in a new season.

1- Lubricate the craft according to the manufacturer's recommendations. Refer to the lubrication chart on Page 4-6. Remove, clean, inspect, adjust, and install the spark plugs with new gaskets (if they require gaskets). Make a thorough check of the ignition system. This check should include: all electrical connections; condition of the wiring; battery electrolyte level; and status of the battery charge.

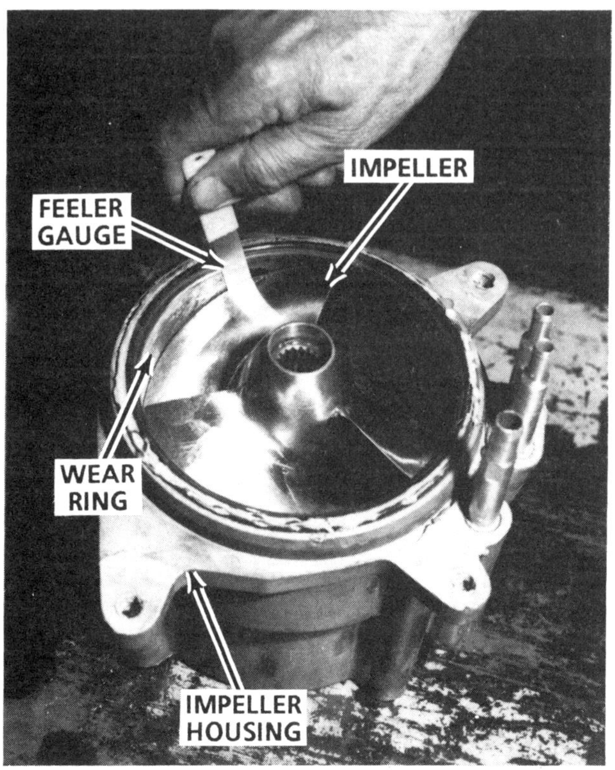

Using a feeler gauge to check the clearance between the impeller blades and the wear ring in the impeller housing, as explained in the text.

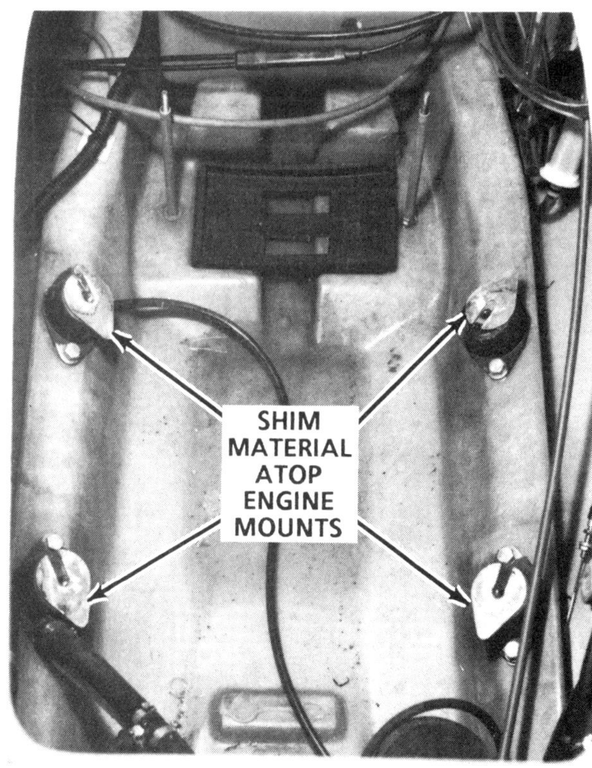

The shim material on top of the engine mounts is determined at the factory and should never need to be changed.

2- Take time to check the gasoline tank and all fuel lines, fittings, couplings, valves, including the flexible tank fill and vent. Turn on the fuel supply valve at the tank. If the fuel was not drained at the end of the previous season, make a careful inspection for gum formation. When gasoline is allowed to stand for long periods of time, gummy deposits form. This gum can clog the filters, lines, and passageways in the carburetor.

GOOD WORDS

All manufacturers recommend the fuel and oil filters be replaced at the start of each season or at least once a year.

3- Close all water drains. Check and replace any defective water hoses. Check to be sure the connections do not leak. Replace any spring-type hose clamps, if they have lost their tension, or if they have distorted the water hose. The new clamps should be the band-type clamps.

4- The engine can be operated with the jet pump in water to flush it. If this is not practical, a flush attachment may be used. On earlier models, this device is attached to the engine water supply hose, between the engine and the pump. On later models, the fitting is located on the aft end of the jet pump thru-hull assembly.

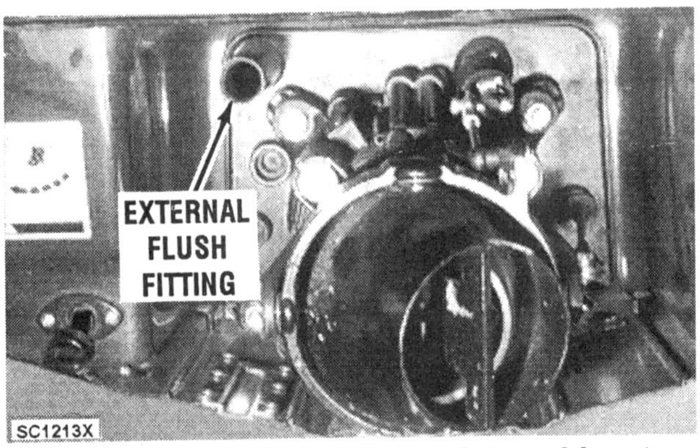

Location of the flush fitting on late model engines. On earlier models, the fitting is located inline, within the engine compartment.

The fuel, oil and cooling connections should be checked to be sure they are fastened securely.

CAUTION

The following procedure MUST be followed whenever the engine is started with a hose attached for cooling the engine. Start the engine, and THEN turn on the water. The engine MUST be started first to prevent water from backing up into the tuned exhaust pipe and into the engine. Water must circulate anytime the engine is operating to prevent overheating -- which could cause damage to moving parts and possibly cause engine seizure.

NEVER allow the engine to operate without cooling water for more than 15 seconds. NEVER operate the engine with a flushing attachment for more than five minutes. The driveline seal is not provided with cooling water if it is not immersed in water. To shut the engine down, first shut off the water supply, and then shut down the engine. If this sequence of tasks is not followed as stated, the engine could be severely damaged.

REMEMBER

Start the engine.
Turn the water on.
Turn off the water.
Shut down the engine.

Check the exhaust outlet for water discharge. Check for leaks.

The flame arrestor should be kept fairly clean and the rubber gasket around the arrestor must remain in good condition.

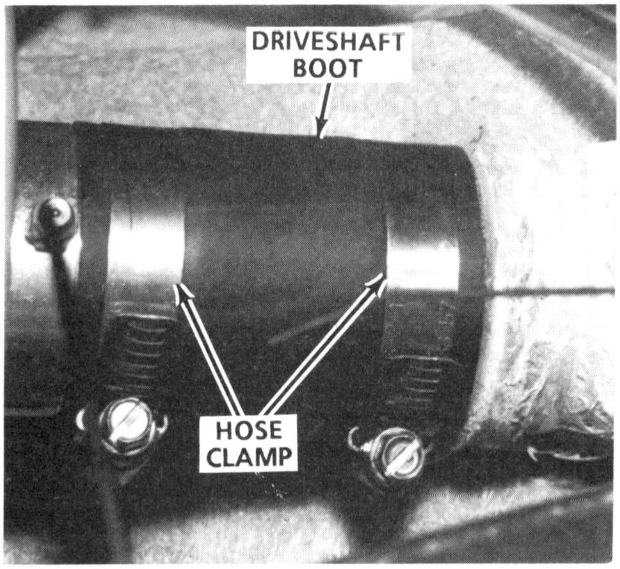

Old style driveshaft boot secured with standard hose clamps.

5- Check the electrolyte level in the battery and the voltage for a full charge. Clean and inspect the battery terminals and cable connections. **TAKE TIME** to check the polarity, if a new battery is being installed. Cover the cable connections with lubricant or special protective compound as a prevention to corrosion formation. Check all electrical wiring and grounding circuits.

6- Check all electrical parts on the engine and lower portions of the hull to be sure they are of the type to prevent ignition of an explosive atmosphere. Rubber caps help keep spark insulators clean and reduce the possibility of arcing. Electric cranking motors and high-tension wiring harnesses should be of a marine type to prevent an explosive mixture from igniting.

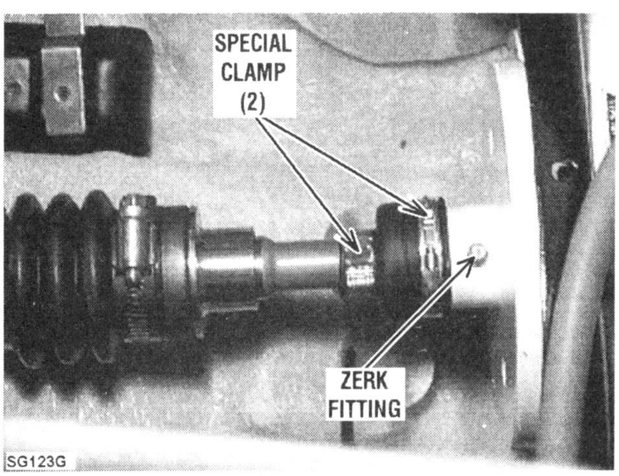

New style driveshaft boot secured with special clamps. A special tool is required to remove/install these clamps as stated in the text.

4-7 SEALANTS, ADHESIVES, LUBRICANTS, AND FUEL STABILIZERS

It is common practice for the larger manufacturers of personal watercraft to market their own line of products for use on their craft. Bombardier chemical engineers have developed such a line available for use with Bombardier personal watercraft and snowmobiles. Throughout this manual, the authors recommend the application of the manufacturer's products as a first choice. All products listed are alternatives of equal value, and may be used with confidence, if the manufacturer's line is not available.

Sealants and Adhesives

Many sealants and adhesives are recommended and they are **NOT** interchangeable. Each is designed to perform under a different set of conditions. Follow the directions on the package for cleaning and preparing surfaces. Sealants and adhesives **MUST** be applied **ONLY** to clean dry sur-

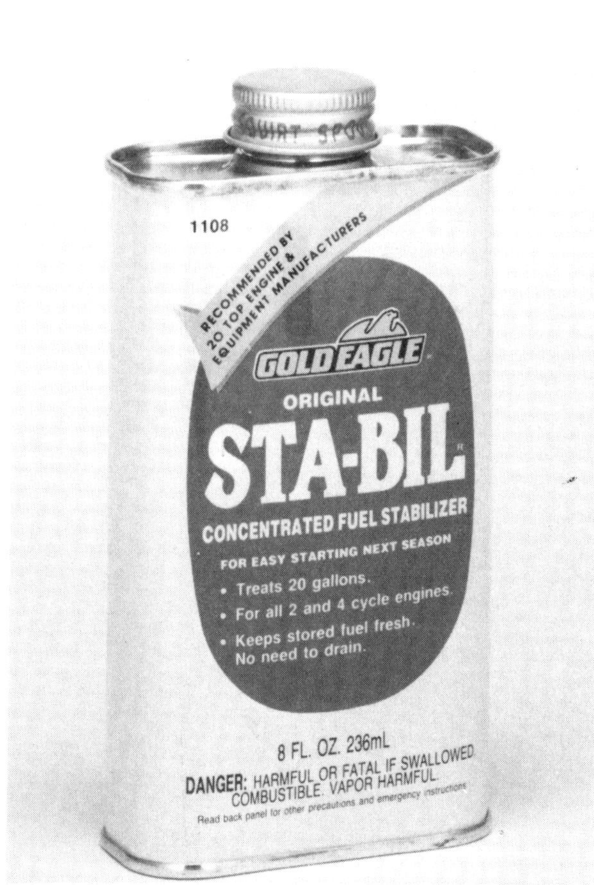

A gasoline stabilizer and conditioner may be used to prevent fuel from "souring" for up to twelve full months.

FACTORY RECOMMENDED LUBRICANTS AND ADHESIVES

PRODUCT	PART NUMBER	USE
Sea-Doo Lube	293-600-006	Spark plug hole threads
Sea-Doo Injection Oil	293-600-005	Oil injection system and rotary valve face.
Sea-Doo Jet pump Synthetic Oil	293-600-011	Jet pump end cap reservoir
Sea-Doo PL-32 Sealant	293-800-001	Outside diameter of all oil seals
Sea-Doo Bearing Grease	293-550-005	Jet Pump thrust bearing and washer
Loctite 157 Stripper	295-000-040	Anywhere Loctite 515 Gasket Eliminator was used.
Loctite 242 (blue)	293-800-002	Cranking motor, rock grate, steering nozzle, engine support, engine mount, exhaust manifold, flame arrestor, rotary valve, and stator plate mounting bolt threads. Also crankshaft taper and all electrical component mounting bolts.
Loctite 414 Superbond	N/A	Drive shaft plug
Loctite 515 Gasket Eliminator	293-800-007	Cylinder head, crankcase, and nozzle-to-impeller housing mating surfaces.
Loctite 517 or 587 Superflex Ultra Blue	293-800-016	Impeller housing mounting holes, ride shoe perimeter, and all oil seal lips.
Loctite 592	N/A	Exhaust outlet water fittings
Loctite PST 567	293-800-013	All other water fittings, and crankcase drain plug threads.
Loctite 648 (green)	290-899-788	Distance sleeve on rotary valve shaft
Loctite 767 Anti-seize	293-550-001	Crankshaft/flywheel threads, drive shaft and impeller shaft threads, and all bearing race exteriors.
Loctite Primer N	293-600-012	Internal aluminum threads
Loctite U Blue	293-800-006	All exposed cable threads
Silicone 732 RTV	N/A	Electrical harness grommets, hull screw threads, and wear ring lip.
Armour All ®	N/A	Vinyl seat

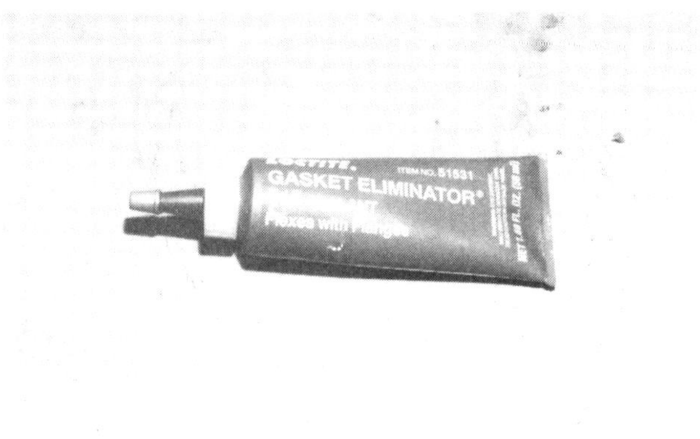

One of the many Loctite products the manufacturer endorses for use in sealing engine parts.

faces. Apply sparingly -- excessive amounts may block oil passageways and cause serious damage.

The accompanying table lists all the required products to lubricate and correctly seal any work or maintenance performed on the craft.

Fuel Stabilizer

"Sta-Bil" Fuel Conditioner and Stabilizer is recommended during engine operation and during the storage period. This fluid absorbs water in the fuel system and protects against corrosion.

If used during operation, this fuel additive will prevent the formation of gum and varnish deposits and greatly extend the period between required carburetor overhauls.

When added to the fuel during storage, the additive will prevent the fuel from "souring" for up to twelve full months.

Bombardier distributes many products under their own label, such as this Multi-Purpose Lubricant.

4-8 FIBERGLASS HULLS

Fiberglass reinforced plastic hulls are tough, durable, and highly resistant to impact. However, like any other material they can be damaged. One of the advantages of this type construction is the relative ease with which it may be repaired. Because of its break characteristics, and the simple techniques used in restoration, these hulls have gained popularity throughout the world.

A fiberglass hull has almost no internal stresses. Therefore, when the hull is broken or stove-in, it retains its true form. It will not dent to take an out-of-shape set. When the hull sustains a severe blow, the impact will be either absorbed by deflection of the laminated panel or the blow will result in a definite, localized break. Repairs are usually confined to the general area of the rupture.

Cleaning, Waxing, and Polishing

Any craft should be washed with clear water after each use to remove surface dirt and any salt deposits from use in salt water. Regular rinsing will extend the time between waxing and polishing. It will also give "pride of ownership", by having a sharp, "squared away" appearance. Elbow grease, a mild detergent, and brush will be required to remove stubborn dirt, oil, and other unsightly deposits.

Avoid harsh abrasives or strong chemical cleaners. A white buffing compound can be used to restore the original gloss to a scratched, dull, or faded area. The finish should be thoroughly cleaned, buffed, and polished at least once each season. Take care when buffing or polishing with a marine cleaner not to overheat the surface being worked, because the area will become "burned".

4-9 SUBMERGED ENGINE SERVICE

A submerged engine is always the result of an unforeseen accident. Once the craft is recovered, special care and service procedures **MUST** be closely followed in order to return the unit to satisfactory performance.

CRITICAL WORDS

Never attempt to start an engine immediately after it has been submerged. If water has entered a cylinder, the piston will not be able to compress the liquid. Any attempt to start the engine will most likely result in a bent connecting rod.

NEVER, again we say **NEVER** allow an engine that has been submerged to stand more than a couple hours before following the procedures outlined in this section and making every effort to get it running. Such delay will result in serious internal damage. If all efforts fail and the engine cannot be started after the following instructions have been performed, the engine should be disassembled, cleaned, assembled, using new gaskets, seals, and **O**-rings, and then started as soon as possible.

Submerged engine treatment is divided into three unique problem areas: submersion in salt water; submerged while engine was running; and a submerged unit in fresh water.

The most critical of these three circumstances is the engine submerged in salt water, with submersion while running, a close second.

Salt Water Submersion

NEVER attempt to start the engine after it has been recovered from salt water submersion. This action will only result in additional parts being damaged and the cost of restoring the engine will be increased considerably. If the engine was submerged in salt water the complete unit **MUST** be disassembled, cleaned, and assembled with new gaskets, **O**-rings, and seals. See Chapter 8 -- **ENGINE.** The corrosive effect of salt water can only be eliminated by the complete job being properly performed.

SUBMERGED WHILE RUNNING

SPECIAL INSTRUCTIONS

If the engine was running when it was submerged, the chances of internal engine damage is greatly increased. Remove the spark plugs to prevent compression in the cylinders. Use a socket wrench on the flywheel nut to rotate the crankshaft. If the attempt fails, the chances of serious internal damage is greatly increased. Serious damage might include, but not be limited to: bent connecting rod/s, bent crankshaft, or damaged cylinder. If the crankshaft cannot be rotated, the engine must be completely disassembled. See Chapter 8 -- **ENGINE.**

Submerged Engine -- Fresh Water

As an aid to performing the restoration work, the following steps are numbered and should be followed in sequence. However, illustrations are not included with the procedural steps because the work involved is general in nature.

1- Recover the craft as quickly as possible.

2- Remove the engine compartment cover and the spark plugs.

3- Remove the carburetor diaphragm cover.

4- Flush the outside of the engine with fresh water to remove silt, mud, sand, weeds, and other debris. **DO NOT** attempt to start the engine if sand has entered any part of the engine. Such action will only result in serious damage to engine components. Sand in the engine means the unit must be disassembled. See Chapter 8 -- **ENGINE.**

CRITICAL WORDS

Never attempt to start an engine immediately after it has been submerged. If water has entered a cylinder, the piston will not be able to compress the liquid. Any attempt to start the engine will most likely result in a bent connecting rod.

5- Remove as much water as possible from the engine. Most of the water can be eliminated by turning the craft over with the spark plug opening facing **DOWN**. With the craft in this position, rotate the flywheel with a socket wrench on the flywheel nut. Rotate the crankshaft through at least 10 complete revolutions. After you are satisfied there is no more water in the cylinders, proceed with the next step.

6- Alcohol will absorb moisture. Therefore, position the craft in an upright position, and then pour alcohol into the carburetor throat. Again rotate the flywheel through several revolutions of the crankshaft.

7- Pour alcohol into the spark plug openings and again rotate the crankshaft.

8- With the craft in the upright position, remove the two crankcase drain plugs -- Allen head bolts -- located under the oil injection pump on the starboard side. Allow all the liquid -- oil and possibly water -- to drain.

9- With the craft still upright, pour approximately one teaspoon of engine oil into each spark plug opening. Rotate the crankshaft to distribute the oil in the cylinders.

10- Install and connect the spark plugs.

11- Install the carburetor diaphragm cover.

Move the craft to a body of water or connect a garden hose to the engine cooling water supply fitting or flush fitting on the cylinder head.

12- Obtain **FRESH** fuel and attempt to start the engine.

CAUTION

The following procedure MUST be followed whenever the engine is started with a hose attached for cooling the engine. Start the engine, and THEN turn on the water. The engine MUST be started first to prevent water from backing up into the tuned exhaust pipe and into the engine. Water must circulate anytime the engine is operating to prevent overheating -- which could cause damage to moving parts and possibly cause engine seizure.

NEVER allow the engine to operate without cooling water for more than 15 seconds. NEVER operate the engine with a flushing attachment for more than five minutes. The driveline seal is not provided with cooling water if it is not immersed in water. To shut the engine down, first shut off the water supply, and then shut down the engine. If this sequence of tasks is not followed as stated, the engine could be severely damaged.

REMEMBER

Start the engine.
Turn the water on.
Turn off the water.
Shut down the engine.

13- If the engine fails to start, determine the cause, electrical or fuel, correct the problem and again attempt to get it running. **NEVER** allow a recovered submerged engine to remain unstarted for more than a couple hours without following the procedures in this section and attempting to get it running. If attempts to start the engine fail, the unit should be disassembled, cleaned, assembled, using new gaskets, seals, and O-rings, as **SOON** as possible.

4-10 WINTER STORAGE

Taking extra time to store the craft properly at the end of each season, will increase the chances of satisfactory service when brought out for the next season's operation. **REMEMBER**, idleness is the greatest enemy of a marine engine. The unit should be run on a monthly basis. The steering and shifting mechanism should also be worked through complete cycles several times each month. The owner who spends a small amount of time involved in such maintenance will be rewarded by satisfactory performance, and greatly reduced maintenance expense for parts and labor.

Proper storage involves adequate protection of the unit from physical damage, rust, corrosion, and dirt.

If a submerged engine is not attended to without delay, the unit may suffer serious damage and keep the owner off the water much longer than necessary, until expensive repairs are completed.

The following steps provide an adequate maintenance program for storing the unit at the end of each season.

1- Empty all fuel from the carburetor. For many years there has been the widespread belief simply shutting off the fuel at the tank and then running the engine until it stops is the proper procedure before storing the engine for any length of time. Right? **WRONG! ! !**

First, it is **NOT** possible to remove all fuel in the carburetor by operating the engine until it shuts down from lack of fuel. Considerable fuel is trapped in the float chamber and other passages and in the line leading to the carburetor. The **ONLY** guaranteed method of removing **ALL** fuel is to take the time to remove the carburetor, and drain the fuel.

Secondly, if the engine is operated with the fuel supply shut off until it quits from lack of fuel, the fuel and oil mixture inside the block is removed, leaving bearings, pistons, rings, and other parts without any protective lubricant.

Remove the spark plugs and pour about an ounce of two-cycle engine oil into each spark plug opening. Wait a minute or two, and then slowly rotate the flywheel through two complete revolutions of the crankshaft. This action will evenly distribute the oil on the cylinder walls. Install the spark plugs.

2- Drain the fuel tank and the fuel lines. Pour approximately one quart (0.96 liters) of benzol (benzine) into the fuel tank, and then rinse the tank and pickup filter with the benzol. Drain the tank. Store the craft with the engine compartment cover **OPEN** to allow air to circulate around the tank. Leave the tank fuel vent **OPEN** to permit air to circulate through the tank.

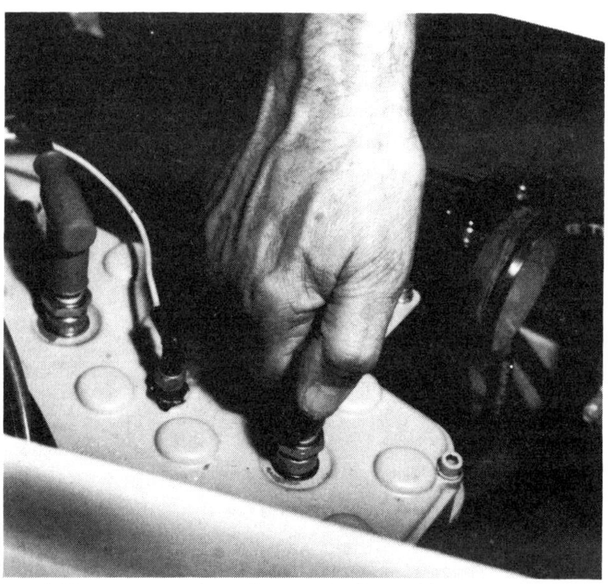

When preparing the craft for off-season storage, remove the spark plugs and insert about an ounce of oil into each opening, as explained in the text.

3- Clean the in-line fuel filter with benzol. See Chapter 6 -- **FUEL**.

4- Lubricate the throttle and shift linkage. Lubricate the steering pivot shaft with water resistant multi-purpose lubricant.

5- Plug the exhaust outlet with a shop towel or similar item to prevent any type of contaminant from entering the exhaust system.

BATTERY STORAGE

Remove the battery from the craft and keep it charged during the storage period. Clean the battery thoroughly of any dirt or corrosion, and then charge the battery to the full specific gravity reading. After charging, store the battery in a clean cool dry place where it will not be damaged or knocked over.

NEVER store the battery with anything on top of it or cover the battery in such a manner as to prevent air from circulating around the filler caps. All batteries, both new and old, will discharge during periods of storage, more so if they become hot than if they are stored in a place to remain cool.

The battery should always be removed from the craft during extended periods of storage. Keep it dry, cool, and charged.

Therefore, the electrolyte level and the specific gravity should be checked at regular intervals. A drop in the specific gravity reading is cause to charge the battery back to a full reading.

One of the largest personal watercraft battery manufacturers recommends a battery in storage should be checked every two weeks and charged if necessary. The electrolyte can also be replenished at this time to prevent sulfation from occurring inside the battery.

In cold climates, **EXERCISE CARE** in selecting the battery storage area. A fully charged battery will freeze at about 60° below zero.

4-11 PRE-SEASON CHECK

Before attempting to start the craft after winter storage, check the steering and throttle action for smooth operation, without any indication of binding. If the battery was removed for storage, check to be sure it has a full charge.

FILLING THE BATTERY

Remove each filler cap using a pair of pliers. Fill each cell to the proper level with distilled water. The level of electrolyte should be between the upper and lower level marks. When filling the battery, fill to the upper line and allow the battery to stand for about twenty minutes. Check the level again and replenish, as necessary.

Cooling water and exhaust system hose connections should be checked as part of the pre-season routine.

Use a temperature corrected hydrometer to test the specific gravity of the electrolyte. At 20°C (68°F), the hydrometer reading should be 1.26 on the scale. If it is necessary to charge the battery, leave the filler caps lightly resting on the cell openings to allow the gases to escape. The charging current should not exceed 1.9 amps for 10 hours.

After charging, push each filler cap into place and wipe the top of the battery clean before installation.

Clean the battery posts and battery cable ends with a wire brush to ensure good, clean connections. Clean the top surface of the battery and identify the **"POS"** or "+" and the **"NEG"** or "-" embossed symbols. Correctly connect the battery cables to the battery observing polarity. If the cables are connected backwards, expensive parts in the ignition system will be **DESTROYED** the first time an attempt is made to start the engine.

Check to be sure the breather pipe is in place on the battery and the pipe was not pinched when the battery was installed.

FIRST TIME START

If the proper maintenance was observed at the end of the previous season, both the carburetor and fuel tank should be completely empty. Fill the gas tank with the proper mixture and open the fuel valve, if equipped. Open the choke. **DO NOT** attempt to crank the engine until it starts, **AT THIS TIME**. Considerable cranking time will be spent before the oil/fuel mixture engine bearings and other moving parts. Ex-

*The bilge inlet and hose should be checked at the beginning of each season to **ENSURE** the system will work on demand in an emergency.*

pensive parts may be damaged due to lack of lubrication and the cranking motor will become excessively hot, which is also **BAD NEWS**.

If the craft cannot be moved to a body of water, connect a garden hose to the flush fitting.

CAUTION

The following procedure MUST be followed whenever the engine is started with a hose attached for cooling the engine. Start the engine, and THEN turn on the water. The engine MUST be started first to prevent water from backing up into the tuned exhaust pipe and into the engine. Water must circulate anytime the engine is operating to prevent overheating -- which could cause damage to moving parts and possibly cause engine seizure.

NEVER allow the engine to operate without cooling water for more than 15 seconds. NEVER operate the engine with a flushing attachment for more than five minutes. The driveline seal is not provided with cooling water if it is not immersed in water. To shut the engine down, first shut off the water supply, and then shut down the engine. If this sequence of tasks is not followed as stated, the engine could be severely damaged.

REMEMBER
Start the engine.
Turn the water on.
Turn off the water.
Shut down the engine.

The correct procedure for starting the engine is to obtain an **AEROSOL** can of WD-40. The product **MUST** be an aerosol canister because the liquid in the container must be atomized. A simple squirt can will not do the job. **DO NOT** use ether or a commercial starting fluid because they will evaporate almost instantly. These products work well with four-stroke engines equipped with an oil pump, but they do not contain a lubricant essential for a two-stroke engine. WD-40 is an extremely flammable oil.

Check to be sure both spark plugs are tight to prevent the mist from the canister from igniting. Spray the WD-40 through the flame arrestor and continue spraying with **NO** throttle advance until the engine starts. As soon as the engine is operating, cease spraying the WD-40. The engine may emit some smoke for a few seconds after it starts. Watch the sediment bowl (if so equipped). As soon as the bowl fills with fuel, the engine should operate on the regular oil/fuel mixture. Work the throttle to keep the engine at idle speed until it reaches operating temperature, then close the choke. Advance the throttle a few times and the engine should then idle smoothly.

If the flame arrestor cover was removed prior to starting the engine, install the cover and check to be sure it is fastened securely to the flame arrestor housing.

Check to be sure the battery electrolyte reaches the upper mark. During engine operation, the electrolyte should remain between the upper and lower level marks.

*Location of the engine flush fitting. If the engine is to be operated out of the water, a garden hose **MUST** be attached to this fitting to provide cooling water, as explained with **CAUTIONS** in the text.*

5
TROUBLESHOOTING

5-1 INTRODUCTION

This chapter is divided into five main sections as follows:

5-2 Mechanical engine problems.
5-3 Fuel system problems.
5-4 Ignition system faults.
5-5 Cranking system failures.
5-6 Charging system malfunctions.
5-7 Troubleshooting Charts

If a change in engine rpm is experienced at normal operating speed, such a change is usually a symptom of an engine problem or trouble in the jet pump. An increase or a decrease in rpm may be an indication of the area to be checked. The following generalizations may be helpful in isolating the problem.

A noticeable change in craft performance, as evidenced by one of the following symptoms, may be attributed to one or more of the areas listed:

Lower than Normal Engine Rpm
Mechanical problems in the engine:
Plugged flame arrestor.
Fouled spark plug/s, incorrect heat range or gap.
Shorted or weak ignition coil.
Contaminated fuel.
Restricted exhaust system.
Inadequate fuel pump pressure.
Loss of compression due to blown head gasket.
Mechanical problems in the jet drive:
Worn or damaged impeller.
Main bearing failure.
Driveshaft spline failure.

Higher than Normal Engine Rpm
Mechanical problem in the engine:
None.
Mechanical problems in the jet drive:
Clogged intake grate -- possibly plastic bag or other debris.
Line, sea weed, etc. entangled in impeller.
Ingested sand.

If a reduction in speed is experienced, accompanied with no loss or gain of engine rpm, the loss of performance can fairly accurately be attributed to some type of drag on the craft or restriction of the water ejected from the jet pump.

Drag on the hull is usually caused by damage to the craft.

Many times, drag from the jet pump is caused by a partially lowered reverse gate, if so equipped, or from dragging debris caught in the intake grate.

A partially lowered reverse gate may be caused by incorrect cable adjustment or by a weak or broken return spring.

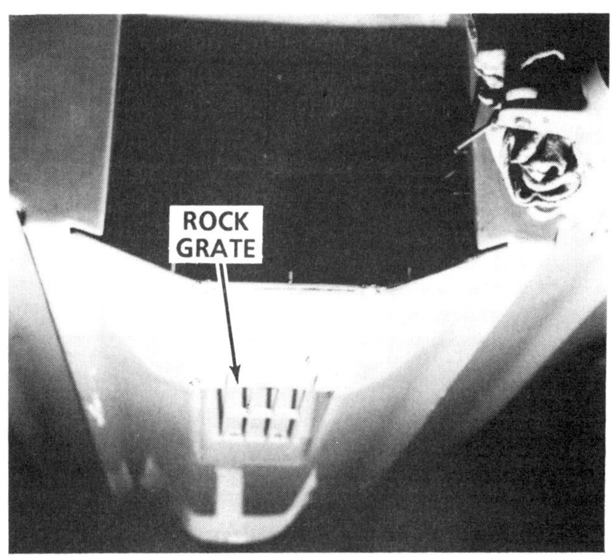

One of the most hazardous conditions for a personal watercraft is restriction of water through the rock grate caused by a plastic bag, plastic wrap, or by some type of sea weed. Even the smallest reduction in water volume will affect jet pump performance and could quickly lead to engine overheating.

Engine Troubleshooting

Troubleshooting must be a well thought-out procedure. Always attempt to proceed with the troubleshooting in an orderly manner. The "shot in the dark" approach will only result in wasted time, incorrect diagnosis, replacement of unnecessary parts, and frustration.

Obviously, if the instructions are to be of maximum benefit as a guide, they must be fully understood and followed in the proper sequence.

When an engine fails to start, the trouble must be localized to one of four general areas: cranking system; ignition system; fuel system; or compression. Each of these areas must be systematically inspected until the trouble is isolated. At that time, detailed tests of the system or area must be made to determine the part causing the problem. The last section of this chapter describes charging system troubleshooting.

Troubleshooting Check

When using this Troubleshooting Check, proceed sequentially through each test until the defect is uncovered. Then skip to the detailed testing procedure and check for that system. For example, if, when using the Troubleshooting Check procedure, the first two systems, cranking and ignition tests OK, but the third test shows there is trouble in the fuel system, then proceed to the detailed test under the Fuel System Troubleshooting Check in Section 5-3.

Cranking System Test

1- Turn the key switch to the **START** position or press the ignition button. The cranking motor should crank the engine at a normal rate of speed.

If the cranking motor cranks the engine slowly or fails to crank it at all, the trouble is either in the battery or in the cranking system. Test the battery first, then if the battery checks out, proceed directly to Troubleshooting the Cranking System in Section 5-5 for detailed testing procedures to isolate the problem.

Ignition System Test

2- Connect a standard spark tester (available at almost any automotive parts store at very nominal cost), between the spark plug and the spark plug lead. Hold the spark plug lead with an insulated holder (to avoid receiving a "shock" when the current passes through). Crank the engine with the cranking motor. If there is **NO** spark, or if the spark is very weak, the trouble is in the ignition system. Proceed directly to the Ignition Troubleshooting in Section 5-4 for

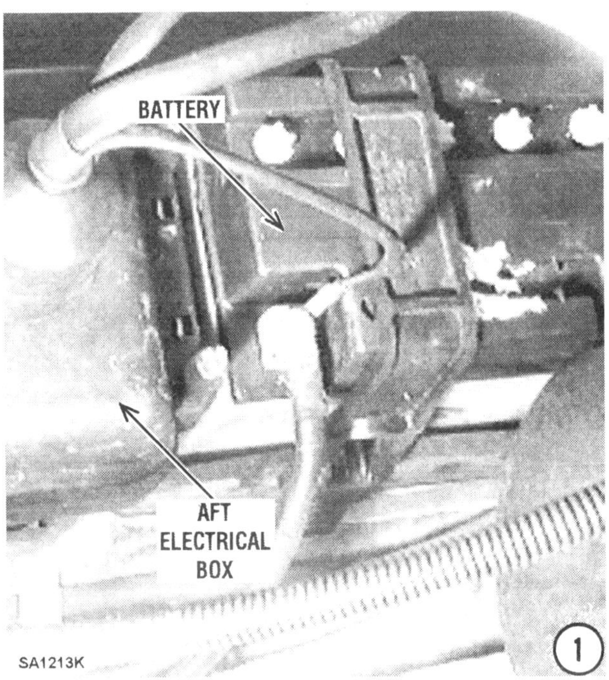

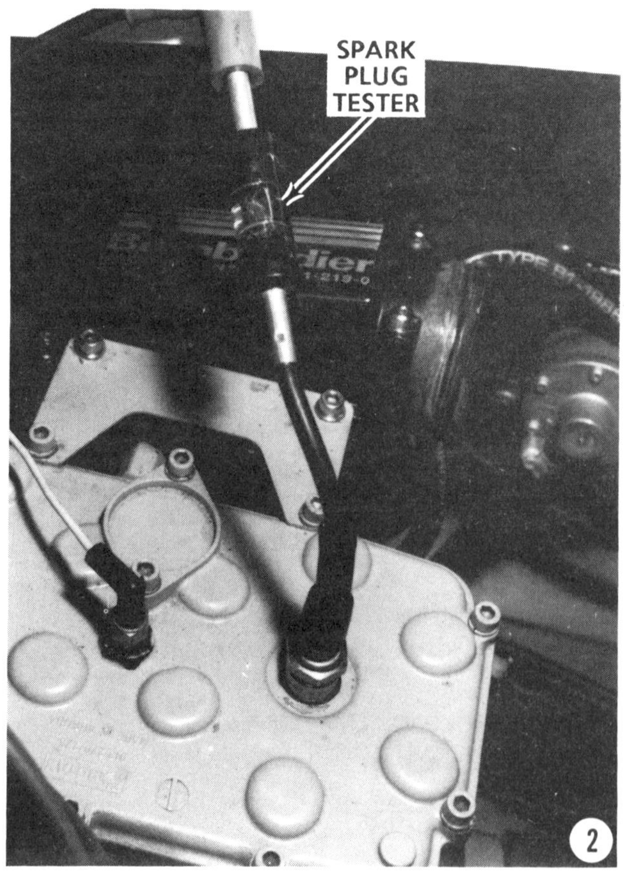

detailed testing procedures to uncover the problem.

3- Remove the flame arrestor. Look down into the throat of the carburetor and at the same time open and close the throttle several times. Observe the throttle valve to verify it moves from the fully closed to the fully open position. Perform the same check with the choke valve.

Attempt to determine the age of the fuel in the fuel tank. Many fuels tend to "sour" in three to four months, especially if an additive such as Sta-Bil has not been added. Other fuels take longer to "sour". In no case should an attempt be made to start the engine if the fuel in the tank is more than 12 months old.

Compression Test

4- Good compression is the key to proper engine performance. An engine with worn piston rings or a blown gasket cannot be made to perform satisfactorily until the mechanical defects contributing to low compression are corrected. Generally, a compression gauge is used to determine the cranking pressure within each cylinder.

To make a compression test, remove both spark plugs and remember or identify from which cylinder they were taken. Evaluation of the spark plug firing end can be most useful in determining how the cylinder is functioning. After the spark plugs have been removed, install a compression gauge into one cylinder and crank the engine.

The throttle valve and choke **MUST** be in the **WIDE OPEN** position in order to obtain maximum readings. Crank the engine through several revolutions and observe the highest reading on the compression gauge.

The cranking compression for a new engine is approximately 140psi. The significance in a compression test is the variation in pressure readings between the cylinders. As long as this variation is between 20-30 psi, cylinder compression is considered as "satisfactory". If a greater variation exists, then the lower reading cylinder should be checked by making a cylinder leak test. A simple leak test can be made by first inserting a teaspoonful of oil into the spark plug opening of the lower reading cylinder, and then cranking the engine a few times to distribute the oil.

Check the compression again to see if inserting the oil caused a change. The oil helps make a temporary seal around the rings and increases the compression. If the reading increased, the compression loss was probably due to worn rings. If the reading did not change appreciably, the loss may be due to a blown head gasket.

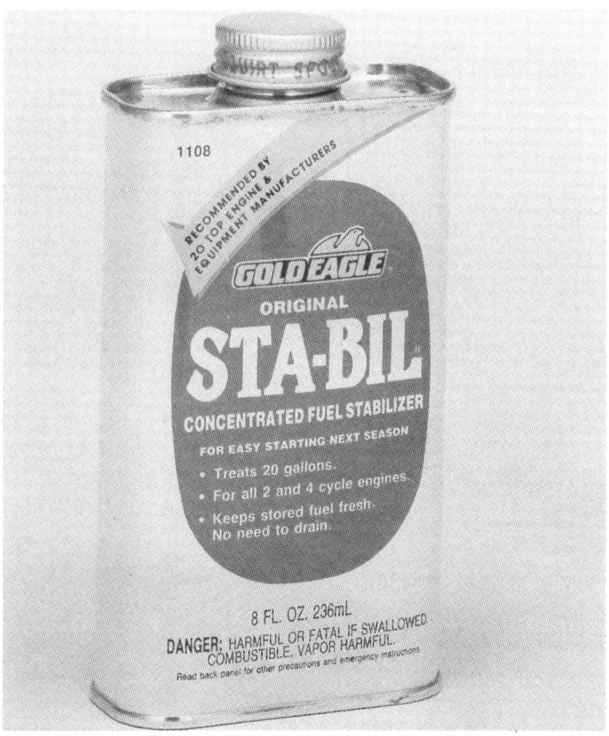

A gasoline stabilizer and conditioner may be used to prevent fuel from "souring" for up to twelve full months.

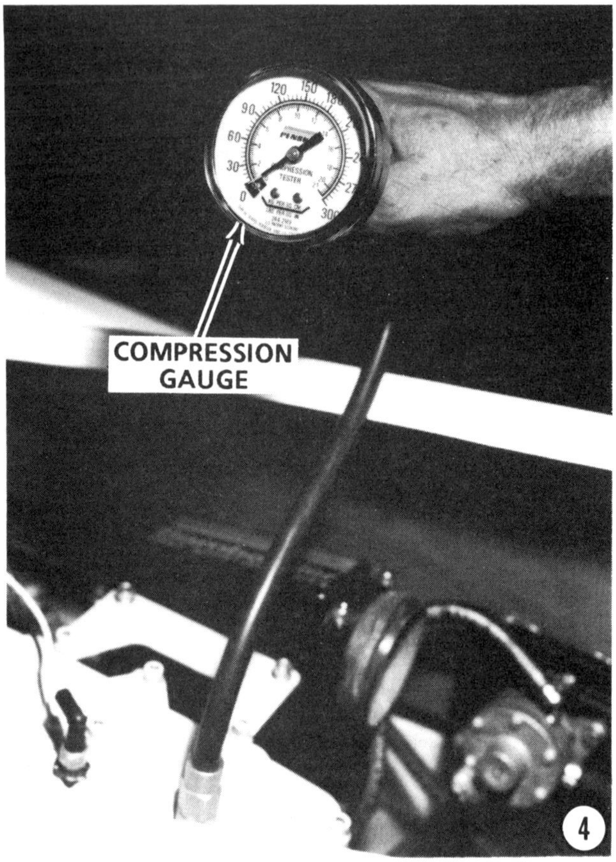

5-2 MECHANICAL ENGINE PROBLEMS

Jumping Waves

Even though "jumping waves" with a personal watercraft may be considered **GREAT** sport, consider the extensive internal engine damage which will surely occur if this type action is consistently practiced.

During wave jumping or using a wet ramp, when the craft clears the water, the engine RPM approaches the limit set by the RPM limiter in an unloaded condition.

Once the craft returns to the water surface, the sudden load on the impeller is tranferred along the driveline. Speed of the rotating pump shaft and the crankshaft is suddenly reduced while the flywheel continues to spin rapidly due to its mass. This condition, especially when repeated over a period of time, often shears the Woodruff key and the flywheel becomes repositioned on the crankshaft.

A very small misalignment will result in a change in ignition timing. The craft operator will notice a drastic loss of power. A worse condition would be the flywheel shifting to the degree engine start would not be possible.

This sudden increase and decrease in engine speed dramatically shortens engine life. Bent crankshafts and broken connecting rods can usually be directly attributed to this type of "fun play".

Easy Check for Twisted Crankshaft

A twisted crankshaft may be detected by first removing the spark plugs and setting up a dial indicator in one of the spark plug openings. If a dial indicator is not available, a pencil, straw, or other small diameter straight object may be used.

With the indicator, or other device, in place, determine the TDC (top dead center) of the No. 1 piston. Now, make two matching marks -- one on the flywheel and the other adjacent on the stator. Next, rotate the flywheel **EXACTLY** 180° -- preferably using a degree wheel on the flywheel. Make a second mark on the stator adjacent to the mark on the flywheel.

NOW, check for TDC of the other piston. The dial indicator should indicate the same reading as for the first piston **AND** the two marks on the stator should be **EXACTLY** opposite each other through the center of the crankshaft. If the indicator is off or if the marks on the stator are not aligned, as described -- the crankshaft is twisted and **MUST** be replaced.

Engine Noises

Engine noises can be generally classified as knocks, slaps, clicks, or squeaks. These noises are usually caused by loose bearings, sloppy pistons, or other moving parts of the engine.

A main bearing knock is usually identified by a dull thud which is noticeable under engine load. Attempting to move the craft

Removal of the crankshaft is not necessary to check for a "twisted" condition, as explained in the text.

under power with it tied to the dock will bring out a main bearing knock. If you pull the spark plug wire from one plug at a time and the noise disappears when one of the wires is removed, then the noise is probably coming from that cylinder. The cause may be either the rod bearing, piston pin, or the piston, which has quieted down because the load has been removed from the part. If a rod bearing has failed, the noise will be loudest when the engine is decelerating. Piston pin noise and piston pin slap are generally louder when a cold engine is first started.

Carbon build up in the combustion chamber can cause interference with a piston.

Many times the source of an unusual sound may be isolated by using a stethoscope or other listening device. One such device is a long shank screwdriver. Allow your hand to extend over the end of the handle, press an ear to your hand, and then probe with the other end of the screwdriver by touching it firmly against the block at each cylinder or noise area. Take care and use good judgment when using this method of attempting to detect the cause of a problem because the noise will travel through other metallic parts of the engine and could lead to a false interpretation of what you are hearing.

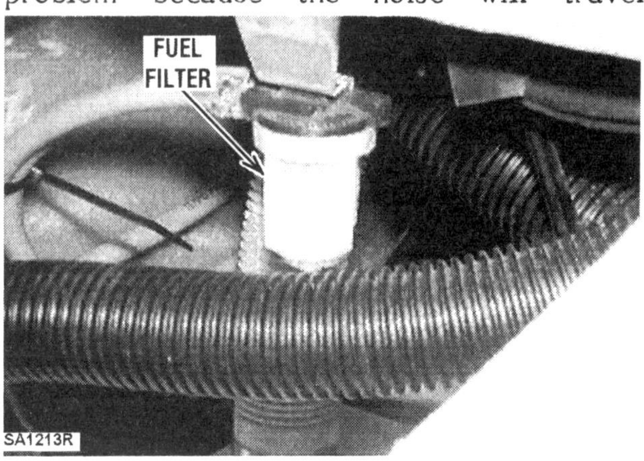

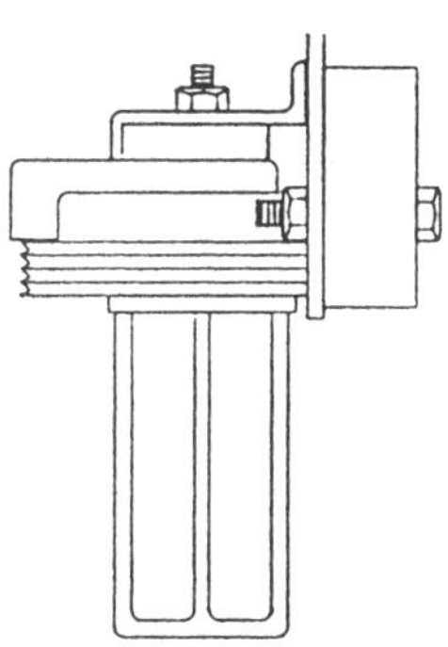

A leaking sealing ring at the fuel filter will draw air into the system. The leak may not be evident until the engine is shut down -- fuel will then leak out past the damaged sealing ring -- top illustration. Check to be sure the filter element is in satisfactory condition -- bottom.

5-3 FUEL SYSTEM PROBLEMS

The following paragraphs provide an orderly sequence of tests to pinpoint problems in the system. It is very rare for the carburetor by itself to cause failure of the engine to start.

Many times fuel system troubles are caused by a plugged fuel filter, a defective fuel pump, or by a leak in the fuel line from the fuel tank to the fuel pump or by a vacuum leak in the line between the pump and the crankcase.

Fuel will begin to sour in three to four months and will cause engine starting problems. A fuel additive such as Sta-Bil may be used to prevent gum from forming during storage or prolonged idle periods.

When the engine is hot, the fuel system can cause starting problems. After a hot engine is shut down, the temperature inside the fuel bowl may rise to 200°F and cause the fuel to actually boil. All carburetors are vented to allow this pressure to escape to the atmosphere. However, some of the fuel may percolate over the high-speed nozzle and overflow into the intake manifold.

In order for this raw fuel to vaporize enough to burn, considerable air must be added to lean out the mixture. Therefore,

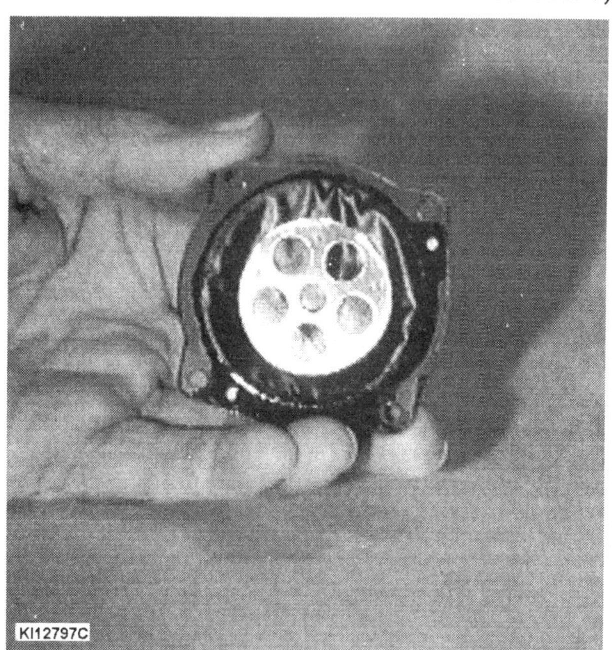

The tiniest "pin" hole in the carburetor diaphragm will seriously affect carburetor performance.

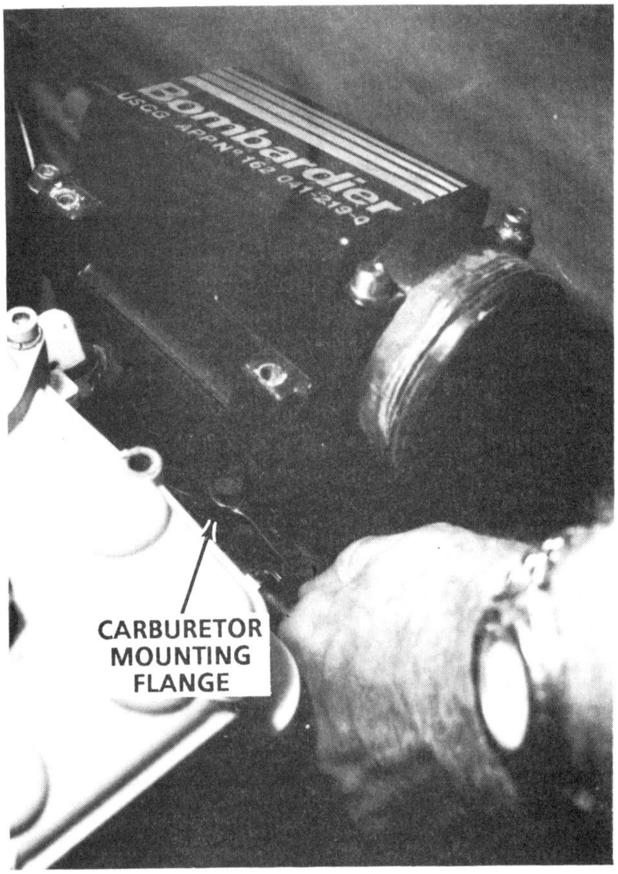

A small leak between the carburetor and the intake manifold, caused by a defective carburetor mounting gasket, will result in rough engine performance at idle speed.

the only remedy is to open the throttle as wide as possible and to crank the engine until enough air is drawn in to provide the proper mixture for the engine to start. **NEVER** move the throttle lever back-and-forth in an attempt to start a hot engine. This action will only compound the problem by adding more fuel to an already too-rich mixture.

A leak between the fuel tank and the pump many times will not appear when the engine is operating because the suction created by the pump sucking fuel will not allow the fuel to leak. Once the engine is turned off and the suction no longer exists, fuel may begin to leak.

Engine Surge

If the engine operates as if the load on the craft was being constantly increased and decreased, even though the operator attempts to maintain a constant engine speed, the problem can most likely be attributed to the fuel pump.

Rough Engine Idle

Problems preventing an engine from running smoothly include: a fouled spark plug; an air leak in the intake manifold; and uneven compression between the cylinders.

Of course any problem in the carburetor affecting the air/fuel mixture will also prevent the engine from operating smoothly at idle speed. These problems usually include: leaking needle valve and seat; a dirty flame arrestor; defective choke; and improper adjustments for idle mixture or idle speed.

5-4 IGNITION SYSTEM FAULTS

WARNING

Check to be sure the engine compartment is well ventilated and free of any gasoline vapors before starting any of the following tests. A spark will be generated creating a potential fire hazard if fuel vapors are present.

1- Check to be sure the battery has a full charge. If not, correct the condition by charging the battery or making a substitution.

2- Check all the terminal connections inside the electrical box. Check all primary and secondary ignition wiring, the engine "kill" switch and the ignition switch. If the craft is equipped with the DESS -- Digitally Enhanced Security System -- circuitry, check to be sure the DESS

1

post (switch) and lanyard are in satisfactory condition.

3- Use a spark tester and check for spark. If a spark tester is not available, hold the plug wire about 1/4" (6.4mm) from the engine. Crank the engine through a few revolutions using the cranking motor and check for spark. A strong spark over a wide gap must be observed when testing in this manner, because under compression a strong spark is necessary in order to ignite the air/fuel mixture in the cylinder. This means it is possible to think a strong spark is present, when in reality the spark will be too weak when the plug is installed. If there is no spark, or if the spark is weak, the trouble is most likely under the flywheel in the stator assembly.

4- Perform the resistance test procedures given in the Ignition chapter to identify the faulty component. The accompanying illustration shows the location of the trigger coil on a model 787 series engine.

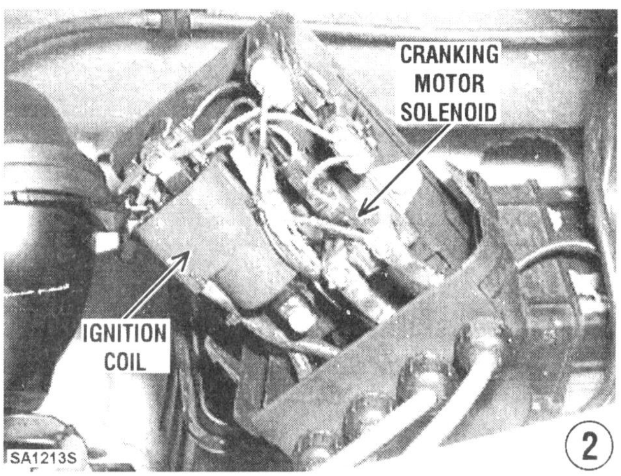

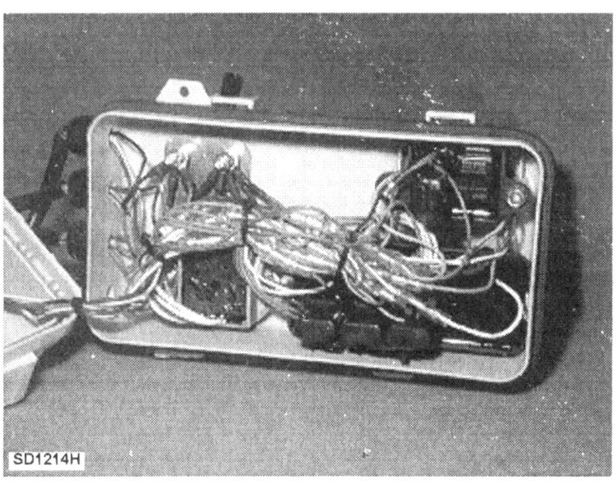

Location of the secondary electrical box, typically located in the forward compartment. The fuses, MPEM and DESS modules, and rectifier/regulator are located within this box on models so equipped.

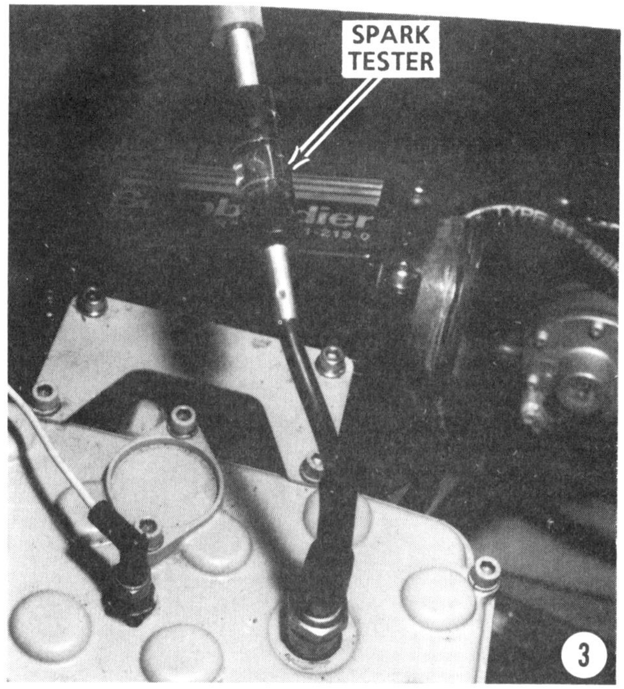

Intermittent or Multiple Problems

Many ignition problems occur only during engine operation, when the component is subject to vibration and/or an increase in temperature. Many of these defective components return to normal once the engine is shut down and allowed to cool.

There is also the possibility of more than one defective component in the ignition system. One component failure may have caused a surge of power and "fried" another component.

Make every attempt to avoid the **BUY** and **TRY** method of troubleshooting. In **MOST** cases electrical components are not returnable, once the item leaves the store.

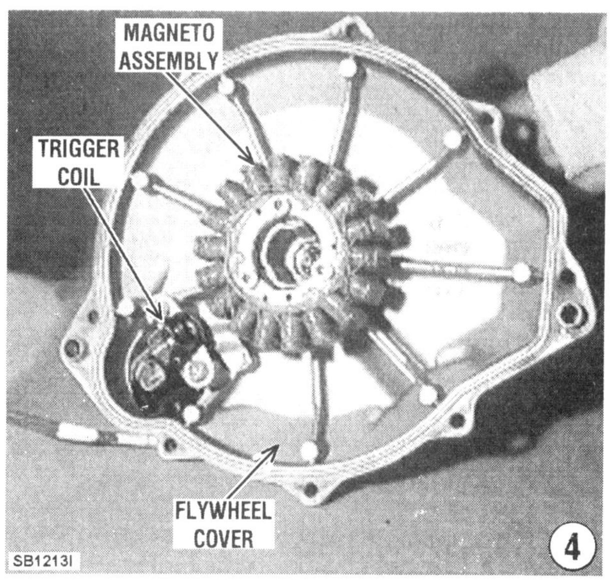

SPARK PLUG EVALUATION

Removal: Remove the spark plug wires by pulling and twisting on only the molded cap. **NEVER** pull on the wire or the connection inside the cap may become separated or the boot damaged. Remove the spark plugs and identify from which cylinder they were taken. **TAKE CARE** not to tilt the socket as you remove the plug or the insulator may be cracked.

Examine: Carefully examine the firing end to determine the firing condition in each cylinder.

Correct Color: A proper firing plug should be dry and powdery. Hard deposits inside the shell indicate the engine is starting to use some oil, but not enough to cause concern. The most important evidence is the light gray color of the porcelain, which is an indication this plug has been running at the correct temperature. This means the plug is one with the correct heat range and also that the air/fuel mixture is correct.

Overheating: A dead white or gray insulator, which is generally blistered, is an indication of overheating and pre-ignition. The electrode gap wear rate will be more than normal and in the case of pre-ignition, will actually cause the electrodes to melt. Overheating and pre-ignition are usually caused by overadvanced timing, detonation from using too-low an octane rating fuel, an excessively lean air/fuel mixture, or problems in the cooling system.

Rich Mixture: A black, sooty condition on both the spark plug shell and the porcelain is caused by an excessively rich air/fuel mixture, both at low and high speeds. The rich mixture lowers the combustion temperature so the spark plug does not run hot enough to burn off the deposits.

Deposits formed only on the shell is an indication the low-speed air/fuel mixture is too rich. At high speeds with the correct mixture, the temperature in the combustion chamber is high enough to burn off the deposits on the insulator.

Too Cool: A dark insulator, with very few deposits, indicates the plug is running too cool. This condition can be caused by low compression or by using a spark plug of an incorrect heat range. If this condition shows on only one plug it is most usually caused by low compression in that cylinder. If both plugs have this appearance, then it is probably due to the plugs having a too-low heat range.

Fouled: A fouled spark plug may be caused by the wet oily deposits on the insulator shorting the high-tension current to ground inside the shell. The condition may also be caused by ignition problems which prevent a high-tension pulse to be delivered to the spark plug.

Carbon Deposits: Heavy carbon-like deposits are an indication of excessive oil consumption. This condition may be the result of worn piston rings.

Example of overheating, heavy load. Use a spark plug with a lower heat rating.

Spark plug fouled by oil. The engine may be in need of an overhaul.

Electrode Wear: Electrode wear results in a wide gap and if the electrode becomes carbonized it will form a high-resistance path for the spark to jump across. Such a condition will cause the engine to misfire during acceleration. If both plugs are in this condition, it can cause an increase in fuel consumption and very poor performance at high-speed operation. The solution is to replace the spark plugs with a rating in the proper heat range and gapped to specification.

Red rust-colored deposits on the entire firing end of a spark plug can be caused by water in the cylinder combustion chamber. This can be the first evidence of water entering the cylinders through the exhaust manifold. This condition **MUST** be corrected at the first opportunity.

5-5 CRANKING SYSTEM FAILURES

Regardless of how or where the solenoid is mounted, the basic circuits of the starting system on all makes of cranking motors are the same and similar tests apply. In the following testing and troubleshooting procedures, the differences are noted.

WARNING
ALWAYS TAKE TIME TO VENT THE BILGE WHEN MAKING ANY OF THE TESTS AS A PREVENTION AGAINST IGNITING ANY FUMES ACCUMULATED IN THAT AREA.

Faulty Symptoms

If the cranking motor spins, but fails to crank the engine, the cause is usually a corroded or gummy Bendix drive. The drive should be removed, cleaned, and given an inspection.

If the cranking motor cranks the engine too slowly, the following are possible causes and the corrective actions that may be taken:

a- Battery charge is low. Charge the battery to full capacity.

b- High resistance connections at the battery, solenoid, or motor. Clean and tighten all connections.

c- Undersize battery cables. Replace cables with sufficient size.

All engine cranking problems fall into one of three problem areas:

1- The cranking motor fails to turn.

2- The cranking motor spins rapidly, but does not crank the engine.

3- The cranking motor cranks the engine, but too slowly.

The following paragraphs provide a logical sequence of tests designed to isolate a problem in the cranking system.

Before wasting too much time troubleshooting the cranking motor circuit, the following checks should be made. Many times, the problem will be corrected.

Powdery deposits have melted and shorted-out this spark plug.

Red, brown, or yellow deposits, are by-products of combustion from fuel and lubricating oil.

a- Battery fully charged.
b- Main fuse is "good" (not blown).
c- All electrical connections clean and tight.
d- Wiring in good condition -- insulation not worn or frayed.

The following troubleshooting procedures are presented in a logical sequence.

Do not operate the cranking motor for more than 15 seconds. Prolonged cranking motor operation will cause overheating and damage the motor.

After each test, allow the cranking motor to cool for a minute or so.

Never depress the **START** button to activate the cranking motor while the engine is operating. Such action will damage the pinion and/or flywheel gears.

GOOD WORDS CRANKING CIRCUIT SYSTEM TESTS

The following "Quick Test" troubleshooting procedures can be performed in a matter of minutes. These steps are presented in a logical sequence and are presented to provide a quick analysis of the cranking system components -- cranking motor, start/stop switch, safety lanyard and solenoid switch.

Performing these simple tasks may identify the problem -- while the components are still in the watercraft. If these steps do not locate the fault, proceed to the "Thorough Tests" Section. Some of these steps will require the aid of an assistant.

"Quick Test"

1- Once again, check to be sure the battery is fully charged. Check all electrical connections. Remove the spark plugs.

2- Obtain a continuity tester. Remove the rubber boot from the positive lead of the crank-

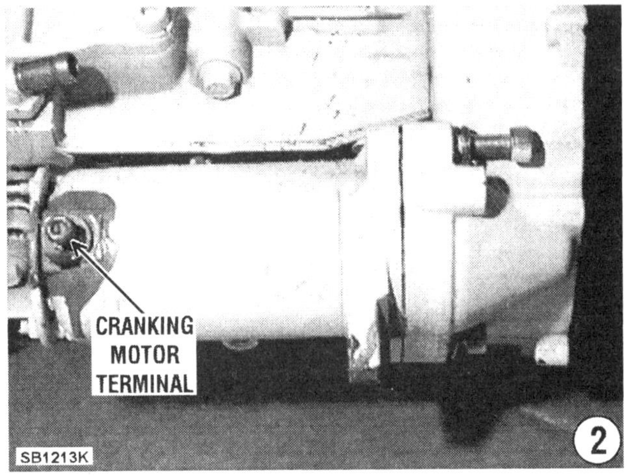

ing motor. Connect the continuity tester lead to the cranking motor terminal. Depress the starter. If the light illuminates and cranking motor does not operate -- or just "clicks" -- the cranking motor must be serviced or replaced.

3- If the continuity tester does **NOT** illuminate, the cranking motor is probably not at fault. Obtain a digital ohmmeter. Disconnect the Red/Yellow start/stop switch lead in the electrical box. Connect a tester lead to the Red/Yellow start switch lead. Connect the other tester lead to ground near the solenoid switch. Depress the start/stop switch and observe the indicated value. The meter should register close to 0 ohm when the switch is depressed. When released, the volt meter should indicate infinity. If both readings are within tolerance, the cranking motor solenoid is probably faulty.

Safety Lanyard Switch Test

Obtain a digital ohmmeter. Disconnect the Black and Black/Yellow leads in the electrical box. Connect the tester leads to the lanyard switch leads. The measured resistance should

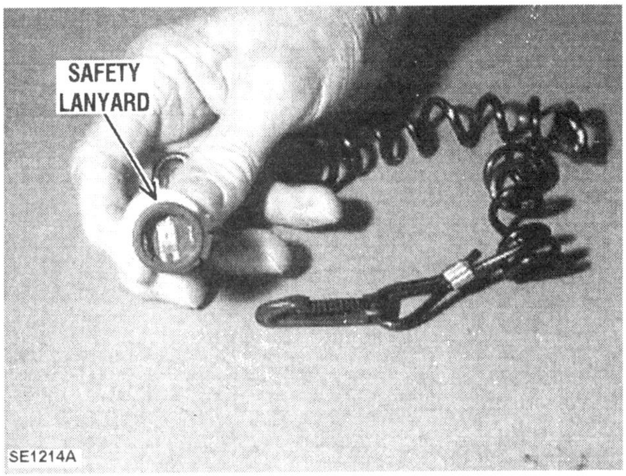

The contact points of the safety lanyard and DESS switch -- post -- must be clean and in satisfactory condition.

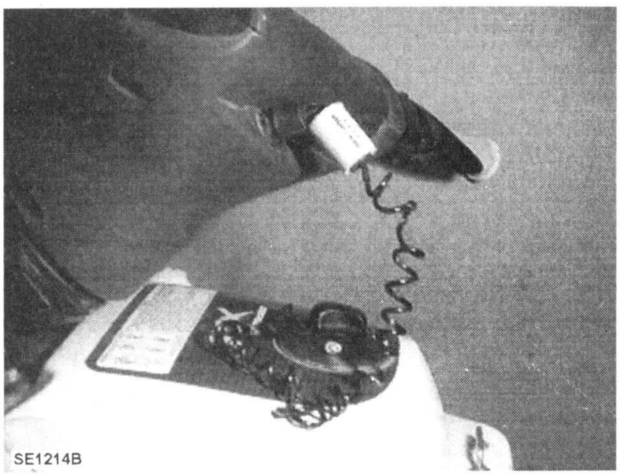

The safety lanyard should seat firmly on the switch or "post". An ohmmeter reading is taken with the lanyard installed, and then removed from the post, as indicated in the test.

be close to 0 ohm when the cap lanyard is securely fastened to the switch -- or "post". The tester should indicate an open circuit when the lanyard is removed.

Cranking Circuit Tests

1- Remove both spark plug leads from the plugs and ground the ends to the engine to prevent accidental engine start. Keep the leads grounded for the following tests.

Obtain a voltmeter and select the Vx1 DC scale. Observe the starter solenoid. Both large terminals have Red leads connected. One terminal has the battery cable connected and the other terminal has the lead from the cranking motor connected. Make contact with the negative Black meter lead to the terminal on the starter solenoid with the battery cable connected.

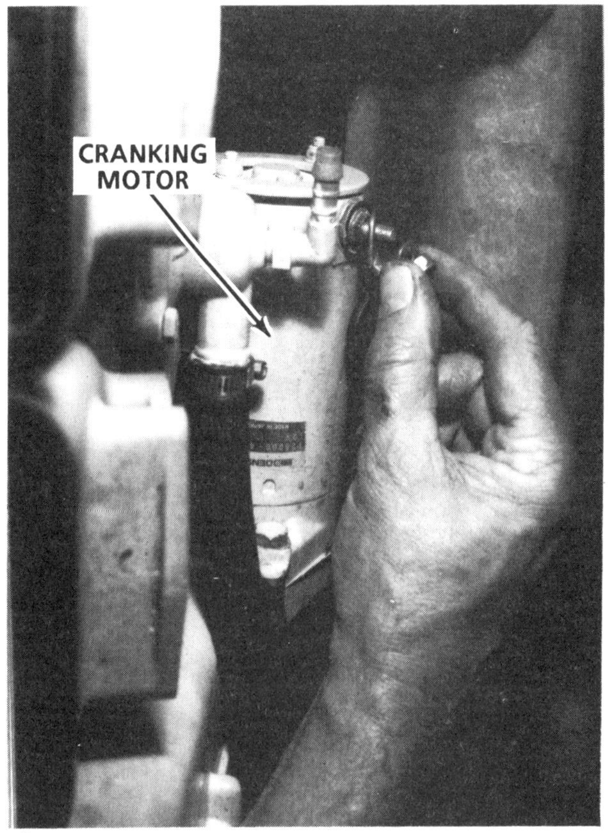

Disconnecting electrical leads in preparation to performing bench tests on the cranking motor, as described in detail -- Chapter 9 -- ELECTRICAL.

Make contact with the positive Red meter lead to the positive battery terminal. The voltmeter should register less then 0.25V DC. If the reading exceeds 0.25V DC, clean the connections at both ends of the positive battery cable and repeat the test. If the reading does not change, replace the cable.

2- Keep the meter on the same scale. Make contact with the negative Black volt-

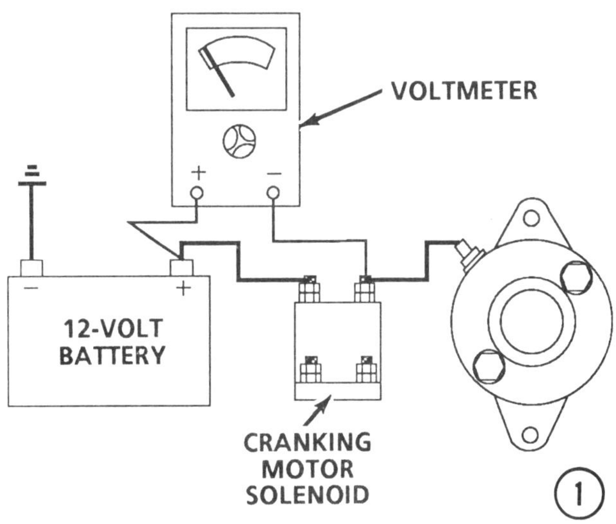

1

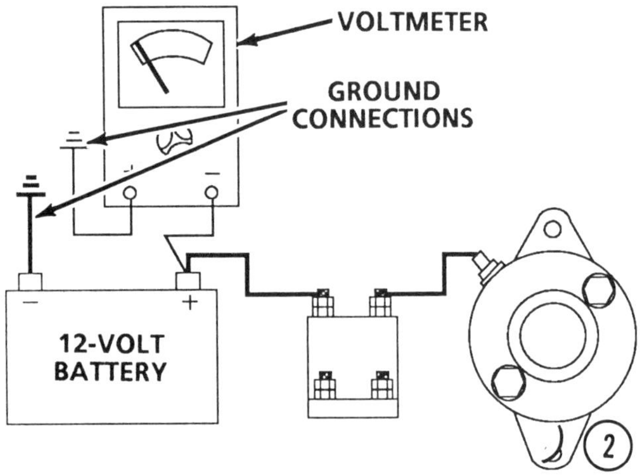

meter lead to the negative terminal on the battery. Make contact with the positive Red voltmeter lead to a good engine ground. The voltmeter should register less then 0.25V DC. If the reading exceeds 0.25V DC, clean the connections at both ends of the negative battery cable and repeat the test. If the reading does not change, replace the cable.

CRITICAL WORDS

During the following tests -- Step 3 and Step 4, both meter leads must be placed on the solenoid terminals **WHILE** the engine is being cranked. **ALSO**, both meter leads must be removed from the solenoid terminals **WHILE** the engine is being cranked and **BEFORE** the engine has stopped cranking. If these precautions are not followed, the voltmeter may be damaged.

3- Keep the meter on the same scale. Obtain the services of an assistant.

Signal the assistant to start cranking the engine. While the engine is being cranked, make contact with the negative Black voltmeter lead to the solenoid terminal which has the lead from the cranking motor connected, and then quickly make contact with the positive Red voltmeter lead to the solenoid terminal which has the battery cable connected.

If the meter reading exceeds 0.25V DC, the starter solenoid must be replaced. Remove the meter leads and signal the assistant to cease cranking.

4- Keep the meter on the same scale and the assistant at the **START** button. Observe the same precautions as before in Step 3. While the engine is being cranked, make contact with the negative Black voltmeter lead to terminal on the cranking motor, and then quickly make contact with the positive Red voltmeter lead to the solenoid terminal which has the Red cranking motor connected.

If the meter reading exceeds 0.25V DC, the connections at both ends of the solenoid-to-cranking motor lead should be cleaned and the test repeated. If the reading does not change, the lead must be replaced.

CRANKING MOTOR SOLENOID TESTING

Description

The cranking motor solenoid is actually a switch between the battery and the engine. The switch cannot be serviced. Therefore, if testing indicates the switch is faulty, it **MUST** be re-

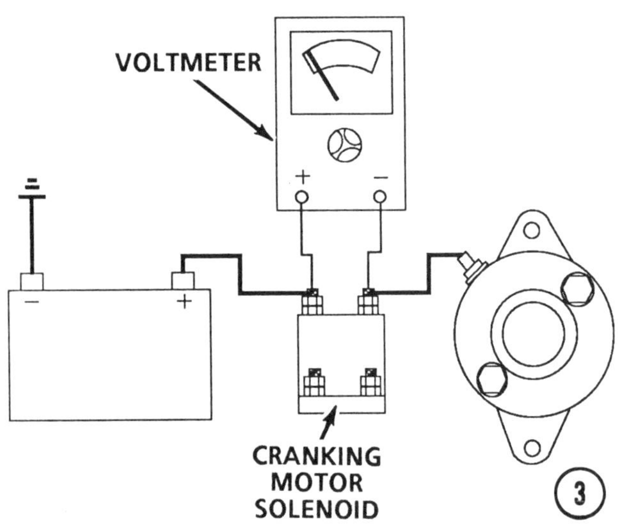

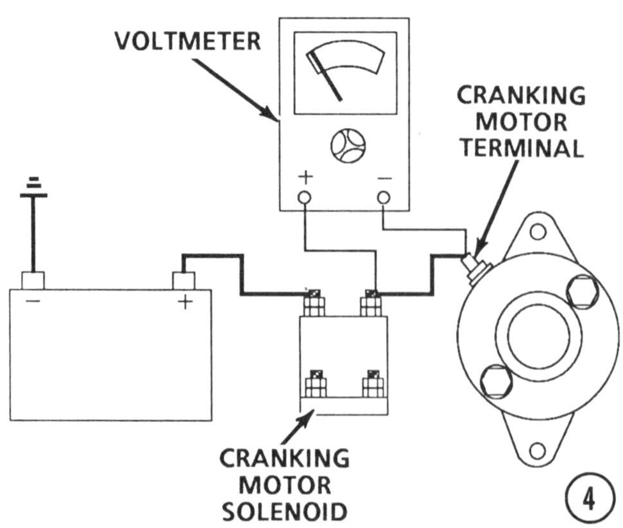

placed. The solenoid is housed inside the aft electrical box on most models covered in this manual.

The solenoid may remain inside the box during testing, but the cover must be removed to make the necessary connections.

Preliminary Tasks

Remove the electrical box cover.

Disconnect the single Red large lead from the upper left terminal.

Disconnect the two or three Red leads from the upper right terminal.

Disconnect the two Black/White leads from the lower left terminal.

Disconnect the Yellow/Red lead from the lower right terminal.

Cranking Solenoid Testing

1- Select the Rx1 ohm scale on the ohmmeter. Make contact with the two meter leads across the two upper solenoid terminals. The meter should read an infinite resistance. If the meter shows a resistance of less than infinity, the solenoid is defective and must be replaced.

2- Keep the ohmmeter on the same scale. Obtain a 12V battery and a pair of jumper leads. Using the set of jumper leads, connect the battery across the two lower terminals. Make contact with the two meter leads across the two upper solenoid terminals. If the solenoid emits a "click" and the ohmmeter registers zero ohms, the solenoid is good. If no "click" is heard, or if the meter registers a high or infinite resistance, the solenoid is defective and must be replaced.

If the solenoid failed the test, it must be replaced. Two bolts secure the solenoid to the electrical box. When installing the replacement solenoid, apply Loctite 242 to the threads of the bolts and install the new solenoid in place. Tighten both bolts securely.

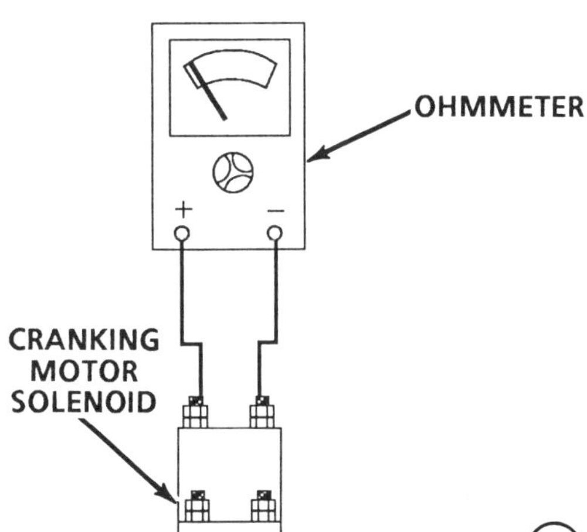

1

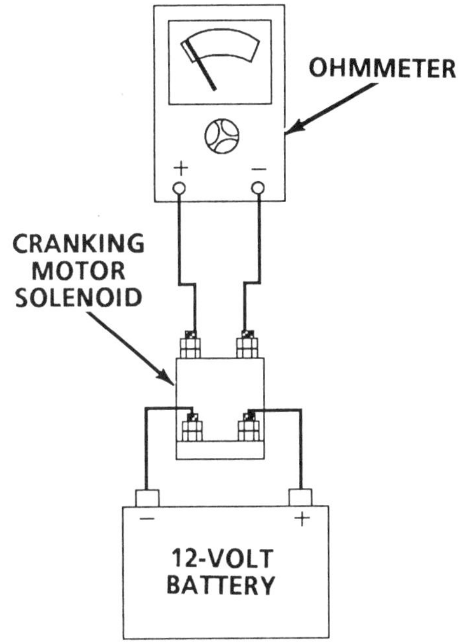

2

Connect the single Red large lead to the upper left terminal.

Connect the two or three Red leads to the upper right terminal.

Connect the two Black/White leads to the lower left terminal.

Connect the Yellow/Red lead from the lower right terminal.

5-6 CHARGING SYSTEM MALFUNCTIONS

The battery charging coil, rectifier/regulator, battery, and the necessary wiring to connect it all together comprise the charging system.

Before the charging system is blamed for battery problems, consider other areas which may be the cause:

1- Operating at low speed for short periods.

2- Voltage losses due to high resistance.

3- Corroded battery cables, connectors, and terminals.

4- Low electrolyte level in the battery cells.

5- Prolonged disuse of the battery causing a self-discharged condition.

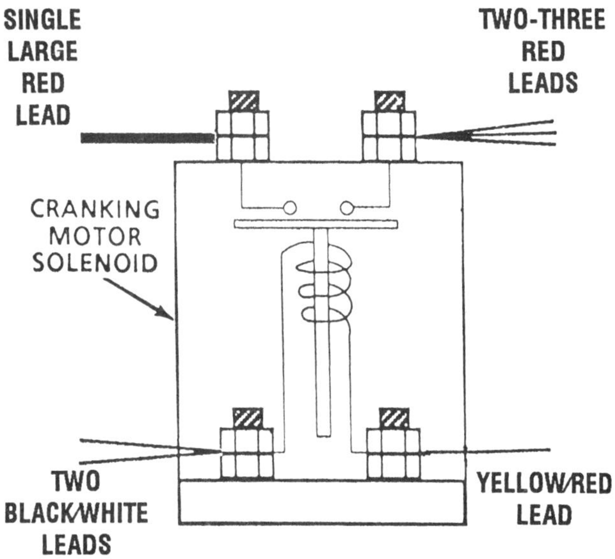

Simplified line drawing to indicate wiring connections on the cranking motor solenoid housed in the rear electrical box.

Service the battery, refer to Section 5-3. If, after charging, the battery fails to give the required specific gravity reading, then the battery will never hold a charge and must be replaced. This problem may have originated with the battery, a shorted cell perhaps, or may be attributed to the failure of the rectifier/regulator or charging coil. Refer to the Ignition chapter for resistance tests on these two components.

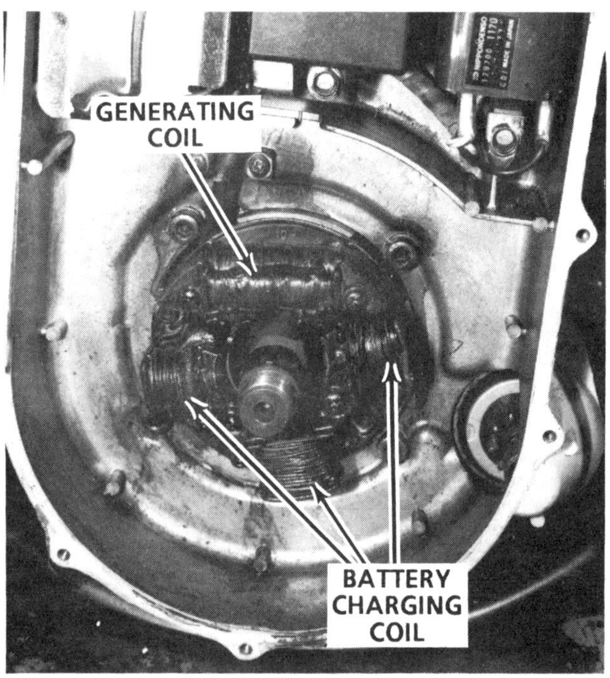

Location of generating coils and battery charging coils, mounted on the magneto assembly. The magneto assembly is secured to the stator plate on model 587 - 717 series engines.

5-7 TROUBLESHOOTING CHARTS

The following six charts are presented as an aid to troubleshooting the engine and jet pump. The uppermost box identifies a specific problem. Start with the problem, then work downward. The probable causes are listed in columns under a specific system, for example: Fuel system, or Electrical system. Once a possible cause is identified, refer to the specific chapter for procedures on testing and service.

When using the charts -- proceed step-by-step through each of the tests or checks until the defect or cause of the problem is uncovered.

FIRST CHART -- Page 5-15 -- Engine fails to start.

SECOND CHART -- Page 5-16 -- Engine starts, then shuts down.

THIRD CHART -- Page 5-17 -- Engine misfires.

FOURTH CHART -- Page 5-18 -- Engine lacks proper power.

FIFTH CHART -- Page 5-19 -- Abnormal engine noises.

SIXTH CHART -- Page 5-20 -- Jet pump problems.

ONE FINAL WORD

Before going out on the water, **TAKE TIME** to verify the drain plug is installed. Countless number of excursions have had a very sad beginning because the craft was eased into the water only to have the hull begin filling with water.

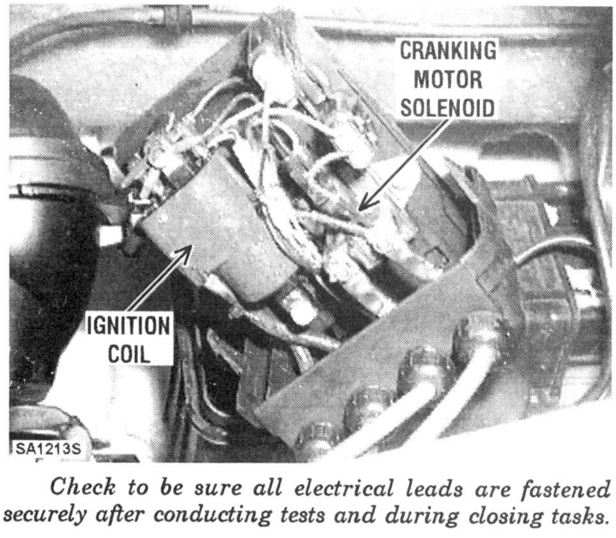

Check to be sure all electrical leads are fastened securely after conducting tests and during closing tasks.

CHART ONE

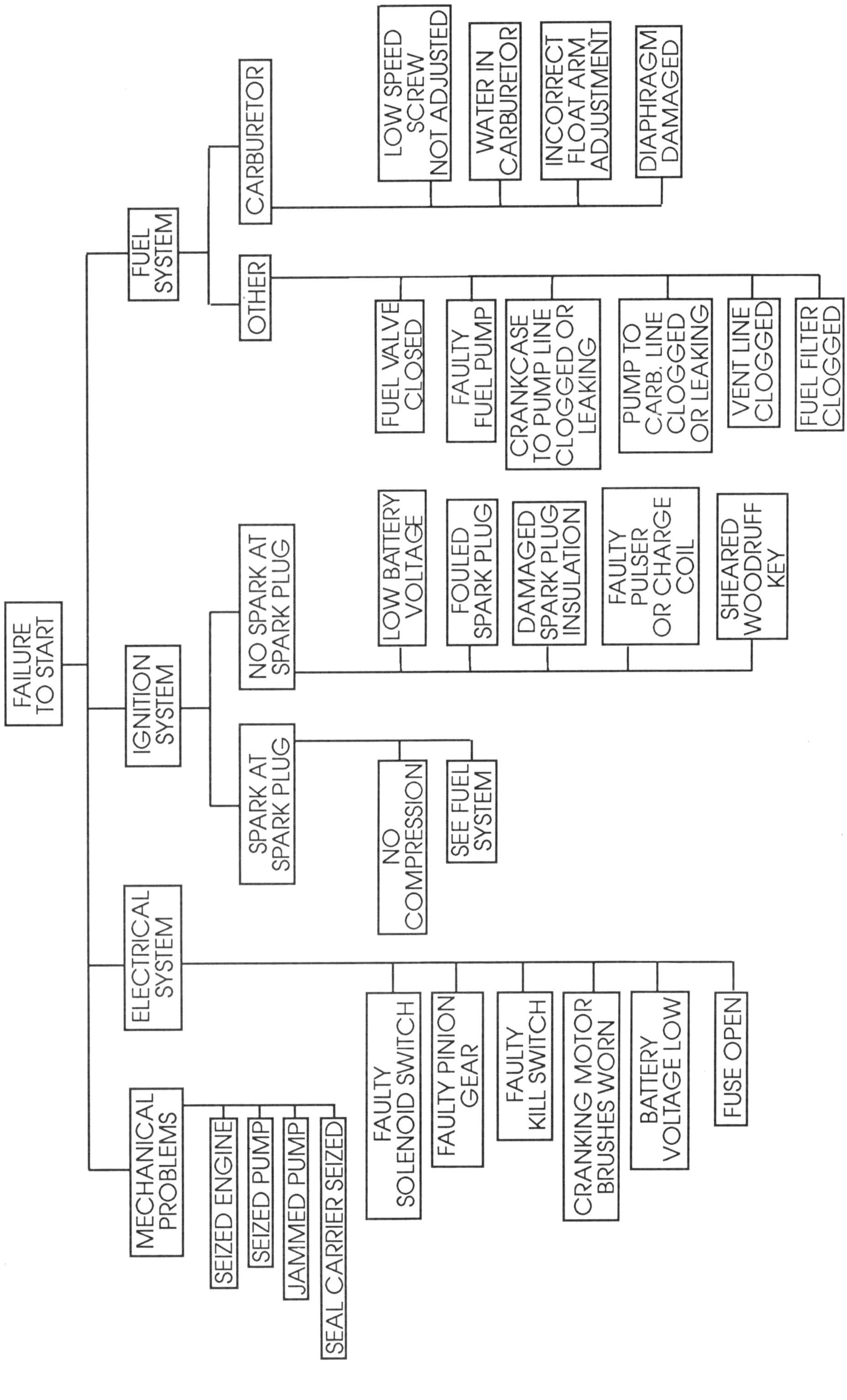
FAILURE TO START
FUEL SYSTEM
CARBURETOR
LOW SPEED SCREW NOT ADJUSTED
WATER IN CARBURETOR
INCORRECT FLOAT ARM ADJUSTMENT
DIAPHRAGM DAMAGED
OTHER
FUEL VALVE CLOSED
FAULTY FUEL PUMP
CRANKCASE TO PUMP LINE CLOGGED OR LEAKING
PUMP TO CARB. LINE CLOGGED OR LEAKING
VENT LINE CLOGGED
FUEL FILTER CLOGGED
IGNITION SYSTEM
NO SPARK AT SPARK PLUG
LOW BATTERY VOLTAGE
FOULED SPARK PLUG
DAMAGED SPARK PLUG INSULATION
FAULTY PULSER OR CHARGE COIL
SHEARED WOODRUFF KEY
SPARK AT SPARK PLUG
NO COMPRESSION
SEE FUEL SYSTEM
ELECTRICAL SYSTEM
FAULTY SOLENOID SWITCH
FAULTY PINION GEAR
FAULTY KILL SWITCH
CRANKING MOTOR BRUSHES WORN
BATTERY VOLTAGE LOW
FUSE OPEN
MECHANICAL PROBLEMS
SEIZED ENGINE
SEIZED PUMP
JAMMED PUMP
SEAL CARRIER SEIZED

CHART TWO

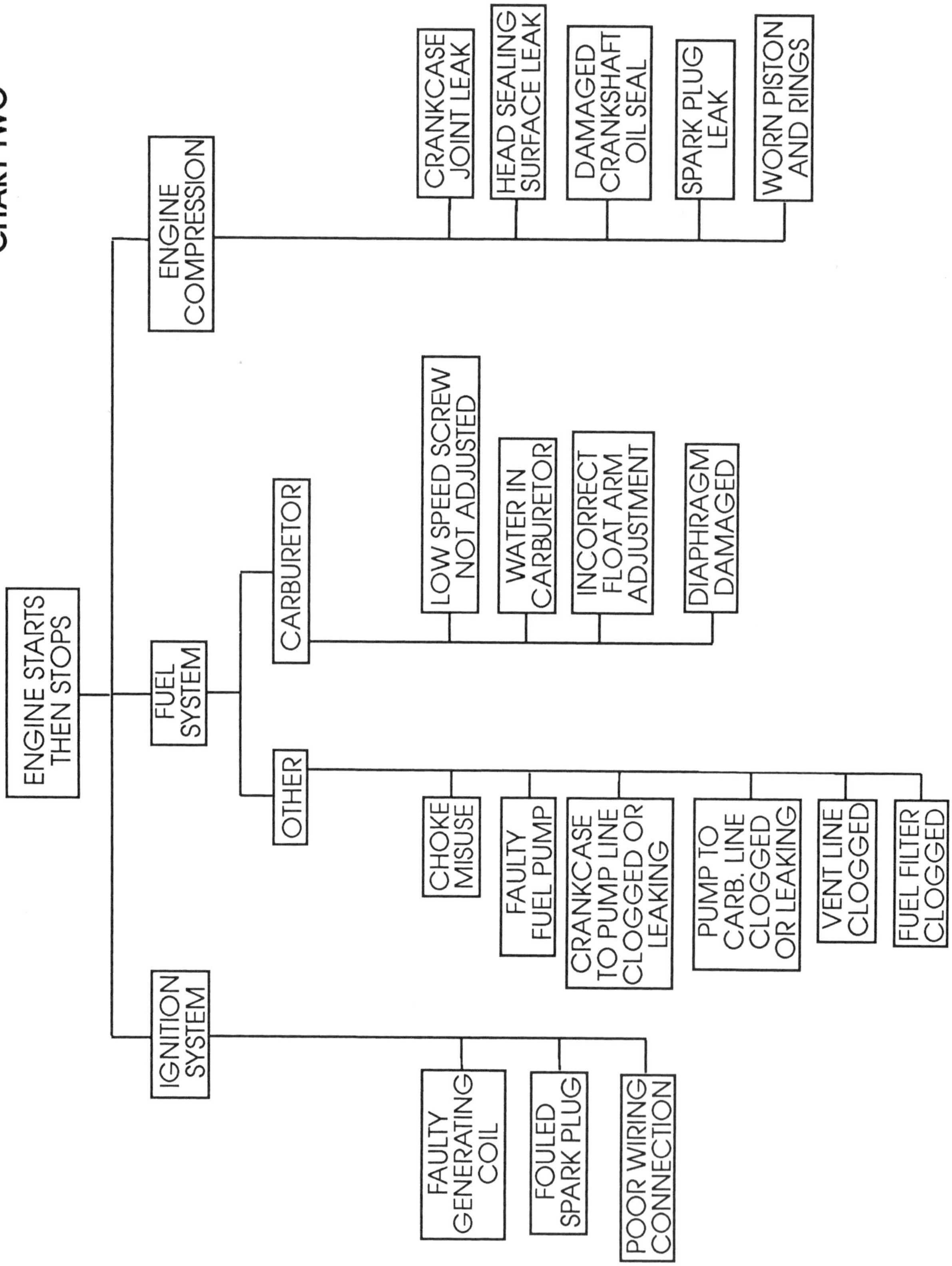
ENGINE STARTS THEN STOPS
ENGINE COMPRESSION
CRANKCASE JOINT LEAK
HEAD SEALING SURFACE LEAK
DAMAGED CRANKSHAFT OIL SEAL
SPARK PLUG LEAK
WORN PISTON AND RINGS
FUEL SYSTEM
CARBURETOR
LOW SPEED SCREW NOT ADJUSTED
WATER IN CARBURETOR
INCORRECT FLOAT ARM ADJUSTMENT
DIAPHRAGM DAMAGED
OTHER
CHOKE MISUSE
FAULTY FUEL PUMP
CRANKCASE TO PUMP LINE CLOGGED OR LEAKING
PUMP TO CARB. LINE CLOGGED OR LEAKING
VENT LINE CLOGGED
FUEL FILTER CLOGGED
IGNITION SYSTEM
FAULTY GENERATING COIL
FOULED SPARK PLUG
POOR WIRING CONNECTION

CHART THREE

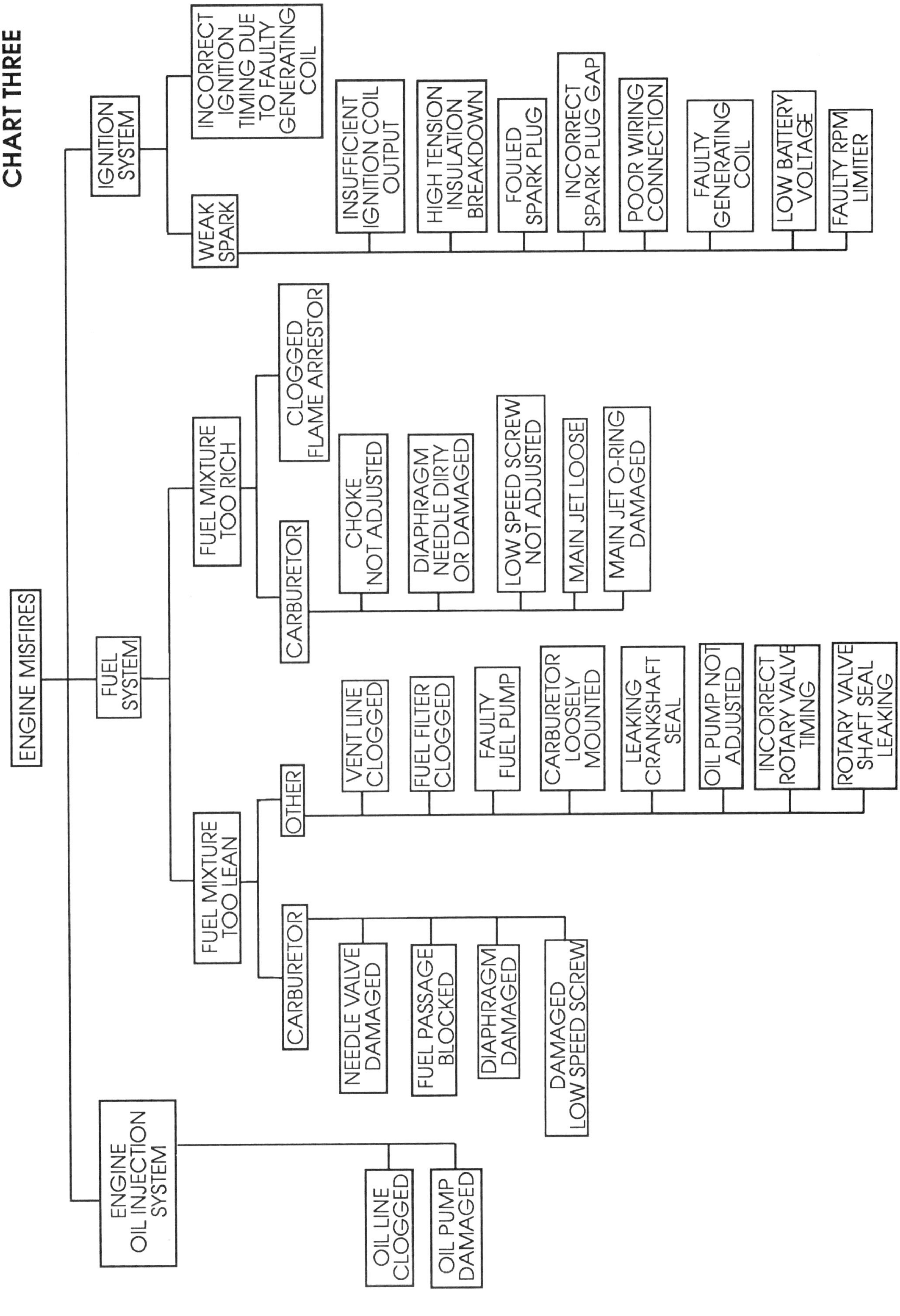
ENGINE MISFIRES
IGNITION SYSTEM
WEAK SPARK
INCORRECT IGNITION TIMING DUE TO FAULTY GENERATING COIL
INSUFFICIENT IGNITION COIL OUTPUT
HIGH TENSION INSULATION BREAKDOWN
FOULED SPARK PLUG
INCORRECT SPARK PLUG GAP
POOR WIRING CONNECTION
FAULTY GENERATING COIL
LOW BATTERY VOLTAGE
FAULTY RPM LIMITER
FUEL SYSTEM
FUEL MIXTURE TOO RICH
CLOGGED FLAME ARRESTOR
CARBURETOR
CHOKE NOT ADJUSTED
DIAPHRAGM NEEDLE DIRTY OR DAMAGED
LOW SPEED SCREW NOT ADJUSTED
MAIN JET LOOSE
MAIN JET O-RING DAMAGED
FUEL MIXTURE TOO LEAN
OTHER
VENT LINE CLOGGED
FUEL FILTER CLOGGED
FAULTY FUEL PUMP
CARBURETOR LOOSELY MOUNTED
LEAKING CRANKSHAFT SEAL
OIL PUMP NOT ADJUSTED
INCORRECT ROTARY VALVE TIMING
ROTARY VALVE SHAFT SEAL LEAKING
CARBURETOR
NEEDLE VALVE DAMAGED
FUEL PASSAGE BLOCKED
DIAPHRAGM DAMAGED
DAMAGED LOW SPEED SCREW
ENGINE OIL INJECTION SYSTEM
OIL LINE CLOGGED
OIL PUMP DAMAGED

CHART FOUR

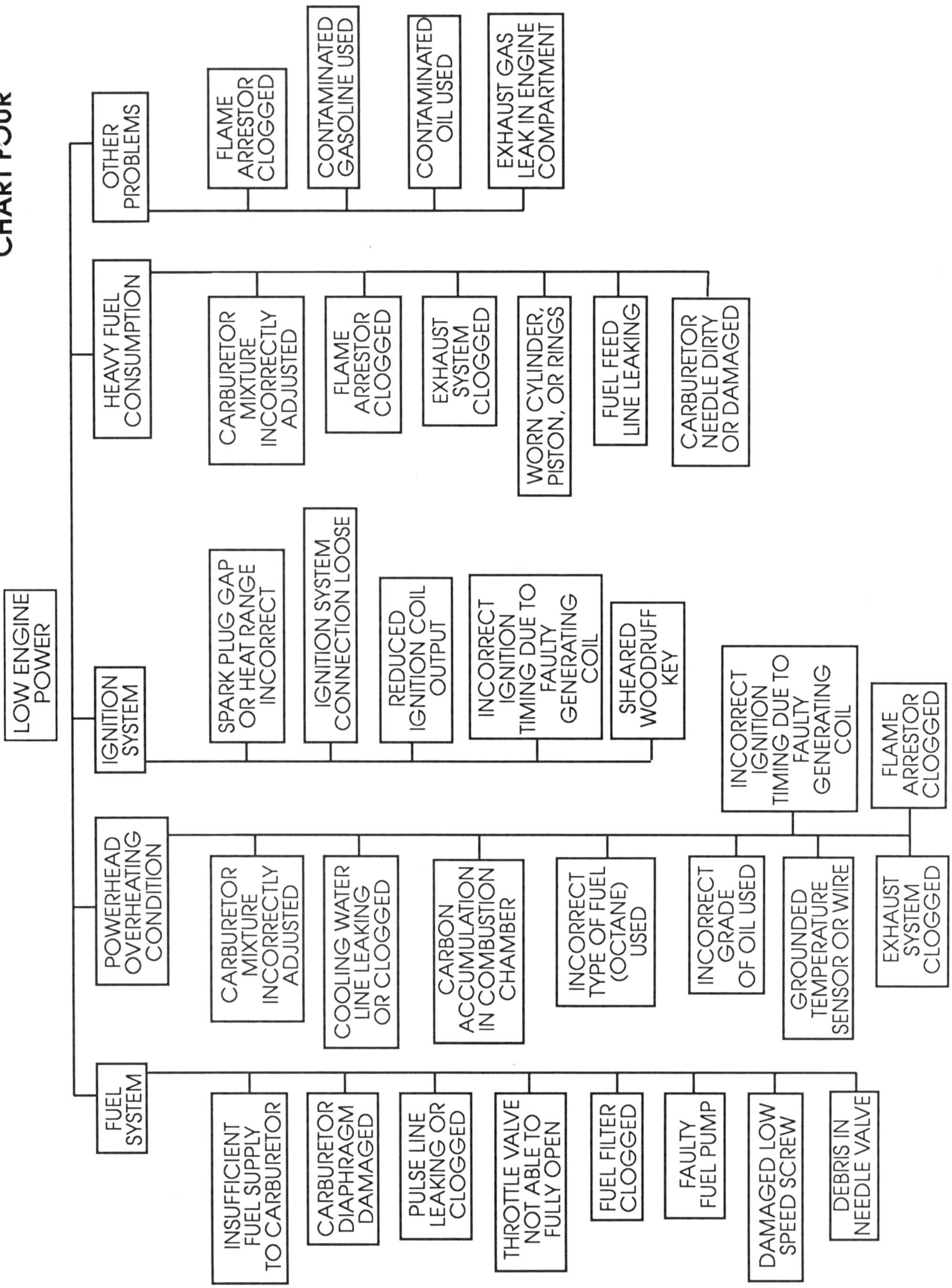
LOW ENGINE POWER
FUEL SYSTEM
INSUFFICIENT FUEL SUPPLY TO CARBURETOR
CARBURETOR DIAPHRAGM DAMAGED
PULSE LINE LEAKING OR CLOGGED
THROTTLE VALVE NOT ABLE TO FULLY OPEN
FUEL FILTER CLOGGED
FAULTY FUEL PUMP
DAMAGED LOW SPEED SCREW
DEBRIS IN NEEDLE VALVE
POWERHEAD OVERHEATING CONDITION
CARBURETOR MIXTURE INCORRECTLY ADJUSTED
COOLING WATER LINE LEAKING OR CLOGGED
CARBON ACCUMULATION IN COMBUSTION CHAMBER
INCORRECT TYPE OF FUEL (OCTANE) USED
INCORRECT GRADE OF OIL USED
GROUNDED TEMPERATURE SENSOR OR WIRE
EXHAUST SYSTEM CLOGGED
INCORRECT IGNITION TIMING DUE TO FAULTY GENERATING COIL
FLAME ARRESTOR CLOGGED
IGNITION SYSTEM
SPARK PLUG GAP OR HEAT RANGE INCORRECT
IGNITION SYSTEM CONNECTION LOOSE
REDUCED IGNITION COIL OUTPUT
INCORRECT IGNITION TIMING DUE TO FAULTY GENERATING COIL
SHEARED WOODRUFF KEY
HEAVY FUEL CONSUMPTION
CARBURETOR MIXTURE INCORRECTLY ADJUSTED
FLAME ARRESTOR CLOGGED
EXHAUST SYSTEM CLOGGED
WORN CYLINDER, PISTON, OR RINGS
FUEL FEED LINE LEAKING
CARBURETOR NEEDLE DIRTY OR DAMAGED
OTHER PROBLEMS
FLAME ARRESTOR CLOGGED
CONTAMINATED GASOLINE USED
CONTAMINATED OIL USED
EXHAUST GAS LEAK IN ENGINE COMPARTMENT

CHART FIVE

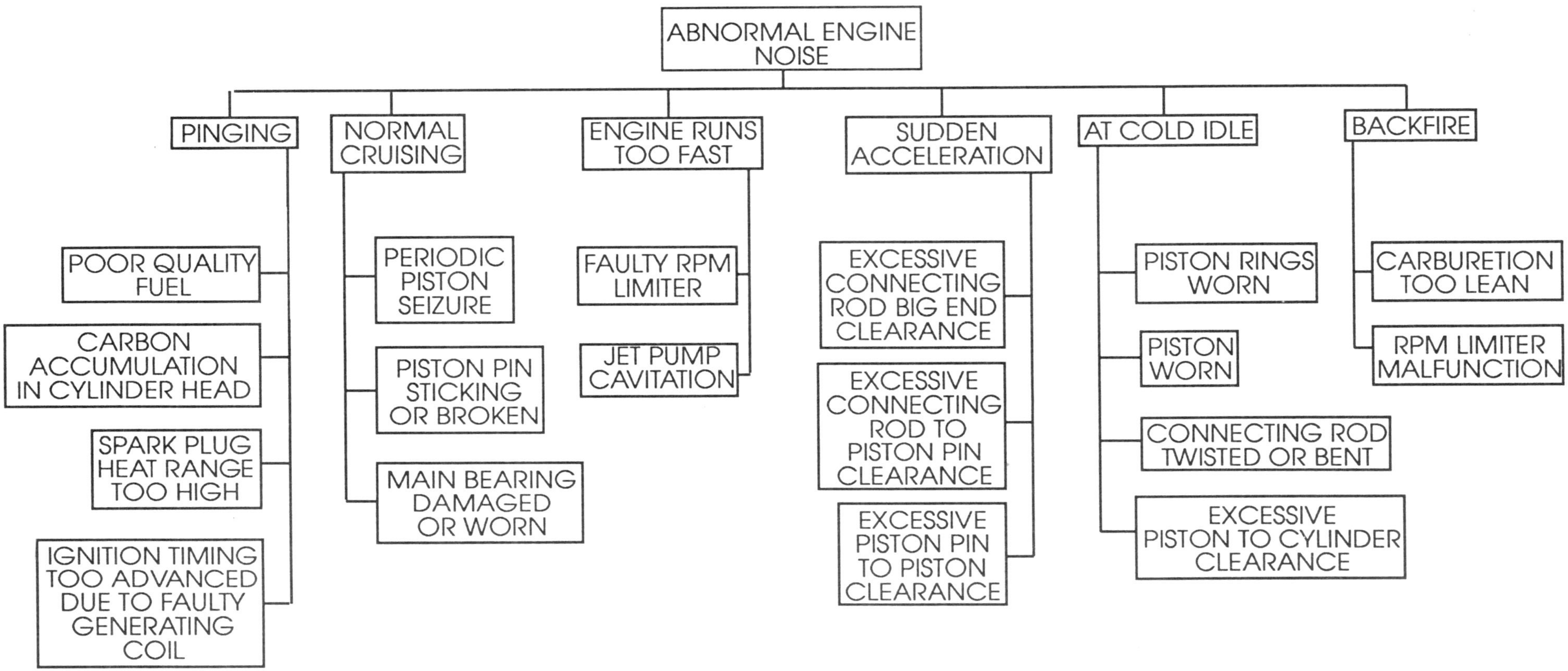
ABNORMAL ENGINE NOISE
PINGING
POOR QUALITY FUEL
CARBON ACCUMULATION IN CYLINDER HEAD
SPARK PLUG HEAT RANGE TOO HIGH
IGNITION TIMING TOO ADVANCED DUE TO FAULTY GENERATING COIL
NORMAL CRUISING
PERIODIC PISTON SEIZURE
PISTON PIN STICKING OR BROKEN
MAIN BEARING DAMAGED OR WORN
ENGINE RUNS TOO FAST
FAULTY RPM LIMITER
JET PUMP CAVITATION
SUDDEN ACCELERATION
EXCESSIVE CONNECTING ROD BIG END CLEARANCE
EXCESSIVE CONNECTING ROD TO PISTON PIN CLEARANCE
EXCESSIVE PISTON PIN TO PISTON CLEARANCE
AT COLD IDLE
PISTON RINGS WORN
PISTON WORN
CONNECTING ROD TWISTED OR BENT
EXCESSIVE PISTON TO CYLINDER CLEARANCE
BACKFIRE
CARBURETION TOO LEAN
RPM LIMITER MALFUNCTION

CHART SIX

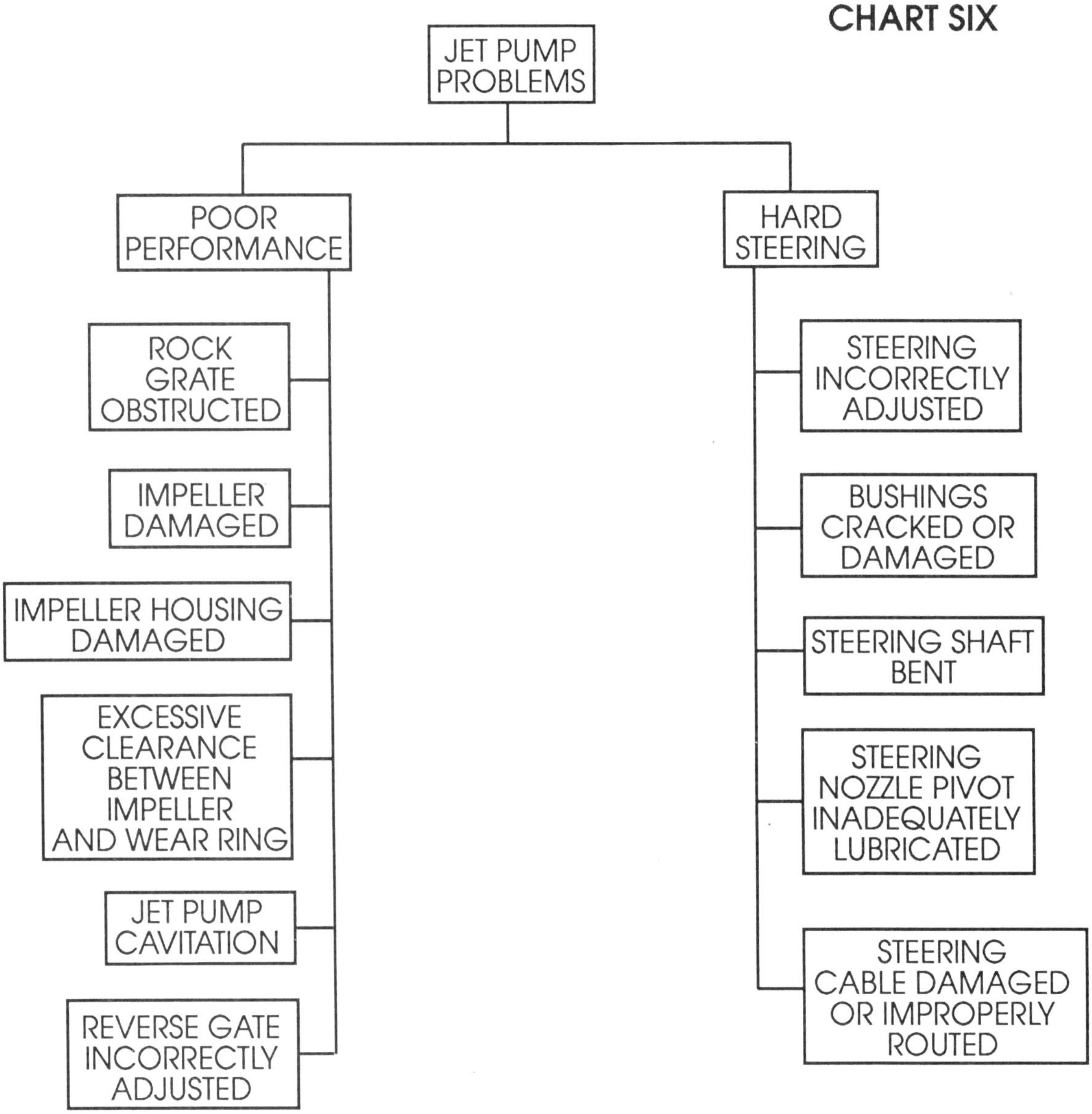
JET PUMP PROBLEMS
POOR PERFORMANCE
ROCK GRATE OBSTRUCTED
IMPELLER DAMAGED
IMPELLER HOUSING DAMAGED
EXCESSIVE CLEARANCE BETWEEN IMPELLER AND WEAR RING
JET PUMP CAVITATION
REVERSE GATE INCORRECTLY ADJUSTED
HARD STEERING
STEERING INCORRECTLY ADJUSTED
BUSHINGS CRACKED OR DAMAGED
STEERING SHAFT BENT
STEERING NOZZLE PIVOT INADEQUATELY LUBRICATED
STEERING CABLE DAMAGED OR IMPROPERLY ROUTED

6
FUEL AND OIL

6-1 INTRODUCTION

The carburetion and ignition principles of two-cycle engine operation **MUST** be understood in order to perform a proper tune-up on a marine engine.

If any doubts exist concerning the reader's understanding of two-cycle engine operation, it would be best to take a few minutes to study the Introduction section in the first portion of Chapter 8 -- **ENGINE**, before tackling any work on the fuel system.

The fuel system includes the fuel tank, fuel pump, fuel filters, carburetor, and the associated parts to connect it all together. Regular maintenance of the fuel system to obtain maximum performance, is limited to changing the fuel filters at regular intervals and using fresh fuel.

If a sudden increase in gas consumption is noticed, or if the engine does not perform properly, a carburetor overhaul, including boil-out, or replacement of the fuel pump may be required.

Making a low speed adjustment with the flame arrestor removed; the craft in a body of water or a flush attachment and garden hose connected; and the engine operating.

6-2 GENERAL CARBURETION INFORMATION

The carburetor is merely a metering device for mixing fuel and air in the proper proportions for efficient engine operation. At idle speed, a marine engine requires a mixture of about 8 parts air to 1 part fuel. At high speed or under heavy duty service, the mixture may change to as much as 12 parts air to 1 part fuel.

DIAPHRAGM CARBURETORS

Description

The only carburetor used on Bombardier Sea-Doo personal watercraft covered in this manual, at press time, is a diaphragm type carburetor. The carburetor installed is manu-

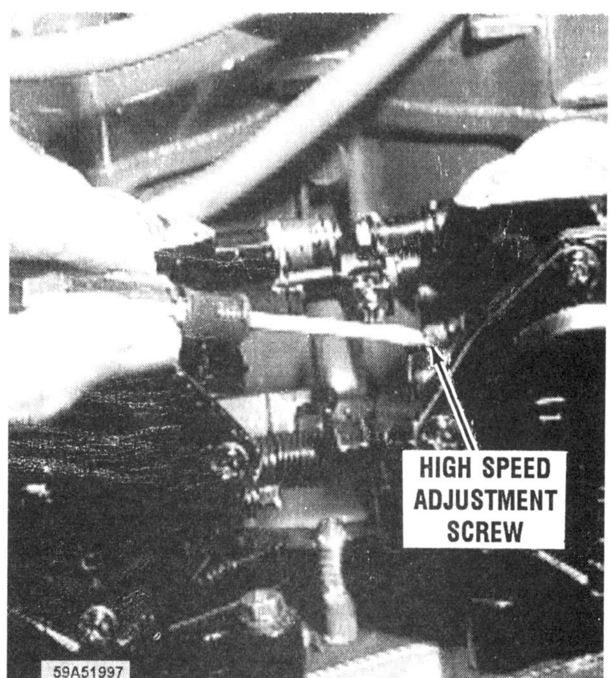

An air/fuel mixture problem in the carburetor will prevent the engine from operating smoothly at idle speed.

Close view of a carburetor typical of those used on watercraft featured in this manual.

factured by Mikuni and identified as a "Mikuni Super BN". The fuel pump serving the fuel system is an integral part of the carburetor.

A diaphragm carburetor does not have a float bowl or a float. Instead, a diaphragm takes the place of these items.

A hinged arm, still called a float arm, rests against the diaphragm and opens and closes the needle and seat assembly.

Operation

Following the fuel through its course, from the fuel tank to the combustion chamber of the cylinder will provide an appreciation of exactly what is taking place and probably "why".

All models covered in this manual have a fuel filter/fuel water separator installed between the fuel tank and the fuel pump inlet on the carburetor.

Fuel is drawn from the tank through the filter by the integral fuel pump and fed directly into the carburetor fuel chamber through the inlet valve. When the choke valve is closed

Close view of the diaphragm assembly removed from the carburetor.

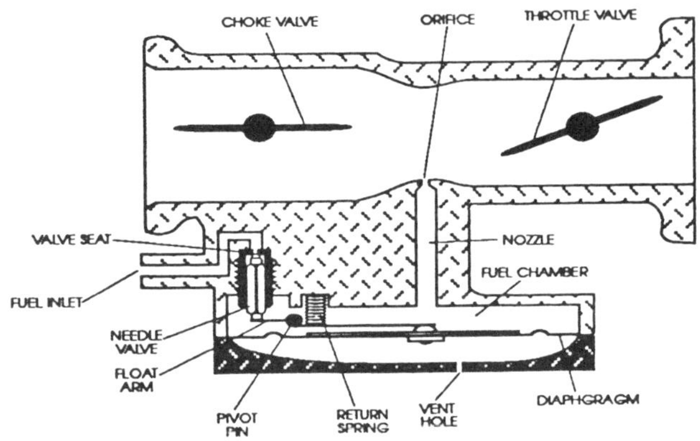

Cross-section drawing of a diaphragm carburetor -- the type used on engines covered in this manual.

and the engine is being cranked, or when the piston is on its upward stroke, the pressure in the carburetor throat orifice and the crankcase is less than atmospheric pressure. The fuel chamber above the diaphragm is connected to the crankcase through a vacuum tube. A nozzle channel connecting the carburetor throat with the fuel chamber also allows the fuel chamber to be at less than atmospheric pressure.

With the two pressure differentials, the diaphragm will flex toward the lowest pressure -- toward the fuel chamber. As the diaphragm flexes, it raises the float arm which is in constant contact with the diaphragm, because of a return spring. The float arm is hinged. As one side rises, the other side lowers and allows the inlet valve to open. At the same time, the needle valve falls away from the valve seat. As soon as the inlet valve opens, fuel flows under pressure from the fuel pump into the fuel chamber.

When the piston is on its downward stroke, the crankcase, the carburetor

Typical installation of the in-line fuel filter on units covered in this manual since 1992.

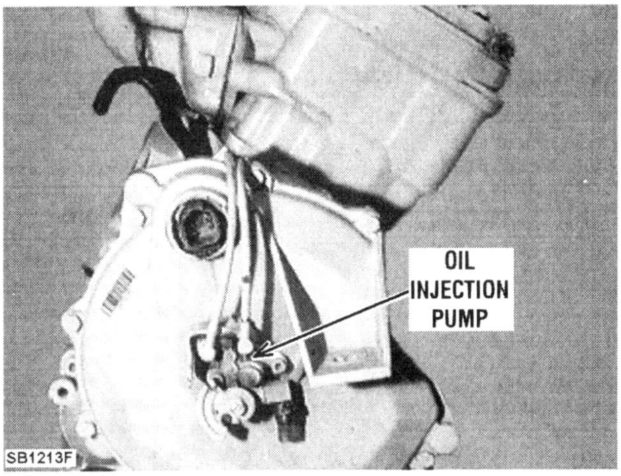

Oil injection pump mounted on the flywheel cover on a Model 787 Series engine. The pump shaft is connected to the end of the crankshaft through a short nylon shaft. Therefore, oil delivery is a function of engine rpm.

throat, and the fuel chamber are all at a higher pressure than atmosphere. The fuel chamber is filled with fuel at a higher pressure. Therefore, the diaphragm flexes toward the side with the lower pressure -- the side vented to atmosphere.

The return spring pushes the float arm down against the diaphragm. The other side of the float arm raises up and pushes the needle valve against its seat, closing the inlet valve.

On the next piston downward stroke, the diaphragm flexes back again forcing the fuel in the fuel chamber to be ejected through a jet into the carburetor throat. On the same stroke, the fuel chamber is replenished with fresh fuel through the open needle valve. This cycle repeats itself, sending a squirt of fuel into the carburetor throat during each downward stroke of the piston.

Inside view of the fuel pump cover. Only the forward carburetor is equipped with the pump. The aft carburetor is serviced through a short coupling from the pump.

In order to obtain the proper air/fuel mixture for all engine speeds, some models have high and low speed jets. These jets have adjustable needle valves which are used to compensate for changing atmospheric conditions. In almost all cases, the high-speed circuit has fixed high-speed jets. These fixed jets cannot be adjusted.

A throttle valve controls the flow of air/fuel mixture drawn into the combustion chambers. A cold engine requires a richer fuel mixture to start and during the brief period it is warming to normal operating temperature. A choke valve is incorporated ahead of the metering jets and venturi. As this valve begins to close, the volume of air intake is reduced, thus enriching the mixture entering the cylinders.

When this choke valve is fully closed, a very rich fuel mixture is drawn into the cylinders.

Fuel/Oil Mixture

All Bombardier Sea-Doo personal watercraft are equipped with an oil injection system.

A new or rebuilt engine must be properly "broken in" if it is to develop maximum performance. An engine which is incorrectly "broken-in" will never be able to achieve its maximum rated horsepower.

During the initial "breaking-in" period, the fuel/oil ratio should be a little "rich" with added oil. This "rich" oil mixture may be obtained by adding one ounce of oil to each gallon of fuel. This "rich" oil mixture should be used for the first 2-1/2 to 3 hours of operation after the unit is brought out of storage, or following an engine overhaul.

Such a ratio will **ENSURE** adequate lubrication of moving parts, which have been drained of oil during storage or service work. Operation of an engine without the proper amount of oil will quickly lead to a significant loss of horsepower.

All fuel pumps on the engines covered in tyhis manual are an integral part of the carburetor. Service procedures for the pump are included in the overhaul inst4ructions for the carburetor, beginning on Page 6-9

On a dual carburetor installation -- most of the units covered -- **ONLY** the forward carburetor has the integral fuel pump. This pump serves both carburetors via a short fuel line between the two.

Use only two-stroke outboard marine oil in the mixture -- **NEVER** automotive oils. Four-stroke automotive engine oil is not designed to burn completely -- only to lubricate. Therefore, automotive oil, if used in a two-stroke engine, will leave an undesirable residue in the combustion chamber.

The oil injection system is covered in Section 6-8 of this chapter, beginning on Page 6-35.

Air/Fuel Mixture

A suction effect is created in the intake manifold each time the piston moves upward in the cylinder. This suction draws air through the throat of the carburetor. A restriction in the throat, called a venturi, controls air velocity and has the effect of reducing air pressure at this point.

The difference in air pressures at the throat and in the fuel chamber, causes the fuel to be pushed out of a metering jet extending down into the fuel chamber. When the fuel leaves the jet, it mixes with the air passing through the venturi. This air/fuel mixture should then be in the proper proportion for burning in the cylinders to provide maximum engine performance.

Throttle and Choke Valves

A throttle valve controls the flow of air/fuel mixture drawn into the combustion chambers. A cold engine requires a richer fuel mixture to start and during the brief period it is warming to normal operating temperature. A choke valve is place ahead of the metering jets and venturi. As the valve begins to close, the volume of air drawn in is reduced, thus enriching the mixture entering the cylinders.

When this choke valve is fully closed, a very rich fuel mixture is drawn into the cylinders aiding engine start.

Engine Revolution Limiter

An engine revolution limiter is a part of the MPEM (Multi-Purpose Electronic Module) and is mounted in the electrical box through 1995. In 1996 and on, the MPEM is housed in a separate box forward of the engine compartment. The function of the limiter is to prevent damage to the engine by operating at too high an rpm figure.

In the very early days of personal watercraft, jumping waves could cause extensive internal

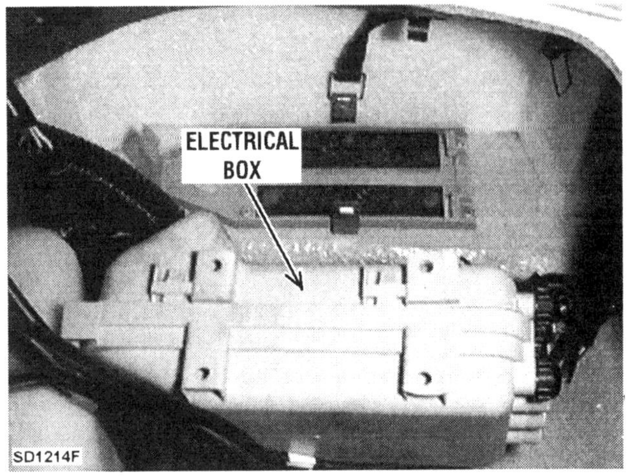

Typical location of the electrical box mounted in the engine compartment. The engine revolution limiter -- a part of the MPEM (Multi-Purpose Electronic Module -- was housed in this box through 1995. Since 1996, the module has been housed in a separate box forward of the engine compartment.

engine damage. When the craft left the water, unchecked engine rpm rapidly increased to as high as 12,000 rpm. As soon as the craft re-entered the water, engine rpm instantly decreased to about 6,000 rpm. This sudden increase and decrease in engine speed dramatically shortened engine life. In those days repeated wave jumping was known to cause bent crankshafts, sheared flywheel Woodruff keys, and broken connecting rods.

Bombardier engineers developed an rpm limiter system to help prevent excessive high engine rpm and an almost "runaway" condition. The engineers answer was an rpm limiter. This limiter restricts engine speed to about 6,500rpm. Thus, wave jumping, using a ramp, and a host of other almost unbelievable maneuvers could be performed without undue stress on the engine. Today, with an rpm limiter system installed, a bent crankshaft or broken connecting rod is almost non-existent.

BELIEVE THESE WORDS

Any attempt to bypass the revolution limiter could have serious and expensive consequences.

6-3 FUEL PUMP

All fuel pumps on the engines covered in this manual are an integral part of the carburetor. Service procedures for the pump are included in the overhaul instructions for the carburetor, beginning on Page 6-9.

On a dual carburetor installation -- most of the units covered -- **ONLY** the forward carburetor has the integral fuel pump. This pump services both carburetors via a short fuel line between the two.

The following five paragraphs describe the pump, operation of the various components, and caution concerning handling of the diaphragm and consequences of the tiniest hole.

Integral Fuel Pump
Description and Operation

The integral fuel pump consists of a series of diaphragms and check valves operated by crankcase pressure. The pump feeds fuel directly into the fuel chamber through the inlet valve.

As the piston moves upward, a suction, or negative pressure, is created in the crankcase. This low pressure acts upon the inner diaphragm through the carburetor. The inner diaphragm is, therefore, pulled inward. When the diaphragm moves inward the check valve opens and fuel is drawn from the bottom of the fuel pump exterior chamber into the fuel pump interior chamber.

As the piston moves downward, a positive pressure is created in the crankcase causing the inner diaphragm to be pushed outward. The delivery check valve opens under pressure and allows fuel to flow into the upper part of the exterior chamber and onto the inlet valve of the carburetor.

The function of the outer diaphragm is to absorb the pulsations of the fuel and allow a smooth uninterrupted fuel flow.

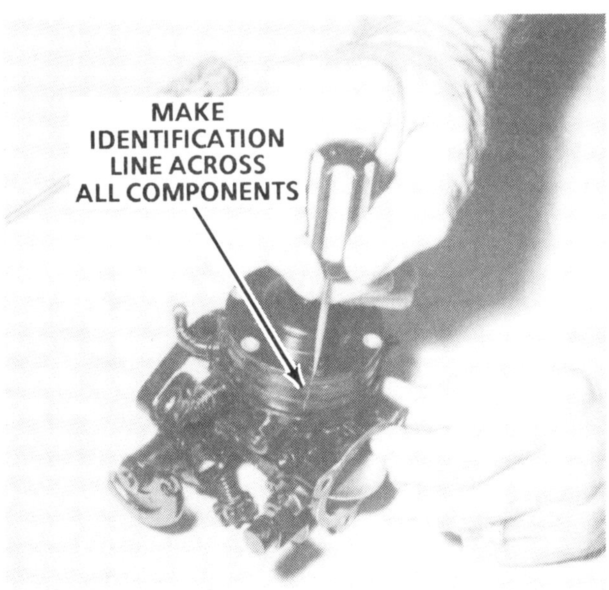

*Prior to disassembling the fuel pump, a mark should be scribed across the edge of all pump components -- eight layers in all -- to **ENSURE** all will be assembled back in their original location and position. This illustration shows marking a "round" body pump, but the mark on a rectangular body is made in the same manner.*

If the fuel pump diaphragm tears or is punctured, even with the tiniest hole, fuel will enter the crankcase through the pulse line causing an overrich mixture. The mixture will foul the plugs and in extreme cases flood the crankcase.

6-4 FUEL IN THE SYSTEM

The fuel system includes the fuel tank, fuel pump, fuel filters, carburetor, connecting fuel lines, and the associated parts to connect it all together. Regular maintenance of the fuel system to obtain maximum performance, is limited to changing the fuel filters at regular intervals and using **FRESH** fuel.

Even with the high price of fuel, removing gasoline that has been standing unused over a long period of time, is still the easiest and least expensive preventive maintenance possible. In most cases, old gas, even if mixed with some oil, can usually be used in an automobile, or other gas driven vehicle using regular gas, without harmful effects.

A lead substitute additive -- octane booster -- can help prevent detonation when unleaded gasoline is used in a personal watercraft.

If a sudden increase in gas consumption is noticed, or if the engine does not perform properly, a carburetor overhaul, including boil-out, or replacement of the fuel pump may be required.

Leaded Gasoline

The manufacturer of the units covered in this manual recommend their engines be operated using unleaded gasoline with a minimum octane rating of 84 or higher.

In the United States leaded gasoline in no longer available. The Environmental Protection Agency (EPA) slated a national phase-out of leaded fuel to become effective back in 1993. For all practical purposes, the phase-out has been completed. Lead in gasoline boosts the octane rating (energy).

Therefore, if the lead is removed, it must be replaced with another agent. Unknown to the general public, many refineries are adding alcohol in an effort to hold the octane rating

Alcohol in gasoline can have a deteriorating effect on certain fuel system parts. Seals can swell, pump check valves can swell, diaphragms distort, and other rubber or neoprene composition parts in the fuel system can be affected.

For over a decade, every effort has been made to use materials having a high resistance to the alcohol being added to fuels for parts coming in contact with the fuel.

Fuels containing alcohol will slowly absorb moisture from the air. Once the moisture content in the fuel exceeds about 1%, the moisture will separate from the fuel taking the alcohol with it. This water/alcohol mixture will settle to the bottom of the fuel tank. The engine will fail to operate. Therefore, storage of this type gasoline for use in marine engines is not recommended for more than just a very short time -- say a few weeks. **USE FRESH FUEL.**

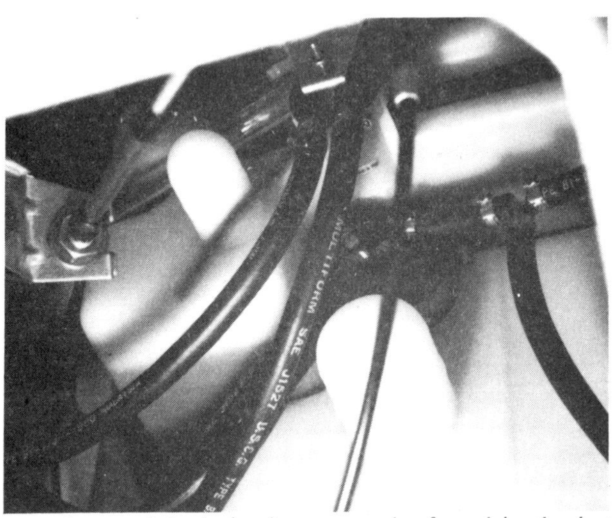

A water/alcohol mixture can be found in the bottom of fuel tanks, when the craft has been operated on a high alcohol content fuel.

Removing Fuel from the System

For many years there has been a popular wide-spread belief that simply shutting off the fuel at the tank and then running the engine until it stops is the proper procedure before storing the engine for any length of time. Right? **WRONG!**

It is **NOT** possible to remove all the fuel in the carburetor by simply operating the engine until it shuts down from lack of fuel. Some fuel is always trapped in passages. The **ONLY** guaranteed method of removing **ALL** fuel, is to take the time to remove the carburetor, and drain the fuel.

BAD NEWS

If the engine is operated with the fuel supply shut off until it stops, the fuel and oil mixture inside the engine is removed, leaving bearings, pistons, rings, and other parts with little protective lubricant, during long periods of storage.

PROPER PROCEDURE
LONG TIME STORAGE

a- Shut off the fuel supply at the tank.
b- Disconnect the fuel at the tank.
c- Operate the engine **ONLY** until it begins to run **ROUGH.**
d- Stop the engine, which will leave some fuel/oil mixture inside.
e- Remove and drain the carburetor.

By disconnecting the fuel supply, all **SMALL** passages are cleared of fuel even though some fuel is left in the carburetor. A light oil should be put in the combustion chamber, as instructed in the Owner's Manual.

SHORT TIME STORAGE

For short periods of storage, the old belief of shutting off the fuel supply and running the engine until it shuts down from lack of fuel may help prevent severe gum and varnish from forming in the carburetor. This is especially true during very hot weather. Again, this procedure is only valid for short periods of time, no more than a couple weeks.

6-5 TROUBLESHOOTING

The following paragraphs provide an orderly sequence of tests to pinpoint problems in the fuel system. It is very rare for the carburetor by itself to cause failure of the engine to start.

FUEL PROBLEMS

Many times fuel system troubles are caused by a plugged fuel filter, a defective fuel pump, or by a leak in the line from the fuel tank to the fuel pump. A defective choke may also cause problems. **WOULD YOU BELIEVE**, a majority of starting troubles which are traced to the fuel system are the result of empty fuel tank, clogged fuel filter, or aged "sour" fuel.

Fuel Filter and Sediment Bowl Service

Some models are equipped with a fuel filter and sediment bowl mounted inside the engine compartment.

WORDS OF CAUTION

Be sure to loosen the fuel filler cap on the fuel tank to relieve any pressure built up inside the fuel system **BEFORE** removing the sediment bowl.

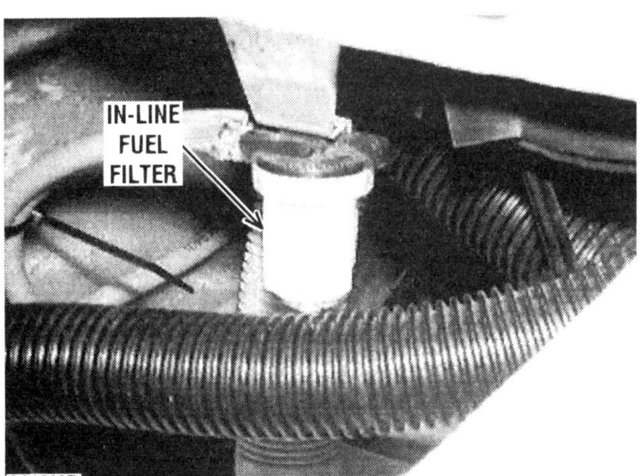

The in-line fuel filter installed in the engine compartment should be checked and cleaned at regular intervals -- especially at the beginning of any fuel troubleshooting.

Have a shop cloth handy to catch any spilled fuel. Unthread the ring nut and pull down the sediment bowl with the fuel filter inside. A large **O**-ring seals the bowl to the filter cap and ring nut. The function of this **O**-ring is to prevent fuel and air leaks. If the **O**-ring is defective, the fuel pump will draw air into the system and the engine will either not start, will falter, or will die. Sometimes, this problem is not apparent with a full fuel tank, because the fuel is gravity fed to the carburetor. When the fuel level drops in the tank, the pump takes over, but its effectiveness is impaired by the air leak at the sediment bowl **O**-ring.

Clean out the sediment bowl and rinse the filter screen in solvent every 25 hours of engine operation.

"Sour" Fuel

Under average conditions (temperate climates), fuel will begin to breakdown in about four months. A gummy substance forms in the bottom of the fuel tank and in other areas. The filter screen filter screen

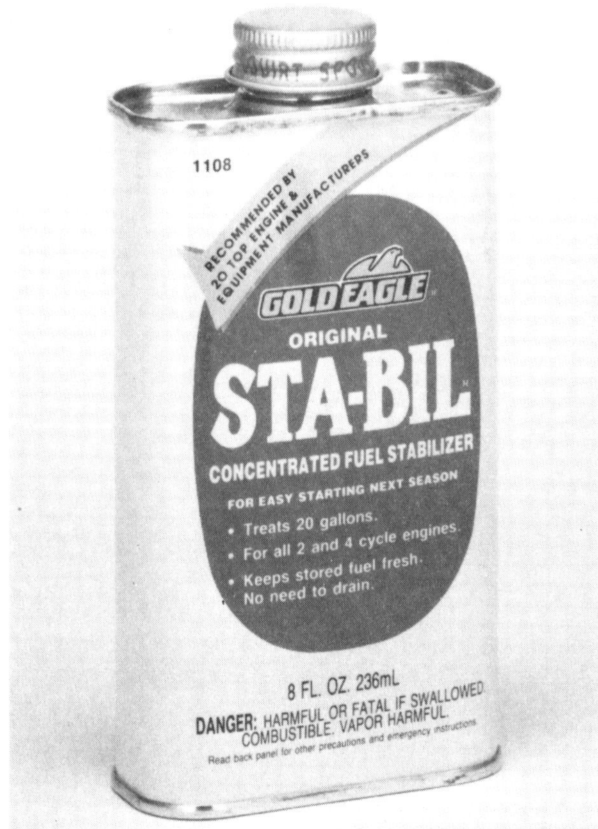

STA-BIL Gasoline Stabilizer and Conditioner -- or equivalent, may be used to prevent fuel from "souring" for up to twelve full months.

between the tank and the carburetor and small passages in the carburetor will become clogged. The gasoline will begin to give off an odor similar to rotten eggs. Such a condition can cause the owner much frustration, time in cleaning components, and the expense of replacement or overhaul parts for the carburetor.

Even with the high price of fuel, removing gasoline that has been standing unused over a long period of time is still the easiest and least expensive preventative maintenance possible. In most cases, this old gas, even with the oil mixture, can be used without harmful effects in an automobile, or other vehicle, using regular gasoline.

A gasoline preservative additive, such as Sta-bil Fuel Conditioner and Stabilizer for 2 and 4-Cycle Engines, will keep the fuel "fresh" for up to twelve months. If this particular product is not available in your area, other similar additives are produced under various trade names.

Choke Problems

When the engine is hot, the fuel system can cause starting problems. After a hot engine is shut down, the temperature inside the fuel bowl may rise to 200°F and cause the fuel to actually boil. All carburetors are vented to allow this pressure to escape to the atmosphere. However, some of the fuel may percolate over the high-speed nozzle.

If the choke should stick in the open position, the engine will be hard to start. If the choke should stick in the closed position, the engine will flood making it very difficult to start.

In order for this raw fuel to vaporize enough to burn, considerable air must be added to lean out the mixture. Therefore, the only remedy is to remove the spark plugs; ground the leads; crank the engine through about ten revolutions; clean the plugs; install the plugs; and again attempt to start the engine. The attempt should be successful.

Rough Engine Idle

If an engine fails to idle smoothly, the most reasonable approach to the problem is to perform a tune-up to eliminate the possibility of fouled or faulty spark plugs.

Other problems preventing an engine from operating smoothly include: an air leak in the intake manifold or uneven compression between the cylinders.

Of course any problem in the carburetor affecting the air/fuel mixture will also prevent the engine from operating smoothly at idle speed. These problems usually include: a weak float arm return spring; leaking needle valve and seat; defective choke; and improper adjustments for idle mixture or idle speed.

Excessive Fuel Consumption

One of two, or both conditions, may cause excessive fuel consumption. One may be inefficient engine operation and the other may be poor boating habits of the operator.

If the fuel consumption suddenly increases over what could be considered normal, then the cause can probably be attributed to the engine or watercraft and not the operator.

Check the fuel system for possible leaks. A leak between the tank and the pump many times will not appear when the engine is operating, because the suction created by the pump drawing fuel will not allow the fuel to leak. Once the engine is turned off and the suction no longer exists, fuel may begin to leak.

6-6 SERVICE MIKUNI SUPER BN CARBURETOR

The Mikuni carburetors installed on all the Sea-Doo watercraft covered in this manual are identified as "Super BN" models.

The carburetor is equipped with an integral fuel pump. On a single carburetor installation, the carburetor has the fuel pump. On a dual carburetor installation, only the forward carburetor has the fuel pump. The pump services both carburetors through a short fuel hose between the two carburetors.

The carburetor linkage may be "horizontal" or "vertical", depending on the model engine installation.

GOOD WORDS

The removal procedures are divided into two minor sections -- one for the Model 587, 657, and 717 Series engines and the other section covering the Model 787 Series engines.

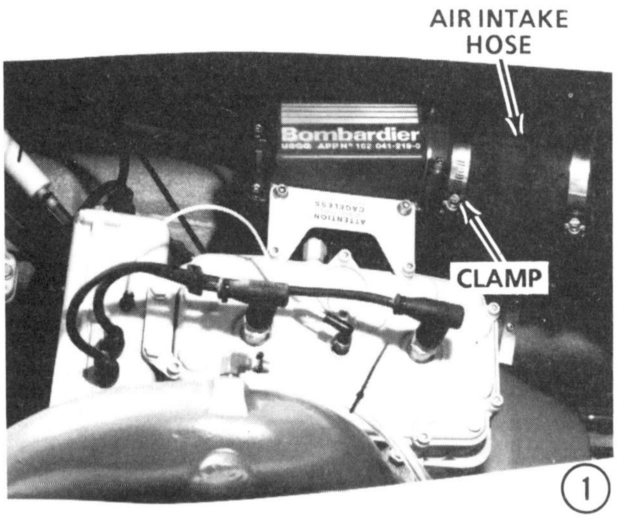

CARBURETOR REMOVAL SINGLE AND DUAL INSTALLATION MODEL 587, 657 AND 717 SERIES

1- Rotate the fuel valve to the **OFF** position. Loosen the clamp securing the air intake hose to the flame arrestor cover and slide the clamp back along the hose. Work the hose back and free of the cover flange.

2- Obtain an M6 Allen wrench and remove the six Allen head bolts and the washers securing the support bracket to the flame arrestor and the engine. Lift off the bracket.

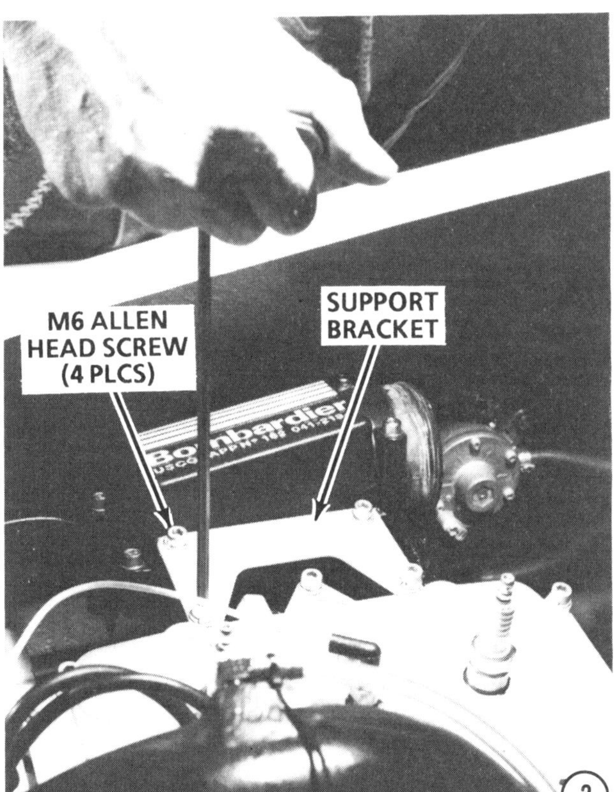

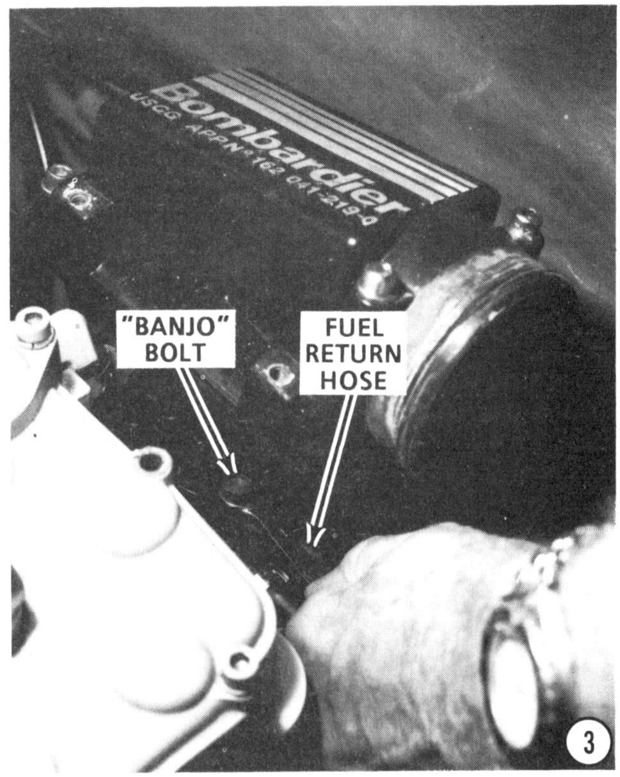

3- Remove the "banjo" bolt and fitting from the carburetor, with the fuel return hose attached. A banjo bolt is a special hollow bolt used to secure a fuel hose to the carburetor. If working on a dual carburetor installation, disconnect the fuel supply hose from the automatic shut-off valve.

4- Disconnect the throttle cable end from the curved guide on the throttle shaft. Remove the two carburetor mounting bolts.

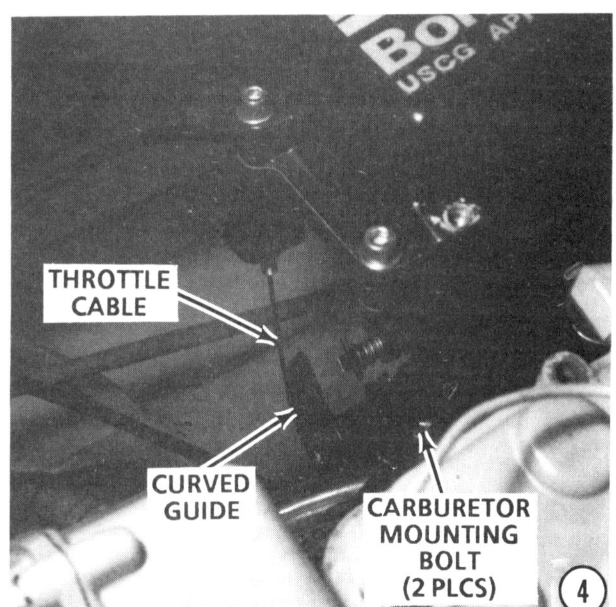

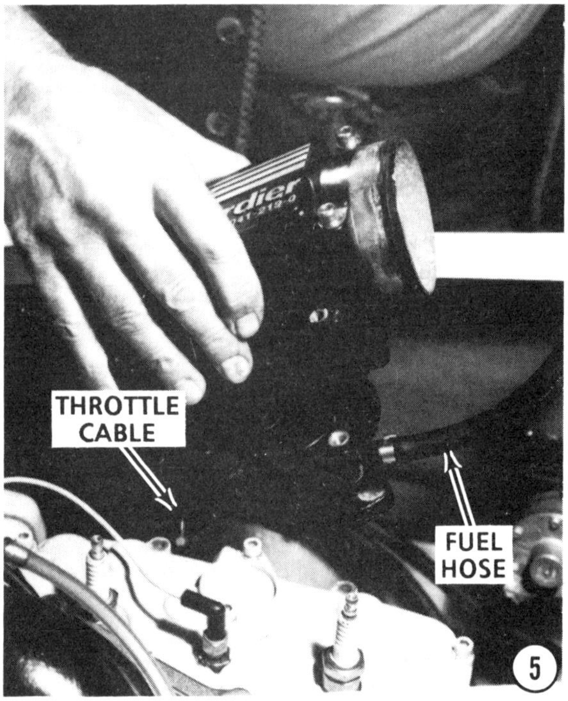

5- Lift the carburetor off the intake manifold, with the choke and throttle cables along with the fuel hose still attached. It is far easier to remove or disconnect these items after the carburetor is free of the manifold.

6- Remove all traces of carburetor gasket material from the carburetor mounting flange and the intake manifold.

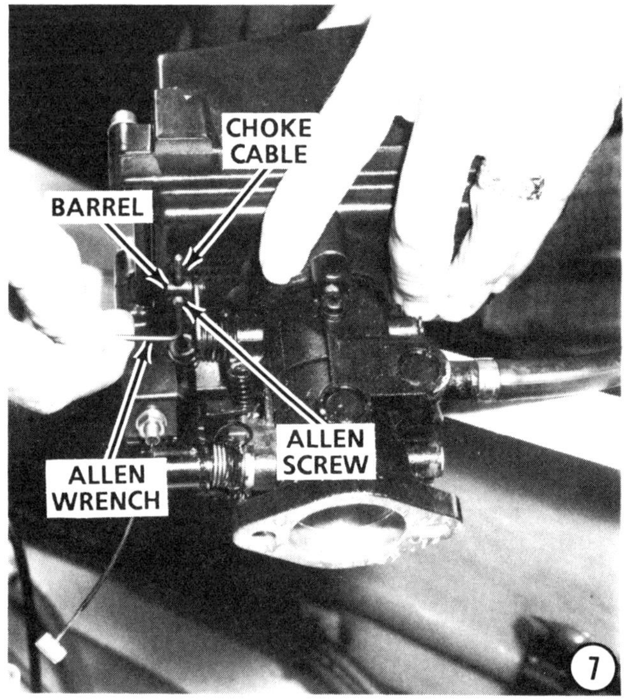

7- BEFORE disturbing the choke cable, obtain some "Whiteout", or some masking tape, or equivalent. Mark the point on the cable where it enters the barrel. Be sure the mark is fairly permanent and will not be disturbed when the cable is removed from the barrel. Marking the cable at this point, at this time, will ensure the cable can be installed back in its original position, with the same amount of cable tension and with the same choke valve setting. After the identifying mark is made, loosen the small Allen screw securing the choke cable to the barrel on the choke shaft.

8- Loosen the two nuts, one on each side, securing the choke cable to the cable

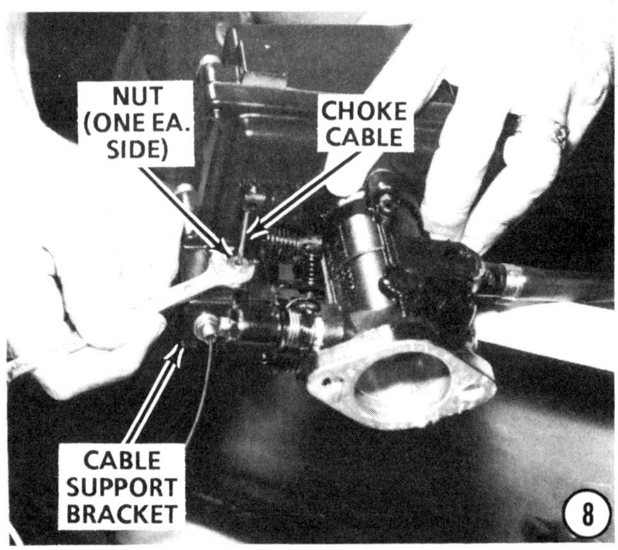

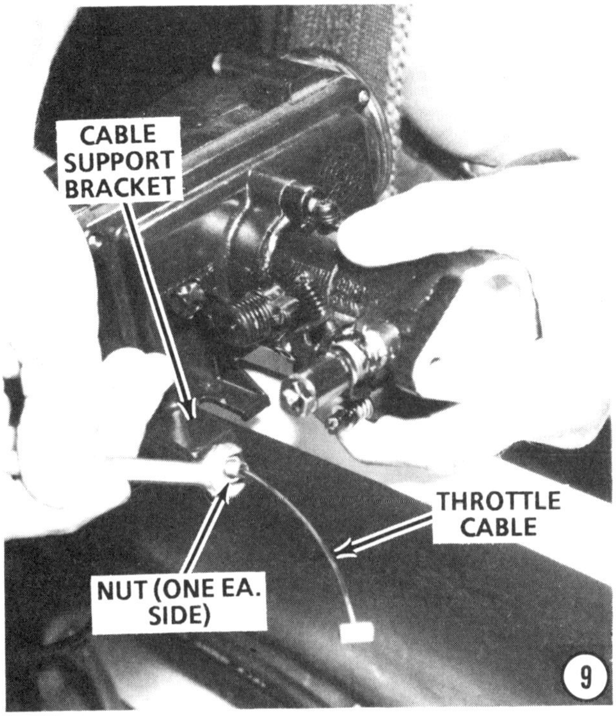

support bracket on the flame arrestor bottom cover. Pull the cable free of the slot and free of the barrel.

9- Release the throttle cable from the cable support bracket in the same manner.

10- Remove the four Allen head bolts securing the top flame arrestor cover.

11- Lift off the top cover and the flame arrestor screen.

12- Loosen the hose clamp and work the fuel supply hose free of the carburetor fitting. Move the carburetor/s to a suitable work surface. **NOW**, remove the single Allen head bolt securing the bottom flame arrestor cover to the carburetor. Lift off the cover.

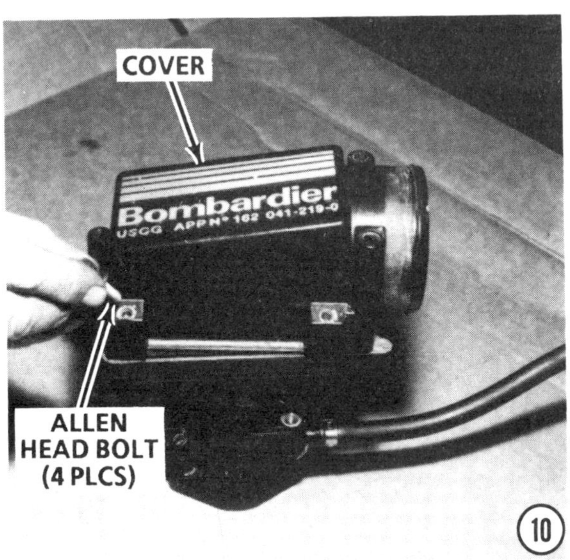

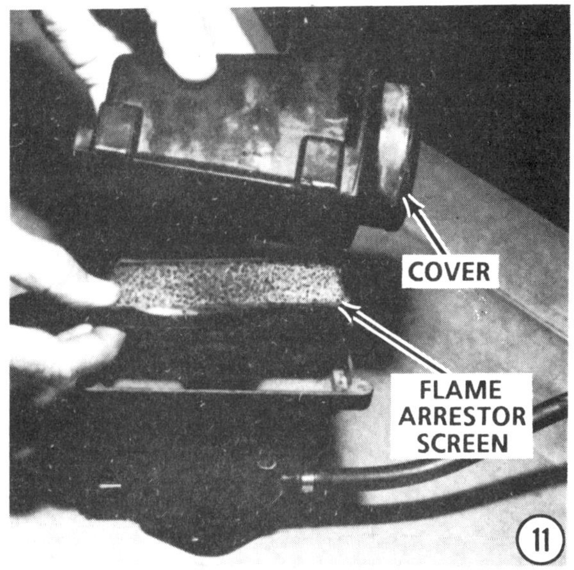

GOOD WORDS

The carburetor with the fuel pump is the **FORWARD** carburetor. The aft carburetor does not have the fuel pump **BUT** it does have the idle speed screw.

Pry off the link connecting the two throttle shafts together from both ball joints. **TAKE CARE** not to alter the length of the link. Normal service may be performed **WITHOUT** separating the carburetors.

INSTALLATION of the carburetor set begins on Page 6-28.

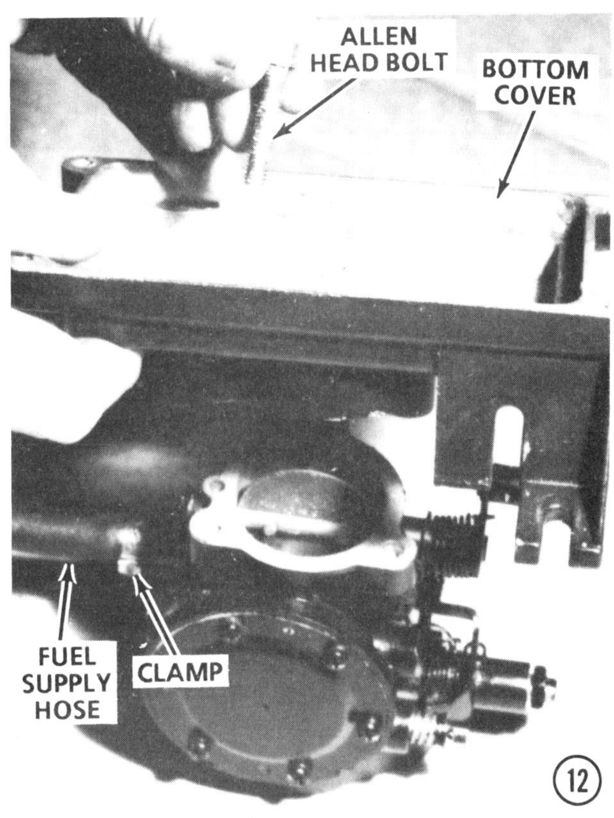

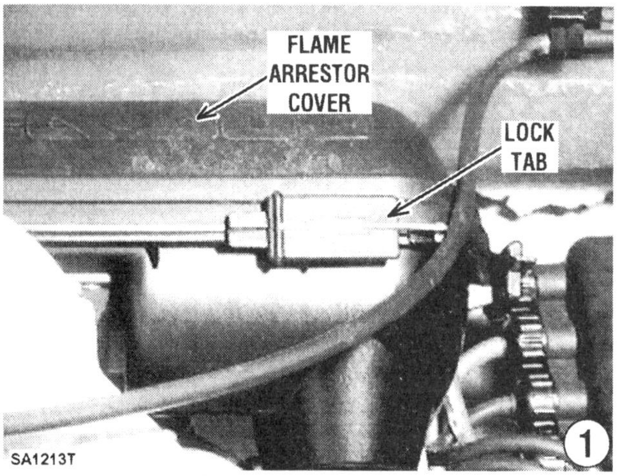

CARBURETOR REMOVAL DUAL INSTALLATION MODEL 787 SERIES

1- Push the flame arrestor cover locking tab down, then sideways to unlock the tab. Perform this procedure for each of the cover lock tabs.

2- Remove the six 5mm Allen head bolts securing the flame arrestor screen assembly to the arrestor pan -- or base.

3- Remove the fastener securing the cooling water bypass hose to the upper portion of the exhaust elbow. Pull the hose free of the fitting. This bypass hose channels water overboard to give the operator a visual indication of coolant water circulating through the engine.

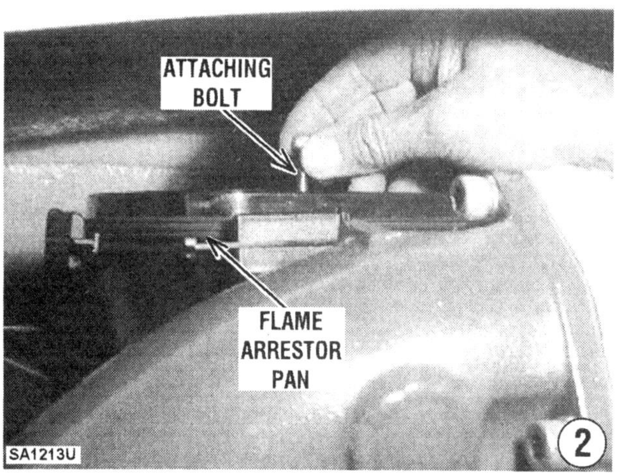

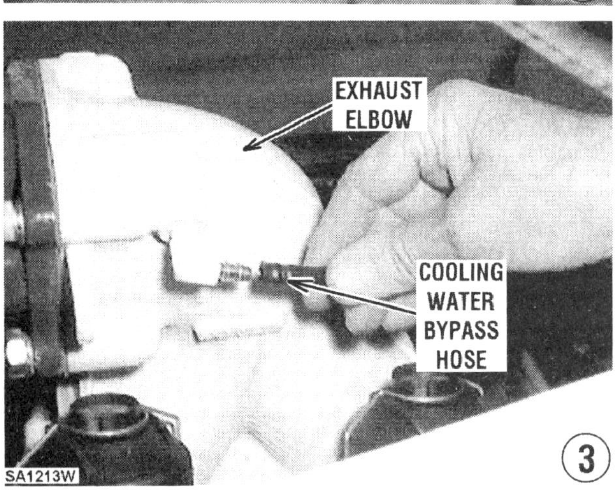

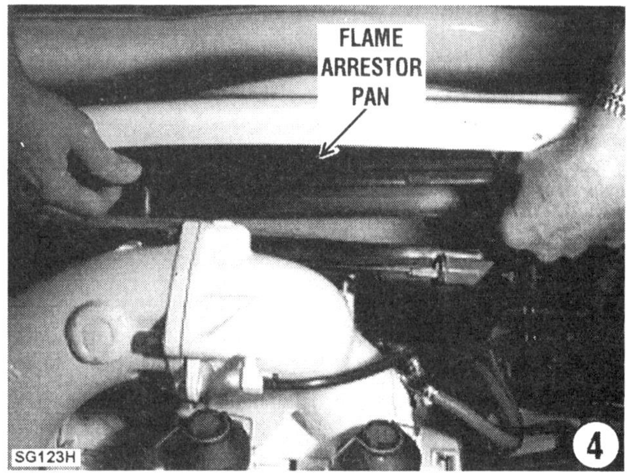

4- Rotate the flame arrestor pan 90 degrees towards the port side, then slide the arrestor pan aft and out of the engine compartment.

5- Pry the arrestor screen away from the bracket assembly using a sharp instrument, and then set the screen aside for further inspection.

6- Remove the four 5mm bolts securing the arrestor stainless steel bracket to the exhaust manifold.

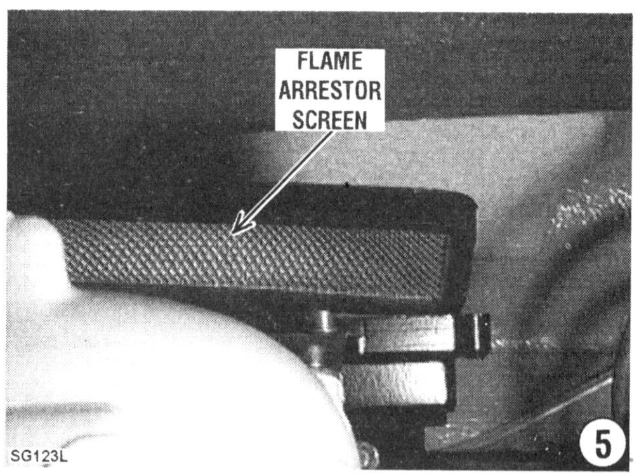

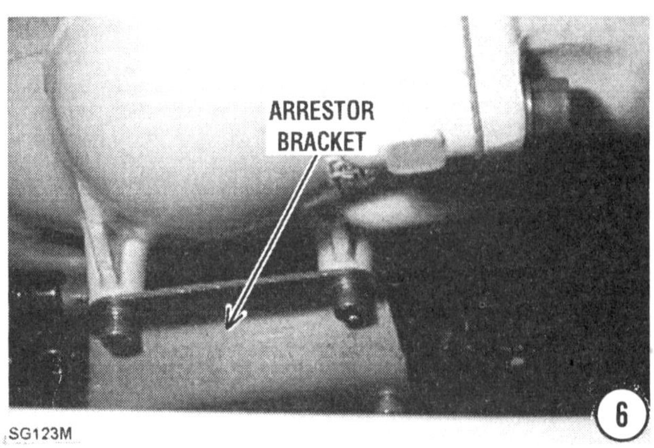

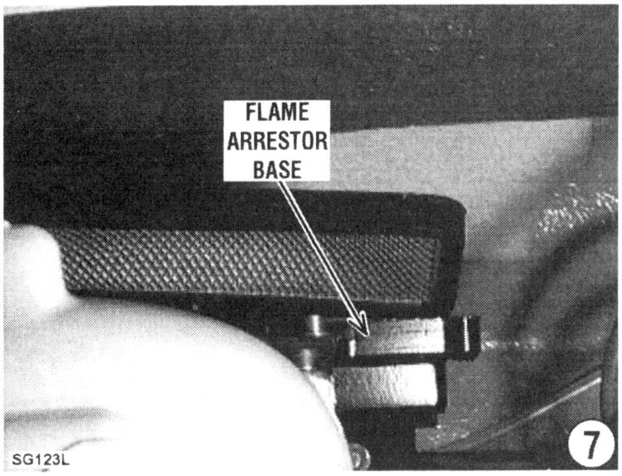

7- Lift the arrestor base bracket free of the carburetors.

8- Loosen the 3mm Allen head locking screw securing the choke cable to the pivot arm. Temporarily remove the 10mm cable locknut and disconnect the choke cable. Install the locknut back on the pivot arm. Remove the throttle in the same fashion.

9- Obtain special tool Sea-Doo P/N 295 000 054 -- Clamp pliers. Squeeze the pliers on the tab of the fuel return hose at the aft carburetor clamp,

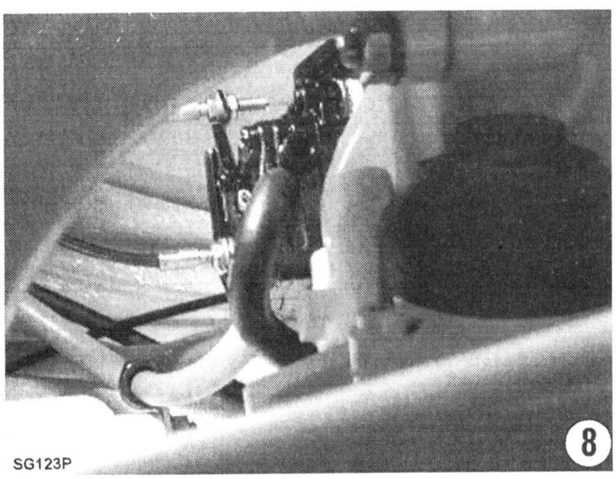

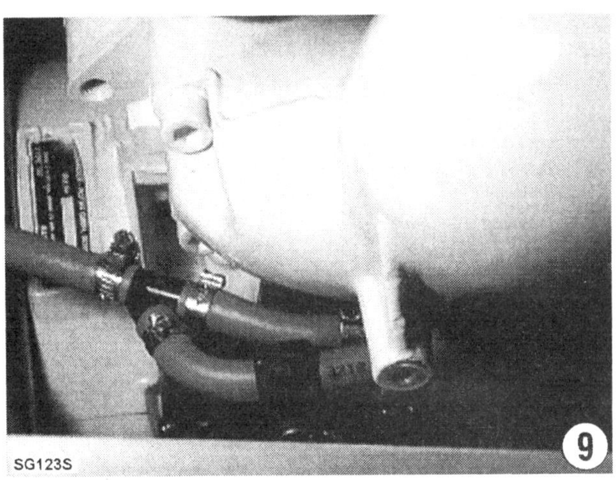

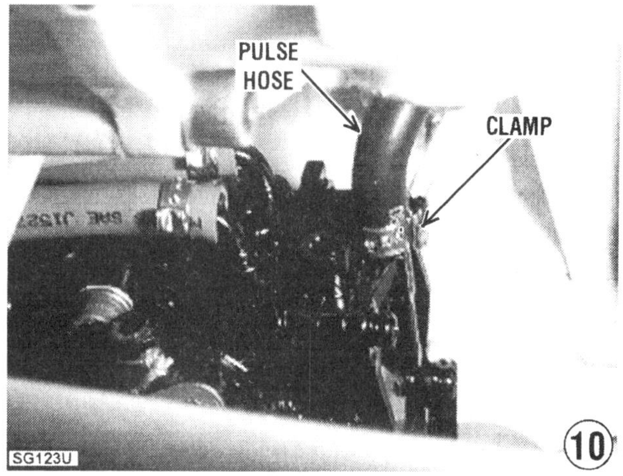

then rotate the pliers 90 degrees to remove the clamp. These clamps may be used again.

10- Cut the pulse hose metal clamp -- **not reusable** -- then remove the pulse hose from the top of the No. 1 -- forward -- carburetor.

11- Remove the restraining strap securing the fuel inlet hose to the forward cranking motor shoulder.

12- Remove the four 6mm Allen head bolts securing the carburetors to the rotary valve cover.

13- Temporarily remove the locknut securing the oil pump cable to the oil injection pump locat-

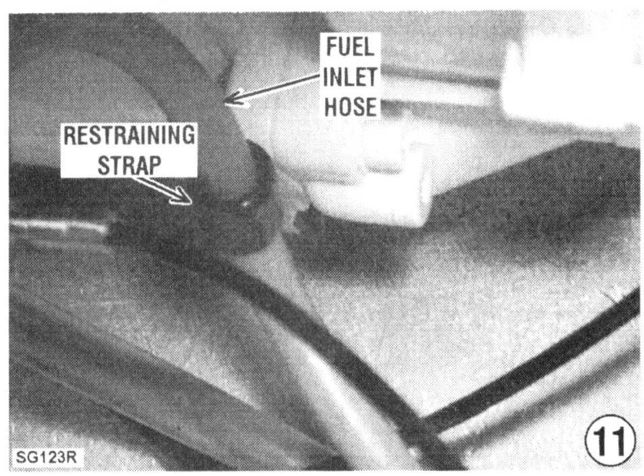

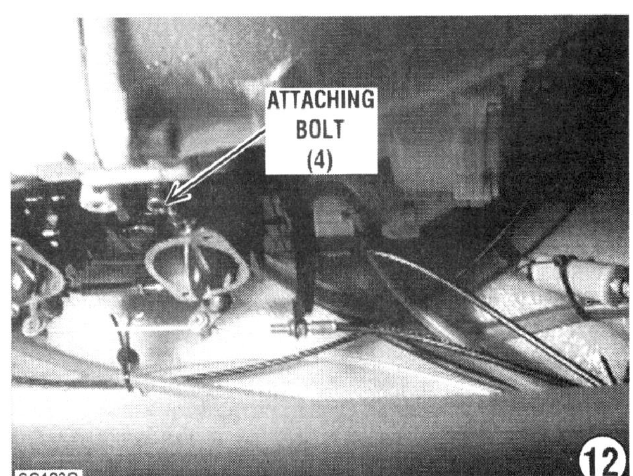

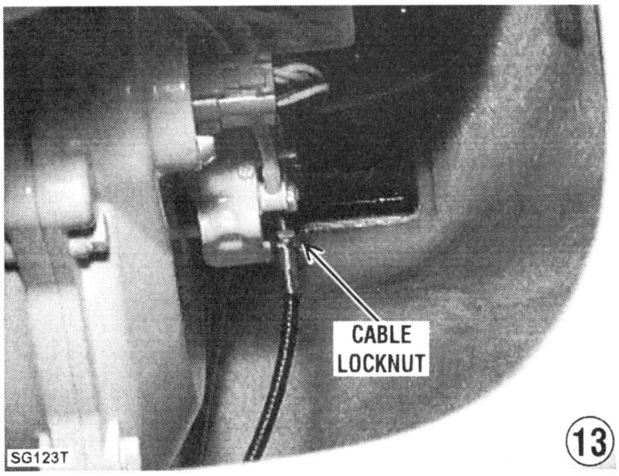

ed on the magneto cover. Disconnect the cable from the pump. Install the locknut after the cable is removed. Do not disconnect the injection pump cable end connected to the carburetor assembly, at this time.

14- Lift the carburetor/s, together as an assembly, from the rotary valve and set it aside for further inspection and cleaning.

Remove the one-time use metal clamp securing the fuel inlet hose to the carburetor. Disconnect the inlet hose.

INSTALLATION of the carburetor set for Model 787 engines, begins on Page 6-31.

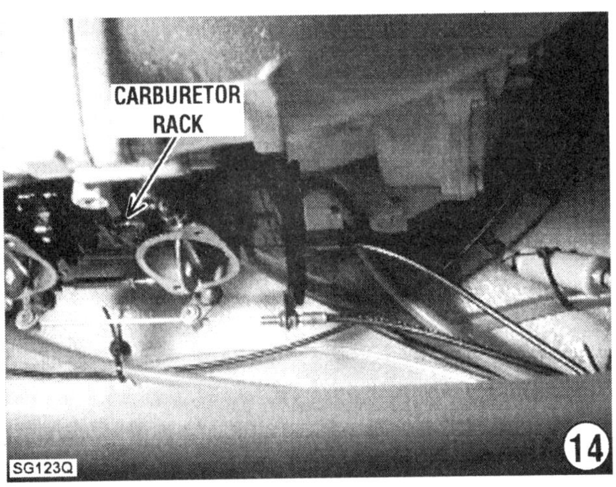

DISASSEMBLING MIKUNI SUPER BN CARBURETOR WITH INTEGRAL FUEL PUMP

Move the carburetors as a set -- to a suitable work surface. Cover the Rotary valve openings with "duct" tape to prevent contamination from entering the block. The carburetors need **NOT** be separated for normal service work. The carburetor in this section was separated for photographic clarity.

SPECIAL FUEL PUMP WORDS

This design fuel pump consists of many layers of components. Some items may be stuck together and appear to be one gasket instead of a stack of different parts. If the fuel pump is to be returned to service without replacing any of the fragile gaskets and check valves made of plastic film, **DO NOT ATTEMPT** to separate any stack of components stuck together. Such action will surely rip or stretch the delicate items. Lay the parts out in the order of removal, as an aid to later assembling.

1- Scribe a mark along the entire thickness of the fuel pump to ensure all components will be positioned correctly during installation. Remove the four Phillips head screws and lockwashers securing the fuel pump cover to the carburetor.

2- Remove the fuel pump cover and gasket.

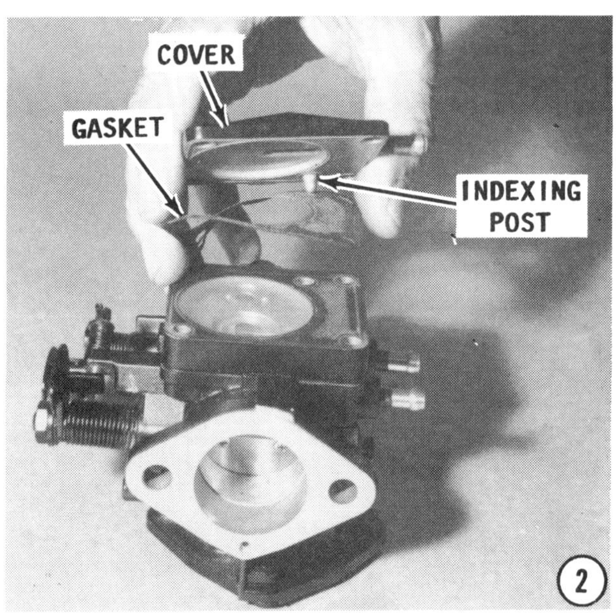

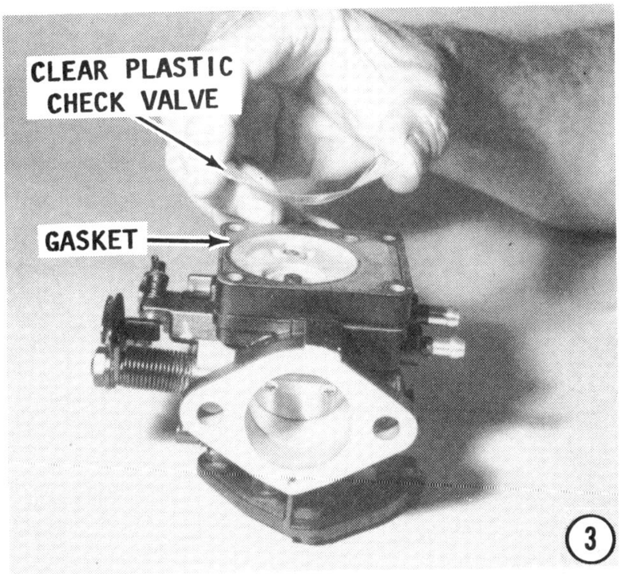

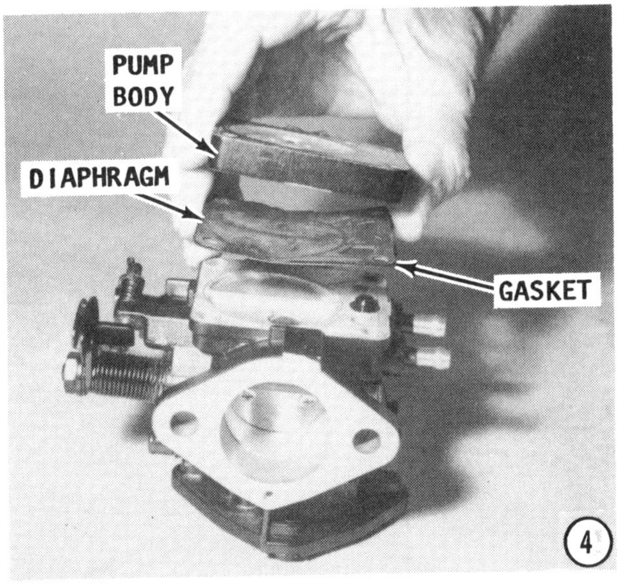

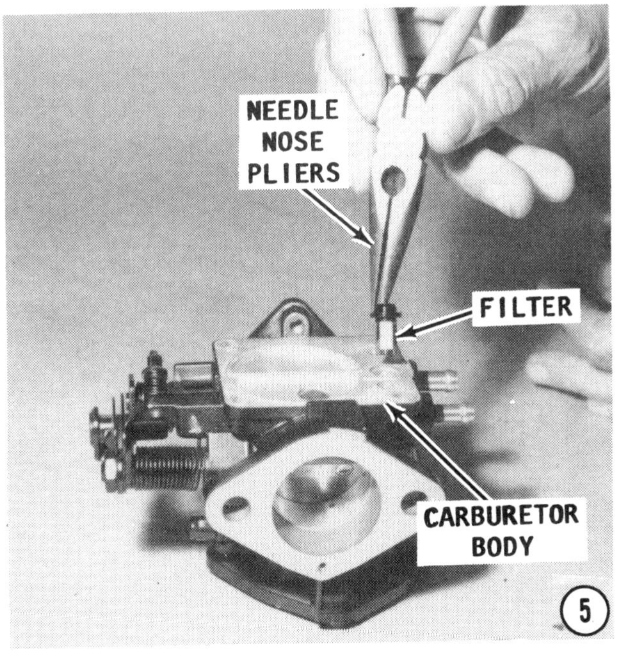

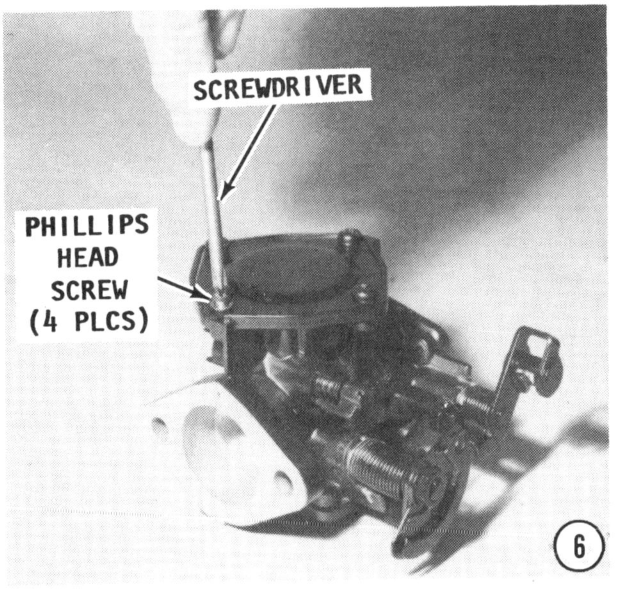

3- Remove the clear plastic check valve and gasket.

4- Lift off the pump body, diaphragm, and gasket.

5- Using a pair of needlenose pliers, withdraw the small fuel filter in the carburetor body.

6- Remove the four Phillips head screws securing the diaphragm cover to the carburetor.

7- Lift off the diaphragm cover.

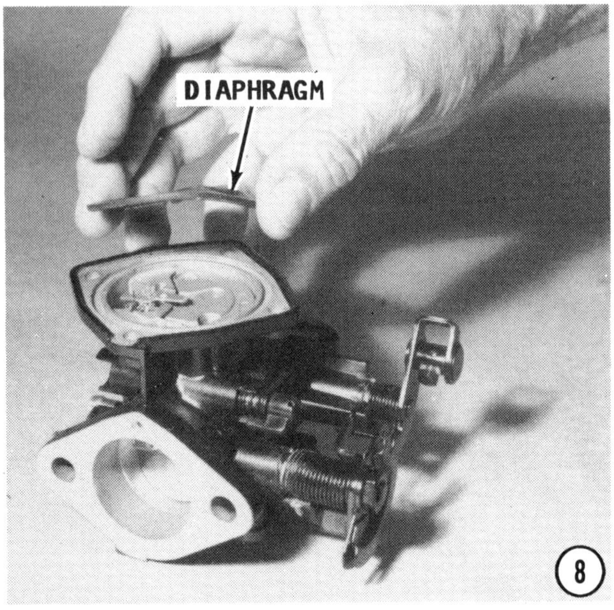

8- Remove the diaphragm from the fuel chamber.

9- Obtain a stubby Phillips head screwdriver and a hammer. Now, tap lightly on the end of the screwdriver while attempting to remove the hinge screw. Do not attempt to remove this screw "cold" -- without tapping the screwdriver -- because the head will almost certainly be stripped!

10- Lift out the float arm and float pin. The needle valve will remain hanging on the

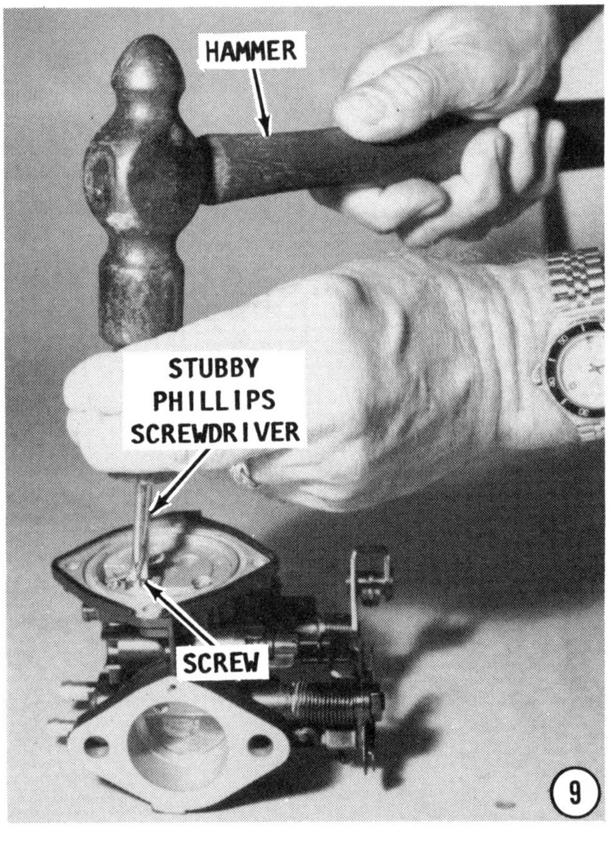

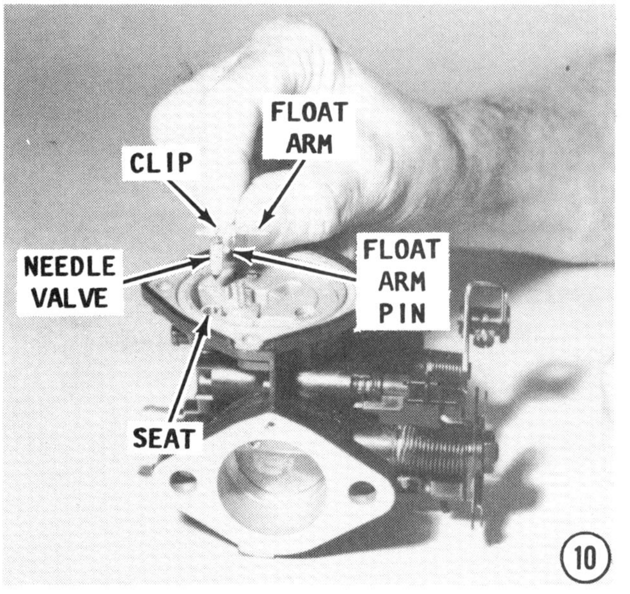

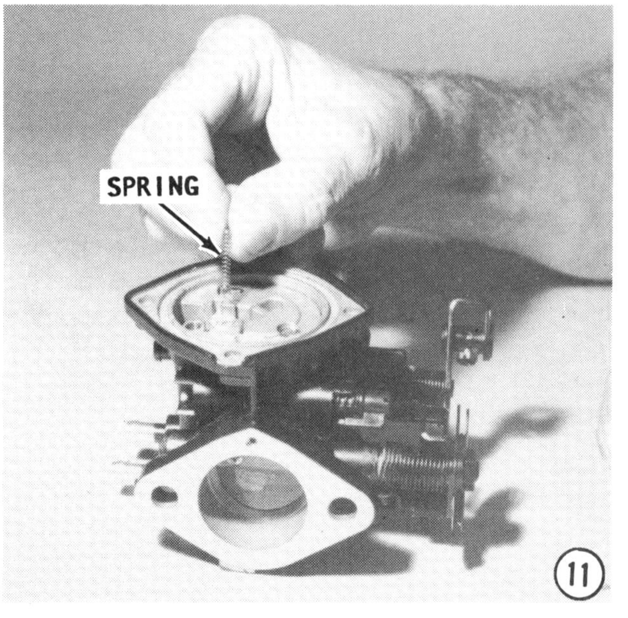

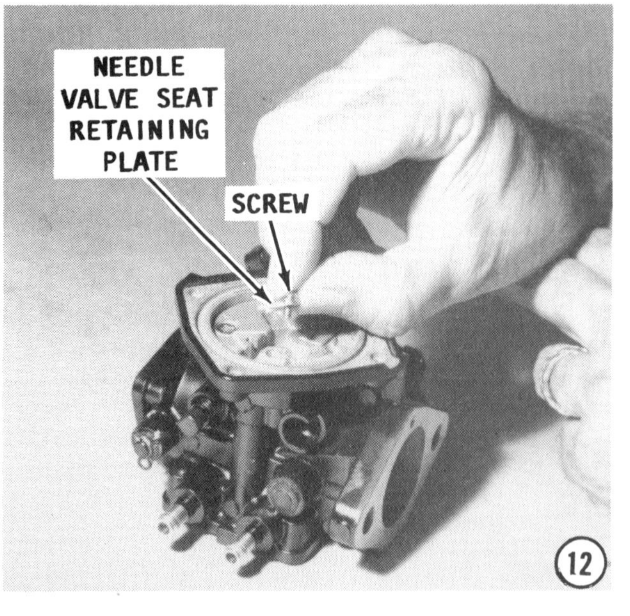

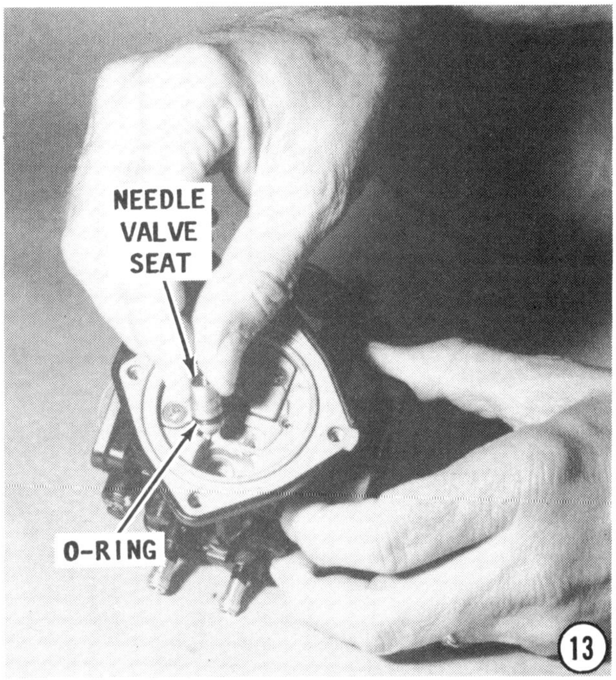

pin by the clip. Lift the valve from the seat.

11- Remove the spring from its recess in the fuel chamber.

12- Use the stubby Phillips head screwdriver and a hammer from Step 15 and lightly tap the end of the screwdriver while attempting to remove the screw securing the retaining plate to the fuel chamber. Remove the screw and plate.

13- Lift out the needle valve seat from the fuel chamber. Remove and discard the O-ring around the seat.

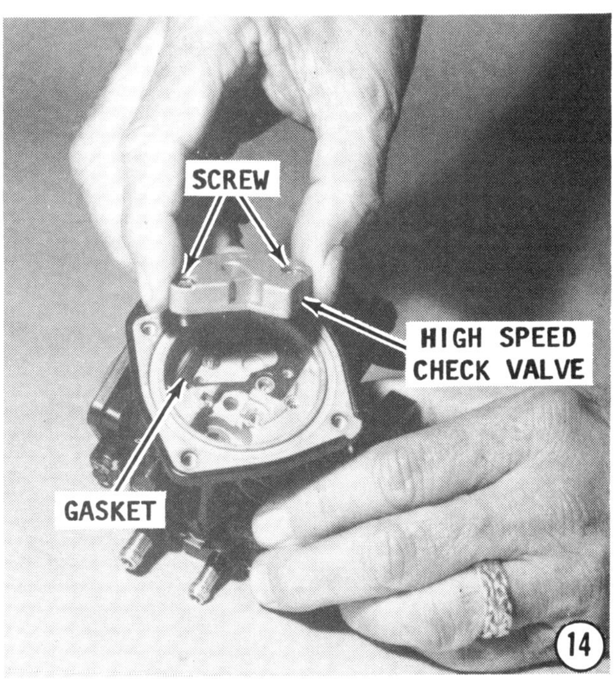

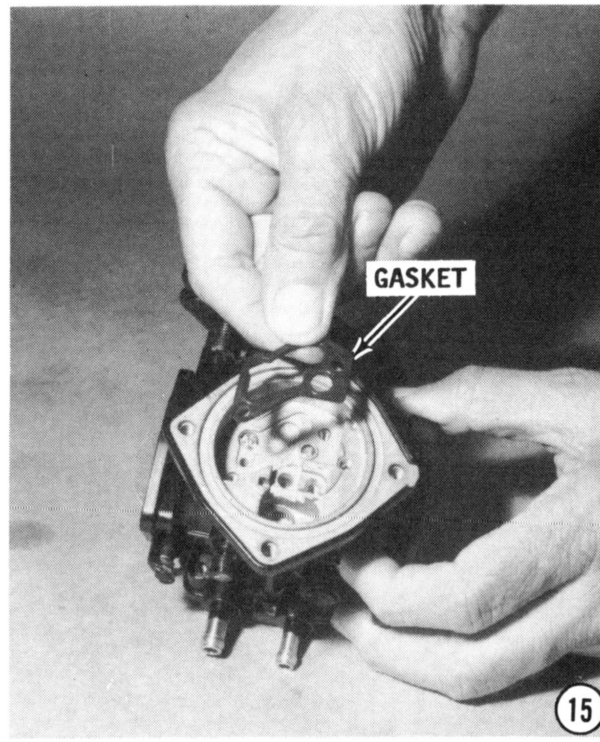

14- If necessary, use the "tapping method" to remove the two small screws securing the high speed check valve housing to the carburetor. Remove the check valve housing.

15- Remove and discard the gasket.

16- Use the correct size screwdriver to remove the main jet (the larger of the two jets), from the fuel chamber.

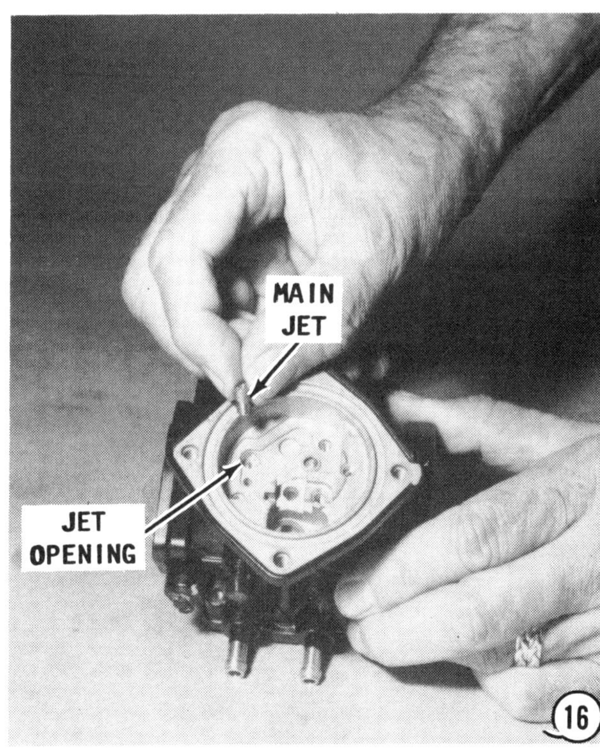

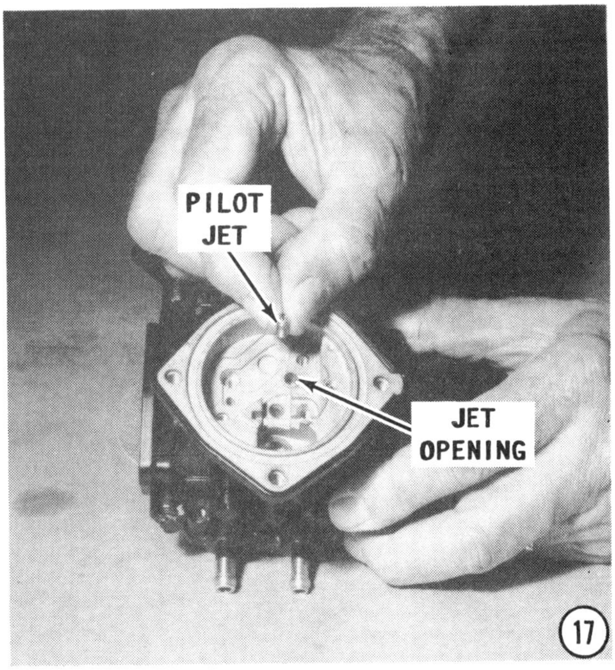

17- Use the correct size screwdriver and remove the pilot jet from the fuel chamber.

18- Remove the small Phillips head screw from the underside of the high speed check valve. Lift off the retainer. Note how a small post on the underside of the retainer indexes into a hole in the housing.

19- CAREFULLY lift off the clear plastic flap. **TAKE CARE** not to lose this item. This flap controls the flow of fuel in one direction and prevents the engine from "faltering" when the throttle is suddenly changed from wide open, to idle, and then back to wide open again.

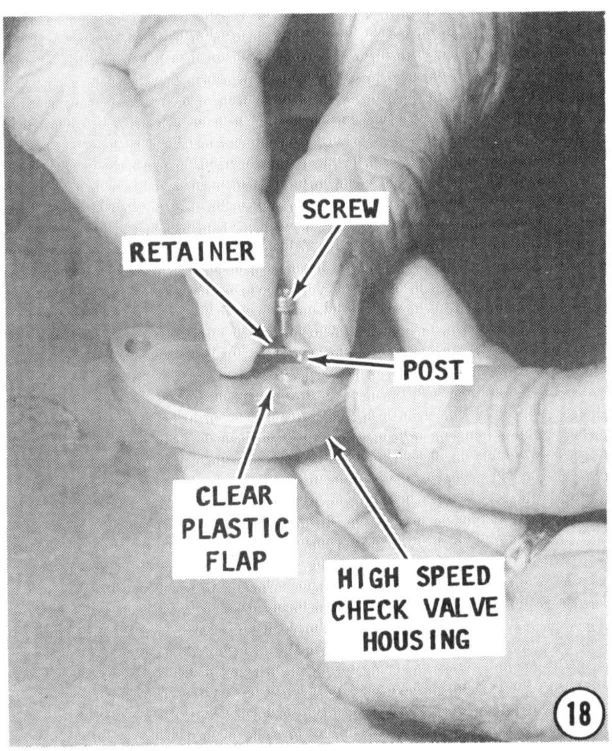

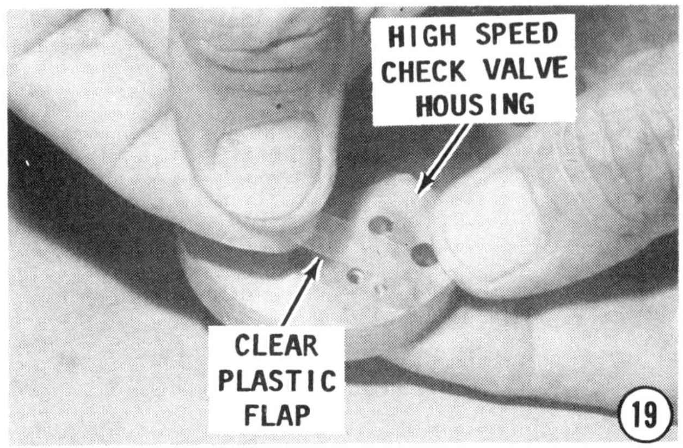

SPECIAL WORDS

Observe the letters **"L"** and **"H"** embossed next to the two mixture adjusting screws. The **"L"** denotes the low speed mixture adjusting screw, and the **"H"** denotes the high speed mixture adjusting screw. Each screw may have a plastic limiter cap installed on the screw head. A tab on the cap indexes into a slot in the collar cast in the carburetor body around the screw. The rotation of the screw is limited by the tab in the slot - hence the name "limiter cap".

20- Pull off the limiter cap on the low speed mixture adjusting screw -- the screw next to the embossed **"L"**.

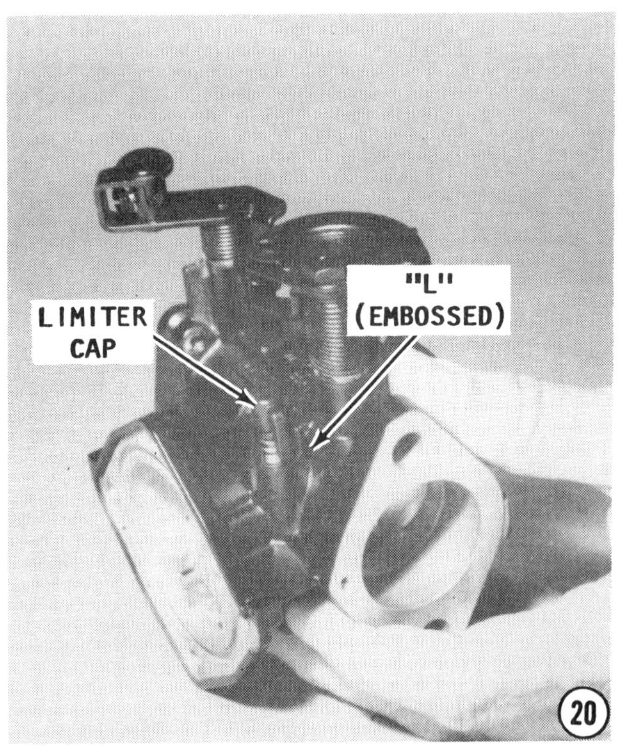

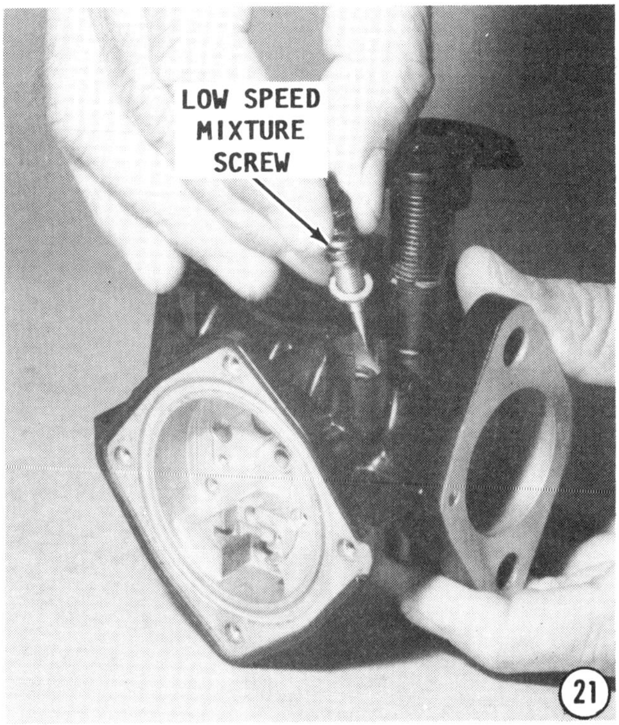

21- Back out the low speed mixture screw.

22- Pull off the limiter cap on the high speed mixture adjusting screw -- the screw next to the embossed **"H"**.

23- Back out the high speed mixture adjusting screw.

CLEANING AND INSPECTING

NEVER dip rubber parts, plastic parts, or diaphragms in carburetor cleaner. These

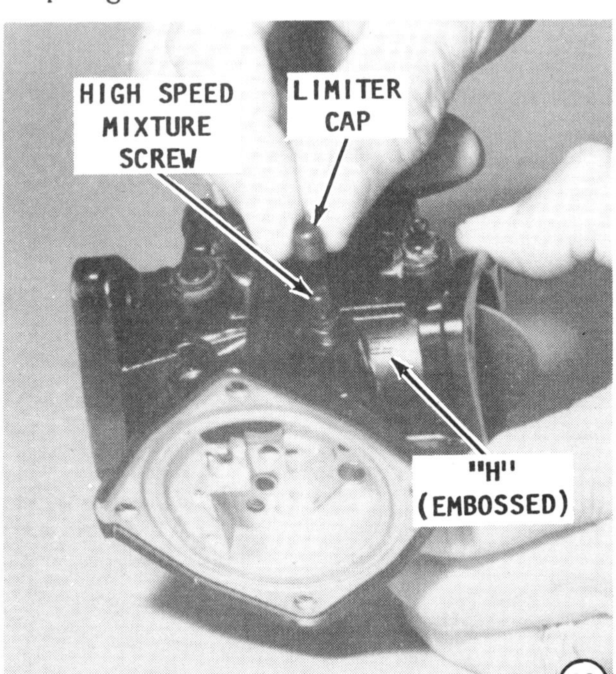

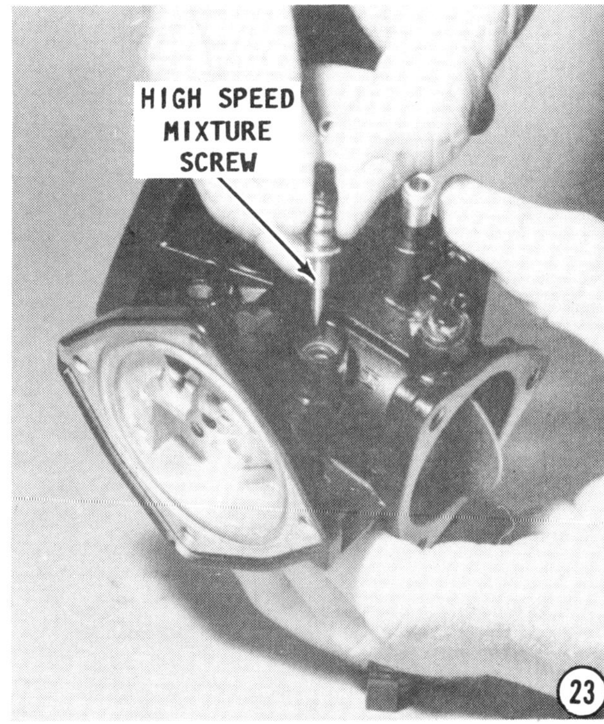

parts should be cleaned **ONLY** in solvent, and then blown dry with compressed air.

Place all metal parts in a screen-type tray and dip them in carburetor cleaner until they appear completely clean, then blow them dry with compressed air.

Blow out all passages in the castings with compressed air. Check all parts and passages to be sure they are not clogged or contain any deposits. **NEVER** use a piece of wire or any type of pointed instrument to clean drilled passages or calibrated holes in a carburetor.

Move the throttle shaft back and forth to check for wear. If the shaft appears to be too loose, replace the complete throttle body because individual replacement parts are **NOT** available.

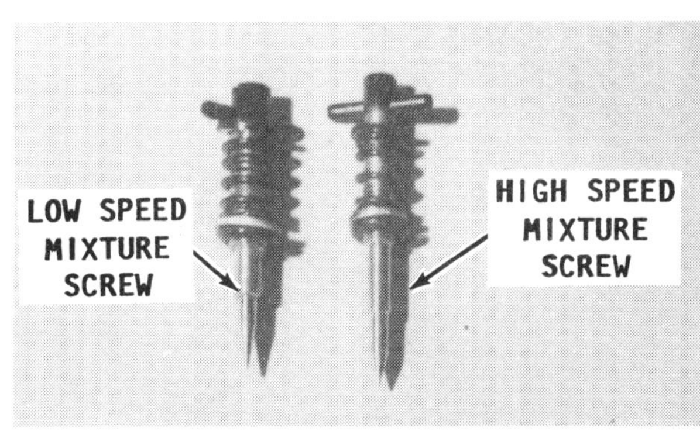

Comparison of the low speed and high speed mixture screws. Notice how the high speed screw is slightly longer than the low speed screw.

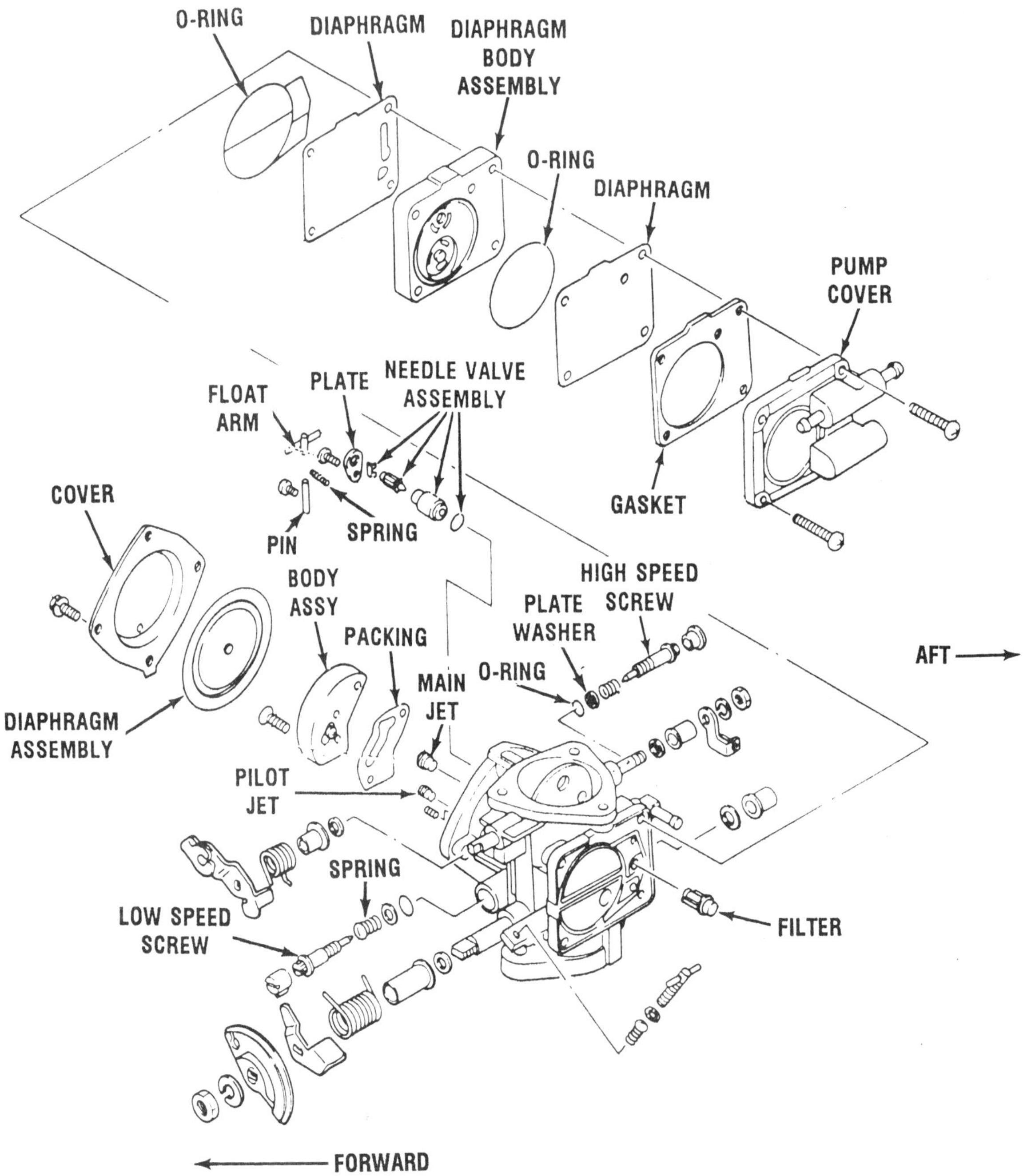

Exploded drawing of a typical Mikuni carburetor used on all Models covered. Major parts have been identified.

CRITICAL AREA

Inspect the condition of both the choke and thottle shafts. Most shaft designs incorporate a cutout portion to enable the butterfly valve to lay flat against the shaft. Two small holes are drilled in this cutout portion to accomodate the small screws securing the valve to the shaft. The weakest point of the shaft is at the screw holes. Small cracks form at the holes due to metal fatigue. These cracks elongate the threaded holes and the small screws shake loose. Once loose, they are sucked into the carburetor, then the intake manifold, and into the crankcase -- eventually finding their way into the combustion chamber. One of these small screws can then cause very expensive damage to pistons, rings and cylinder walls. Therefore, check the cutout in the shaft **CLOSELY** for any evidence of a small crack.

Inspect the main body, airhorn, and venturi cluster gasket surfaces for cracks and burrs which might cause a leak. Check to be sure the float arm return spring has not been stretched. Check the float arm needle contacting surface and replace the diaphragm if this surface has a groove worn in it.

Inspect the tapered section of the low and high speed adjusting needles and replace any that have developed a groove.

As previously mentioned, most of the parts which should be replaced during a carburetor overhaul are included in overhaul kits available from your local marine dealer. One of these kits will contain a matched fuel inlet needle and seat. This combination should be replaced each time the carburetor is disassembled as a precaution against leakage.

Make a thorough inspection of the fuel chamber and fuel pump diaphragms for the tiniest pin hole. If one is discovered, the hole will only get bigger. Therefore, the diaphragms must be replaced in order to obtain full performance from the engine.

Inspect the two vent holes in the first pump body. These holes **MUST** be clean. If these two holes are not perfectly clear, the diaphragm in the fuel pump will not function properly -- full fuel volume will not be pumped.

ASSEMBLING

Special Words on Mixture Screws

Observe the low and high speed mixture screws. They appear to be very similar, but they are **NOT** interchangeable. To identify the correct screw, place them side by side as shown in the lower right illustration on Page 6-17. The low speed screw is longer than the high speed screw. The low speed screw is installed on the linkage side of the carburetor -- next to the embossed "**L**". The high speed screw is installed on the opposite side -- close to the fuel supply and return fittings -- next to the embossed "**H**".

Both screws are adjusted later at the completion of the assembling procedures.

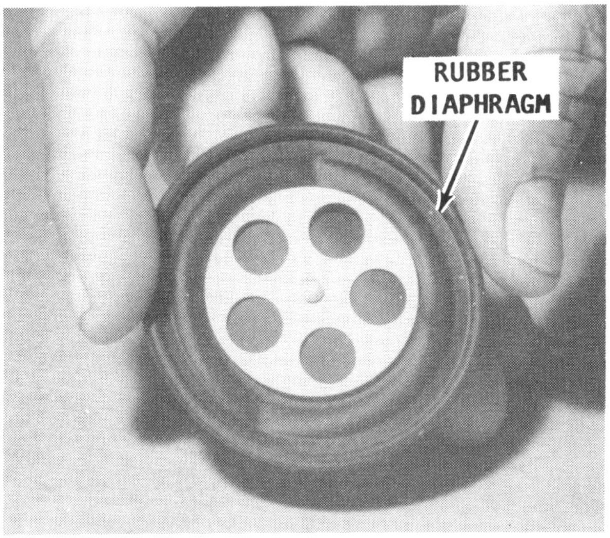

The pump diaphragm must be carefully inspected and replaced if any damage is discovered or even suspected. The tiniest "pin" hole is sufficient cause for replacement.

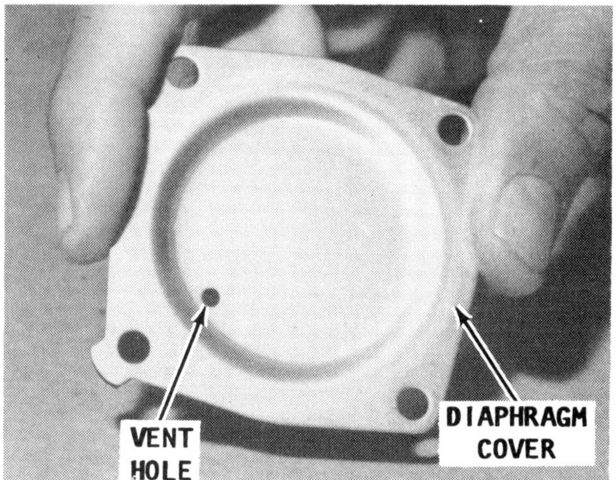

Carefully clean the vent hole in the diaphragm cover. One side of the rubber diaphragm is vented to the atmosphere thru this hole. Diaphragm action is seriously impaired if the hole becomes blocked -- even by the smallest obstruction.

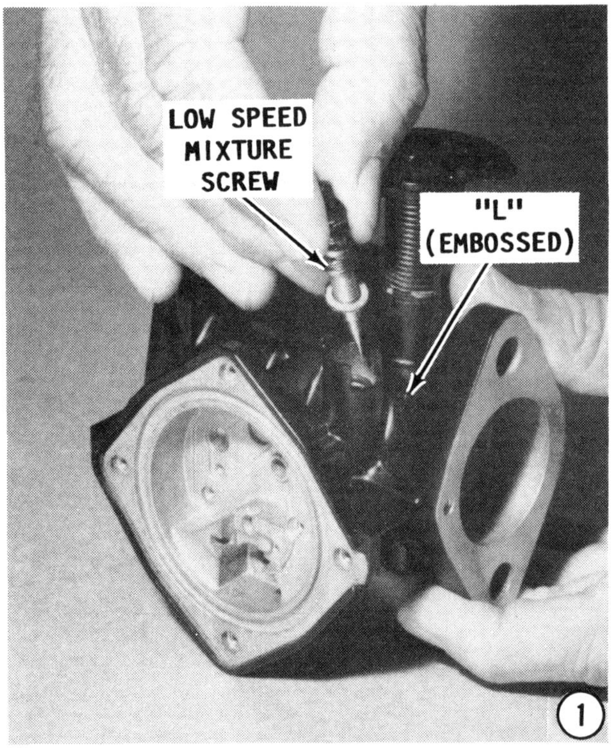

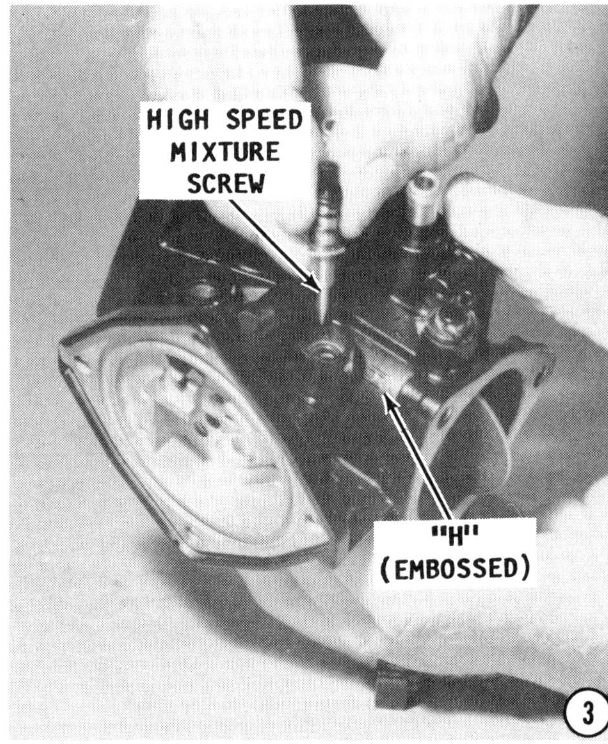

Therefore, thread each screw in just a few turns at this time.

1- Install the low speed mixture screw just a few turns on the linkage side of the carburetor -- next to the embossed "L".

2- Index the tab on the limiter cap into the slot in the collar and snap the cap squarely down over the screw head.

3- Install the high speed mixture screw just a few turns on the same side as the fuel supply and return fittings -- next to the embossed "H".

4- Index the tab on the limiter cap into the slot in the collar and snap the cap squarely down over the screw head.

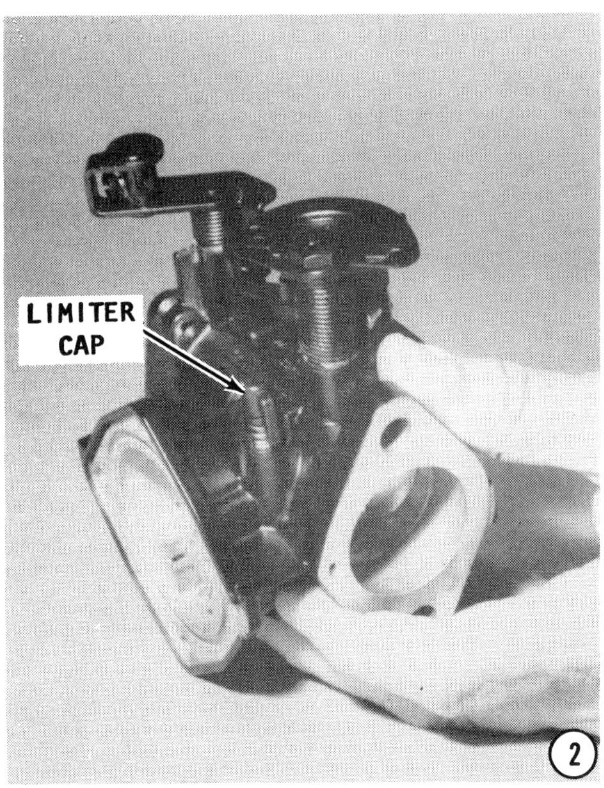

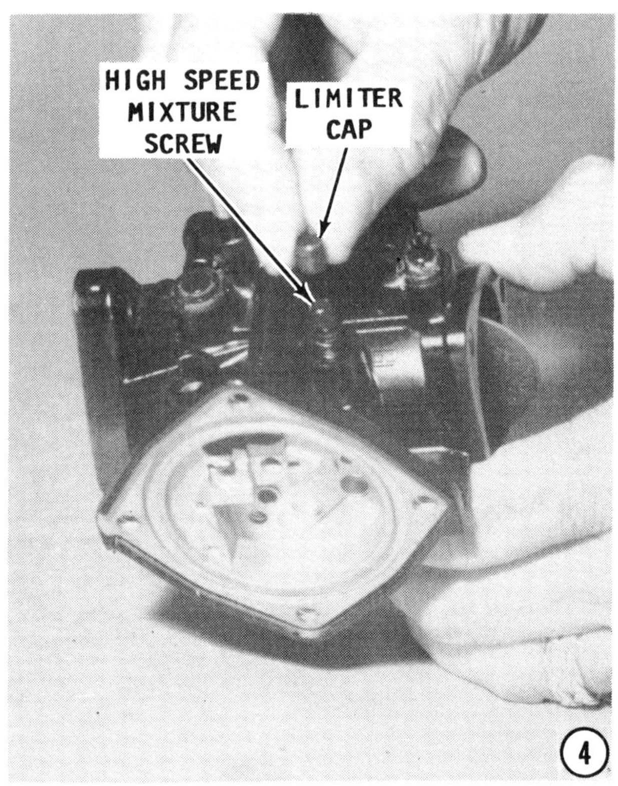

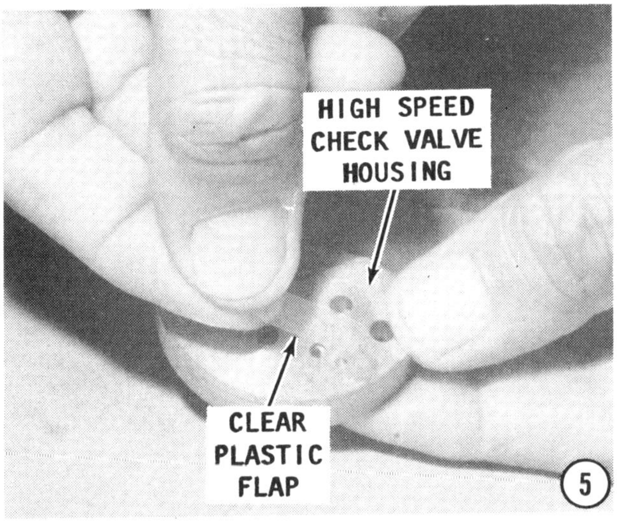

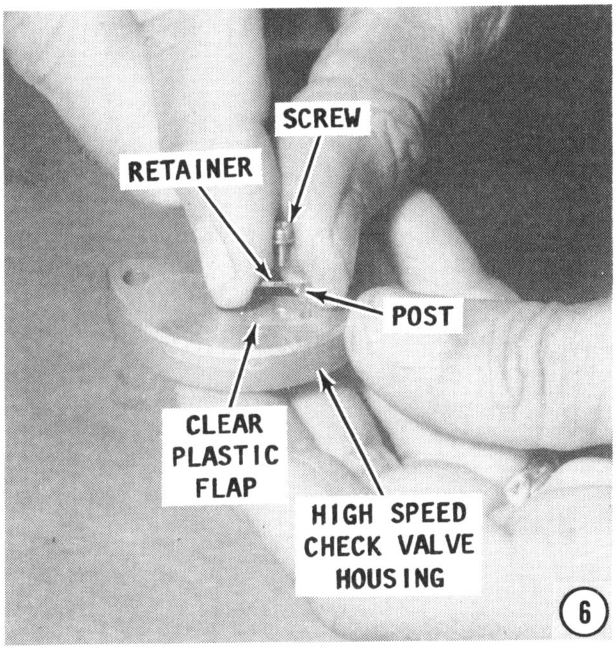

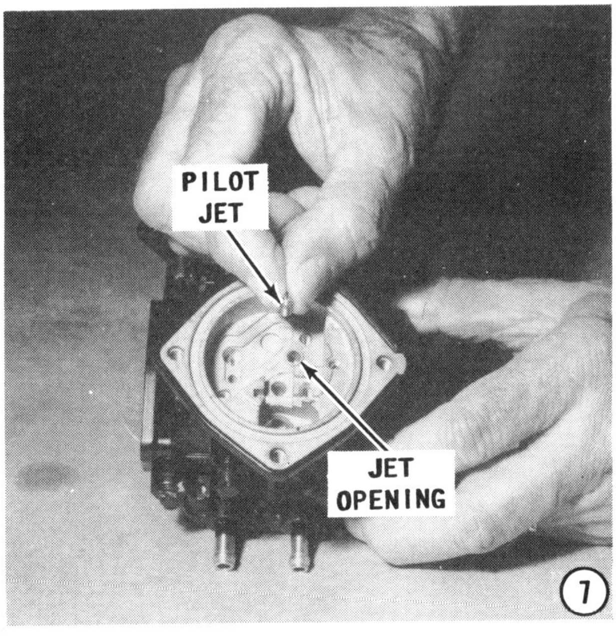

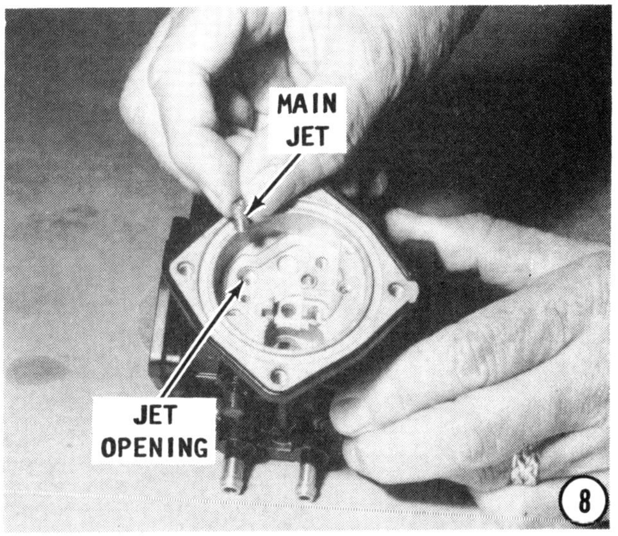

5- Align the two holes in the clear plastic flap with the two holes on the underside of the high speed check valve housing. If the four holes align, the flap is installed properly. If the holes will not align, the flap is upside-down.

6- Position the retainer over the plastic flap, with the post on the underside of the retainer indexed into both the hole in the plastic flap and the hole in the housing. Install and tighten the screw securing both parts to the housing.

7- Install the pilot jet (the smaller of the two), into the jet opening in the fuel chamber. Tighten the jet securely using the proper size screwdriver.

8- Install the main jet into the jet opening in the fuel chamber. Again, using the proper size screwdriver tighten the jet securely.

9- Position the high speed check valve housing gasket into the recess in the fuel chamber.

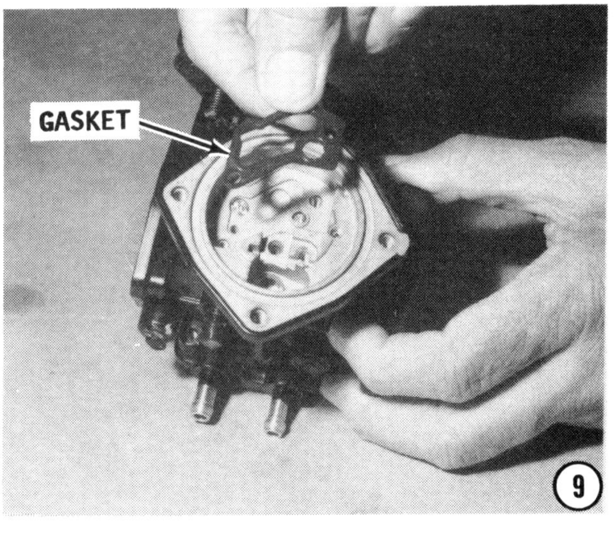

10- Place the high speed check valve over the gasket. Secure it in place with two Phillips head screws. Tighten the screws to a torque value of 1.5 ft lb (2Nm).

11- Install a new **O**-ring around the needle valve seat. Lower the seat into the recess in the fuel chamber.

12- Install the seat retainer over the seat. Secure the seat in place with the Phillips head screw. Tighten the screw to a torque value of 0.7 ft lb (1Nm).

13- Lower the spring into the recess in the fuel chamber.

14- Slide the needle valve clip over the square tab on the float arm. Slide the float arm over the pin, as shown. Lower the assembly down into place and at the same time: guide the needle valve into the needle

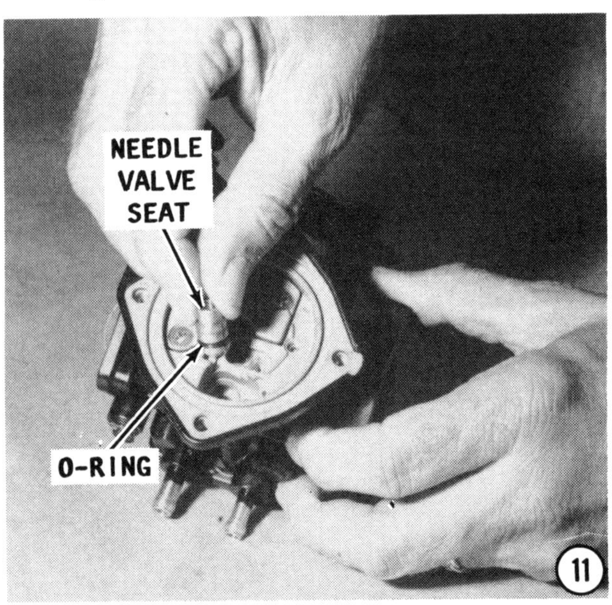

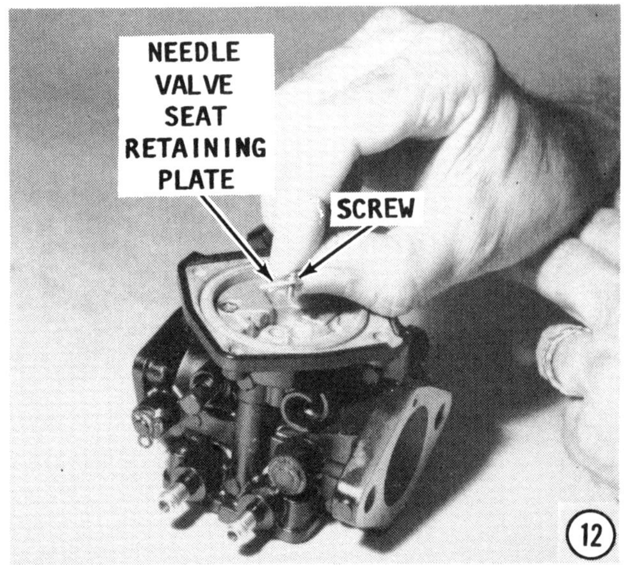

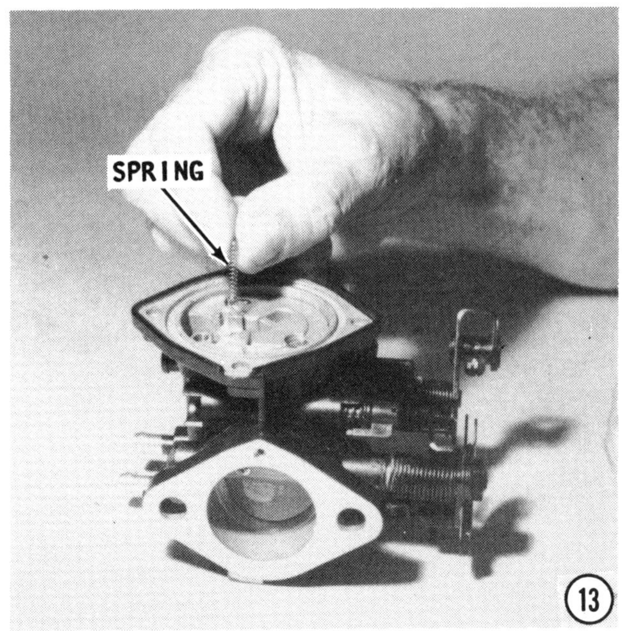

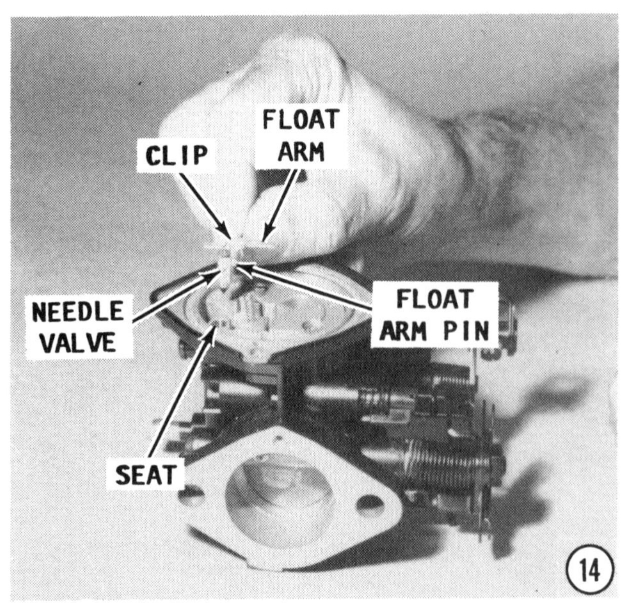

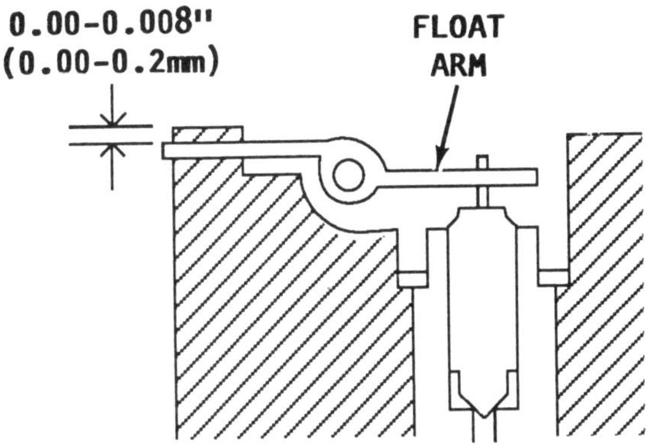

15

seat; place the float pin into the elongated slot in the fuel chamber; and position the dimple in the float arm over the spring. Secure the pin to the carburetor with the small Phillips head screw. Tighten the screw to a torque value of 0.7 ft lb (1Nm).

Float Arm Adjustment

15- The float arm adjustment is correct when three conditions are satisfied:

a) The top of the float arm is parallel to the fuel chamber surface.

b) The distance between the top surface of the float arm and the diaphragm contact surface of the carburetor is between 0.00-0.008" (0.0-0.2mm).

c) The float arm makes contact with the needle valve, but does not apply any pressure to the valve.

If any of these conditions are not satisfied, then replace the float arm and/or return spring. Do not attempt to bend the

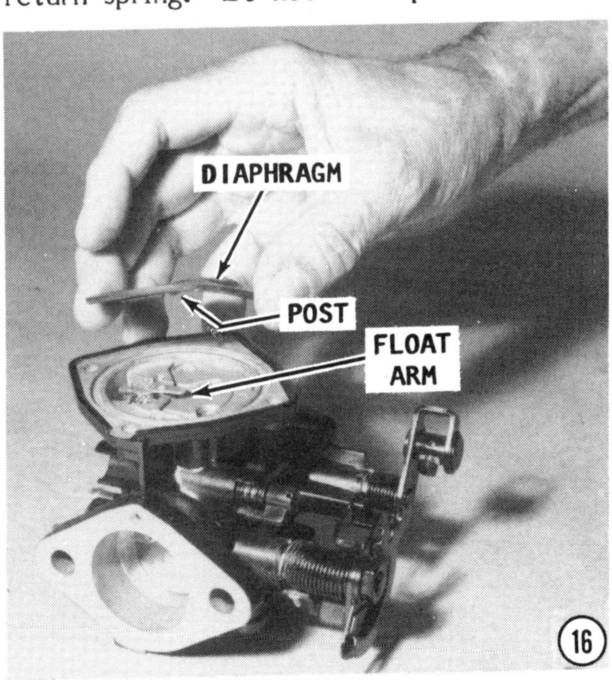

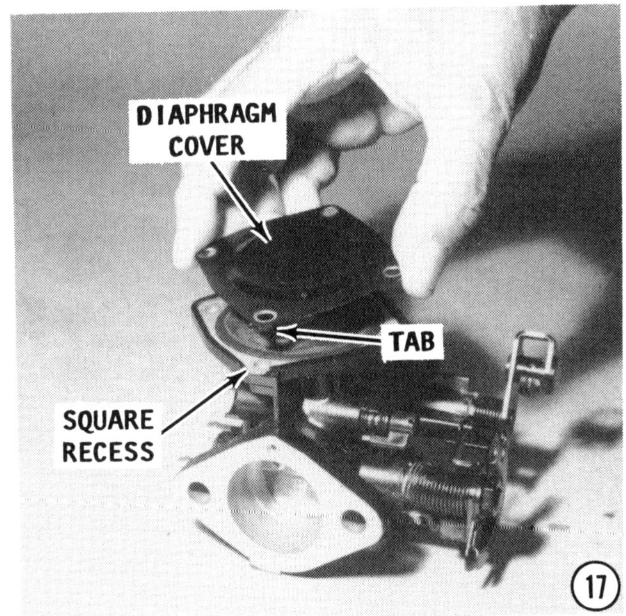

float arm or to change the characteristics of the old return spring.

Depress the rounded end of the float arm. The needle valve should lift smoothly and return inside the seat when the arm is released.

16- Place the diaphragm over the fuel chamber with the float post on the underside facing **DOWN** to rest against the float arm.

17- Position the diaphragm cover over the diaphragm with the tab in the cover corner indexed into the square recess in the carburetor body.

18- Secure the cover to the carburetor with the four Phillips head screws. Tighten

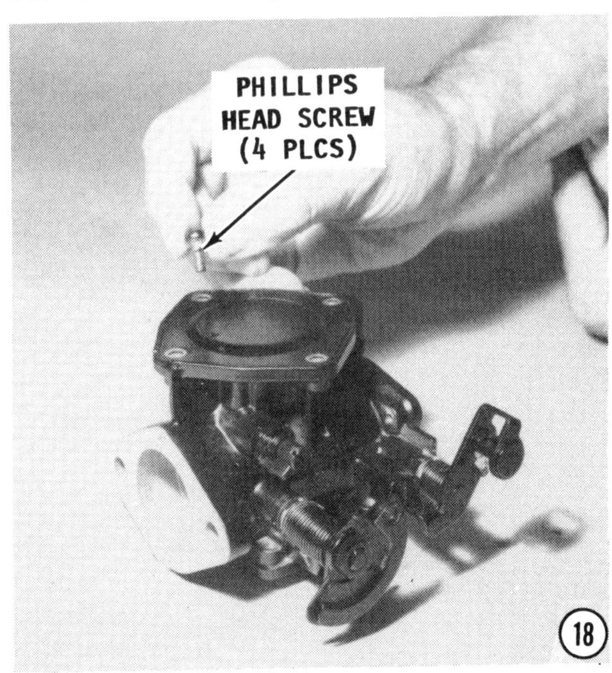

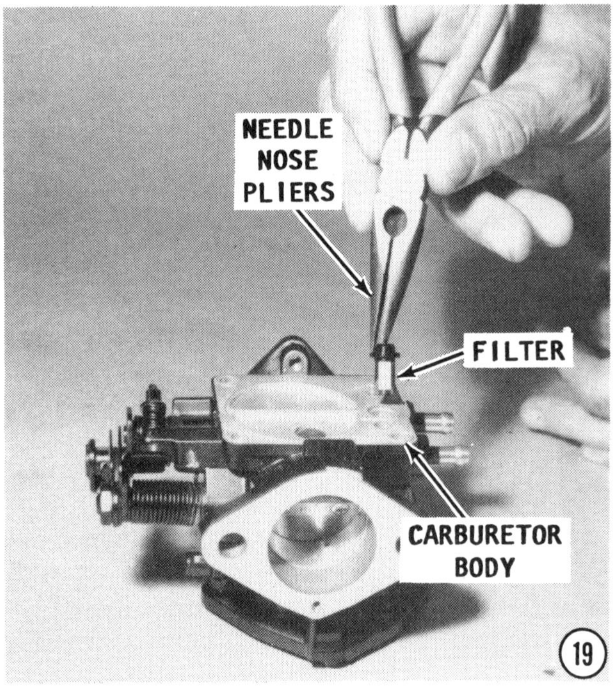

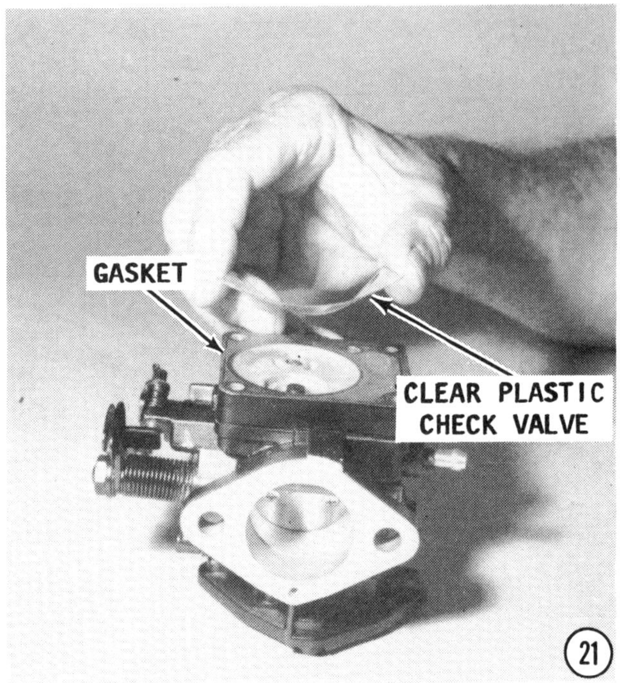

the screws alternately and evenly to a torque value of 3 ft lb (4Nm).

19- Lower the fuel filter into the recess in the carburetor body.

Fuel Pump Assembling

First these words: Take care to install all **seven** layers of diaphragms, gaskets, and check valves (made from plastic film), with their tabs facing **UPWARD** toward the choke valve. The fuel pump components must be carefully stacked and held together until the securing screws may be passed through all layers and threaded into the fuel pump body.

If a gasket, diaphragm or check valve is askew, the slightest pressure from the end of the securing screw will tear the component and affect the performance of the fuel pump.

20- Install the gasket, diaphragm and pump body over the carburetor.

21- Place the gasket, followed by the clear plastic check valve, over the pump body.

22- Position the gasket over the plastic check valve. Finally, place the fuel pump cover over the gasket with the post, on the underside of the cover, indexed into the

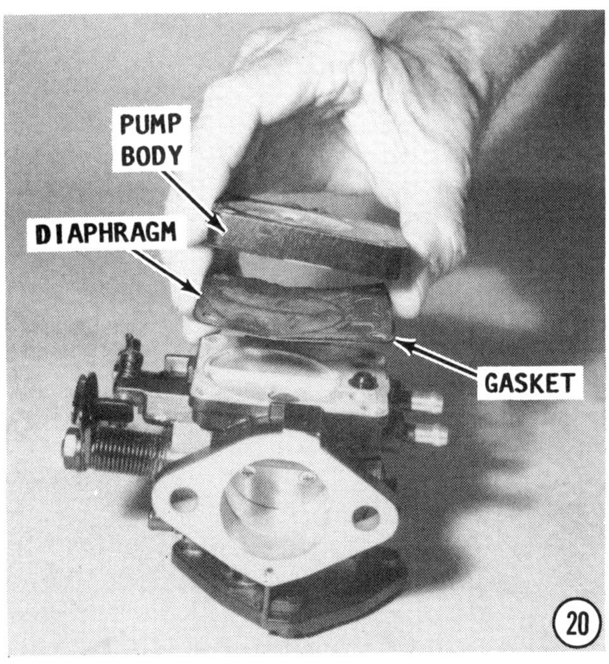

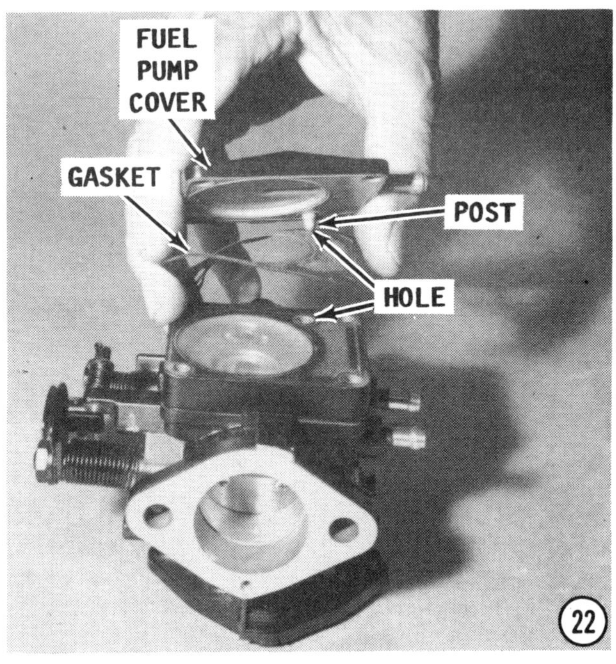

holes in the stack of components between the cover and the pump body. Check to be sure the layers are aligned with all tabs facing **UPWARD.**

23- Visually inspect the screw holes to verify all components are squarely in place and the path for each screw is clear. Continue to hold the stack of components tightly together until all four screws are started into the fuel chamber body.

After the screws are started, tighten them in a diagonal pattern to a torque value of 3 ft lb (4Nm).

BENCH ADJUSTMENTS

Set the low speed and high speed adjustment screws the correct number of turns "out" from a lightly seated position, as listed in the table in the Appendix.

SPECIAL WORDS

If the following adjustments are attempted while the carburetor/s are installed on the

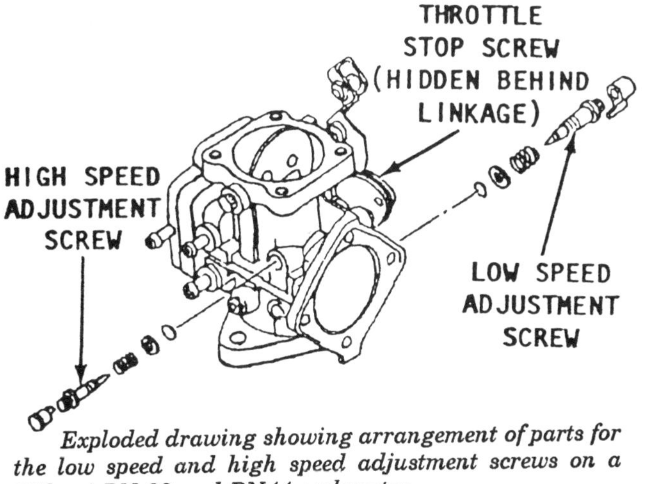

Exploded drawing showing arrangement of parts for the low speed and high speed adjustment screws on a Mikuni BN 38 and BN44 carburetor.

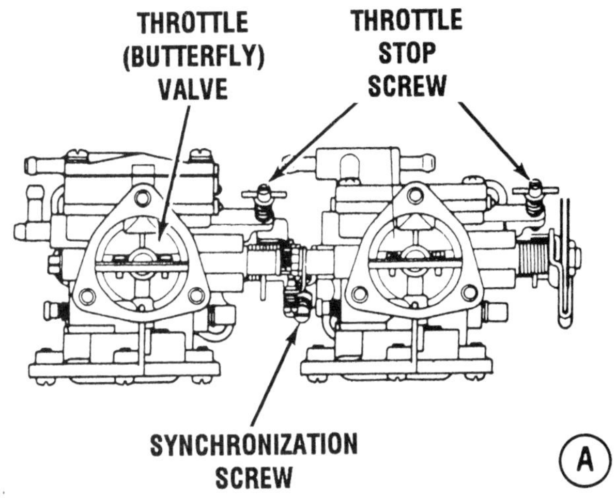

manifold, the flame arrestor cover and the flame arrestor screens must first be removed to enable the choke/throttle (butterfly) valve to be observed.

Low Speed and Idle Mixture Adjustment

Move the craft to a test tank, to a body of water, or connect a flush attachment and hose to the engine. For maximum performance, the idle mixture and the idle rpm should be adjusted under actual operating conditions. Connect a tachometer to the engine.

Lightly seat the low speed adjustment screw. From this position, back the screw out 3/4 to 1 full turn as a preliminary adjustment, at this time.

Start the engine and allow it to warm to operating temperature.

CAUTION

Water must circulate through the jet pump -- to and from the engine, anytime the engine is operating. Circulating water will prevent overheating -- which could cause damage to moving engine parts and possible engine seizure.

NEVER, AGAIN NEVER, operate the engine at high speed with a flush device attached. An engine operating at high speed with such a device attached, might very likely **RUNAWAY** from lack of a load on the impeller shaft, causing extensive damage.

The throttle stop screw is located on the throttle linkage on the opposite side of the carburetor from the fuel fittings

With the engine operating, slowly rotate the throttle stop screw until the engine idles at 1,200 to 1,300 rpm. Next, rotate the low speed screw inward or outward 1/8th of a turn at a

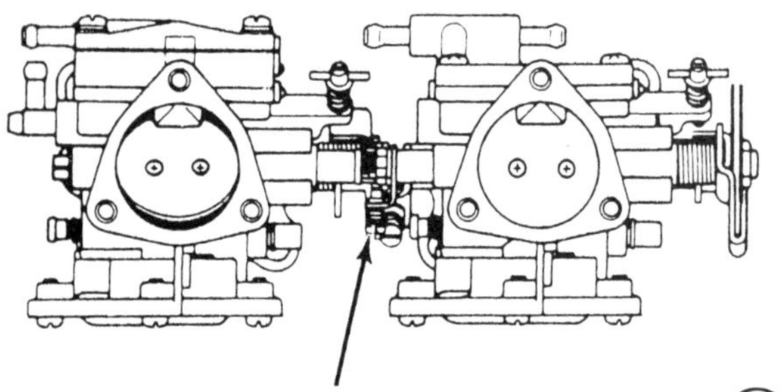

B

time until the cylinders fire evenly and engine rpm increases. Readjust the throttle stop screw to reduce the rpm to specified idle speed.

Synchronize Throttle Valves (Illustration "A")

Back off the throttle stop screws on both carburetors until the end of the screws just barely clear the throttle lever.

Make some sort of record as to the position of the throttle stop screws. Consider this as the "set" position. Check to be sure the throttle valve of the No. 2 carburetor is fully closed.

Now, turn the synchronization screw inward or outward until the No. 1 throttle valve is fully closed.

Finally, turn the throttle stop screws on both carburetors to the "set" position, as noted above.

Synchronize Choke Valves (Illustration "B")

Rotate the synchronization screw on the No. 1 carburetor inward or outward to bring both choke valves into the fully closed position when the choke lever is turned to the closed position.

CARBURETOR INSTALLATION -- ALL ENGINES EXCEPT MODEL 787 SERIES

Preparation Tasks

Place the two carburetors on the workbench with the **FORWARD** carburetor -- the one with the fuel pump -- on the right and the aft carburetor to the left. Both ball joints on the throttle levers should be facing **UP**. Apply just a "dab" of lubricant to both ball joints. Install the link over the ball joints. Install the short piece of fuel hose between the two carburetors.

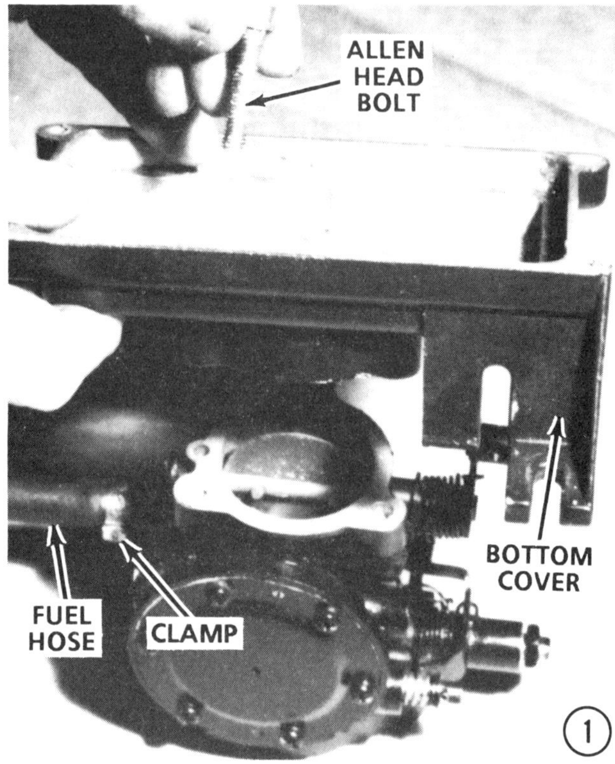

1

1- Move the set of carburetors into position and secure the rack to the intake manifold with the attaching bolts.

Work the fuel supply hose onto the fuel pump fitting on the forward carburetor. Slide the hose clamp down over the hose and secure the hose in place, by tightening the clamp securely.

Now, position the bottom flame arrestor cover in place over the carburetors, with the cable support bracket on the same side as the carburetor diaphragm cover. Apply a light coating of Loctite

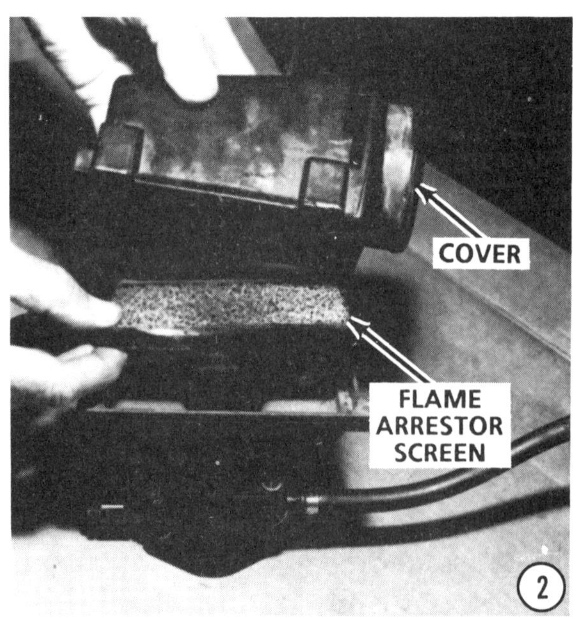

2

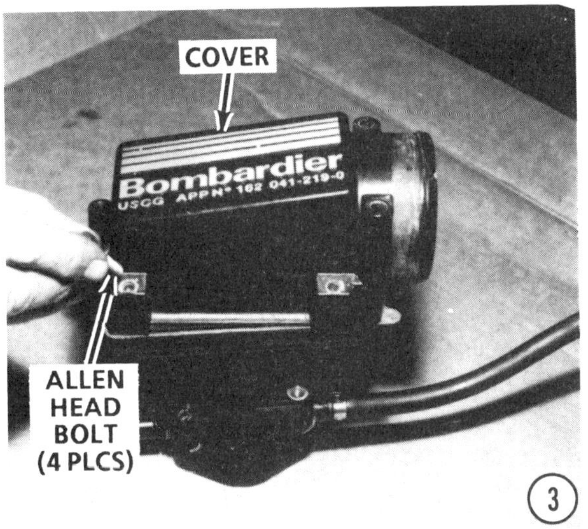

3

242 onto the threads of the Allen head bolts. Install and tighten the bolt securely.

2- Place a **CLEAN** flame arrestor screen into the bottom cover and install the top cover over the flame arrestor.

3- Install and tighten the four Allen head bolts securing the two covers together. Tighten the bolts to a torque value of 8.8 ft lb (10Nm).

4- Slide the throttle cable into the slot in the cable support bracket, with a nut on each side of the slot, as shown. Tighten the two nuts against each other to secure the cable in place.

5- Slide the choke cable into the other slot on the bracket with a nut on each side of the slot **AND** at the same time, guide the cable end into the hole in the barrel on the choke shaft. Tighten the two nuts against each other just **SNUG,** to secure the cable in place.

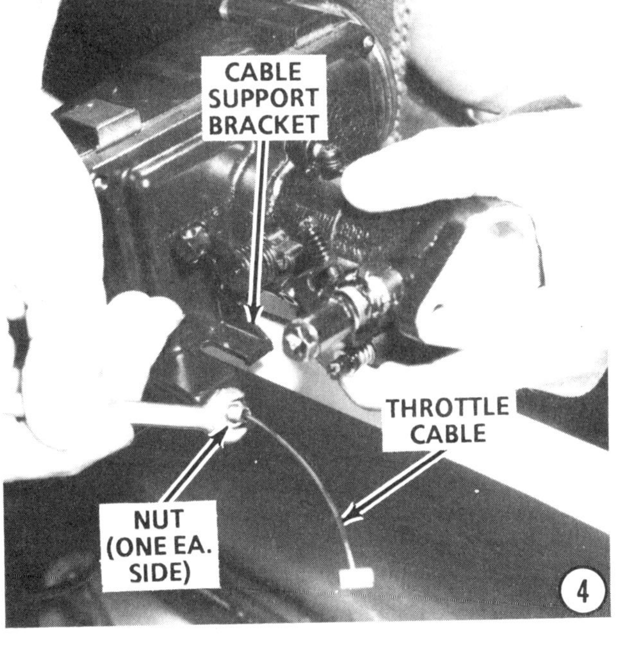

4

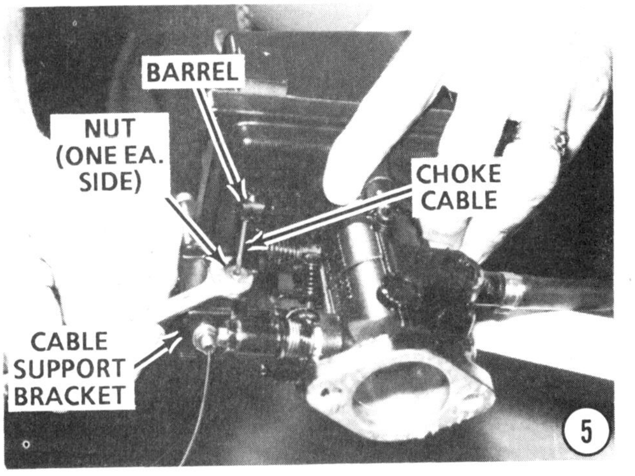

5

6- Adjust the choke cable until the mark on the cable is at the position where the cable just enters the barrel. Maintain this position and tighten the small Allen head screw on the barrel to secure the cable. If necessary, adjust the slack in the cable with the two nuts at the support bracket, and then tighten them securely.

7- Position a new carburetor mounting gasket over the intake manifold, and then lower the carburetor/s down over the manifold studs.

8- Apply a light coating of Loctite 242 to the threads of the carburetor mounting studs. Install lockwashers, and then the nuts. Tighten the nuts to a torque value of 18 ft lb (25Nm). Insert the throttle cable end into the hole in the curved guide and route the cable into the groove in the guide.

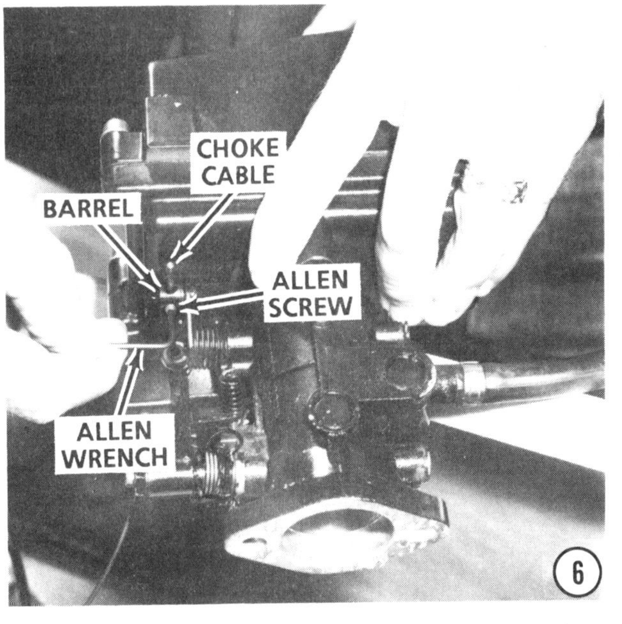

6

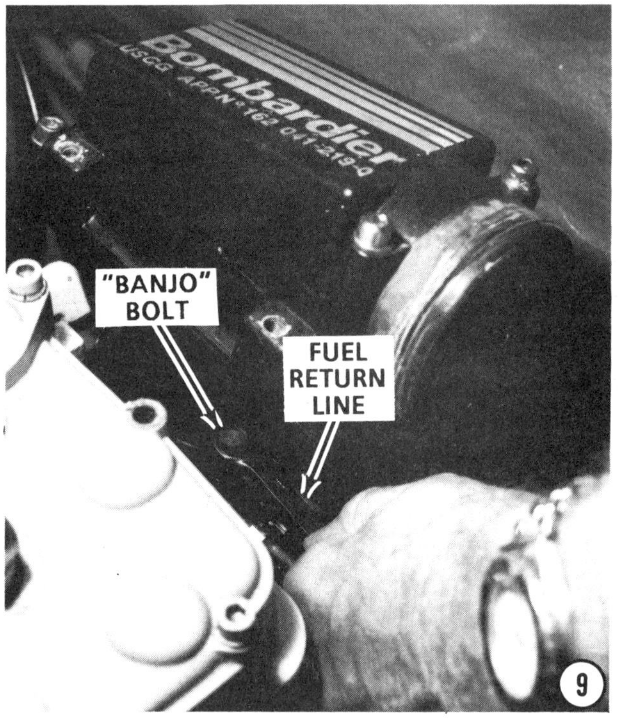

Carburetor Priming

The carburetor must be primed with fuel before attempting to start the engine to save excessive drain on the battery and overheating of the cranking motor.

Place a shop cloth under the fuel return fitting (the fitting above the installed fuel supply line). Blow through the disconnected fuel return hose (the hose attached to the banjo fitting), until fuel begins to drip from the fuel return fitting on the carburetor.

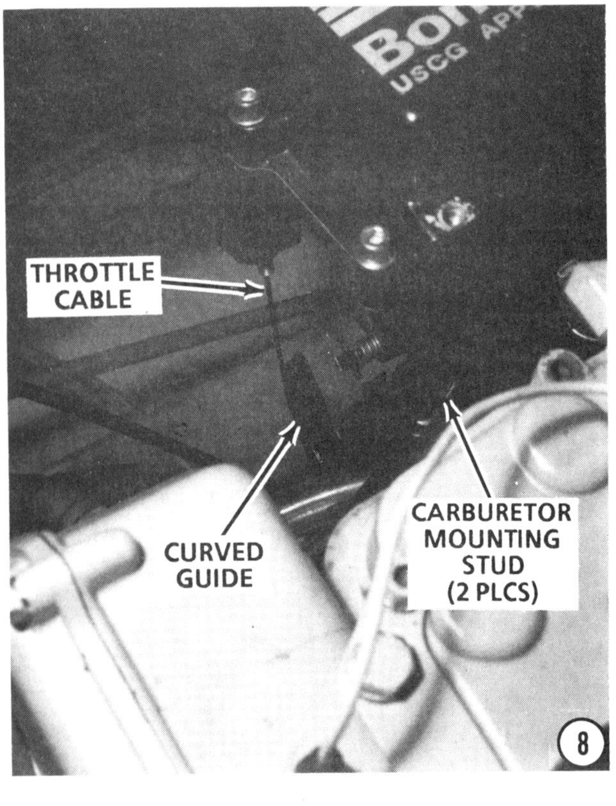

When the fuel drips, the carburetor is full of fuel and the engine is ready for cranking after all other tasks are completed.

9- Position the "banjo" bolt, with the fuel return line attached, over the hole above the installed fuel supply hose. Install and tighten the "banjo" bolt and washer securely.

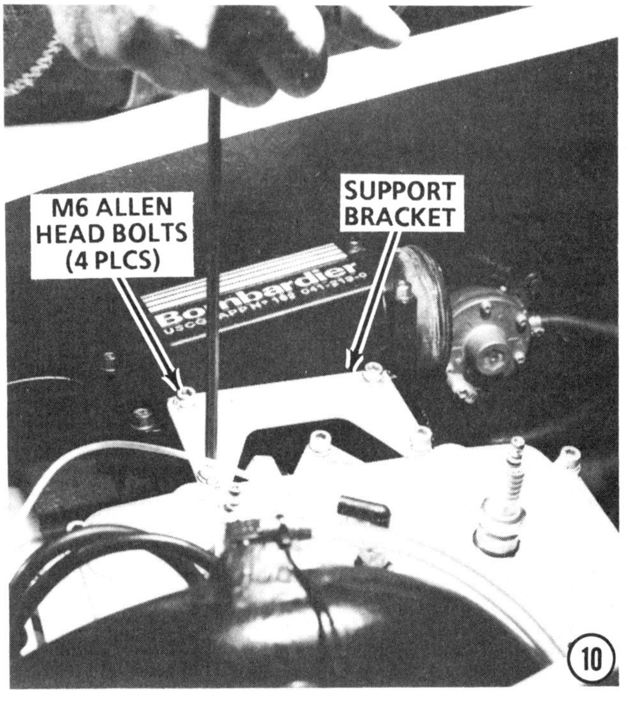

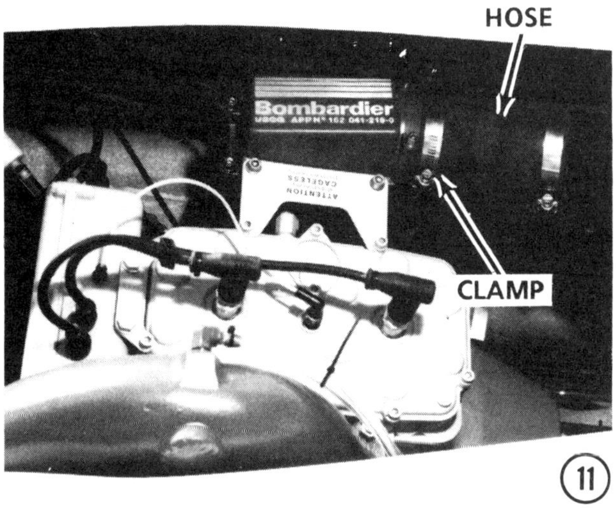

10- Install the support bracket over the flame arrestor cover and engine. Apply a light coating of Loctite 242 to the threads of the four M6 Allen head bolts. Install and tighten the four bolts securely.

11- Work the air intake hose onto the flame arrestor cover flange. Install and tighten the hose clamp to secure the hose to the flange.

INSTALLATION -- DUAL CARBURETORS MODEL 787 SERIES ONLY

Preparation Tasks

Place the two carburetors on the workbench with the **FORWARD** carburetor -- the one with the fuel pump -- on the right and the aft carburetor to the left. Both ball joints on the throttle levers should be facing **UP**. Apply just a "dab" of lubricant to both ball joints. Install the link over the ball joints. Install the short piece of fuel hose between the two carburetors.

1- Install a new rubber carburetor base gasket on each rotary valve port in the cover. Lower the

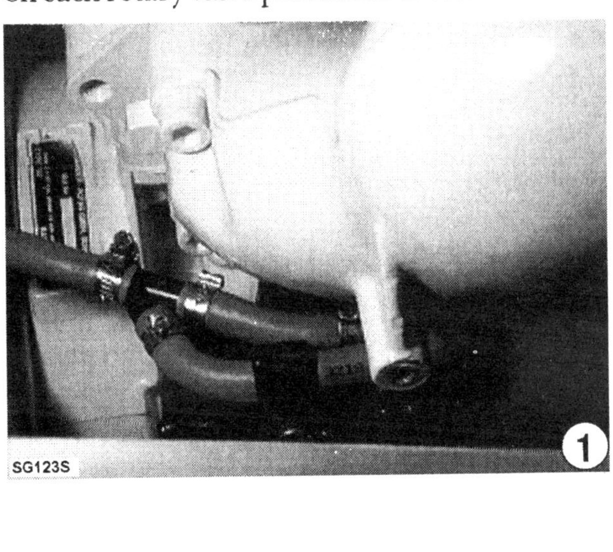

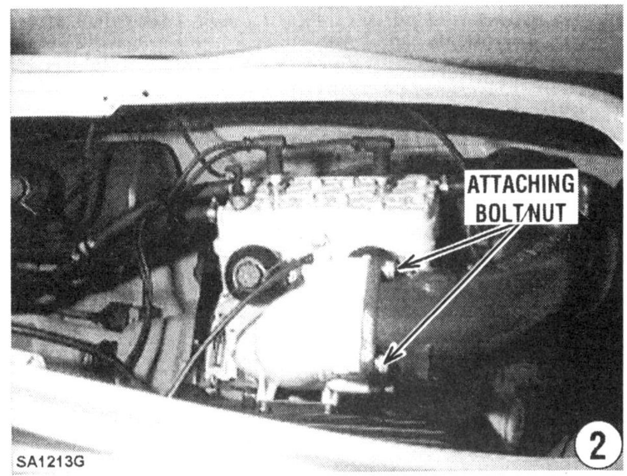

carburetor rack into position on the rotary valve cover. Install the four attaching bolts and tighten them securely. Cover the carburetors with a clean shop towel to prevent debris from contaminating the carburetors.

Connect the fuel inlet hose to the No. 1 lower end of the -- forward -- carburetor. Connect the pulse hose to the top side of the No. 1 carburetor. Connect the fuel return hose to the aft end of the No. 2 -- aft -- carburetor. Secure each hose with a zip tie -- except the fuel return hose. This hose uses a special clamp. Observe the hose bracket, located on the forward engine mount bracket. Wrap a zip tie around the fuel return hose and the oil injection cable and secure them to this bracket.

2- Connect the throttle and choke cables to the No. 1 carburetor. Due to the angle of the engine, care and patience will go a long way when connecting these cables. While the choke locknuts are loose, slide the cable in place in the lever arm, then tighten the tension on the choke cable. Next, tighten the locknuts on each side. Tighten the 3mm Allen head set screw on the No. 1 choke lever arm first, then tighten the No. 2 cable set screw. Test the throttle and choke levers to be sure the cable moves freely without binding.

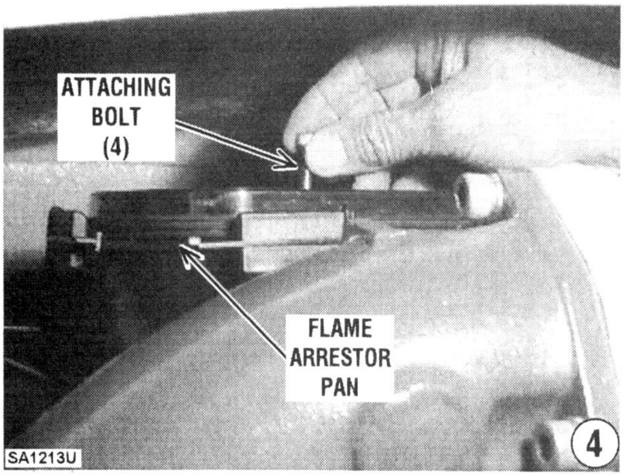

3- Connect the oil pump cable to the oil pump. Rotate the oil pump lever **COUNTER-CLOCKWISE** until it almost makes contact with the bracket -- say with 1/4" clearance. When the lever is rotated it should be possible to verify there is spring pressure against the lever. Without spring pressure, the lever arm will not actuate. Operate the throttle to be sure the oil injection pump lever functions properly.
should be possible to verify there is spring pressure against the lever. Without spring pressure, the lever arm will not actuate. Operate the throttle to be sure the oil injection pump lever functions properly.

4- Remove the shop towels placed over the carburetors. Lower the flame arrestor base into position on the carburetor rack. Install the four 5mm attaching bolts. Tighten the bolts securely. Place the shop towel over the arrestor base, to prevent any foreign object from falling into the carburetor throat.

5- Position the stainless steel bracket securing the flame arrestor pan to the exhaust manifold. Install the four lockwashers, then the 5mm attaching bolts and tighten them securely. Remove the shop towel from the arrestor base. Lower the

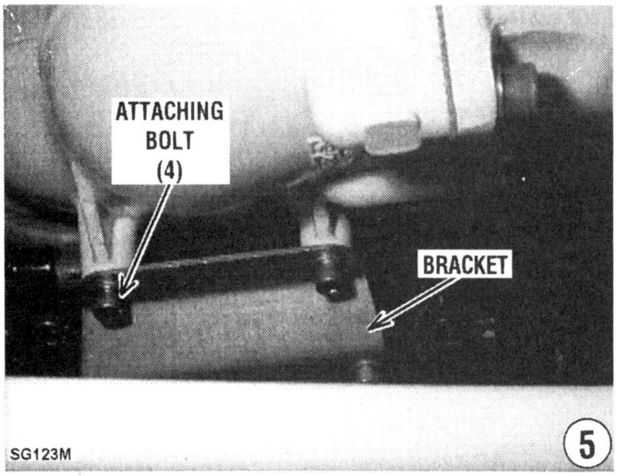

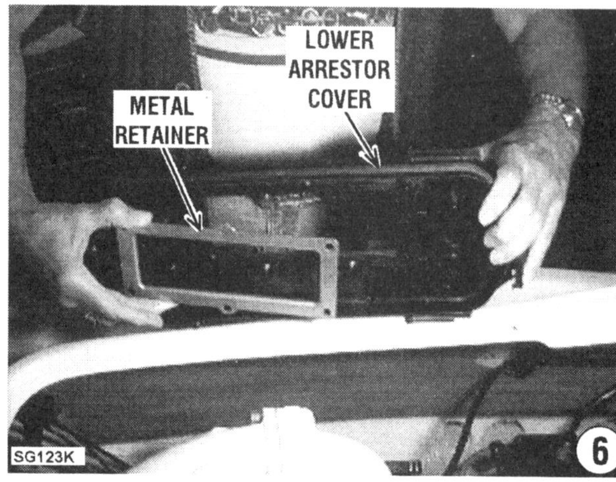

stainless steel flame arrestor screen into place. Be sure the cambered top -- bowed portion of screen -- faces **UP**.

6- Move the large lower arrestor cover, with the metal retainer, into position on the arrestor base -- over the arrestor screen. Secure the cover in place with six 5mm Allen head bolts. Tighten the bolts securely.

6-7 FUEL ADJUSTMENTS

Five fuel adjustments may be made on the Mikuni carburetor, for both a single and dual carburetor installation -- for this manual they will be all dual installations -- as follows:

a- Main Jet selection.

b- Throttle Cable Adjustment -- **ALWAYS** followed by oil injection synchronization.

c- Low Speed Mixture Adjustment.

d- Idle Speed Adjustment – **ALWAYS** followed by oil injection pump synchronization.

e- Choke Cable Adjustment.

Each of the listed adjustments are explained in detail in the following paragraphs with individual headings identifying each adjustment.

Main Jet Selection

Several main jet sizes are available to compensate for variations in altitude and temperature. The out-the-door stock main jet is intended for operation at sea level. The jet size is stamped on one face. The following main jet sizes are available: 70, 75, 80, 85, 90, 95, and 100. When operating the craft at higher altitudes, try changing the main jet in increments of 0.005" or 0.010", until an acceptable level of performance is achieved.

Operating the engine on an excessively lean mixture may cause overheating, pre-ignition, and a decrease in performance.

Any time the main jet size is altered, keep a close eye on the condition of the spark plug tip. If the tip begins to show signs of overheating, change to a colder heat range. Check with the local dealer for the correct number spark plug. The heat range of a spark is identified by a number. However, the numbers are not consistant. Some spark plug manufacturers use a higher number for colder plug, and others use a lower number for colder plug.

The main jet is located externally. Therefore, the carburetor need not be removed to change a main jet. The jet is threaded into the fuel chamber, on the same side as the fuel supply hose. When removing the main jet, look for the **O**-ring, which will remain in the jet bore. When installing a new main jet, install the **O**-ring onto the jet first, and then install the jet. **DO NOT** overtighten the jet. Tighten it just enough to prevent fuel from escaping. If working on a twin carburetor installation -- both jets **MUST** be the same size.

Throttle Cable Adjustment

When no throttle is applied, the idle speed screw must contact the stopper on the carburetor. When the throttle lever is fully applied -- actually making contact with the handlebar grip -- the stop lever on the throttle linkage must contact the other side of the stopper on the carburetor. If these two conditions are not met, an adjustment on the throttle cable is necessary at the cable support bracket. Adjust the throttle cable length by loosening or tightening the two nuts -- one on each side of the support bracket.

After a throttle cable adjustment is made, **ALWAYS** perform a check to ensure the oil injection pump is synchronized with the carburetor. If the oil injection pump is not properly synchronized with the carburetor, serious damage may be caused to the engine due to lack of adequate lubrication.

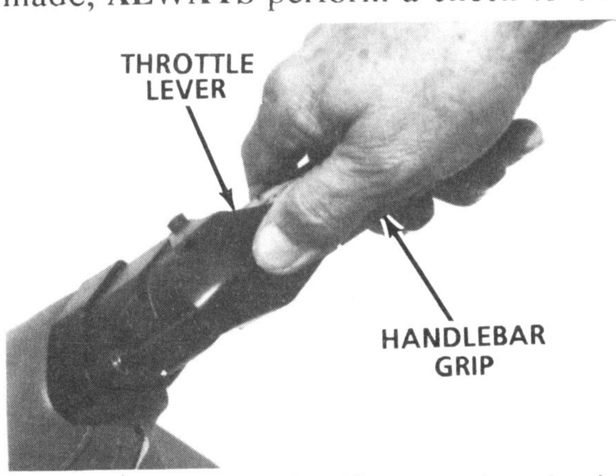

The throttle valve is wide open when the throttle lever is fully applied -- making contact with the handlebar grip.

Low Speed Mixture Screw

The low speed mixture screw must **NEVER** be used to alter engine rpm. Such action could result in severe engine damage. This screw is used to smooth engine operation at a set idle speed. The idle speed is changed by adjustment of the idle speed screw, covered in the following paragraph. If working on a twin carburetor installation, perform the required work on each carburetor in turn.

Now, with the engine shut down, tighten the low speed mixture screw, identified in the accompanying illustration, until the screw is **LIGHTLY** seated, and then back it out 1-1/2 to 2 full turns as a preliminary setting.

Start the engine and allow it to warm to operating temperature.

CAUTION

The following procedure MUST be followed whenever the engine is started with a hose attached for cooling the engine. Start the engine, and then IMMEDIATELY turn the water on. The engine MUST be started first to prevent water from backing up into the tuned exhaust pipe and into the engine. Water must circulate anytime the engine is operating to prevent overheating -- which could cause damage to moving parts and possibly cause engine seizure. NEVER al-

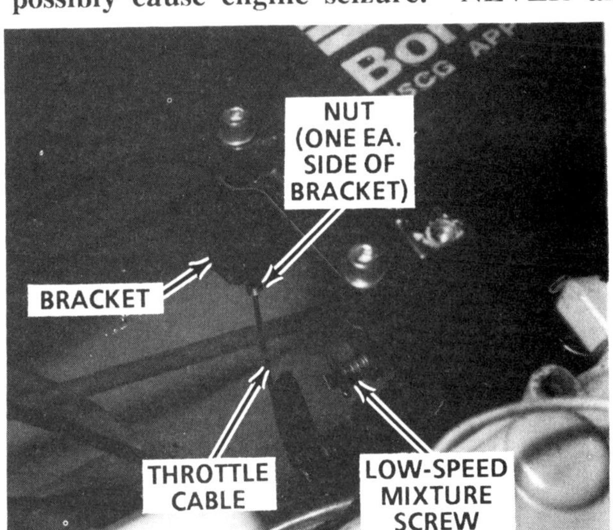

Typical location of the low-speed mixture screw and the throttle cable adjustment nuts.

low the engine to operate without cooling water for more than 15 seconds. NEVER run the engine with a flushing attachment for more than 5 minutes. The driveline seal is not provided with cooling water if it is not immersed in water. To shut the engine down, first shut off the water supply, and then shut down the engine. If this sequence of tasks is not followed as stated, the engine could be severely damaged.

REMEMBER
Start the engine.
Turn the water on.
Turn off the water.
Shut down the engine.

To lean the mixture -- rotate the low speed mixture screw **CLOCKWISE**.

To enrichen the mixture -- rotate the low speed mixture screw **COUNTERCLOCKWISE**.

Rotate the low speed mixture screw in the direction which will result in the smoothest engine operation.

Idle Speed Adjustment

The idle speed adjustment screw is located on the throttle shaft linkage. If working on a twin carburetor installation, only the aft carburetor is equipped with an idle speed screw. A link rod connects the two throttle shafts together. Therefore, when an adjustment is made at the idle speed screw on the aft carburetor, the adjustment holds good for both carburetors.

When no throttle is applied, the tip of the screw rests against a stopper cast into the carburetor. To set the idle speed, obtain an induction type tachometer.

To increase engine rpm -- rotate the idle speed screw **CLOCKWISE**.

To decrease engine rpm -- rotate the idle speed screw **COUNTERCLOCKWISE**.

Start the engine and allow it to warm to operating temperature.

CAUTION

The following procedure MUST be followed whenever the engine is started with a hose attached for cooling the engine. Start the engine, and then IMMEDIATELY turn the water on. The engine MUST be started first to prevent water from backing up into the tuned exhaust pipe and into the engine. Water must circulate anytime the engine is

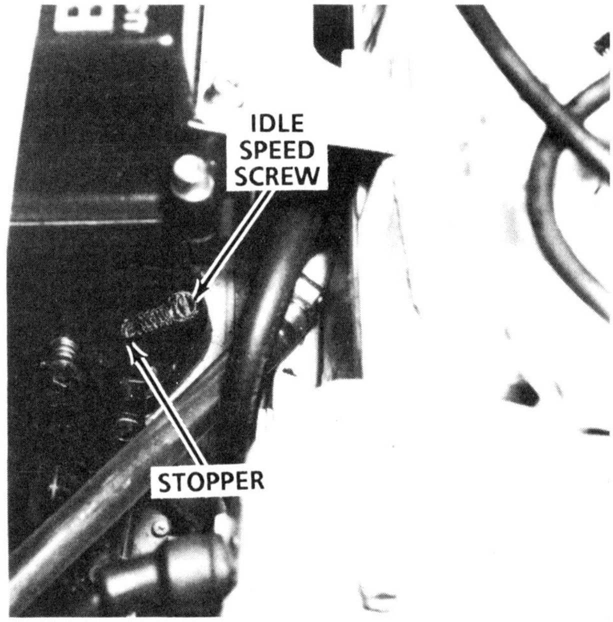

Typical location of the idle-speed adjustment screw on a Mikuni carburetor.

operating to prevent overheating -- which could cause damage to moving parts and possibly cause engine seizure. NEVER allow the engine to operate without cooling water for more than 15 seconds. NEVER run the engine with a flushing attachment for more than 5 minutes. The driveline seal is not provided with cooling water if it is not immersed in water. To shut the engine down, first shut off the water supply, and then shut down the engine. If this sequence of tasks is not followed as stated, the engine could be severely damaged.

REMEMBER
Start the engine.
Turn the water on.
Turn off the water.
Shut down the engine.

Rotate the idle speed screw until the engine idles at 1500 rpm -- if using a flush attachment, or 1200 rpm, if the craft is in the water.

After a throttle cable adjustment is made, **ALWAYS** perform a check to ensure the oil injection pump is synchronized with the carburetor. If the oil injection pump is not properly synchronized with the carburetor, serious damage may be caused to the engine due to lack of adequate lubrication.

Choke Cable Adjustment

Push the choke knob inward. Observe the choke pivot arm on the side of the carburetor. The arm should be in the position shown in the accompanying illustration, with minimum slack in the choke cable. If the choke needs adjustment: Loosen the small screw securing the choke cable end to the barrel on the choke lever. Push the choke knob inward until it rests against the control panel. Set the choke pivot arm in the fully open position, and then tighten the screw securing the cable to the barrel.

Choke Cable -- Excessive Slack

If there is too much slack in the cable, loosen the top nut around the cable at the cable support bracket and tighten the bottom nut until only a small amount of slack remains.

If the cable is tight -- with no slack -- loosen the bottom nut around the cable at the cable support bracket and tighten the top nut until there is just a small amount of slack in the cable.

If the cable has stretched, or has become kinked or frayed, or tends to stick, the entire cable should be replaced.

6-8 OIL INJECTION ALL ENGINES

An oil injection system is standard equipment on all Bombardier Sea-Doo personal watercraft covered in this manual. This oil injection system replaces the age old method of manually mixing oil with the fuel for lubrication of internal moving parts of a two-stroke engine. Two slightly different model pumps are used. One, mounted on the intake manifold, is installed on all engines **EXCEPT** the model 787 Series. The oil injection pump on the 787 Series, is mounted on the flywheel (magneto) cover. Both pumps operate in the same manner and are serviced with the same procedures except for the removal and installation instructions.

Bombardier engineers designed and developed their oil injection system as an independent unit -- separate from the fuel system until the lubricant enters the intake manifold. This system injects oil directly into the intake manifold between the carburetor and the rotary valve. Such an arrangement provides immediate oil delivery response to changes in rpm and load demands. Because the oil bypasses the carburetor, gummy deposits in the carburetor passages are avoided.

Oil Mixture

Fill the oil tank to the top line embossed on the oil tank dipstick. Use only two-cycle NMMA engine oil **OR** a Low Ash oil with a BIA certified rating of TCW-3. This oil is suitable for use through a temperature range of 14°F to 140°F (-10°C to +60°C).

The rate of oil injection is controlled by engine rpm. The oil pump on all engines **EXCEPT** Model 787 is directly coupled to the rotary valve shaft driven by the crankshaft. The oil pump on the Model 787 Series engines is coupled to the forward end of the crankshaft. The pump at both locations is capable of delivering the same amount of oil under any engine operating condition.

After the initial break-in period, oil is injected at a rate giving approximately a 40:1 mixture at low and idle speed. The mixture increases as the throttle is advanced to WOT (wide open throttle).

In order to obtain extra lubrication for a new, rebuilt, or engine just out of long storage, during the first 10 hours of break-in period, the manufac-

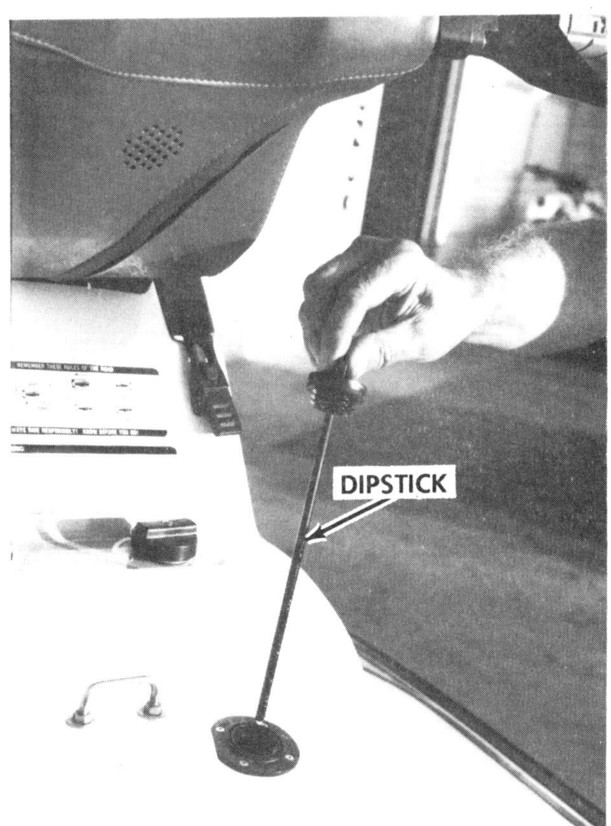

Fill the oil tank to the top line embossed on the oil tank dipstick.

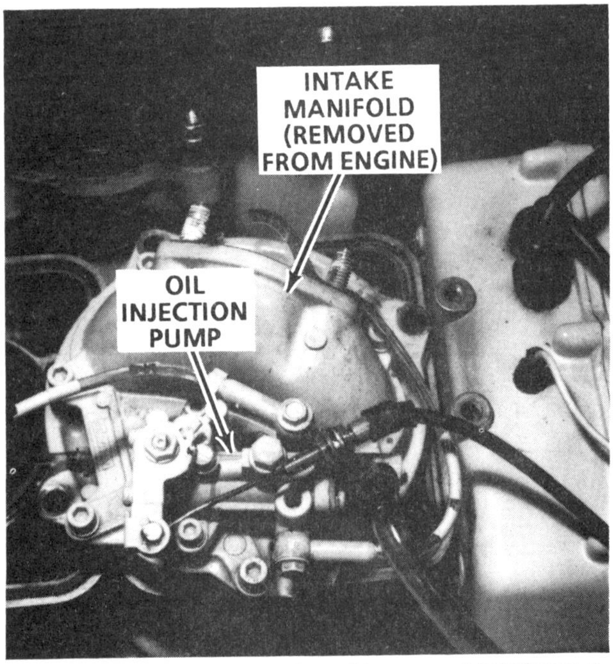

On all engines covered in this manual ***EXCEPT*** *the Model 787 Series, the oil injection pump is mounted on the intake manifold on the starboard side of the engine. A gear on the pump shaft indexes with a gear on the rotary valve shaft, making oil delivery a function of rpm.*

turer recommends oil be added directly into the fuel tank to give a mixture of 25:1. (Translated, this means adding 16oz of oil per tank full -- 10 to 12 gallons of fuel.)

Professional mechanics, who own oil injected units, constantly add an extra ounce of oil per gallon of fuel after the break-in period. The addition of this extra oil may prevent engine seizure in the event of an oil pump failure or clog in an oil delivery line.

Mechanics are quite cost conscious. They realize all too well how much cheaper replacing fouled spark plugs can be when weighed against an engine overhaul to replace expensive parts damaged from lack of adequate lubrication.

Use only outboard marine oil in the mixture, **NEVER** automotive oils. Four-stroke automotive engine oil is not formulated to burn completely, only to lubricate. Therefore, automotive oil, if used in a two-stroke engine, will leave undesirable residue.

SYSTEM COMPONENTS

Principle components of the oil injection system include:

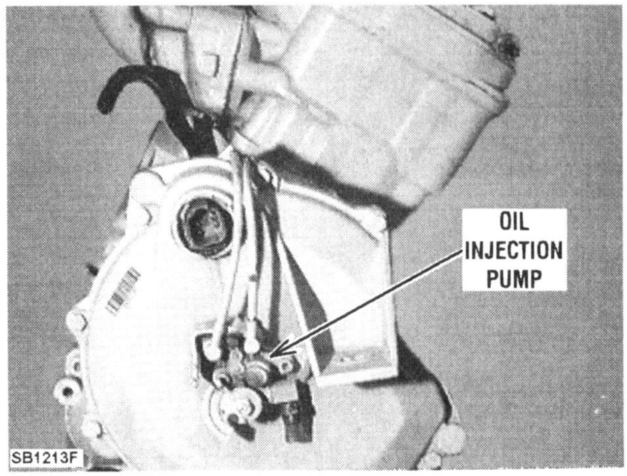

On a Model 787 Series engine, the oil injection pump is mounted on the flywheel cover. The pump shaft is connected to the end of the crankshaft through a short nylon shaft. Therefore, oil delivery is a direct function of engine rpm.

Main oil tank, with dipstick.

A one-way check valve.

An oil injection pump.

An in-line oil filter.

A network of connecting hoses.

Oil Tank

The oil tank is mounted inside the engine compartment, with the filler easily accessible when the engine cover is removed. The capacity of this tank is 95 fl oz (2.8L).

Oil supply to the oil injection pump is gravity fed from the oil tank. Therefore, a

The oil tank is mounted next to the fuel tank. The usual capacity of the oil tank is about 95 fl oz (2.8L).

one-way check valve is mounted on top of the tank. The other end of the valve is open to atmosphere. The check valve allows air to enter the tank replacing oil used, but will not allow oil to flow out of the tank -- thus the tank is "spill-proof".

Oil Injection Pump

The oil injection pump is a positive displacement type unit driven by the rotary valve shaft on all engines **EXCEPT** the Model 787 Series. On the 787, the pump is driven through a coupling to the forward end of the crankshaft. With each revolution of the crankshaft, a plunger cam moves up and down, pumping oil through two transparent hoses to the intake manifold.

System Inspection

Remove the oil tank filler cap. Inspect the condition of the seal beneath the cap. Replace the seal if it is worn or cracked.

Disconnect the ventilation hose from the fitting on the oil tank.

To test operation of the check valve:

Pull the two hoses free of the fittings. Now, blow through one of the fittings, and then the other fitting. Air should pass through the valve **ONLY** in one direction.

A faulty check valve will starve the engine of lubrication and quickly cause extensive and expensive damage.

Troubleshooting

Unfortunately, most problems with the oil injection system are only discovered after it is too late. If either of the two hoses becomes clogged, kinked, or restricted, oil delivery to the cylinders will be reduced. Insufficient oil delivery to the engine will cause the oil level to drop slower than normal; the engine will overheat due to inadequate lubrication; and moving internal parts will wear more quickly or be severely damaged.

Excessive oil delivery to the cylinders will cause the oil level to drop at a much faster rate than normal. The engine will smoke -- especially at idle speed -- and the spark plugs will become fouled, causing the engine to misfire.

First Checks
Oil Delivery

If excessive or reduced oil delivery occurs, as described above, one or more of the following areas may require **IMMEDIATE** attention:

- Defective oil pump.
- Clogged oil tank check valve.
- Kinked oil hose.
- In-line oil filter requires cleaning.
- Blocked oil hose.
- Air in the system.
- Leak in the system -- fitting or hose.
- Improper idle fuel adjustment.
- Poor quality fuel.
- Poor quality two-stroke oil.

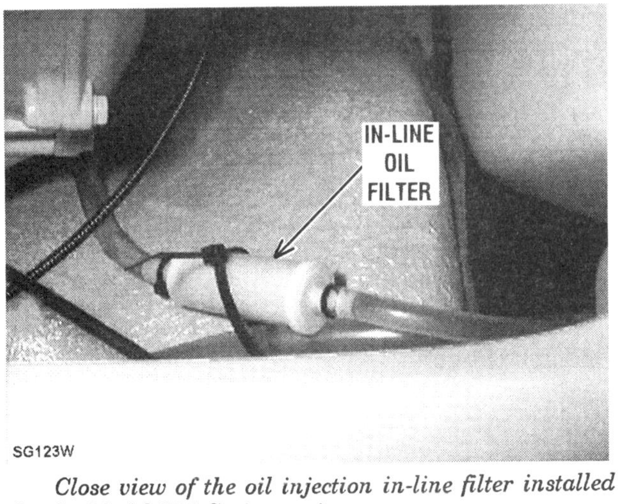

Close view of the oil injection in-line filter installed for the Model 787 Series engines.

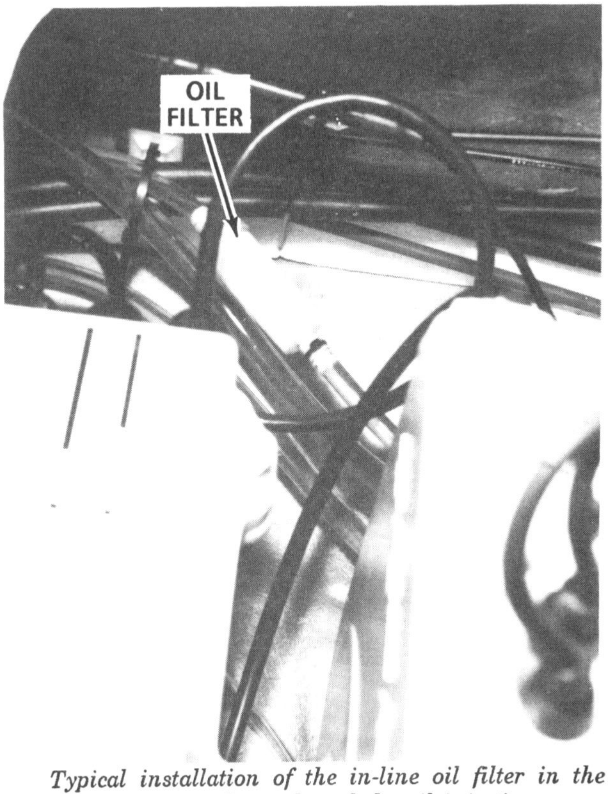

Typical installation of the in-line oil filter in the supply line between the tank and the oil injection pump for all engines in this manual ***EXCEPT*** *Model 787 Series.*

BAD NEWS

An oil pump gasket set is the only kit available to overhaul an oil pump. No spare parts are sold for the oil pump used on these units. Therefore, a defective oil pump **MUST** be replaced.

Systematically check each oil line and connection. A free flow of oil through an unrestricted line is **CRITICAL** for adequate engine lubrication.

CAUTION

Any time an oil tank hose is disconnected, the oil injection pump **MUST** be purged ("bled") of any trapped air. Failure to "bleed" the system could lead to engine seizure due to lack of adequate lubrication.

Instruction for "bleeding" -- purging -- air from the system follows.

Purging Air from Oil Injection System

The following procedures are to be performed any time the oil injection system has been opened (other than to add oil to the tank), and air has entered the system.

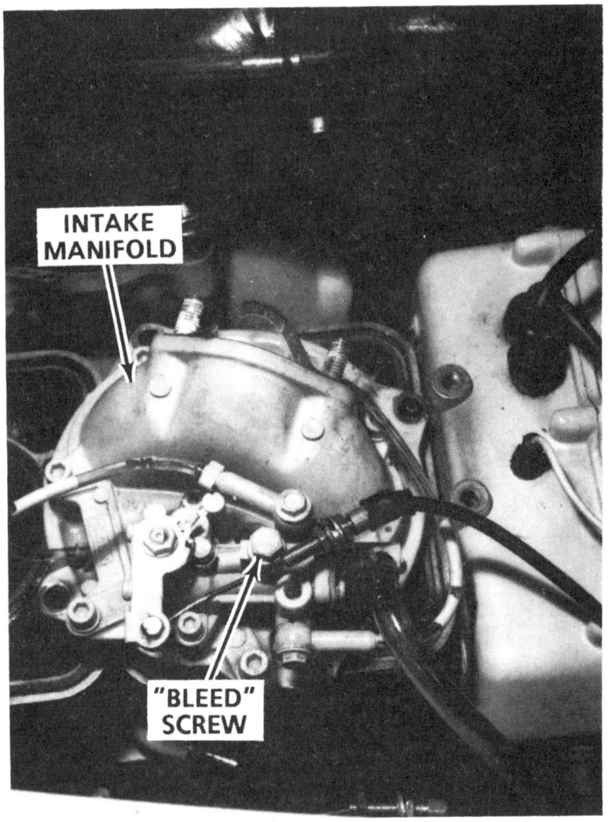

*Typical location of the oil injection pump "bleed" screw for all engines covered in this manual **EXCEPT** the Model 787 Series engines.*

Prepare about a ten minute supply of 40:1 premix using two-stroke oil. "Jerry-rig" a setup to provide this mixture to the fuel pump during this purging procedure. The oil/fuel mixture prepared will provide adequate lubrication while the system is being purged. Check the level of oil in the oil tank and replenish as necessary.

CAUTION

The following procedure MUST be followed whenever the engine is started with a hose attached for cooling the engine. Start the engine, and then IMMEDIATELY turn the water on. The engine MUST be started first to prevent water from backing up into the tuned exhaust pipe and into the engine. Water must circulate anytime the engine is operating to prevent overheating -- which could cause damage to moving parts and possibly cause engine seizure. NEVER allow the engine to operate without cooling water for more than 15 seconds. NEVER run the engine with a flushing attachment for more than 5 minutes. The driveline seal is not provided with cooling water if it is not immersed in water. To shut the engine down, first shut the water supply off, and then shut down the engine. If this sequence of tasks is not followed as stated, the engine could be severely damaged.

REMEMBER

Start the engine.
Turn the water on.
Turn off the water.
Shut down the engine.

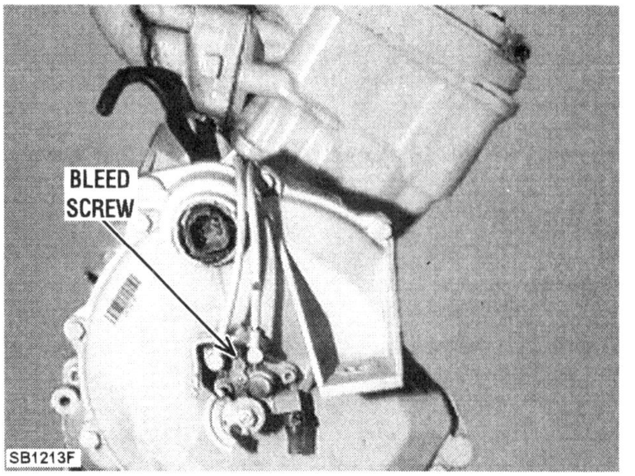

Location of the oil injection pump "bleed" screw on the Model 787 Series engines.

SPECIAL WORDS

Because the fuel and oil premix is being fed through the carburetor, in addition to oil from the oil injection system also being fed directly into the manifold, the engine will be operating on a heavy mixture of oil. This condition will cause the engine to smoke heavier than normal.

Obtain a small container and have a shop cloth handy to catch any oil drips. Loosen the bleed screw on the oil pump a few turns and wait until a slow pulsing flow of oil emerges without any air bubbles. Observe the flow of oil for four full minutes to verify the pump is functioning properly and all air has been expelled from the system.

The two small transparent hoses between the pump and the intake manifold **MUST** be full of oil. If not, operate the engine at idle speed while holding the pump lever in the fully open (**UP**) position. **DO NOT** activate the throttle lever at the handlebar until the two hoses are full of oil. Tighten the bleed screw securely.

Replenish the oil tank with two-stroke oil.

Oil Injection Pump Synchronization

Oil injection pump adjustment is dependent on throttle cable position. Therefore, any time the throttle cable is adjusted, the oil injection pump **MUST** be synchronized. The throttle cable from the handlebar splits into two cables. One cable controls engine rpm at the carburetor. The other cable is attached to the oil injection pump. With this arrangement, the rate of oil injected into the intake manifold is con-

Observe the mark on the pump lever and the mark on the pump boss -- see the accompanying photographic illustration for all engines **EXCEPT** the Model 787 Series. See the accompanying line drawing illustrating the alignment marks for the Model 787 Series.

Adjust the length of the throttle cable to bring these two marks into alignment. This is accomplished using the two adjustment nuts -- one on each side of the slot in the cable support bracket on the pump. A small mirror would be most helpful to see these marks, because there is very little clearance between the oil injection pump and the starboard engine compartment bulkhead.

Oil Injection Pump Operational Test

Remove the four bolts and washers securing the rotary valve housing to the upper and lower crankcase. The intake manifold and the oil pump are attached to the valve housing and will come away with the housing.

Remove the six Allen head bolts and the washers securing the intake manifold to the rotary valve housing. The oil injection pump will remain on the manifold.

Lift the intake manifold free of the rotary valve housing. **TAKE CARE** not to

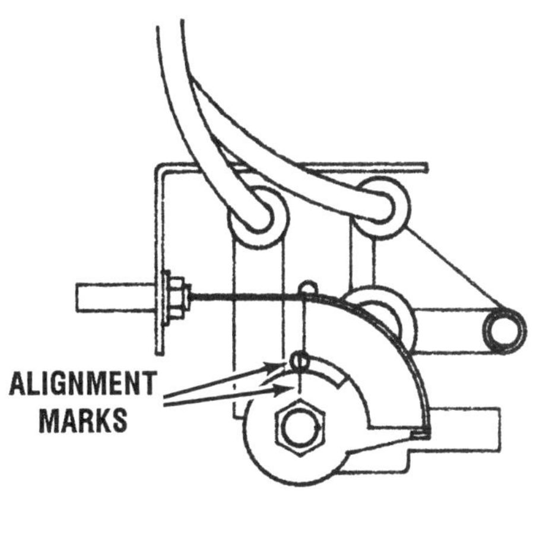

Simple line drawing to depict the marks to be aligned when synchronizing the oil pump on a Model 787 Series engine. The text details the alignment procedure.

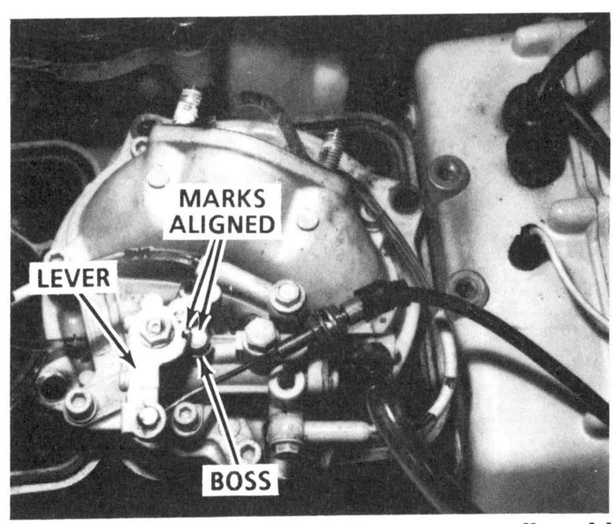

*Close view of the oil injection pump on all model engines covered in this manual **EXCEPT** the Model 787 series to indicate location of the alignment marks when synchronizing the oil pump. The text details procedure.*

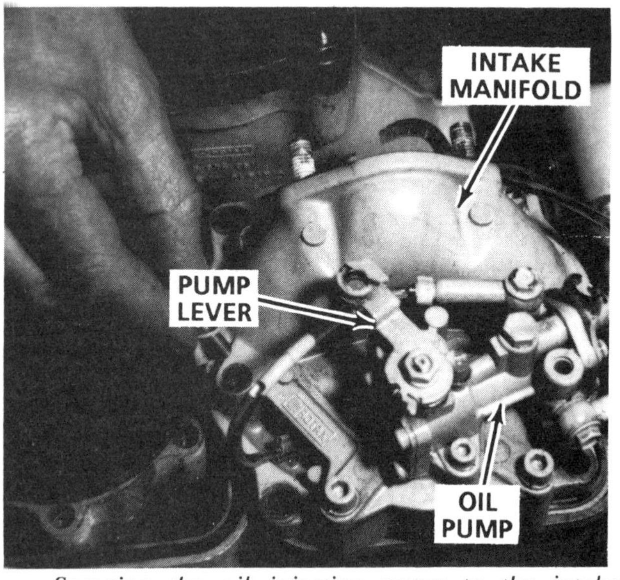

*Securing the oil injection pump to the intake manifold. The throttle cable is installed into the pump lever **AFTER** the rotary valve housing has been installed on the engine.*

lose the single dowel pin from one of the mating surfaces. The dowel pin may remain in the housing or in the manifold.

Obtain Gear Holder P/N 290-277-905 and use it to prevent the oil pump gear from rotating while the nut is loosened. If this tool is not available, hold the gear with a shop cloth, as shown. Remove the nut and washer, gear, and another washer.

Obtain an electric drill capable of **COUNTERCLOCKWISE** rotation (reverse). Insert the oil pump shaft into the drill chuck and tighten the chuck just **SNUG** -- not enough to damage the shaft threads.

Place a shop cloth inside the rotary valve housing under the discharge ports of the two transparent hoses. Check to be sure the drill selector is in the **REVERSE** position. Now, hold the pump lever in the fully open position and at the same time turn the drill motor on. Oil should drip from both discharge ports. If not, the pump is defective and **MUST** be replaced.

GOOD WORDS

Individual replacement parts are **NOT** available. However, if the pump is leaking, the condition may be corrected by installing a gasket kit, which is available. If the gasket set does not solve the problem, the pump **MUST** be replaced. Removal procedures follow.

Oil Pump Removal
All Engines EXCEPT
Model 787 Series

Begin by removing the flame arrestor cover, screen and base. Next, obtain hose pincer tool P/N 295-000-076. Clamp the tool around the oil return hose. If this tool is not available, make arrangements to plug the disconnected hose when it is pulled free of the fitting in the next step.

Now, pull the oil return hose free of the fitting above the rotary valve housing. If the hose was not pinched to halt the flow of oil, quickly plug the end of the hose to prevent losing oil into the bilge.

Remove the four bolts and washers securing the rotary valve housing to the upper and lower crankcase. The intake manifold and the oil pump are attached to the valve housing and will come away with the housing. Remove and discard the sealing ring.

Install a hose pincer tool onto the oil supply hose, and then remove the hose from the fitting at the base of the rotary valve housing. If this tool is not available, make adequate preparations to catch oil when the hose is disconnected and to plug the hose immediately.

Next, disconnect the throttle cable from the oil injection pump by disengaging the cable end from the slotted hole in the lever, and loosening the two nuts -- one on each side of the support bracket.

Remove the six Allen head bolts and the washers securing the intake manifold and the rotary valve housing together. Lift the intake manifold free of the rotary valve housing. The oil injection pump will remain on the manifold. **TAKE CARE** not to lose the single dowel pin from on of the mating surfaces. The dowel pin may remain in either surface. Remove and discard the sealing ring.

Obtain gear holder too P/N 290-277-905 and use it to prevent the oil pump gear from rotating while the nut is loosened. If the tool is not available, hold the gear with a shop cloth, as shown in the accompanying illustration.

Remove the nut and washer, gear, and another washer. Remove the two mounting bolts. Lift the pump off the intake manifold, and then remove and discard the **O**-ring.

GOOD WORDS AND
BAD NEWS

In some cases, a leaking oil pump can be overhauled with a gasket kit.

HOWEVER, No individual parts are available. Therefore, if the problem cannot be corrected with a gasket kit, the pump **MUST** be replaced. A defective pump will quickly cause damage to internal parts from lack of adequate lubrication.

Oil Injection Pump
Installation -- All Engines
EXCEPT Model 787 Series

Begin by installing a new **O**-ring over the oil pump shaft. Install the pump onto the intake manifold with the two bolts and washers. Tighten the bolts to a torque value of 53in lbs (6Nm). Install the two "banjo" bolts and fittings, with the oil lines attached, to the pump. Apply Loctite 242 to the pump shaft threads.

Now, inside the manifold: Install the washer, gear, another washer, and then the nut onto the oil pump shaft. Obtain Gear Holder tool P/N 290-277-9-5 or a shop cloth. Hold the gear firmly to prevent it from rotating while the nut is tightened to a torque value of 89 in lbs (10Nm).

Fit a new sealing ring into the groove of the rotary valve cover. Install the intake manifold onto the rotary valve cover, with the manifold indexed over the single dowel pin in the cover.

Next, install the six bolts and washers securing the intake manifold to the rotary valve cover. Tighten the bolts alternately and evenly to a torque value of 80 in lbs (9Nm).

Hook the end of the throttle cable into the pump lever. Push the cable into the slot with an adjustment nut on each side of the slot.

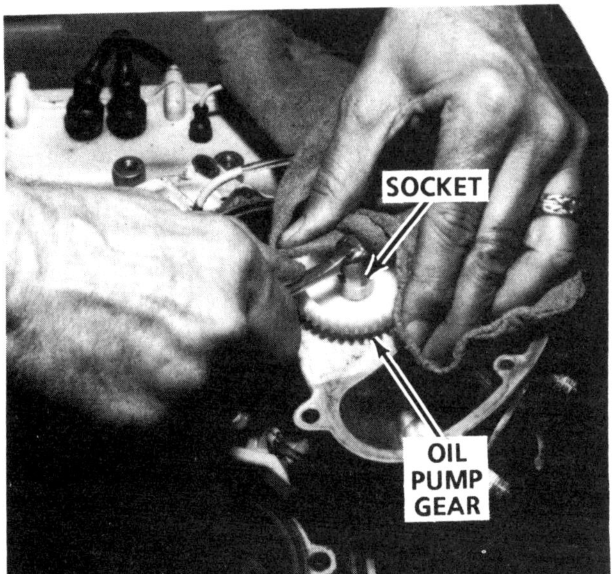

Installation of the washer, gear, another washer, and nut onto the pump shaft from inside the intake manifold.

Observe the mark on the pump lever and the mark on the pump boss. These two marks **MUST** align for oil pump synchronization. Adjust the position of the two nuts to change the length of the cable, and therefore, the position of the lever until the two marks align.

Install the rotary valve cover by first sliding a clamp over the end of the oil supply hose. Remove the pincer tool (or plug in the end of the hose), and the quickly work the hose onto the fitting at the base of the rotary valve. Tighten the clamp securely.

Fit a new sealing ring into the groove in the rotary valve housing. Install the housing against the upper and lower crankcase. Install the four securing bolts and washers. Tighten the bolts to a torque value of 15 ft lbs (20Nm).

Finally, slide a clamp over the end of the oil return hose. Remove the pincer tool (or plug in the end of the hose), and then quickly work the hose onto the fitting at the top of the rotary valve. Tighten the clamp securely.

Oil Pump Removal
Model 787 Only

An oil pump gasket kit, available from the local Bombardier dealer, may solve a leaking oil pump problem.

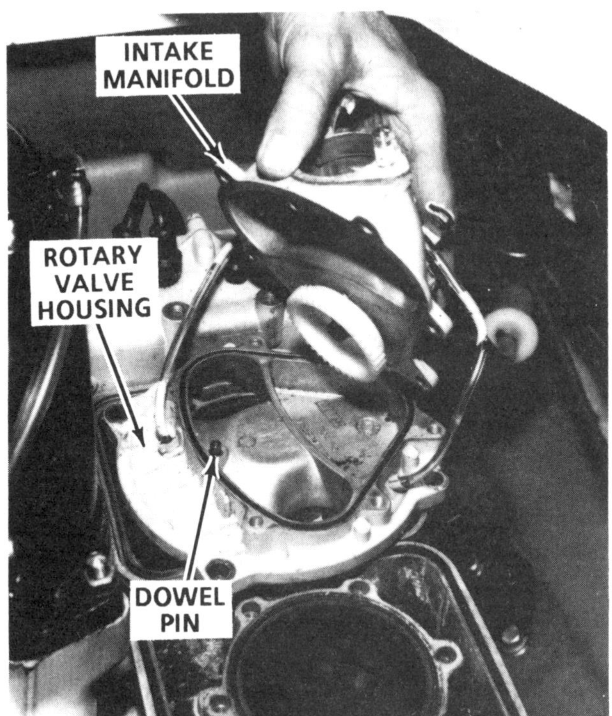

Installation of the intake manifold, with the oil injection pump attached onto the rotary valve housing. A hole in the intake manifold must index over a single dowel pin in the rotary valve housing.

HOWEVER, No individual parts are available. Therefore, if the problem cannot be corrected with a gasket kit, the pump **MUST** be replaced. A defective pump will quickly cause damage to internal parts from lack of adequate lubrication. Connect the oil injection pump cable at the pump.

Remove the two Allen head bolts with flat washers securing the oil injection pump to the flywheel cover (magneto cover). Withdraw the pump and shaft from the cover. Disconnect the oil hoses from the fittings on the pump.

Oil Pump Installation

Model 787 Series Only

Slide the nylon connector shaft through the flywheel cover (magneto cover) and into the end of the crankshaft.

Fit a **NEW O**-ring into the recess of the flywheel cover.

Bring the oil injection pump up to the cover, and then into place against the surface of the flywheel cover. It may be necessary to rotate the **AFT** flywheel slightly to permit the "spade" of the pump shaft to index into the slot in the end of the nylon connector shaft.

Secure the pump in place with the two bolts and washers tightened to a torque value of 26 in lbs (3Nm).

Connect the oil injection pump cable to the fitting and tighten the nut to a torque value of 40 in lbs (4.5Nm).

SYNCHRONIZING OIL PUMP

Check to be sure the oil system has been "bled" -- purged of air -- as described in detail earlier in this section -- before operating the engine. There should be **NO** evidence of an air bubble in the system. Verify the oil reservoir is properly filled.

CRITICAL WORDS

Proper adjustment of the oil pump is critical to **ENSURE** adequate lubrication to internal engine parts during all phases of engine operation -- rpm and load.

CAUTION

The following procedure MUST be followed whenever the engine is started with a hose attached for cooling the engine. Start the engine, and THEN turn on the water. The engine MUST be started first to prevent water from backing up into the tuned exhaust pipe and into the engine. Water must circulate anytime the engine is operating to prevent overheating -- which could cause damage to moving parts and possibly cause engine seizure.

NEVER allow the engine to operate without cooling water for more than 15 seconds. NEVER operate the engine with a flushing attachment for more than five minutes. The driveline seal is not provided with cooling water if it is not immersed in water. To shut the engine down, first shut off the water supply, and then shut down the engine. If this sequence of tasks is not followed as stated, the engine could be severely damaged.

Start the engine. Check to be sure the idle speed is properly adjusted.

Verify the carburetors are synchronized, on a twin carburetor installation.

Remove any throttle cable "free play" by depressing the throttle lever until a slight resistance is felt. With the throttle lever in this position, marks on the pump body must be aligned.

GOOD WORDS

Regardless of the engine being serviced, a small mirror attached to some type of rod will prove most useful in observing the alignment marks, because of the limited space between the pump and the engine compartment bulkhead.

If required, rotate the cable adjustment screw to obtain proper pump alignment of the marks. The accompanying two illustrations clearly indicate location of the alignment marks.

Once alignment is obtained, tighten the jam nut to hold the alignment.

7
IGNITION

7-1 INTRODUCTION AND CHAPTER COVERAGE

The less a marine engine is operated, the more care it needs. Allowing a marine engine to remain idle will do more harm than if it is used regularly. To maintain the engine in top shape and always ready for efficient operation at any time, the engine should be operated every 3 to 4 weeks throughout the year.

The carburetion and ignition principles of two-cycle engine operation **MUST** be understood in order to perform a proper tune-up on a marine engine.

If you have any doubts concerning your understanding of two-cycle engine operation, it would be best to study the operation theory section in the first portion of Chapter 8 -- **ENGINE,** before tackling any work on the ignition system.

A Capacitor Discharge Ignition (CDI) is used on all engine models covered in this manual.

Two generating coils and three battery charging coils are mounted on the stator plate behind the flywheel on all Model 587 thru 717 Series engines.

IMPORTANT WORDS DC CDI IGNITION 787 ENGINES ONLY

The Direct Current Capacitor Discharge Ignition system -- DC CDI -- on 787 engines only -- is a state-of-the-art electrical system. Engines equipped with DC CDI may be **CHECKED** for proper timing, however, adjustments to ignition timing on these systems must be performed using a special programmer module. This module actually writes a timing code within the MPEM -- Multi-Purpose Electronic Unit -- used for controlling ignition. This device is sold to dealers **ONLY** -- it is not available for retail sale.

The "programming" numbers for adjusting ignition timing on all engines equipped with a DC CDI system are listed in this chapter. These are for reference only -- they cannot be used for timing adjustment **UNLESS** the special programmer module is available.

Normally, ignition timing adjustment should not be necessary on engines equipped with DC CDI. Timing has been set at the factory and should remain in calibration since the system is "fixed" -- there are no moving parts. The only

The magneto assembly on all Model 787 Series engines is located inside and attached to the magneto cover.

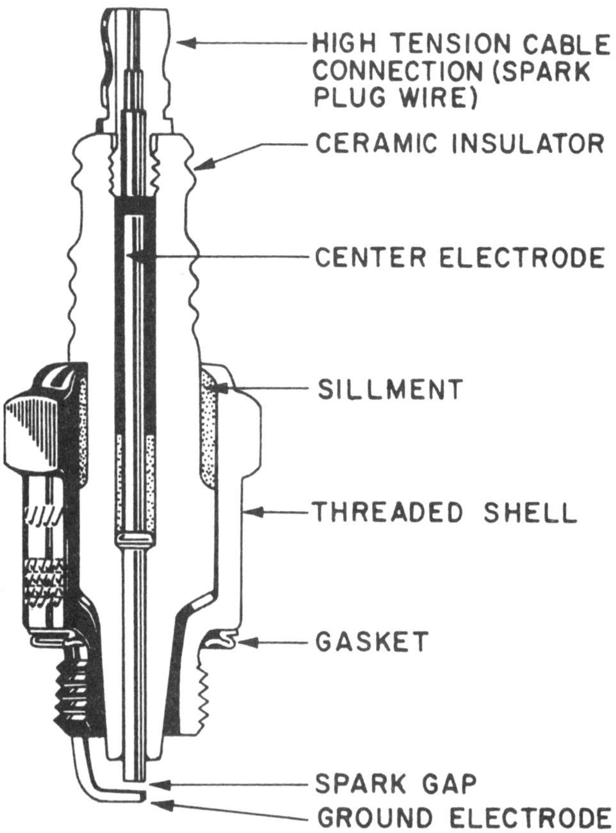

Cross-section drawing of a typical spark plug, with major parts identified.

time ignition adjustment may be necessary would be during replacement of the crankshaft, magneto rotor, trigger coil, or the MPEM unit.

If the timing is found to be incorrect on these models, the crankshaft should be checked for proper alignment. Improper timing may be the result of a twisted crankshaft.

7-2 SPARK PLUG EVALUATION

Removal

Remove the spark plug leads by pulling and twisting on only the molded cap. **NEV-**

The spark plug is foul from operating with an overrich air/fuel mixture, possibly caused by an improper carburetor adjustment.

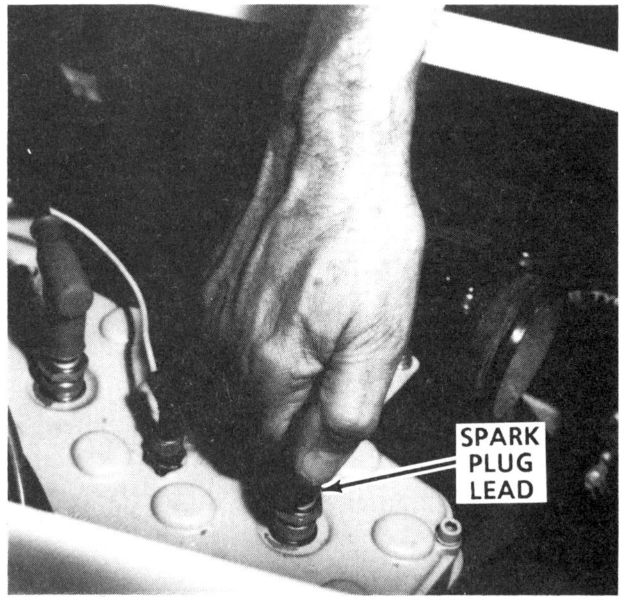

***ALWAYS** remove the spark plug high tension lead by pulling and twisting with a good grip on the molded cap.*

ER pull on the lead or the connection inside the cap may become separated or the boot damaged. Remove the spark plugs. **TAKE CARE** not to tilt the socket as you remove the plug or the insulator may be cracked.

Examine

Carefully examine both plugs to determine the firing conditions in the cylinder. If the side electrode is bent down onto the center electrode, the piston is traveling too far upward in the cylinder and striking the spark plug. Such damage indicates the piston pin or the rod bearing is worn excessively.

In most cases, an engine overhaul is required to correct the condition. To verify the cause of the problem, rotate the flywheel by hand. As the piston moves to the full up position, push on the piston crown

Buildup of combustion deposits is caused primarily from the burning of oil and/or fuel additives during normal combustion. These deposits can result in misfire.

with a screwdriver inserted through the spark plug hole, and at the same time rock the flywheel back-and-forth. If any play in the piston is detected, the engine must be rebuilt.

Correct Color

A proper firing plug should be dry and powdery. Hard deposits inside the shell indicate too much oil is being mixed with the fuel. The most important evidence is the light gray to tan color of the porcelain, which is an indication this plug has been running at the correct temperature. This means the plug is one with the correct heat range and also that the air/fuel mixture is correct.

Rich Mixture

A black, sooty condition on both spark plug shells and the porcelain is caused by an excessively rich air/fuel mixture, both at low and high speeds. The rich mixture lowers the combustion temperature so the spark plug does not run hot enough to burn off the deposits.

Deposits formed only on the shell is an indication the low-speed air/fuel mixture is too rich. At high speeds with the correct mixture, the temperature in the combustion chamber is high enough to burn off the deposits on the insulator.

Too Cool

A dark insulator, with very few deposits, indicates the plug is running too cool. This condition can be caused by low compression or by using a spark plug of an incorrect heat range. If this condition shows on only one plug it is most usually caused by low compression in that cylinder. If both plugs have this appearance, then it is probably due to the plugs having a too-low heat range.

Damaged spark plugs. Notice the broken electrode on the left plug. The missing part MUST be found and removed before returning the engine to service. If overlooked, the part could cause extensive and expensive damage to internal parts.

Fouled

A fouled spark plug may be caused by the wet oily deposits on the insulator shorting the high-tension current to ground inside the shell. The condition may also be caused by ignition problems which prevent a high-tension pulse from being delivered to the spark plug.

Carbon Deposits

Heavy carbon-like deposits are an indication of excessive oil in the fuel. This condition may be the result of worn piston rings or excessive ring end gap.

Overheating

A dead white or gray insulator, which is generally blistered, is an indication of overheating and pre-ignition. The electrode gap wear rate will be more than normal and in the case of pre-ignition, will actually cause the electrodes to melt. Overheating and pre-ignition are usually caused by overadvanced timing, detonation from using too-low an octane rating fuel, an excessively lean air/fuel mixture, or problems in the cooling system.

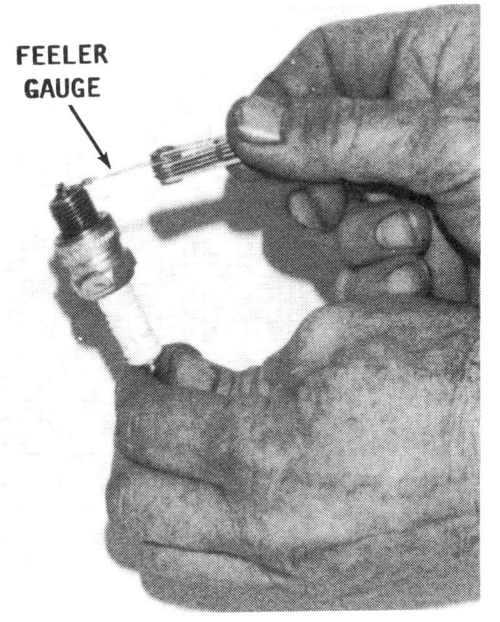

The spark plug gap should always be checked with a wire-type feeler gauge before installing new or used spark plugs.

Electrode Wear

Electrode wear results in a wide gap and if the electrode becomes carbonized it will form a high-resistance path for the spark to jump across. Such a condition will cause the engine to misfire during acceleration. If both plugs are in this condition, it can cause an increase in fuel consumption and very poor performance at high-speed operation. The solution is to replace the spark plugs with a rating in the proper heat range and gapped to specification.

Red rust-colored deposits on the entire firing end of a spark plug can be caused by water in the cylinder combustion chamber. This can be the first evidence of water entering the cylinders through the exhaust manifold because of an accumulation of scale. This condition **MUST** be corrected at the first opportunity. Refer to Chapter 8 -- **ENGINE.**

7-3 CDI (CAPACITOR DISCHARGE IGNITION) AND CHARGING SYSTEM 587 THRU 717 ENGINES

A Capacitor Discharge Ignition (CDI) system is used on all Bombardier Sea-Doo watercraft equipped with a Model 587 thru 717 Series engine. Models equipped with 787 series engine feature the DC CD system -- the components vary from "conventional" CDI systems. Since the CDI system is more readily serviceable, following is a description of the charging and ignition system used on the CDI system.

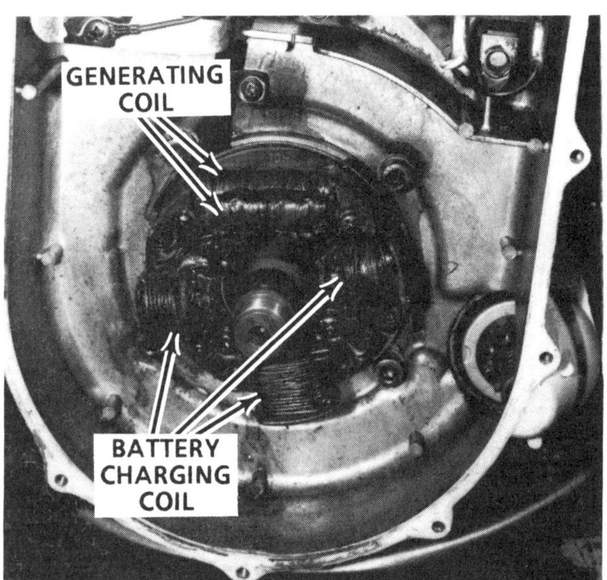

Although both generating coils and all three battery charging coils are mounted on the magneto assembly behind the flywheel, the flywheel does not have to be "pulled" in order to test these coils, as explained later in the text.

This system incorporates an automatic electronic spark advance, which should not require adjustment **UNLESS** the magneto assembly becomes misaligned or an ignition component is replaced.

On early models, about half of the ignition circuit was housed behind the flywheel and the remainder of the components were covered by the magneto housing, or located in an electrical box. Since the technology changes rapidly, new components are added to the system, and are located in an additional box.

SPECIAL WORDS

Five coils are mounted on the magneto assembly. A group of three coils together are the battery charging coils. The other two coils, stacked one on top the other, are the generating coils. The magneto assembly is mounted on the stator plate behind the flywheel. Therefore, the charging system is presented in this chapter and in this section because once the flywheel is removed the components for both systems are exposed.

DESCRIPTION AND OPERATION IGNITION CIRCUIT

The following paragraphs describe -- in layman's language -- the CDI system and how it functions. Every effort has been made to keep "technical" terms to a minimum.

Three major components are used in the CDI system.

1- Two stacked generating coils, are mounted on the magneto assembly behind the flywheel. These coils are sometimes called trigger coils, depending on the manufacturer of the ignition system. The generating coils provide a source of AC current and also time the electrical pulses.

2- The ignition module -- mounted in the high amperage **OR** "conventional" electrical box -- is used to convert AC to DC current. The module stores the DC current, then releases it and automatically controls the timing.

3- The ignition coil, also located in the electrical box, next to the ignition module, boosts the DC voltage instantly to approximately 20,000 volts to the spark plugs.

Two basic circuits are used with the CDI system -- a charge circuit and an ignition circuit.

Charge Circuit

The charge circuit consists of the flywheel magnet, battery charging coils, a diode, and a rectifier/regulator. As the flywheel magnet passes the charging coils, a voltage is induced in the coils. As the flywheel continues to rotate and the coils are no longer influenced by the magnet, the magnetic field collapses. Therefore, an alternating current is produced at the charging coils. This AC current is changed to DC by diodes inside the rectifier/regulator.

Ignition Generating Circuit

The ignition generating circuit has its own flywheel magnet, a generating coil, a diode, and a thyristor. A thyristor is a solid state electronic switching device which permits voltage to flow only after it is triggered by another voltage source.

At this point in time, when the ignition timing marks align, an alternating current is induced in the generating coil, in the same manner as previously described for the charging coil. This current is then passed to a second diode located in the CDI unit where it becomes DC current and flows on to the thyristor. This voltage triggers the thyristor to permit the voltage stored in the capacitor to be discharged. The capacitor voltage passes through the thyristor and on to the primary windings of the ignition coil.

In this manner, a spark at the plug may be accurately timed by the timing marks on the flywheel relative to the magnets in the flywheel and to provide as many as 100 sparks per second for an engine operating at 6,000 rpm.

Timing Advance

All Bombardier Personal Watercraft covered in this manual are equipped with an automatic advance type CDI system with no moving or sliding parts. Therefore, adjustment is not required. Instead, ignition advance timing is accomplished electronically and signal output increases proportionately to engine speed increase.

The timing is correctly set at the factory. As mentioned earlier, the timing should never require adjustments **UNLESS** the magneto assembly is misaligned or an ignition component has been replaced.

Detailed ignition timing procedures are presented in Section 7-7 of this chapter, beginning on Page 7-20.

Charging System

The charging system consists of the flywheel magnets, three battery charging coils, a rectifier/regulator, and the battery. As the flywheel magnets rotate, voltage is induced in the charging coil mounted on the magneto assembly under the flywheel. The alternating current passes through a series of four diodes inside the rectifier/regulator and is transformed to DC current.

The rectifier/regulator allows a maximum of 14 volts to pass on to the battery. Any excess voltage is passed to a SCR gate and back through the charging coil. In this manner, overcharging the battery is avoided. The life of components in the charging circuit is greatly extended due to the absence of power surges.

TROUBLESHOOTING IGNITION CDI SYSTEM

Always attempt to proceed with the troubleshooting in an orderly manner. The "shot in the dark" approach will only result in wasted time, incorrect diagnosis, replacement of unnecessary parts, and frustration.

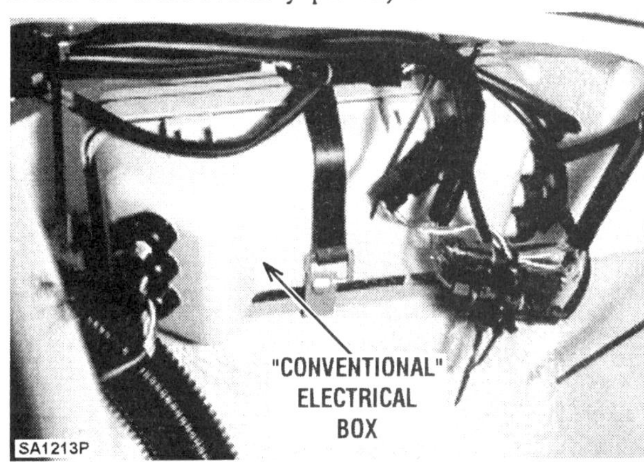

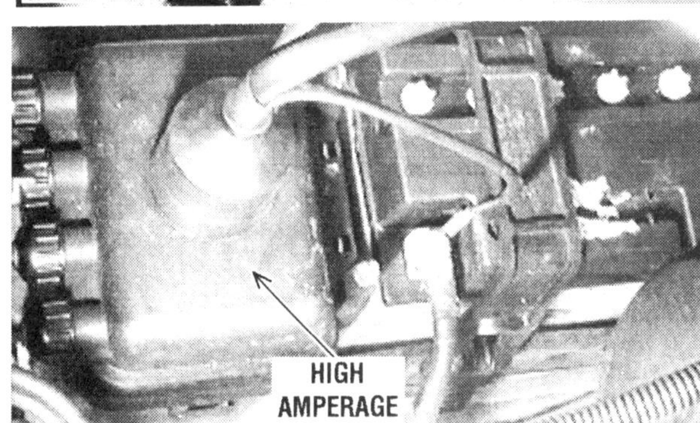

The rectifier/regulator and MPEM are located in the "conventional" electrical box -- top. On some models, a dedicated high amperage box houses the ignition coil and cranking motor solenoid -- bottom.

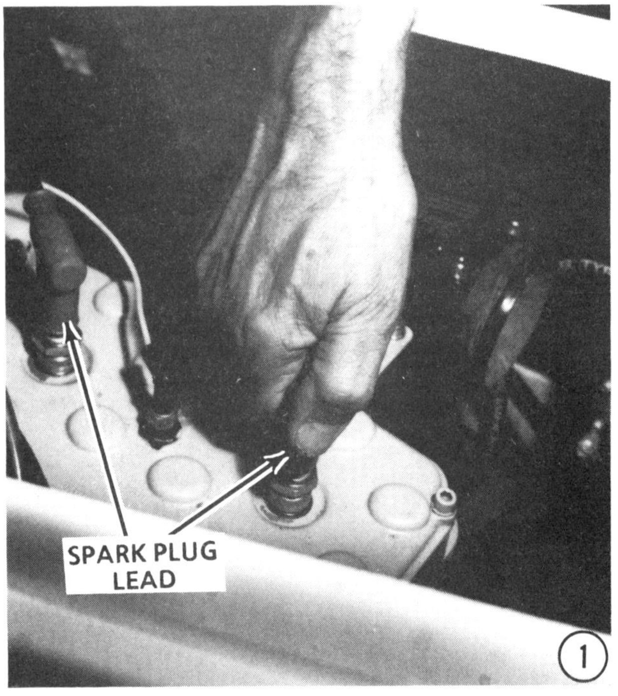

Begin the ignition system troubleshooting with the spark plug and continue through the system until the source of trouble is located.

Spark Plugs

1- Check the plug leads to be sure they are properly connected. Check the entire length of the leads from the plugs to the magneto assembly on the stator plate. If the leads are to be removed from the spark plug, **ALWAYS** use a pulling and twisting motion as a precaution against damaging the connection.

Attempt to remove the spark plug by hand. This is a rough test to determine if the plug is tightened properly. The attempt to loosen the plug by hand should fail. The plug should be tight and require the proper socket size tool. Remove the spark plug and evaluate its condition as described in Section 7-2.

2- Use a spark tester and check for spark. If a spark tester is not available, hold the plug wire about 1/4" (6.4mm) from the engine. Crank the engine through a few revolutions using the cranking motor and check for spark. A strong spark over a wide gap must be observed when testing in this manner, because under compression a strong spark is necessary in order to ignite the air/fuel mixture in the cylinder. This means it is possible to think a strong spark is present, when in reality the spark will be too weak when the plug is installed. If there is no spark, or if the spark is weak, the trouble is most likely behind the flywheel in the magneto assembly.

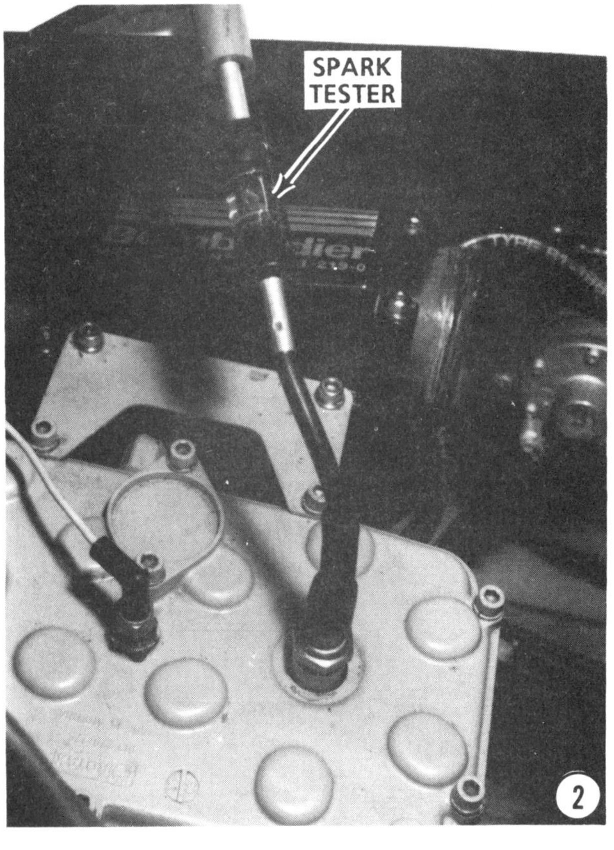

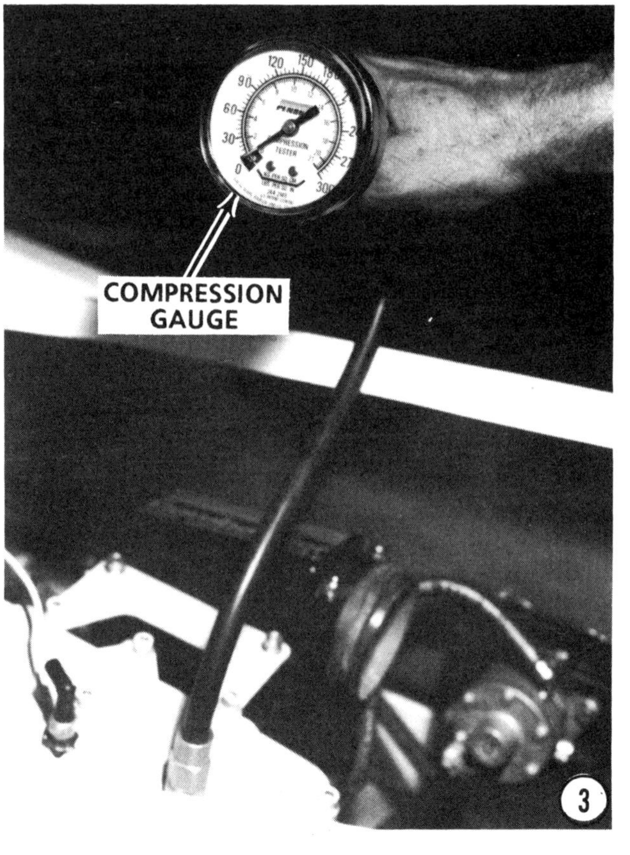

Compression

3- Before spending too much time and money attempting to trace a problem to the ignition system, a compression check of both cylinders should be made. If a cylinder does not have adequate compression, troubleshooting and attempted service of the ignition or fuel system will fail to give the desired results of satisfactory engine performance.

Remove the spark plug lead by pulling and twisting **ONLY** on the molded cap. **NEVER** pull on the lead because the connection inside the cap may be separated or the boot may be damaged. Remove the spark plug. Insert a compression gauge into the cylinder spark plug hole. Ground the spark plug leads to prevent damage to the ignition coil. If a lead is not grounded, the coil will attempt to match the demand created by the spark trying to jump from the electrode to the nearest ground.

Crank a cold engine through several revolutions with the electric cranking motor. Note the compression reading.

An acceptable pressure reading for an engine covered in this manual is 142 psi (981 kPa) or more. The readings between cylinders should not vary by more than 10%.

TROUBLESHOOTING CHARGING SYSTEM

As mentioned earlier, the charging coil is mounted on the magneto assembly located under and protected by the flywheel. The charging coil seldom causes problems in the charging circuit. Most problems can be traced to the rectifier/regulator, the battery, or to the connecting wiring.

The following three conditions will cause damage to the rectifier/regulator.

a- Battery cables connected in reverse.

b- Operating the engine with one or both battery cables disconnected.

c- A loose or broken connection in the charging circuit.

The voltage generated by the charging coil is intended to charge the battery. Anytime the circuit is incomplete, the voltage has actually no where to go. As a result, the voltage will actually reverse itself. Such reversed voltage will damage the internal structure of a diode in the circuit. The diode will then have a scorched or burned appearance.

Perform a visual inspection of the following areas before making any resistance tests on charging circuit components.

a- Check to be sure the battery cables are connected to the correct terminals.

b- Check condition of the battery and recharge, if necessary.

c- Inspect the wiring and the connections between the charging coil, the rectifier/regulator, and the battery. Replace any damaged or deteriorated cable insulation. Pay particular attention for loose, corroded, or disconnected terminals or connections.

7-4 TESTING IGNITION COMPONENTS

General Information

Due to the complex nature of the circuitry and high energy output pulses of microsecond duration, conventional testing devices such as a Volt/Ohm/Ammeter will not measure electrical output with the degree of accuracy required. A Bombardier Ignition Tester, as used by the local Bombardier dealer is an electrical device capable of measuring the peak energy output of capaci-

NEVER attempt to verify the charging circuit by operating the engine with the battery disconnected. Such action will damage the rectifier/regulator.

tor discharge ignition units, as well as the charging and generating coils. This instrument was designed to troubleshoot the CDI ignition system. Unfortunately this tester may not be easily accessible. Therefore, this chapter includes only those tests which may be performed with instruments normally and easily available.

WORDS ON TESTERS

The following static tests may be performed with a digital multitester (high-sensitivity ohmmeter) P/N 529 022 000 -- an instrument which is readily available. In almost all cases, it is only necessary to make contact with the meter probes to the leads or terminals listed for the test. Refer to the Resistance tables located on pages 7-10 and 7-11, for components to be tested, lead colors, and resistance values.

In 1996, Bombardier introduced an inexpensive magneto wiring harness adapter, allowing parts to be tested without removing individual component leads. This adapter is available in a four-pin configuration -- P/N 295 000 131 -- or a six-pin configuration -- P/N 295 000 136 -- to be used on all engines equipped with a harness quick disconnect fitting. Be **SURE** to obtain the proper harness adapter for the engine being serviced.

Dynamic testing, requiring Bombardier Ignition Tester P/N 295 000 008 may not be readily available. For this reason, only static tests requiring a digital multimeter will be covered in this section. If the Bombardier Ignition Tester is available, follow the instructions provided with the unit.

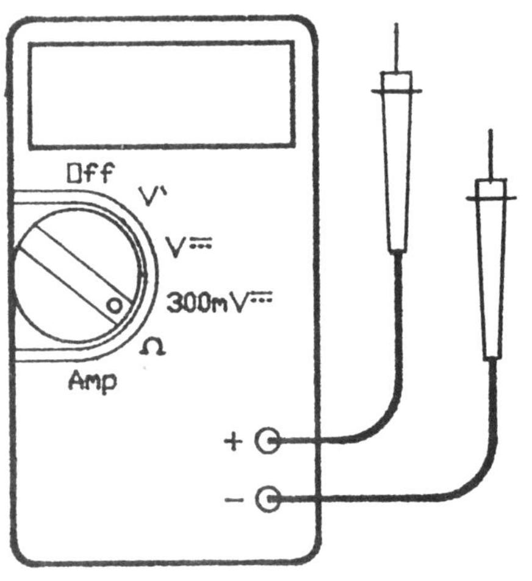

Test results listed throughout this chapter were obtained using a Fluke 73 -- or equivalent -- digital multitester.

WORDS FROM EXPERIENCE

During the tests, if a reading is slightly different from the specifications, but the engine still operates, then there is no real need to replace the affected component, until it actually fails. Bear in mind, in **MOST** cases electrical components are not returnable, once the item leaves the store. Therefore, make every attempt to avoid the **BUY** and **TRY** method of troubleshooting. Such a practice will lead only to wasted time, unneeded cash outlay, and frustration.

Intermittent or Multiple Problems

Many ignition problems occur only during engine operation -- when the component is subject to vibration, the engine is operating under a load, and/or there is a marked increase in temperature -- from the engine operating.

SPECIAL WORDS ON IGNITION AND CHARGING CIRCUIT TESTING

Most of the components in the ignition and charging circuits have leads with quick-disconnect fittings or harness plugs. Some components in these circuits are housed inside the electrical box -- others are housed on the magneto assembly behind the flywheel.

If a component inside the electrical box or on the 787 series engine stator plate is found to be defective, it may be replaced immediately -- usually through attaching hardware. If a component behind the flywheel -- battery charging coil or generating coil -- is found to be defective, the flywheel must be "pulled" in order to remove and replace the item. See Section 7-6, beginning on Page 7-14 for instructions to "pull" the flywheel.

CRITICAL WORDS

Never attempt to verify the charging circuit by operating the engine with the battery disconnected. Such action would force current -- normally directed to charge the battery -- back through the rectifier and damage the diodes in the rectifier.

Magneto Assembly Resistance Tests All Engines except 787

Remove the bolts and washers securing the magneto housing to the crankcase. **TAKE**

CARE not to damage the sealing ring, if so equipped.

Generating Coil Test

Typically, two generating coils -- one a high-speed coil, the other a low-speed coil -- are stacked on top of each other and secured to the magneto plate behind the flywheel. Some engines use a single-coil system design. In either case, it is not necessary for these coils to actually be visible in order to test them. Refer to the Resistance tables provided on the pages 7-10 and 7-11, and locate the proper component lead colors and resistance values for the engine being serviced.

If the magneto harness adapter is available, disconnect the magneto harness connector. Install the magneto harness adapter. If the adapter is not available, identify the appropriate colored leads -- obtained from the Resistance tables -- between the magneto assembly harness and ignition coil. Disconnect the leads at the quick disconnect fitting. Obtain an ohmmeter and measure the resistance across the given colored leads.

Verify the resistance values for both the high-speed and low-speed generating coils, if so equipped. If the meter registers resistance values within specifications, the coil is satisfactory. If servicing a dual-coil system, if either coil fails the resistance test, both coils **MUST** be replaced, because they share a common lead. If the coil passes the resistance test, but the problem is still determined to be in the ignition portion of the magneto assembly, the cause may be a loss of magnetism in the flywheel magnets, requiring replacement of the flywheel.

To replace the coil, refer to the procedure directly after Step **9** on Page 7-16 -- removing the magneto assembly from the stator plate.

Battery Charging Coil Test

The battery charging coil consists of three individual coils, mounted in a semi-circle on the magneto assembly, behind the flywheel. It is not necessary for these three coils to actually be visible in order to test them. Refer to the Resistance tables provided on pages 7-10 and 7-11, to obtain the proper component lead colors and resistance values for the engine being serviced.

If the magneto harness adapter is available, disconnect the magneto harness connector. Install the magneto harness adapter. If the adapter is not available, identify the appropriate colored leads -- obtained from the Resistance tables -- between the magneto assembly harness and ignition coil. Disconnect the leads at the quick disconnect fitting. Obtain an ohmmeter and measure the resistance across the given colored leads.

If the reading is less than expected, a short probably exists in the coil and the coil **MUST** be replaced. If the meter indicates an infinite resistance, an open circuit exists in the coil and the coil **MUST** be replaced.

If the coil passes this resistance test, but the problem is still determined to be in the charging portion of the magneto assembly, the cause is most likely a loss of magnetism in the flywheel magnets. If such is the case, the flywheel **MUST** be replaced.

To replace the battery charging coil, refer to the procedures on Page 7-17 -- after the magneto assembly has been removed from the magneto housing.

Ignition Coil Test
All Engines

The ignition coil is mounted in the high amperage or "conventional" electrical box. To check **PRIMARY** resistance: First, refer to the Resistance Table provided on Pages 7-10 and 7-11, to obtain the lead colors and resistance values for the engine being serviced. Disconnect the leads at their quick disconnect fitting. Obtain an ohmmeter and set the Rx1 ohm scale. Measure the resistance between the two leads and make a note of the reading.

To check the **SECONDARY** resistance: Refer to the Resistance Table listed on Pages 7-10 and 7-11 and obtain the Secondary windings resistance value for the engine being serviced. Insert the meter probes into the high tension lead boots and make contact with the spark plug caps. Check the meter reading against the specifications from the table.

If the ignition coil fails either test, it is defective and **MUST** be replaced.

If the reading is questionable, but the engine operates, install a spark plug tester and check the strength of the spark.

If the spark is acceptable, it is not necessary to replace the ignition coil. However, keep this area in mind for possible future trouble.

If the spark is weak, it will get weaker. Therefore, replace the ignition coil.

Text Continues on Page 7-12

IGNITION SYSTEM RESISTANCE TESTS

Component	Lead Color	Resistance (ohms)	Notes
92 -- All Models			
Generating Coil	Black and Black/Red	40 - 76	
Secondary Winding	End of each spark lead, caps removed	9 - 15 k	
Spark Plug Cap		4.48 k ohms	
Charging Coil	Yellow and Yellow/Black	0.05 - 0.6	
93 -- Single Generating Coil			
Generating Coil	Black and Black/Red	40 - 76	
Secondary Winding	End of each spark lead, caps removed	9 - 15 k	
Spark Plug Cap		4.48 k ohms	
Battery Charging Coil	Yellow and Yellow/Black	0.05 - 0.6	
93 -- Dual Generating Coil			
High Speed Generating Coil	White and Black/Red	2.8 - 4.2	
Low Speed Generating Coil	Black and Black/Red	120 - 180	
Primary Winding	Black and White/Blue	0.23 - .043	
Secondary Winding	End of each spark lead, caps removed	9 - 15 k	
Insulation	Check each wire of Primary and Secondary windings with ground for short circuit	open circuit (infinity)	
Battery Charging Coil	Yellow and Yellow/Black	0.21 - 0.31	
Insulation	Check each wire of Primary and Secondary windings with ground for short circuit	open circuit (infinity)	
94 -- Single Generating Coil			
Generating Coil	Black and Black/Red	40 - 76	
Secondary Winding	End of each spark lead, caps removed	9 - 15 k	
Spark Plug Cap		4.48 k ohms	
Battery Charging Coil	Yellow and Yellow/Black	0.05 - 0.6	
94 -- Dual Generating Coil			
High Speed Generating Coil	White and Black/Red	2.8 - 4.2	
Low Speed Generating Coil	Black and Black/Red	120 - 180	
Primary Winding	Black and White/Blue	0.23 - .043	
Secondary Winding	End of each spark lead, caps removed	5.85 - 6.5 k	
Insulation	Check each wire of Primary and Secondary windings with ground for short circuit	open circuit (infinity)	
Battery Charging Coil	Yellow and Yellow/Black	0.21 - 0.31	
Insulation	Check each wire of Primary and Secondary windings with ground for short circuit	open circuit (infinity)	

IGNITION SYSTEM RESISTANCE TESTS

Component	Lead Color	Resistance (ohms)	Notes
95 -- 587 - 717 Series Engines			
Generating Coil	Black and Black/Red	40 - 76	
Secondary Winding	End of each spark lead, caps removed	9 - 15 k	
Spark Plug Cap		4.48 k ohms	
Battery Charging Coil	Yellow and Yellow/Black	0.05 - 0.6	
96 -- 587 - 717 Series Engines			
Generating Coil	Black and Red/Blue	40 - 76	1
Primary Winding -- All Models		N/A	
Secondary Winding	End of each spark lead, caps removed	9 - 15 k	
Spark Plug Cap		4.48 k ohms	
Battery Charging Coil	Yellow and Yellow/Black	0.05 - 0.6	1
787 Engine Only			
Primary Winding	Primary terminals of coil	0.34 - 0.62	
Secondary Winding	End of each spark lead, caps removed	8.4 - 15.6 K	
Trigger Coil	Adapter White lead Adpater Black/Yellow lead	190 - 300	2
Stator	Two Yellow leads	0.1 - 1.0	2
97 -- 587 - 717 Series Engines			
Generating Coil	Black and Red/Blue	40 - 76	1
Primary Winding -- All Models Except SP, GTS and HX	Primary terminals of coil	0.34 - 0.62	
Primary Wndg -- SP, GTS and HX	N/A	N/A	
Secondary Winding SP, GTS and HX Models only	End of each spark lead, caps removed	9 - 15 k	
Secondary Winding All other Models	End of each spark lead, caps removed	8.4 - 15.6 K	
Spark Plug Cap		4.48 k ohms	
Battery Charging Coil	Yellow and Yellow/Black	0.05 - 0.6	1
787 Engine Only			
Primary Winding	Primary terminals of coil	0.34 - 0.62	
Secondary Winding	End of each spark lead, caps removed	8.4 - 15.6 K	
Trigger Coil	Adapter White lead Adapter Black/Yellow lead	190 - 300	2
Stator	Two Yellow leads	0.1 - 1.0	2

Notes:

1 = Requires 4-pin magneto harness adapter P/N 295-000-131

2 = Requires 6-pin magneto harness adapter P/N 295-000-136

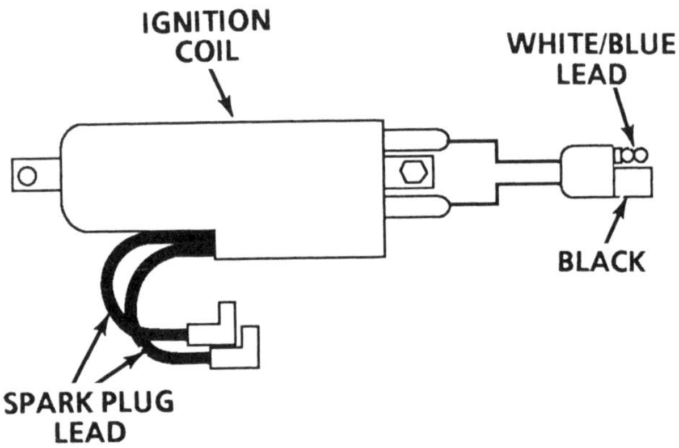

Detailed instructions are presented in the text for testing the primary and secondary resistance of the ignition coil. Lead colors may vary by year and model. Consult the electrical schematic for the model being serviced.

Trigger Coil Test
787 Series Engine Only

The trigger coil is mounted on the inside of the flywheel -- magneto -- cover, adjacent to the charging coil assembly, also mounted on the magneto cover. To check the condition of the trigger coil, first disconnect the magneto electrical harness connector. Install the six-pin electrical harness adapter -- P/N 295-000-136. Connect a ohmmeter lead to the adapter White lead. Connect the other tester lead to the harness adapter Black/Yellow lead. Measure the resistance between the two leads.

The reading should be between 190 - 300 ohms. If the indicated value is not within this range, the trigger coil must be replaced.

7-5 CHARGING CIRCUIT

DESCRIPTION AND OPERATION

The introduction of the 787 engine series brought with it radical design changes in the ignition circuitry. The following description however, covers the basics of a typical charging circuit, applicable even to the 787 engine.

The charging circuit consists of the following items:

- Permanent magnets -- attached to inside perimeter of the flywheel.
- Charging coil -- installed on the magneto assembly.
- A rectifier/regulator -- located inside the battery case cover.
- An external battery.
- Necessary wiring -- to connect the components together.

The **NEGATIVE** side of the rectifier/regulator is always grounded. The **POSITIVE** side is always connected to the battery. The negative terminal on the battery must be connected to a good ground on the engine.

The alternating current generated in the charging coil windings passes to the rectifier/regulator. The rectifier portion changes the alternating current (AC) to direct current (DC) to charge the 12-volt battery. A second function of the rectifier is to smooth the pulses of current to form a steady DC current flow.

The regulator portion of the rectifier/-regulator limits the current produced and sends any excess current back through the charging coil windings.

Electronic components in the charging circuit, including the battery, will last much longer if subjected to a low steady voltage. A sudden voltage surge could overheat and damage charging circuit components.

The magneto assembly is located behind and protected by the flywheel. Therefore the magneto assembly, including the charging coil, seldom causes problems in the charging circuit. When problems do occur, they can usually be traced to the rectifier/regulator or to the battery. Due to its internal circuitry, no static tests may be performed to check the condition of the unit.

If either the charging coil or the rectifier/regulator fails troubleshooting tests, the defective unit cannot be repaired, it **MUST** be replaced.

Testing Charging Circuit Output

Before testing the charging circuit, verify the condition of the battery -- see Chapter 9 -- **ELECTRICAL.**

Move the craft to a test tank or secure it in a body of water next to a float or in a "slip". Secure the craft fore and aft to prevent movement during the following tests.

CAUTION

Water must circulate through the jet pump -- to and from the engine, anytime the engine is operating. Circulating water will prevent overheating -- which could cause damage to moving engine parts and possible engine seizure.

NEVER allow the engine to operate without cooling water for more than 15 seconds.

Amperage Output Test

Obtain an ammeter and set it to the highest DC amp scale. The current flow is much higher when the engine is first started and could very well damage the meter if it were set directly to the operating range scale.

Disconnect the positive battery cable and connect the meter between the disconnected lead and the battery --positive meter terminal to the positive battery terminal and the negative meter terminal to the disconnected battery lead. Connect a tachometer to the engine.

CAUTION

The following tests **CANNOT** be made with a flush attachment connected to the unit. The craft **MUST** be moved to a body of water or to a test tank.

Start the engine and allow it to warm to operating temperature, with the rpm rate stabilized. Now, select the Ax1 amp scale on the meter. Advance the throttle to 5,500 rpm. The meter should register approximate 4 amps -- 587 through 717 engines -- and approximately 5 amps -- 787 engines only, if the rectifier/regulator is functioning correctly. If the rectifier/regulator is faulty, the engine may not be able to reach the required rpm and the ammeter will fail to indicate the proper amperage. If such is the case, check the magneto output -- as indicated below -- prior replacing the rectifier/regulator.

Shut down the engine and disconnect the ammeter.

Voltage Output Test

With the craft in a test tank or body of water, obtain a voltmeter capable of indicating up to 25 DC volts. Connect the voltmeter across the battery terminals, negative-to-negative -- positive-to-positive. Connect a tachometer to the engine.

Start the engine and allow it to warm to operating temperature with the rpm rate stabilized. Now, advance the throttle to 5,500 rpm. If the meter registers over 15 VDC, the rectifier/regulator is defective and **MUST** be replaced. If the meter registers less than 12 VDC, inadequate voltage is being produced to charge the battery.

Check the battery charging coil behind the flywheel -- see Page 7-10 for resistance tests. If the coil is satisfactory, replace the rectifier/regulator.

Ignition Module Test

No resistance test values are provided by the manufacturer for testing the ignition module. If all ignition components are found good, and an ignition malfunction is still suspected, then the module should be replaced.

Replacing a questionable component with a known good one will eliminate a possible problem area, which is difficult to trace, in a defective ignition system.

Stator Test
787 Engine Only

The stator assembly is mounted on the inside of the flywheel -- magneto -- cover. To check the condition of the stator, first disconnect the magneto electrical harness connector. Install the six-pin electrical harness adapter -- P/N 295-000-136. Check the resistance between the two harness adapter Yellow leads with an ohmmeter. The reading should be between 0.1 to 1.0 ohm.

Install either meter lead into the remaining adapter Yellow lead and observe the resistance reading.

If the indicated values are not within 0.1 through 1.0, the stator assembly **MUST** be replaced.

7-6 FLYWHEEL AND MAGNETO

The following short section lists the procedures required to "pull" the flywheel and remove the stator plate in order to service the ignition system. Cleaning and Inspecting procedures in addition to proper assembling and installation instructions are also included.

Alternate Flywheel Holder

If the special flywheel holder listed is not available, a tool may be fabricated from a metal bar and two long bolts. The longer the bar, the more leverage gained, but still limited by the confines of the engine compartment. The bolts should be of sufficient length to allow a socket and a torque wrench to be placed over the flywheel nut and still clear the bar. Drill two holes, large enough for the bolts to pass through, at the end of the bar, with the same distance between as the two holes in the flywheel. Secure the bolts to the bar.

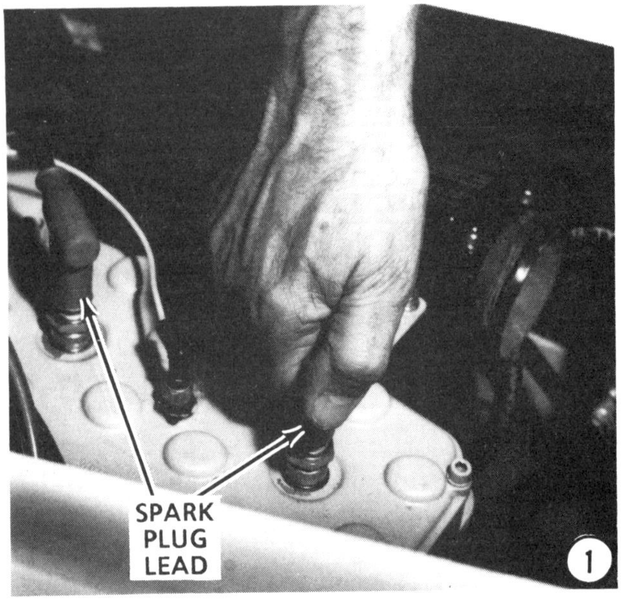

Alternate Flywheel Puller

If the special flywheel puller listed is not available, a similar puller, shown in the accompanying illustration, may be used **PROVIDED** it will pull from the bolt holes in the flywheel and **NOT** from around the perimeter of the flywheel. **NEVER** attempt to use a puller which would put a strain on the outside edge of the flywheel.

"PULLING" THE FLYWHEEL

1- Remove the spark plug high tension leads, and then remove the spark plugs.

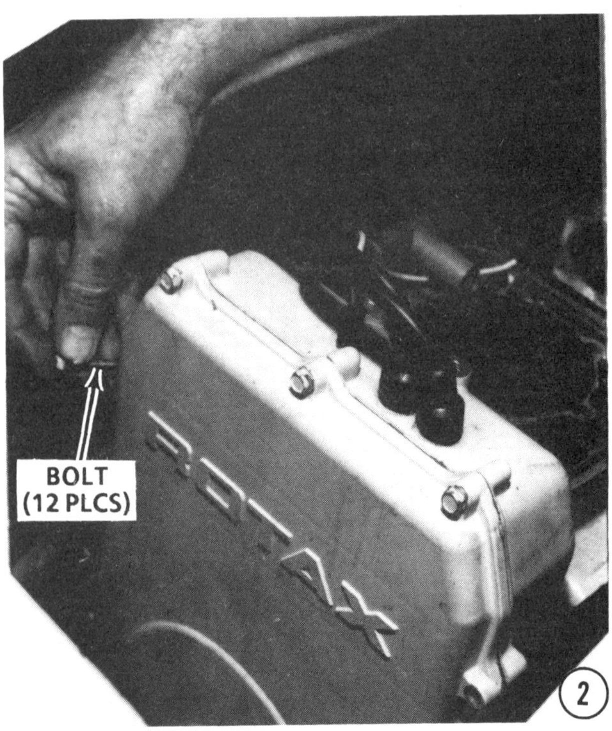

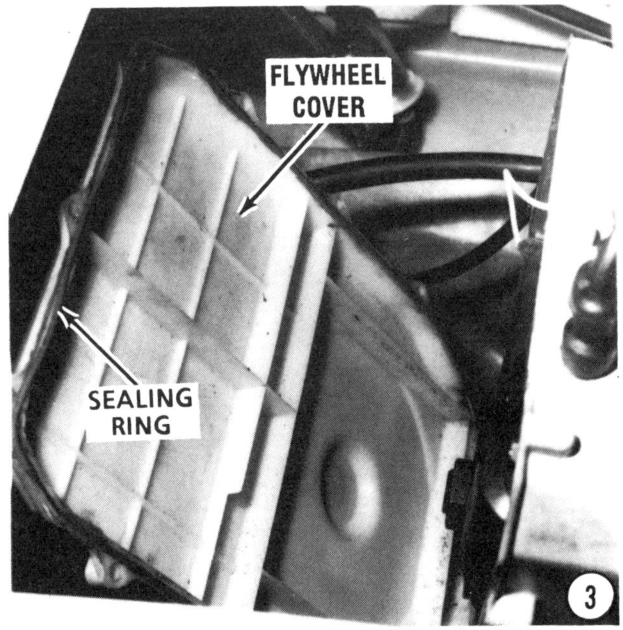

AUTHORS' WORDS

For photographic clarity, the illustrations in this section were made while the work was being performed on a work bench. The ignition system components are housed within a separate electrical box on all models covered. The following procedures are valid for all models covered, and may be performed while the engine is in the craft.

Flywheel Cover Removal

2- Remove the 10 bolts securing the flywheel cover to the magneto housing.

3- Lift off the cover. Remove and discard the sealing ring. This type ring cannot be used a second time -- once used, it will never seal properly again.

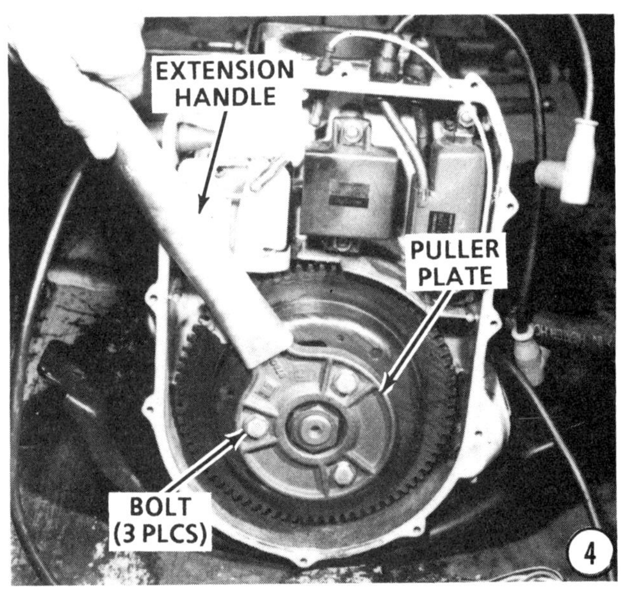

4- Obtain Puller Plate P/N 290-876-080, Extension Handle P/N 290-000-003, and three M8x25mm bolts. Install the plate onto the flywheel using the three bolts. Tighten the bolts securely.

Install the extension handle onto the puller plate arm.

5- Install a 22mm thin-walled socket over the flywheel nut. The socket **MUST** fit inside the puller plate hole.

If the only available socket will not fit into the puller plate hole, grind down the outside diameter of the socket until it will fit into the hole. **DO NOT** grind on the internal threads of the puller plate.

Place a breaker bar into the socket and rotate it **COUNTERCLOCKWISE** to "break" the flywheel nut free. Remove the nut and washer.

6- Obtain Puller and Puller Bolt P/N 290-876-065. Thread the bolt into the puller until just a couple threads are engaged. Thread the puller into the internal threads of the puller plate. Thread the bolt into the puller until the bolt makes contact with the end of the crankshaft.

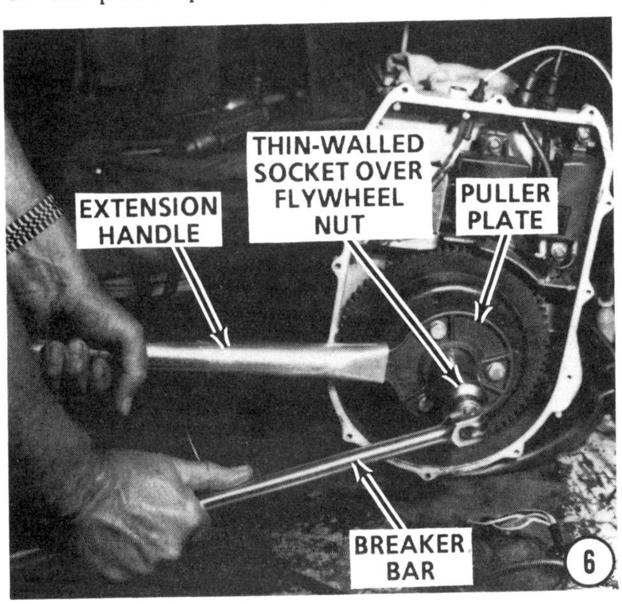

7- Obtain the correct size wrench or socket and place it over the puller bolt.

Now, continue to tighten the bolt and at the same time, shock the crankshaft with a gentle to moderate tap on the end of the bolt, with a soft head mallet. This shock will assist in "breaking" the flywheel loose from the crankshaft.

DO NOT "pound" on the bolt. The force of such action would be transmitted to the seals and bearings along the crankshaft and cause damage. **NEVER** strike the flywheel directly, because this type of shock would surely cause the flywheel magnets to loose their magnetism.

8- Pull the flywheel free of the crankshaft. Remove and **SAVE** the Woodruff key from the crankshaft keyway.

SPECIAL WORDS ON WOODRUFF KEYS

Inspect the condition of the Woodruff key and its keyway in the crankshaft. If the key shows any sign of shearing, it is probably caused by jumping waves with the watercraft and the effect will be a loss of power.

When the craft is clear of the water, the engine rpm approaches the limit set by the rpm limiter in an unloaded condition.

When the craft returns to the water surface, the sudden load on the impeller is transferred along the driveline. Speed of the rotating driveshaft and the crankshaft is suddenly reduced while the flywheel continues to spin rapidly, due to its mass. This condition, especially when repeated over a period of time, often shears the Woodruff key and the flywheel becomes repositioned on the crankshaft.

A very small misalignment will result in a change in ignition timing. The craft operator will notice a drastic loss of power. A worse condition would be the flywheel shifting to the degree engine start would not be possible.

Stator Plate and Magneto Assembly Removal

9- The securing screws pass through elongated slots in the magneto assembly. The assembly could be installed as much as 15° off its original position -- drastically affecting the timing characteristics of the engine. Therefore, at the elongated slot in the magneto assembly, make a set of aligned punch marks -- one on the magneto assembly -- the other on the stator plate. These marks will serve as an aid during installation to **ENSURE** correct positioning of the magneto assembly on the stator plate.

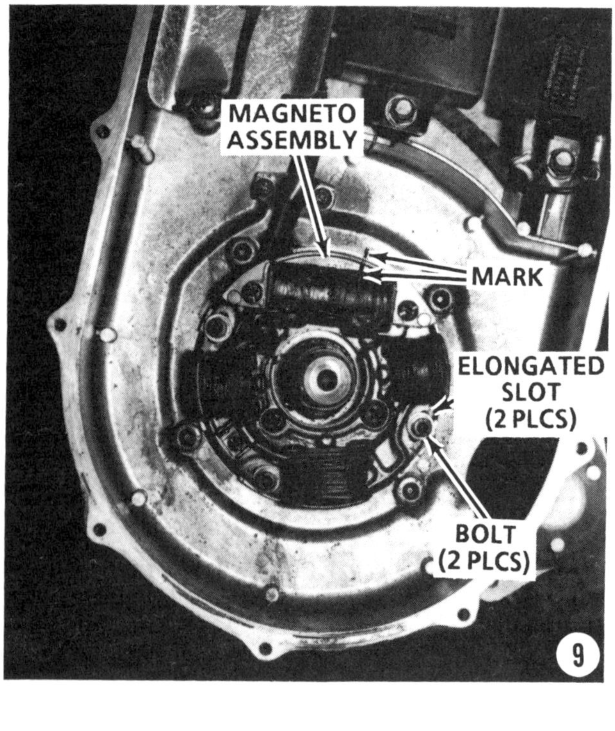

After the stator plate is removed from the engine, all internal and attached components are easily removed and installed.

Remove the two screws securing the magneto assembly to the stator plate. These screws must be removed because they are threaded into the crankcase.

Generating Coil Removal and Installation

Lift the magneto assembly off the crankshaft and turn it over. Obtain a blow torch and heat the two screw holes, shown in the accompanying illustration, to break the Loctite bond. **TAKE CARE** not to melt the protective sleeve on the magneto assembly harness lead.

Remove the two securing screws. Uncrimp and unsolder the Black/Red lead from the coil end tab.

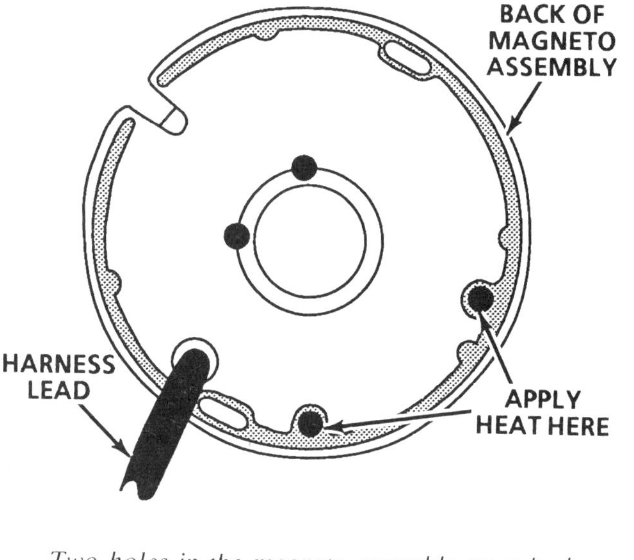

Two holes in the magneto assembly must be heated in order to "break" the Loctite bond and enable removal of a generating coil. This simple line drawing depicts the back side of the magneto assembly with the two holes to be heated identified. Text explains all.

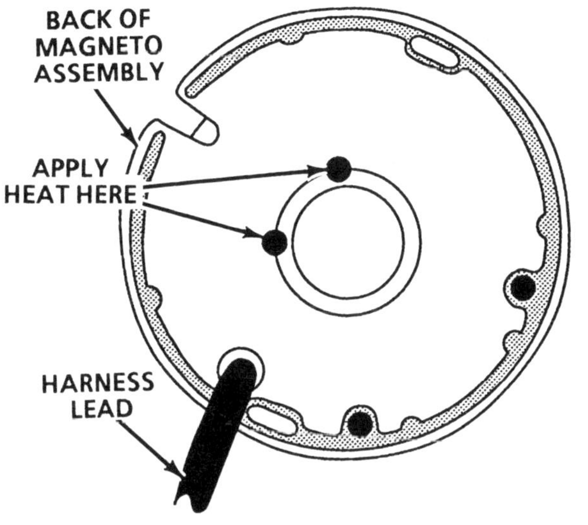

Simple line drawing to depict the back side of the magneto assembly. The two holes to be heated in order to "break" the Loctite bond and remove a battery charging coil, are identified.

GOOD WORDS
COIL CENTERING

Obtaining the magneto coil centering tool -- P/N 290-876-922 is **STRONGLY** recommended. This device **PREVENTS** the coils from moving out of proper alignment during installation.

Remove the shipping nuts from the new generating coil and install the coil over the plate. Strip the end of the old lead, and then crimp and solder the new coil. Apply Loctite 242 onto the threads of the two securing screws. Align the marks made during assembly. Obtain magneto coil centering tool -- P/N 290-876-922 -- and install it to fit around the armature plate before tightening the screws. Now, tighten the screws very **SECURELY**. Remove the coil centering tool.

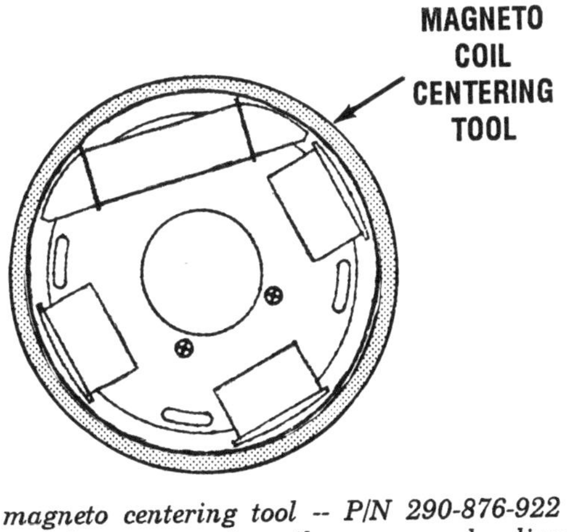

The magneto centering tool -- P/N 290-876-922 -- MUST be used to ensure the coils are properly aligned during assembly.

Battery Charging Coil
Removal and Installation

Use a blow torch and heat the two screw holes, shown in the accompanying illustration to break the Loctite bond. **TAKE CARE** not to melt the protective sleeve on the magneto assembly harness lead. Remove the two securing screws.

On some models, plastic tubes are used to protect the coil leads. Remove the wire retaining clip from the plastic tube. Pull back the White and Black plastic tubes from the spliced connections. Make a note indicating the positioning of each coil lead color and location. Uncrimp and unsolder the Yellow and Yellow/Black leads from the coil. Uncrimp and unsolder the Black ground lead from the coil core. Disconnect the leads from the remaining coils. Remove the three coils.

To install the new coils, begin by simply placing the three coils over the magneto assembly plate, with the bolt holes aligned. Next, crimp and solder the Yellow lead from the new coil to the connector with the White plastic tube. After the soldering is complete and has cooled, pull the tube over the splice. Crimp and solder the Yellow/Black lead from the new coil to the connector with the Black plastic tube. After the solder has cooled, pull the tube over the splice. Crimp and solder the Black ground lead to the proper connector.

Install the lead retaining clip over the tube. Check to be sure the ground lead is attached to the retaining clip. Apply Loctite 242 onto the threads of the two securing screws. Obtain magneto coil centering tool -- P/N 290 876 922 -- and install it to fit around the armature plate before tightening the screws. Now, tighten the screws very **SECURELY**.

CLEANING AND INSPECTING
ALL ENGINES

Inspect the flywheel for cracks or other damage, especially around the inside of the center hub. A damaged flywheel can be **EXTREMELY** dangerous. The forces exerted on the flywheel while it is spinning at high rpm are sufficient to shatter an unbalanced damaged flywheel. Inspect the magnets to see if they are held tightly around the outer circumference of the flywheel rim. Check the surface of the magnets for fractures and grooves.

Check the flywheel ***CAREFULLY****, to be sure metal particles have not become stuck to the magnets.*

Grooves around the inner perimeter would indicate a component mounted on the stator plate is not aligned properly or the component is loose. Fractured or damaged magnets cannot be replaced. Both the magnets and flywheel share the same part number, therefore attempting to remove the magnets would be a pointless task. If magnets or the flywheel is defective, the entire flywheel must be replaced.

Check to be sure small metal particles have not become attached to the magnets. Verify each magnet has good magnetism by using a screwdriver or other suitable tool.

Thoroughly clean the inside taper of the flywheel and the taper on the crankshaft to prevent the flywheel from "walking" on the crankshaft during operation.

Test the magneto assembly to verify it is not loose. Attempt to lift each component on the plate. There should be little or no evidence of movement.

SPECIAL WORDS ON FLYWHEEL MAGNETS

The outer ends of any magnet are called poles. One end is the north pole and the other end is the south pole. The magnetic field surrounding a magnet is concentrated around these two poles.

Some flywheel magnets are fairly long and curved around the outer perimeter of the underside of the flywheel. Others are short and are mounted around the center hub, depending on the location of the coils mounted on the stator. Magnets are usually installed in pairs with the north pole of one adjacent to the south pole of its neighbor and so on. In this manner continuous magnetic field surrounds the inside of the flywheel.

If a flywheel is accidently dropped, not only could the teeth be damaged, but the impact will weaken the magnetic strength of all the magnets housed in the flywheel.

If one or more of the magnets should break or fracture, two new magnetic poles will be created. A long magnet with two poles will become two short magnets with four poles. The new poles will possess only half the magnetic strength of an original pole. The overall magnetic field will be altered. The new field of the shorter magnets will not extend to cover the area of the flywheel.

Serious consequences apply to a CDI type ignition system in the event of a flywheel magnet fracture. Charging and generating coils are energized by the concentrated magnetic field at the magnet poles. If new poles are suddenly created, the ignition module will receive conflicting signals from the magnets and may even attempt to fire the cylinder twice in one revolution.

All these reasons require the flywheel to be handled with **CARE.**

If there is evidence of oil on the magneto assembly, the oil seal/s at the forward main bearing inside the crankcase may have failed. Refer to Chapter 8, **ENGINE.**

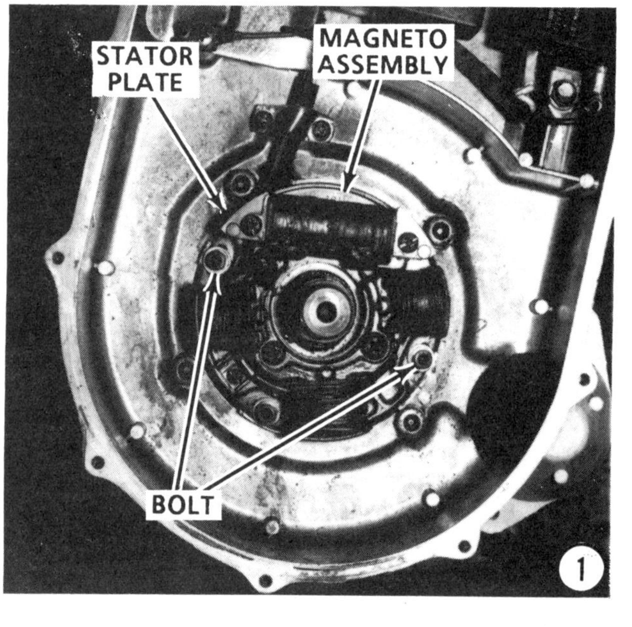

ASSEMBLING AND INSTALLATION

Magneto Assembly

1- Coat the threads of the two attaching Allen head bolts with Loctite 242. Install the magneto assembly onto the stator plate. Align the marks made during disassembling. Install the bolts, washers, and lockwashers. Tighten the bolts alternately and evenly to a torque value of 16 ft lb (22Nm). Check to be sure the two marks remained aligned while the bolts were tightened.

Flywheel Installation

2- Place a dab of thick lubricant onto the curved surface of the Woodruff key to hold it in place while the flywheel is being installed. Press the Woodruff key into place in the crankshaft recess. Wipe away any excess lubricant to prevent the flywheel

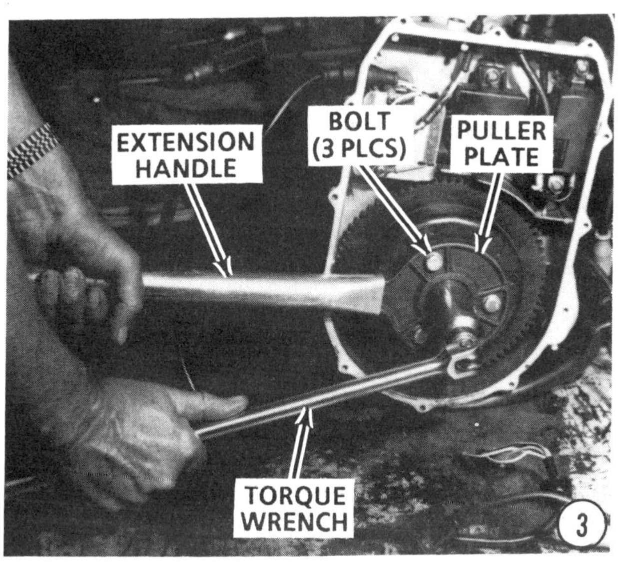

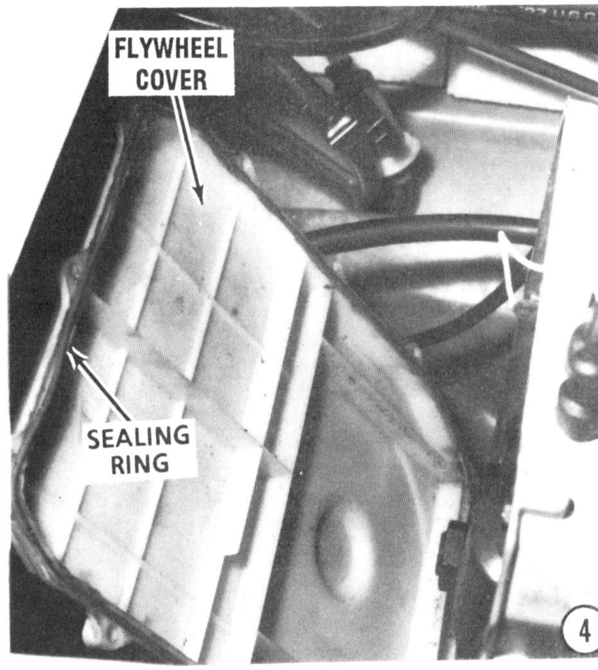

from "walking" during engine operation. Check the flywheel magnets to ensure they are free of any metal particles. Double check the taper in the flywheel hub and the taper on the crankshaft to verify they are clean and free of any oil.

Now, slide the flywheel onto the crankshaft, with the Woodruff key in place. Rotate the flywheel **COUNTERCLOCKWISE** to be sure it does not contact any part of the magneto assembly or wiring on the stator plate.

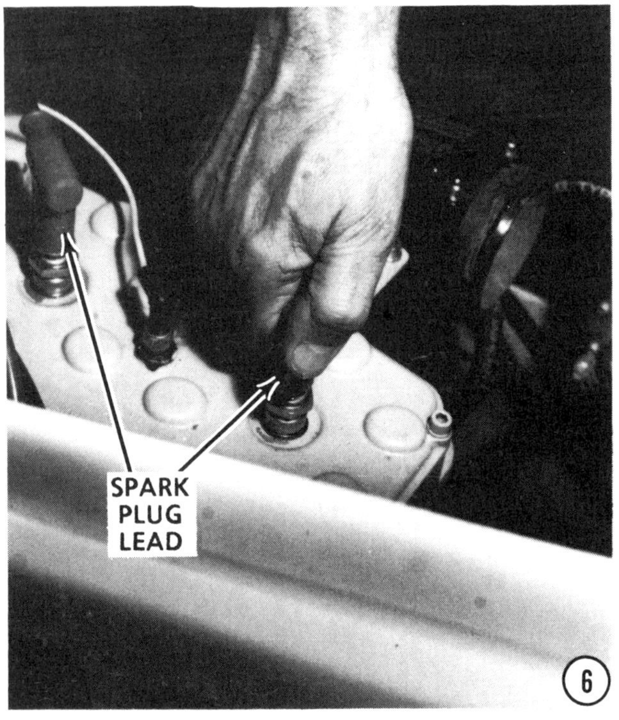

3- Attach Puller Plate P/N 209-876-065 to the flywheel with the three M8x25mm bolts. Install the extension handle onto the puller arm. Apply Loctite 242 onto the crankshaft taper and threads. Install the flywheel nut and washer. If servicing an engine manufactured in 1992 or 1993, tighten the nut to a torque value of 77 ft lbs (105Nm). If servicing an engine manufactured after 1993, tighten the nut to a torque value of 107 ft lbs (145Nm).

4- Apply Loctite 242 to the threads of the ten 10mm magneto housing cover securing bolts. Install a new sealing ring into the groove of the housing cover.

5- Install the cover and secure it in place with the attaching bolts. If servicing an engine manufactured from 1992 through 1994 inclusive, tighten the bolts to a torque value of 44 in lbs (5Nm). If servicing an engine manufactured after 1994, tighten the bolts to a torque value of 80 in lbs (9Nm).

6- Attach the high tension leads to the spark plugs.

7-7 IGNITION TIMING ADJUSTMENTS

The electronic ignition systems used in all Bombardier watercraft covered in this manual have no moving or sliding parts. Typically, adjustment is not required. Ignition advance timing is accomplished electronically. Signal output increases proportionally to engine speed. The timing is set at the factory and should never require adjustment **UNLESS** the magneto assembly becomes misaligned, a flywheel is replaced, or an ignition component has been replaced.

Therefore, the purpose of this section is to give instruction on how to verify the flywheel timing mark position and to give procedures to correctly time a new flywheel which has been purchased and does not have a timing mark.

A new flywheel will always be indexed with the crankshaft, because it is installed over the Woodruff key -- forward flywheel only. **HOWEVER**, the timing mark on the magneto assembly may not match with the characteristics of the new flywheel.

IGNITION WORDS 787 ENGINE ONLY

If timing ignition adjustment is necessary, the procedures **MUST** be performed with a special MPEM programmer unit. This device is **NOT** available as a retail item -- it is available to dealers only. Therefore, if the ignition requires adjustment, the craft must be taken to a dealer. If the special programmer is available, follow the instructions included with the unit.

Craft Location for Timing Adjustments

The watercraft **MUST** be moved to a test tank or to a slip in a body of water where it can be restrained from moving with adequate mooring lines.

Determining TDC (Top Dead Center)

Remove the forward spark plug. Install a dial indicator into the spark plug opening and determine TDC of the piston.

If a dial indicator is not available, use a pencil, straw, or equivalent, resting on top of the piston dome. Determine the highest point to which the inserted object rises when the crankshaft is **SLOWLY** rotated.

Timing Marks

Remove wingnuts securing the aft flywheel cover to the mounting bracket. Remove the flywheel cover, and then install the timing mark pointer tool P/N 295-000-130 -- all engines except 787 series -- using the wing nuts previously removed. Tighten the wingnuts securely, to ensure the pointer tool bracket will remain stationary during the dynamic test.

If servicing a 787 engine, remove the middle bolt or screw securing the engine to the aft engine mount. Install the pointer tool P/N 295-000-135 with the screw/bolt just removed.

Refer to the Tuneup Adjustments Tables in the Appendix for BTDC data on the engine being serviced. With the piston at TDC -- as explained in the previous paragraphs -- zero the dial indicator gauge. Rotate the flywheel **COUNTERCLOCKWISE** about 1/4 revolution, then rotate the flywheel **CLOCKWISE** until the reading on the dial indicator is at the value obtained from the ignition timing table. For example, a 1997 SP model is set at .102" (2.59mm) BTDC. Now, scribe a thin mark on the **AFT** flywheel in the middle of the alignment tool slot -- all but 787 engines -- or **EXACTLY** aligned with the timing mark pointer tool -- 787 engines only. This mark will be used as the reference mark while using a stroboscopic timing light.

Remove the dial indicator from the spark plug hole. Install the spark plug and tighten it to a torque value of 17 ft lbs (24Nm). Install the spark plug cap, check to be sure it is fastened securely.

If the marks align -- proceed to the paragraph under the heading "Adjusting Ignition Timing".

If the marks do not align scribe a new mark on the flywheel which does align with the timing mark on the guard. Obliterate the old flywheel mark, leaving only the new mark. Do not rely on the engine having correct timing just because the marks align. The timing **MUST** be verified dynamically as the final check for mark alignment. With the engine in a static condition -- not operating -- shut down -- the marks may be aligned, but the magneto assembly may have changed position on the stator plate. Such a change makes the mark on the flywheel guard meaningless.

The timing mark on the flywheel guard and the timing mark on the flywheel will be compared during the following timing procedure. **THEREFORE,** make any new marks as close together as practical -- extend the embossed mark, if necessary.

Dynamic Testing

Obtain the Bombardier timing light P/N 529-031-900. If this device is not available, be **SURE** to use **ONLY** a stroboscopic light rated up to 6000 rpm, otherwise the reading will surely be inaccurate -- all but 787 series engine, timed at 3500 rpm.

Connect the timing light to the battery and hook the pickup around the forward spark plug high tension lead. Check to be sure the direction of the embossed arrow on the timing light pickup is **TOWARDS** the spark plug.

Connect a Bombardier inductive tachometer -- P/N 295-000-100 -- to the other spark plug high tension lead. If the manufacturers' tachometer is not available, follow the instructions supplied with the alternate device. The tachometer leads may need to be connected across the battery charging leads -- refer to the Resistance Tests chart on Pages 7-10 and 7-11.

SPECIAL WORDS
IGNITION TIMING

The ignition components are affected by temperature, therefore, it is **IMPERATIVE** to check the timing when the engine is cold. This is true of all watercraft equipped with a CDI ignition system. The manufacturer recommends idling for no more than 20 seconds before conducting the ignition check.

In a test tank or with the craft in a body of water and restrained with adequate mooring lines to prevent the craft from moving, start the engine and point the light perpendicular and straight in line with the timing tool slot. Bring the engine to 6000 rpm -- all engines except the Model 717D Series. On the 717D Series only, bring the up to 6500 rpm.

Aim the timing light at the timing alignment tool. If the mark on the flywheel aligns with the pointer device, the timing is set correctly. **HOWEVER**, if the mark on the flywheel is not

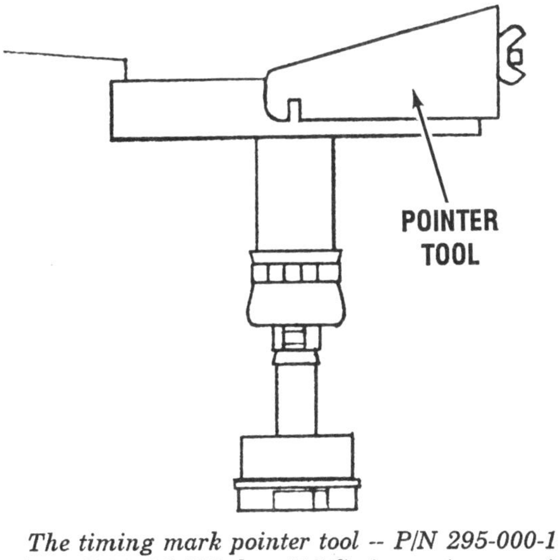

The timing mark pointer tool -- P/N 295-000-102 -- used on all Model 587 thru 717 Series engines provides a stable locator for checking timing alignment.

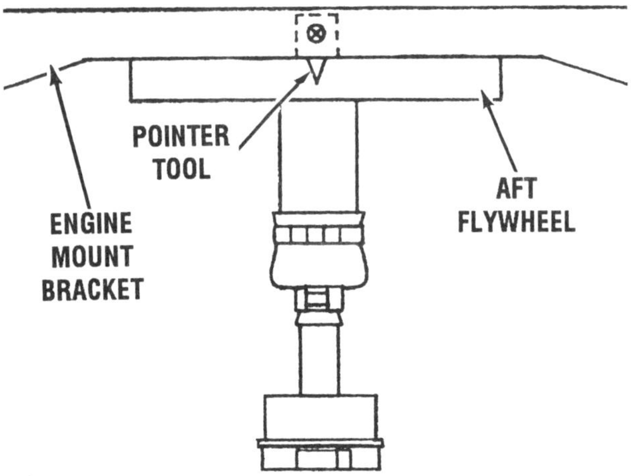

The pointer tool P/N 295-000-135 used on Model 787 Series engines only. The tool is secured to the aft engine bracket using the existing bolt. Check to be sure the pointer is fastened securely.

aligned with the stationary mark -- a timing adjustment is required. Shut down the engine and perform the following procedures outlined under "Adjusting Ignition Timing".

Adjusting Ignition Timing
All Engines Except 787

Ignition timing adjustment is performed by altering the position of the magneto assembly in relation to the stator plate. The magneto assembly is secured to the stator plate with two screws passing through elongated slots. The elongated slots permit the magneto assembly to be rotated within a small limit.

Unfortunately, the flywheel must be removed to gain access to the two screws and alter the position of the magneto assembly. See Section 7-6, beginning on Page 7-14, to remove the flywheel.

After the flywheel has been removed, loosen but **DO NOT REMOVE** the three magneto assembly bolts. If this is the first work performed since the unit left the factory, these bolts are almost impossible to loosen "cold". While attempting to rotate the screwdriver **COUNTERCLOCKWISE**, tap the end of the screwdriver with a hammer to jar the bolt loose.

Once the bolts have been loosened slightly, very **CAREFULLY** rotate the magneto assembly either clockwise or counterclockwise, as necessary to correct the timing. Move the magneto assembly the same amount as the aft flywheel would be moved to align the marks.

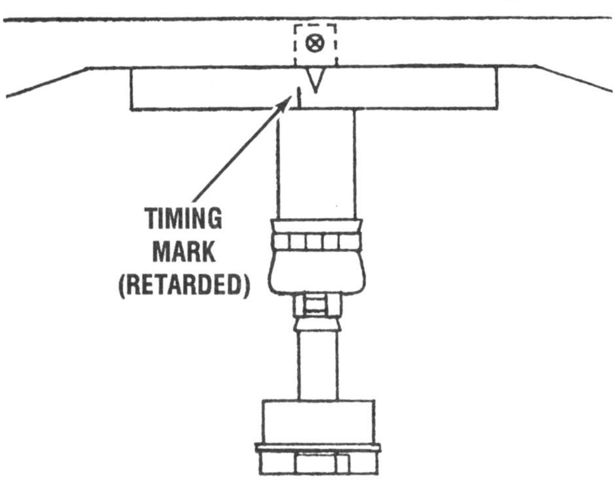

*If the timing mark is **LEFT** of the pointer slot, the timing is **RETARDED**. The magneto must be rotated **COUNTERCLOCKWISE** to advance the timing.*

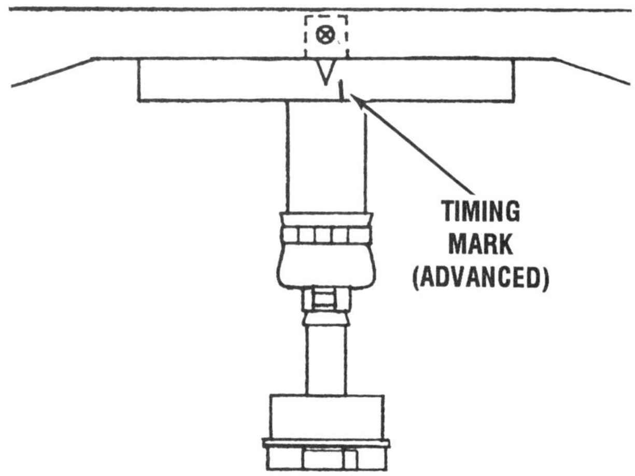

*If the mark is **RIGHT** of the pointer slot, the timing is **ADVANCED**. The magneto must be rotated **CLOCKWISE** to retard the timing.*

*The magneto screws must be loosened, but **NOT** removed prior to rotating the magneto on the stator plate. Rotate the magneto the same amount the aft flywheel would be moved to align the mark.*

Rotate the magneto assembly **CLOCKWISE** to retard spark advance. Rotate the assembly **COUNTERCLOCKWISE** to advance the spark.

SPECIAL WORDS ON SPARK ADVANCE

The ignition module used on all Bombardier Sea-Doo engines allows for a timing decrease in advance as engine speed increases. However, at 6000 rpm, when the marks align, and this procedure is "trial and error", tighten the three attaching bolts very **SECURELY**. If these screws should happen to vibrate loose, they could jam between the rotating flywheel magnets and the coils on the magneto assembly -- causing very serious and expensive damage. Install the flywheel. Refer to Steps 2 thru 6, beginning on Page 7-18.

WORDS FROM EXPERIENCE ALL ENGINES

The two most important items in establishing and maintaining perfect timing on an engine are the magneto assembly securing screws and the crankshaft Woodruff key.

Magneto Assembly Securing Screws

The magneto assembly is mounted on top of the stator plate and has two elongated slots for the securing screws to pass. If the magneto assembly is allowed to rotate just a small amount because the securing screws have worked loose, the ignition timing may be off as much as 15°. The symptoms of this misalignment are loss of power and difficult starting.

Crankshaft Woodruff Key

The same symptoms apply to a sheared Woodruff key allowing the flywheel -- actually the flywheel magnets -- to become out of position on the crankshaft.

During wave jumping or using a wet ramp, when the craft clears the water, the engine RPM approaches the limit set by the RPM limiter in an unloaded condition.

Once the craft returns to the water surface, the sudden load on the impeller is transferred along the driveline. Speed of the rotating pump shaft and the crankshaft is suddenly reduced while the flywheel continues to spin rapidly due to its mass. This condition, especially when repeated over a period of time, often shears the Woodruff key and the flywheel becomes repositioned on the crankshaft.

A very small misalignment will result in a change in ignition timing. The craft operator will notice a drastic loss of power. A worse condition would be the flywheel shifting to the degree engine start would not be possible.

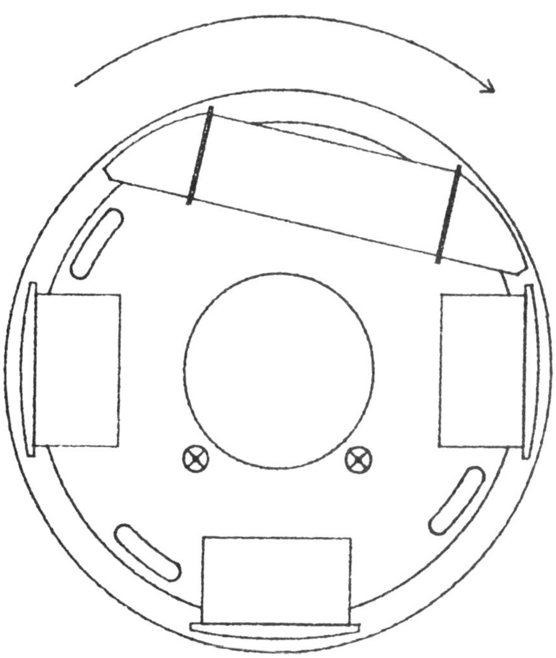

*Rotate the magneto assembly **CLOCKWISE** to retard the ignition timing.*

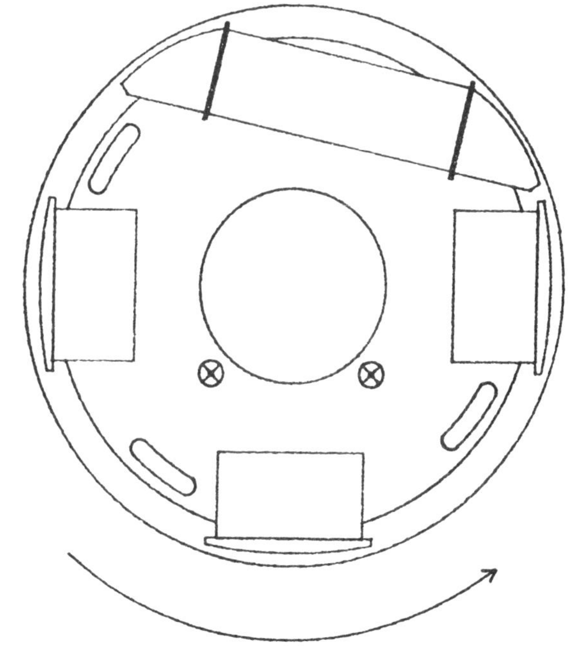

*Rotate the magneto assembly **COUNTERCLOCKWISE** to advance the ignition timing.*

DESS System

The Digitally Encoded Security System -- DESS -- is a unique feature providing anti-theft protection. The basic components of this system include the MPEM, safety lanyard cap, and safety lanyard switch, or DESS post.

Featured on several models, this system may be installed as a kit on any Bombardier Sea-Doo personal watercraft equipped with the Multi Purpose Electrical Module -- MPEM -- unit. The safety lanyard has a cap containing a magnet and Read-Only-Memory (ROM) chip, which stores a unique code. Without the safety lanyard, the watercraft will not start. Due to the processes involved in activation, the watercraft **MUST** be taken to a dealer for the initial set up of a kit installation.

The memory of the DESS "unit" -- actually the MPEM -- is permanent. If the battery is disconnected, the security information remains intact. If the MPEM must be replaced, be **SURE** to inform the dealer the replacement MPEM is used in conjunction with a DESS system.

Two self-diagnostic tests are available for checking the status of the DESS and MPEM system. The first test -- Basic -- may be performed at any time and is listed below. The second test however, -- Advanced -- may only be performed with the special MPEM programmer unit -- not available for retail sale -- therefore, the watercraft must be taken to a dealer to conduct this test.

Basic DESS Self-Diagnostic Test

SIGNAL	PROBLEM	SOLUTION
2 short beeps while installing the safety lanyard on the switch -- DESS post	The safety lanyard is recognized by the MPEM	Normal condition, ignition is authorized.
1 long beep while installing the safety lanyard on the DESS post or pressing the start/stop button	Bad connection between the safety lanyard cap and the DESS post	Remove, then install the safety lanyard on the switch until 2 short beeps are heard to indicate the system will allow ignition.
	Unprogrammed or defective safety lanyard	Be sure the lanyard unit is programmed for the specific watercraft. If it does not work after verification, check to be sure the inside of cap is clean and dry.
	Improper operation of MPEM or defective wiring harness	The unit should be taken to a dealer for advanced diagnostics.
8 short beeps	Defective MPEM	Replace the MPEM unit.
Continuous beep	Engine overheating	Refer to Chapter 5 -- TROUBLESHOOTING

8
ENGINE

8-1 INTRODUCTION AND CHAPTER ORGANIZATION

This chapter includes three basic Rotax engines powering the Sea-Doo Personal Watercraft covered in this manual. Two of the engines -- the 587 and the 717 are almost identical two cylinder engines. Both are installed in the craft in what might be considered "a normal upright" position.

The third engine is the Rotax 787 two-cylinder unit. This power plant is installed at an approximate 30° angle to the port side of the craft. One of the advantages of canting the engine at an angle is lowering the center of gravity, resulting in a more stable ride for the occupants. Naturally, the engineers have a host of reasons for the unique installation.

The chapter is divided into five main working sections as follows:

8-2 Description and operation of a two-stroke engine.

8-3 Complete service -- Removal, Disassembly, Assembly and Installation of a 587 or 717 engine.

8-4 Complete service -- Removal, Disassembly, Assembly, and Installation of a 787 engine.

8-5 Cleaning and Inspecting -- all engines.

8-6 Sealants, Adhesives, Lubricants, and Fuel Stabilizers.

The carburetion and ignition principles of two-stroke engine operation must be understood in order to perform service work on the marine engine covered in this manual.

REPAIR PROCEDURES

Service and repair procedures will vary slightly between individual models, but the basic instructions are almost identical. Any variations between models are clearly indicated. Special tools may be called out in certain procedures. These tools may be purchased through the local dealer. In certain instances, an alternate tool may be obtained or fabricated.

TORQUE VALUES

All torque values given in the text must be met to ensure the overhaul work will be completed for the desired results. Torque values for various parts of the engine are given in the text and a complete list is presented in the Appendix.

A torque wrench is essential to correctly assemble the engine. Never attempt to assemble an engine without a torque wrench. Attaching bolts must be tightened to the required torque value in two progressive sequences, following the specified order. On the first sequence, tighten the bolts to half the torque value. On the second sequence, tighten the bolt to the full torque value.

OTHER ENGINE COMPONENTS

Service procedures for the carburetor, oil injection system, fuel pump, cranking motor and other components are presented in separate chapters. See the Table of Contents. Due to the unique design of these engines, a rotary valve is used in place of a reed valve system.

CLEANLINESS

Make a determined effort to keep parts and the work area as clean as possible. Parts **MUST** be cleaned and thoroughly inspected before they are assembled, installed, or adjusted. Use proper lubricants, or their equivalent, whenever they are recommended.

8-2 TWO-STROKE ENGINE DESCRIPTION AND OPERATION

Two-stroke engines utilize an arrangement of port openings to admit fuel into the combustion chamber and to purge the exhaust gases after burning has been completed. The ports are located in a precise pattern in order for them to be opened and closed at an exact moment by the piston as it moves up and down in the cylinder bore. The exhaust port is located slightly higher than the fuel intake port. This arrangement opens the exhaust port first as the piston starts downward and therefore, the exhaust phase begins a fraction of a second before the intake phase.

LUBRICATION

A two-stroke engine is lubricated by mixing oil with the fuel. Therefore, various parts are lubricated as the mixture passes through the crankcase and the cylinder. All Bombardier Sea-Doo engines are equipped with an oil injection system which delivers oil directly into the intake manifold.

Oil Injection Pump

On all engines, **EXCEPT** the Model 787 Series, the oil injection pump is mounted on the rotary valve housing and driven by a rotary

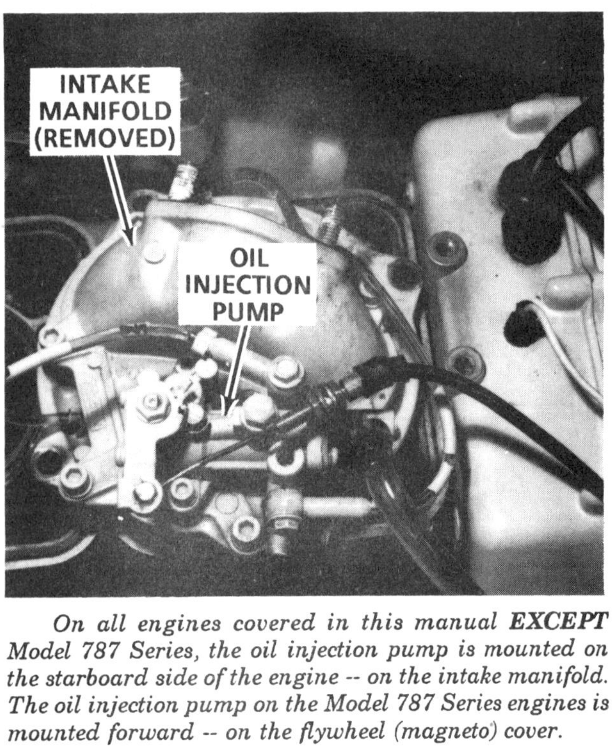

On all engines covered in this manual ***EXCEPT*** *Model 787 Series, the oil injection pump is mounted on the starboard side of the engine -- on the intake manifold. The oil injection pump on the Model 787 Series engines is mounted forward -- on the flywheel (magneto) cover.*

valve shaft which meshes with a gear on the crankshaft.

On the Model 787 Series engines, the pump is mounted on the flywheel cover and driven by a short coupling indexed into the forward end of the crankshaft.

In both cases, the rate of oil injection is controlled by engine rpm.

Physical Laws

The two-stroke engine is able to function because of two very simple physical laws.

1- Gases will flow from an area of high pressure to an area of less pressure. A tire blowout is an example of this principle. The high pressure air escapes rapidly if the tire is punctured.

2- If a gas is compressed into a smaller area, the pressure increases, and if a gas expands into a larger area, the pressure decreases.

If these two laws are kept in mind, the operation of the two-stroke engine will be more easily understood.

ACTUAL OPERATION TWO-STROKE ENGINE WITH ROTARY VALVE

On a four-stroke engine, the four strokes are: intake, compression, power, and exhaust. On a two-stroke engine: the first stroke shares intake and compression, and the second stroke shares power and exhaust. Beginning with the piston at the start of the upward stroke -- the intake stroke -- the exhaust port is closed by the piston; the rotary valve window is over the intake port; and the oil/air/fuel mixture enters the crankcase/combustion chamber.

As the piston travels up the bore, approaching TDC, the spark plug fires and the compressed mixture is ignited -- when the compression stroke is completed. As the piston passes TDC and begins to move downward -- on the power stroke -- the combustion chamber is filled with burning gases. The exhaust port is uncovered. The gases, which are under tremendous pressure, escape rapidly through the exhaust port. When the piston reaches 52° ATDC, the intake port is completely closed by the rotary valve. The piston completes the exhaust stroke when it reaches BDC (bottom dead center). At this instant, both intake and exhaust ports are closed.

As soon as the piston reaches 132° BTDC, the rotary valve window begins to open the intake port. The intake port remains fully or partially open for 184° (132° plus 52°) of crankshaft rotation -- more than half the two-stroke cycle. The intake port is fully open by the time the piston completes its intake portion of the upward stroke.

The cycle then repeats, as described.

TIMING TWO-STROKE ENGINES

The exact time of spark plug firing depends on engine rpm. At low rpm, the spark is retarded, fires later than when the piston is at or beyond TDC. Engine timing is built into the unit at the factory. At high rpm, the spark is advanced, fires earlier than when the piston is at TDC. No adjustment to this timing is necessary, unless the magneto assembly has been removed.

8-3 COMPLETE SERVICE -- MODEL 587, 657 AND 717 ENGINE

Before commencing any work on the engine, an understanding of two-stroke engine operation will be most helpful. Therefore, it would be well worth the time to study the principles of two-stroke engine operation, as outlined briefly in the previous section -- Section 8-2. A Polaroid, or equivalent instant type camera is an extremely useful item, providing the means of accurately recording the arrangement of parts and wire connections before the disassembly work begins. Such a record is most valuable during assembling.

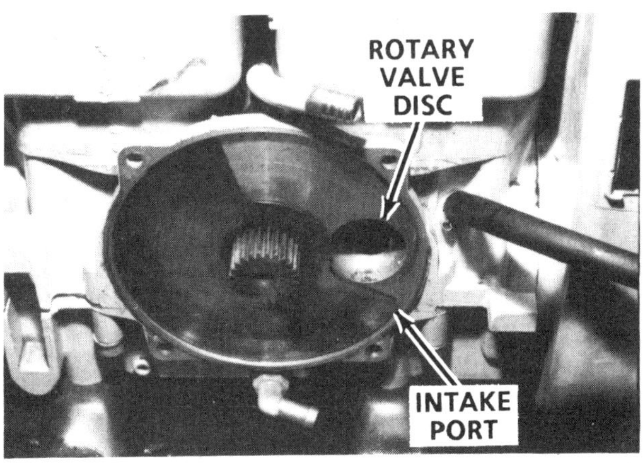

The rotary valve is a very unique feature to the Rotax engine powering the Bombardier Sea-Doo watercraft. A detailed description and explanation of the operating principles of this valve is presented in the text.

PRELIMINARY TASKS

Disconnect the negative and positive leads from the battery. Disconnect the throttle and choke cables from the carburetor.

Disconnect the fuel hoses and remove the carburetor. See Chapter 6 -- **FUEL.**

Remove the driveshaft from the jet pump. See Chapter 10 -- **JET PUMP.**

ILLUSTRATION WORDS

The illustrations presented throughout this section represent a typical installation for the engines covered. The size, shape, and location of certain components will vary depending on the model engine being serviced. **HOWEVER,** the procedures are valid and the differences are usually mentioned in the text.

ENGINE REMOVAL

FIRST, THESE WORDS

The cylinder head, pistons, cylinder bores, oil pump, stator plate, rotary valve disc, and the flywheel may all be serviced while the engine remains in the craft. **HOWEVER,** to service the rotary valve shaft and the crankshaft, the engine must be removed.

The following series of photographs were taken while the engine was removed after all the "hang-on" items were removed. The hull cavity is too cramped an environment to show clearly with photographs the necessary work to be performed. Therefore, depending on the parts requiring service, the engine may remain in place and the unnecessary steps in the following procedures may be skipped.

Briefly stated: perform only the necessary work to return the engine to satisfactory operation. Unnecessary work may include leaving bearings, seals, O-rings, etc. in place, if they are fit for further service.

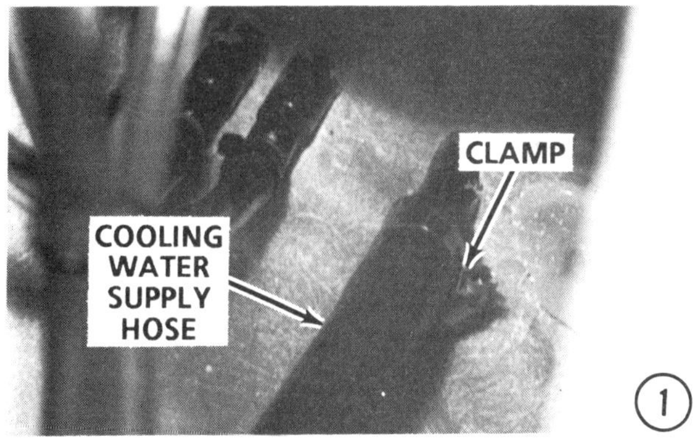

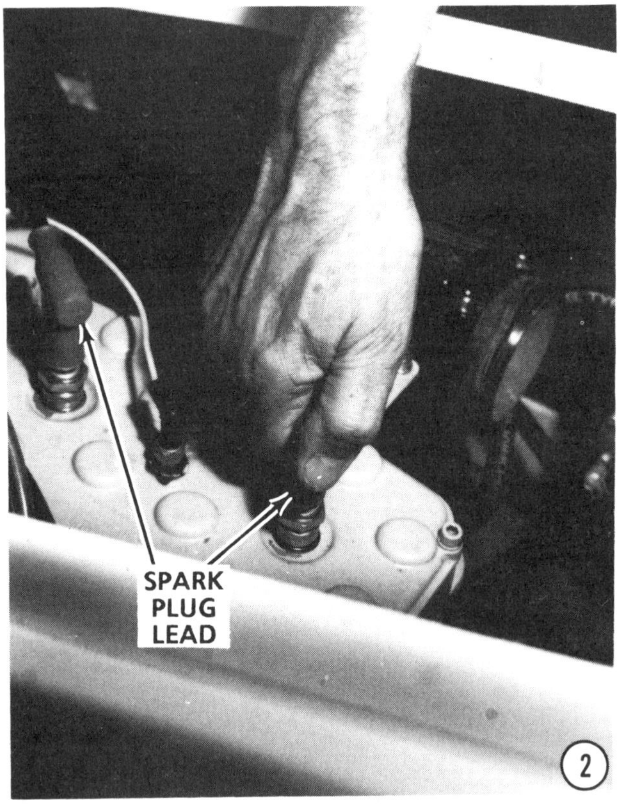

1- Loosen the clamp on the cooling water hose, and then slide it up the hose. Ease the hose free of the fitting on the jet pump.

Cylinder Head Removal

2- Remove both spark plug leads.

3- Remove both spark plugs. Remove the lead from the temperature sensor mounted on the cylinder head cover.

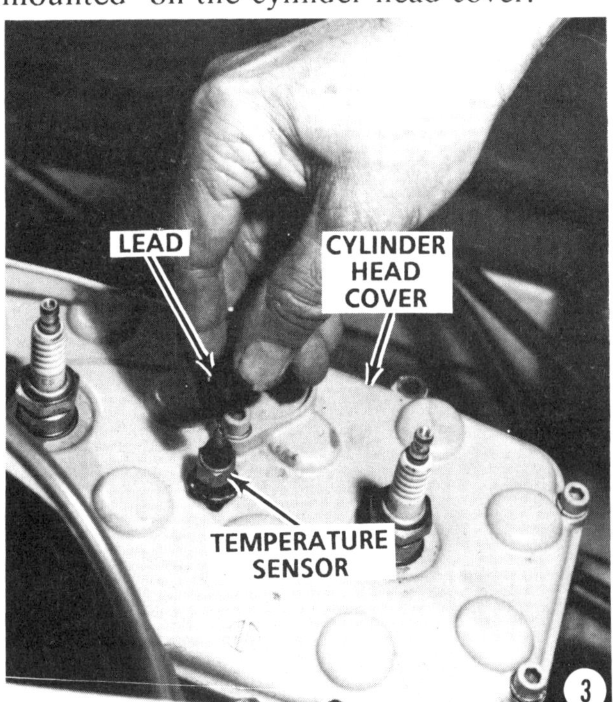

4- On some models, it may be necessary to remove the two bolts and washers securing the support bracket to the flywheel cover.

5- Remove the nine M6 Allen Head bolts and washers securing the cylinder head cover to the cylinder head. The two forward bolts may also secure the stator plate support bracket to the cylinder head cover. Remove the bracket, if

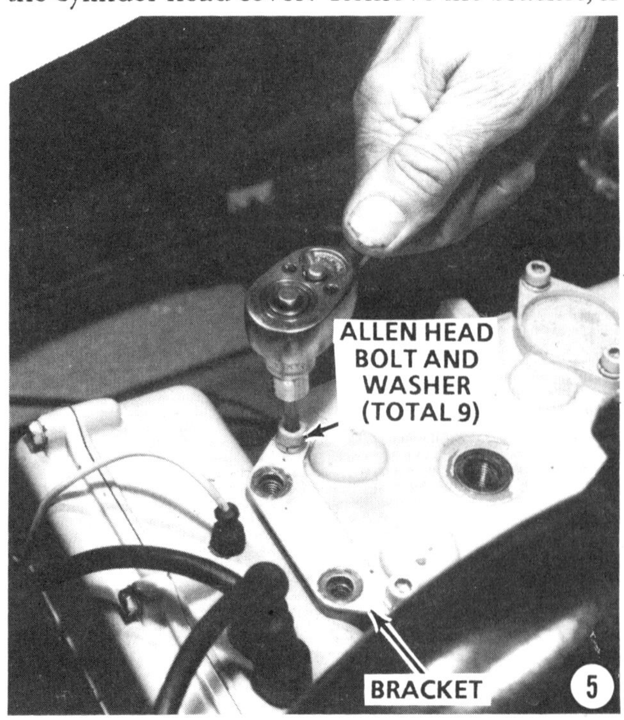

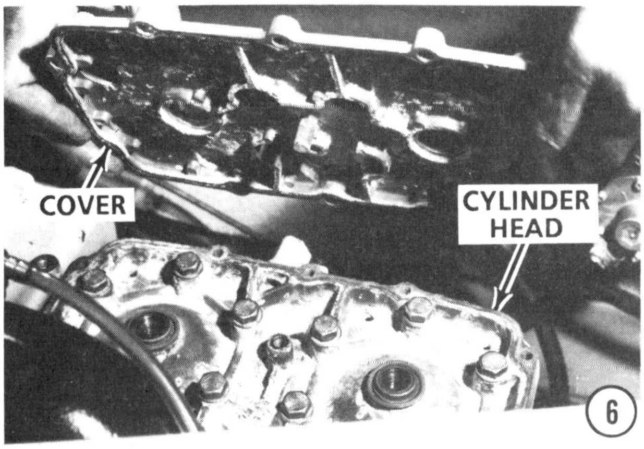

installed. (Two M6 Allen head bolts were removed when the flame arrestor was removed as part of carburetor removal.)

6- Lift off the cylinder head cover. If the cover seems to be stuck in place, use a soft head mallet and tap alternately and evenly on both sides to jar the cover free. Remove and discard the sealing ring.

7- Remove and discard the single small O-ring in the center of the head and the two larger O-rings around the spark plug openings. Remove the twelve bolts and washers securing the cylinder head to the two cylinders.

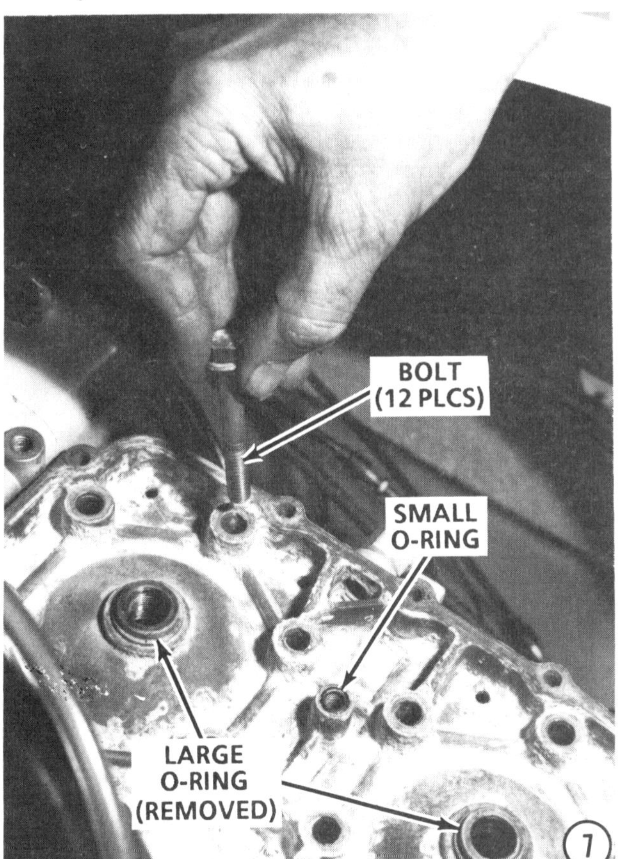

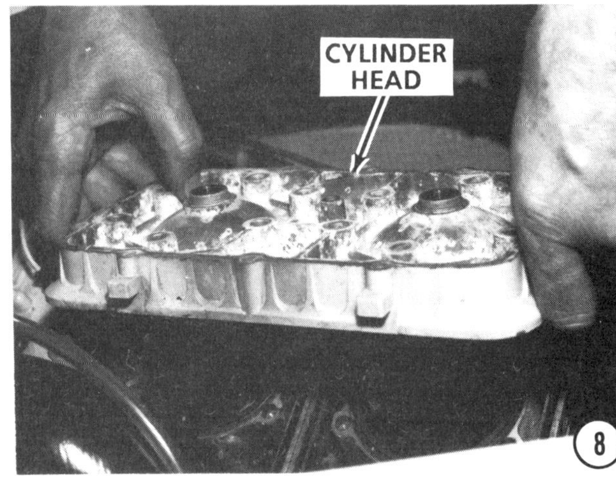

8- Lift off the cylinder head from the cylinders. If the head is stuck, use a soft head mallet and tap alternately and evenly on both sides to jar the head loose.

9- Remove and discard the four sealing rings, one ring around each cylinder bore and one ring around each water jacket. Remove all traces of sealing compound from the sealing ring groove in the cylinders and the corresponding grooves in the head.

Rotary Valve and Oil Injection Pump Removal

10- Obtain hose pincer tool P/N 295-000-076. Clamp the tool around the oil return hose. If this tool is not available, make arrangements to plug the disconnected hose when it is pulled free of the fitting in the next step.

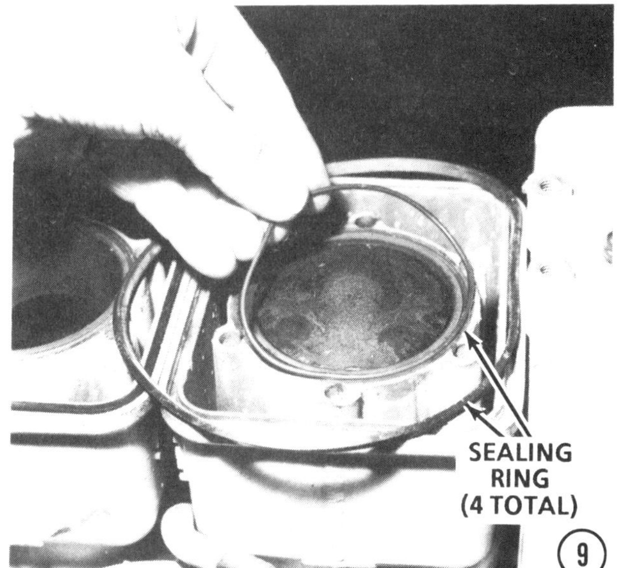

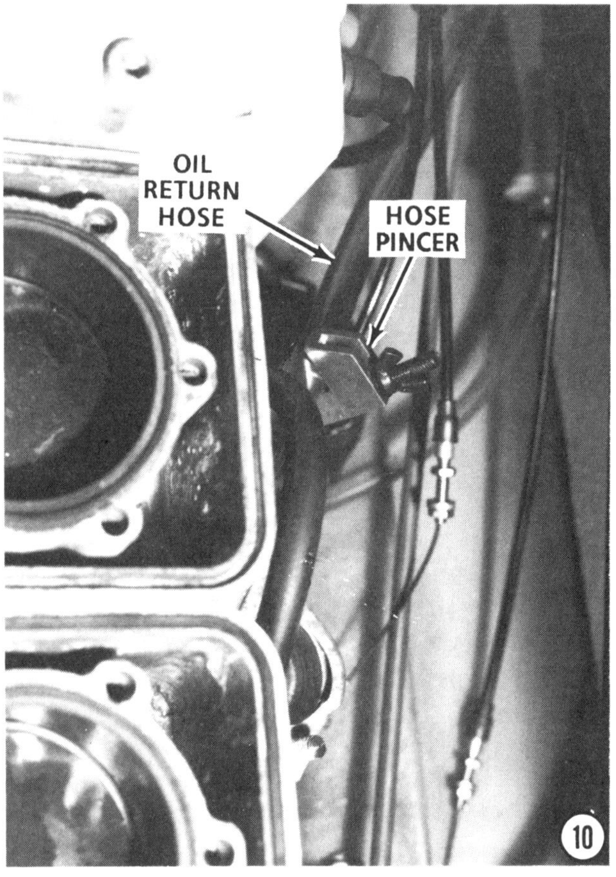

11- Pull the oil return hose free of the fitting above the rotary valve housing. If the hose was not pinched to halt the flow of oil, quickly plug the end of the hose to prevent losing oil into the bilge.

Remove the four bolts and washers securing the rotary valve housing to upper and lower crankcase. The intake manifold and the oil pump are attached to the valve housing and will come away with the housing. Remove and discard the sealing ring.

12- The rotary valve disc is now accessible. If the disc is to be replaced, make a note of the alignment of the disc window edges in relation to the housing, **BEFORE** disturbing the disc. If the disc is suspected of not being properly timed, proceed to the section following Step 27, on Page 8-33, to properly time the disc. If the disc is merely being replaced with a new one, install the new disc in **EXACTLY** the same position as the disc removed. If this is the only service work to be performed, proceed directly to the appropriate area of assembling.

Install a hose pincer tool onto the oil supply hose, and then remove the hose from the fitting at the base of the rotary

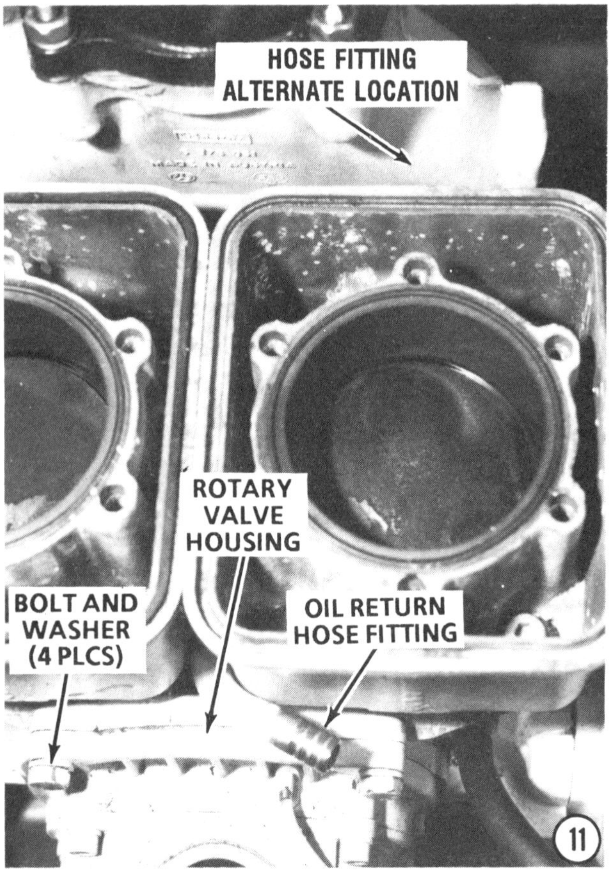

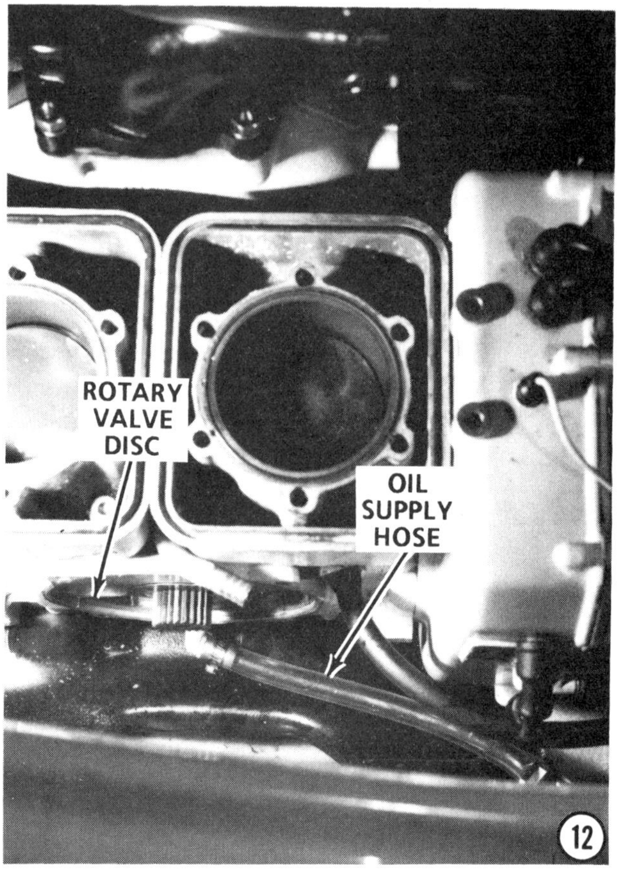

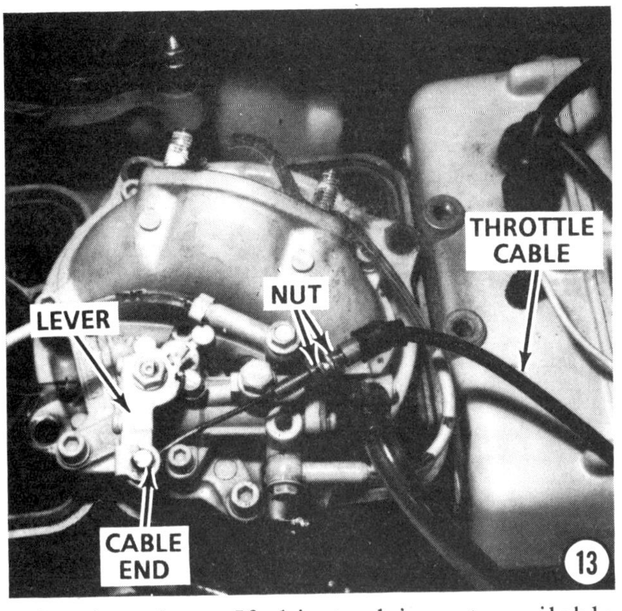

valve housing. If this tool is not available, make adequate preparations to catch oil when the hose is disconnected and to plug the hose immediately.

13- Disconnect the throttle cable from the oil injection pump by disengaging the cable end from the slotted hole in the lever, and loosening the two nuts -- one on each side of the support bracket.

14- Remove the six Allen bolts and the washers securing the intake manifold to the rotary valve housing. The oil injection pump will remain on the manifold.

15- Lift the intake manifold free of the rotary valve housing. **TAKE CARE** not to lose the single dowel pin from one of the mating surfaces. The dowel pin may remain in either surface. Remove and discard the sealing ring.

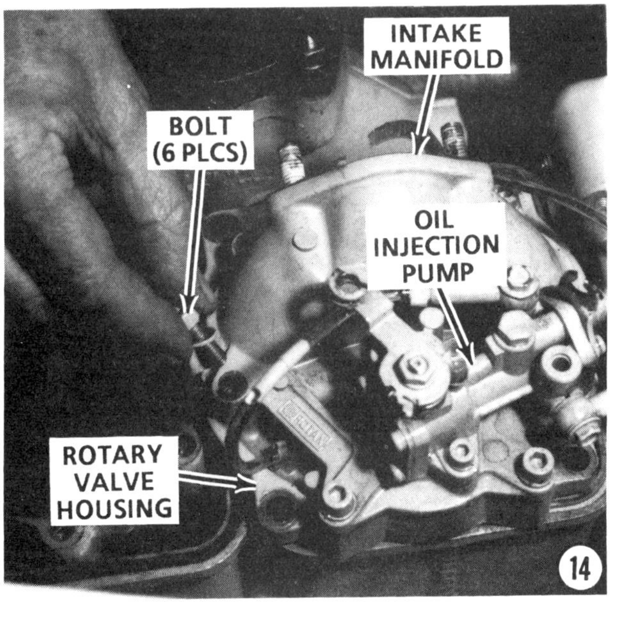

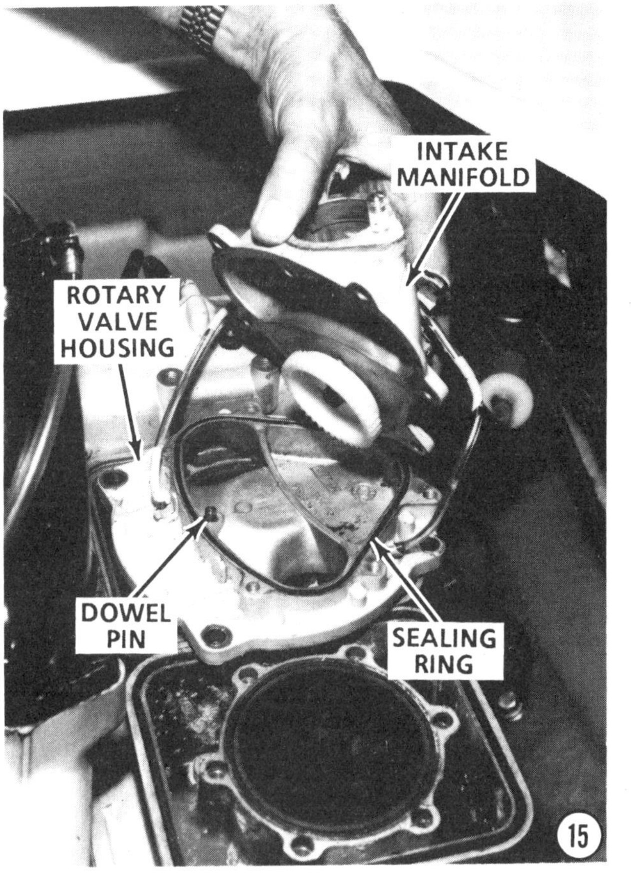

16- Obtain gear holder P/N 290-277-905 and use it to prevent the oil pump gear from rotating while the nut is loosened. If this tool is not available, hold the gear with a shop cloth, as shown. Remove the nut and washer, gear, and another washer. Remove the two mounting bolts. Lift the pump off the intake manifold, and then remove and discard the O-ring.

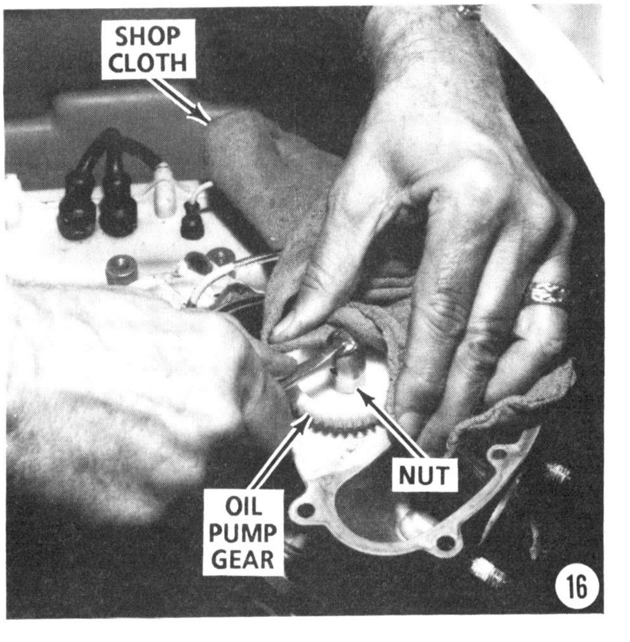

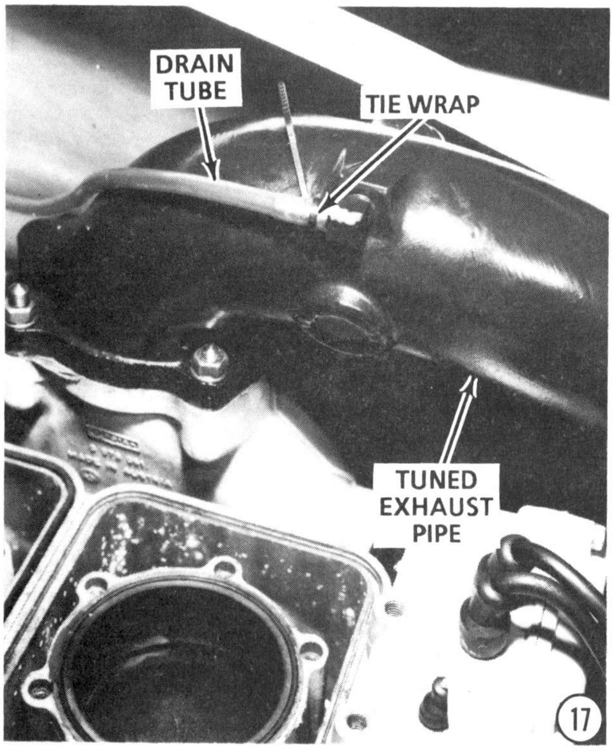

GOOD WORDS

No individual parts are available if the pump is not functioning properly. **HOWEVER**, a leaking oil pump can be overhauled with a gasket kit. If the problem cannot be corrected with the gasket kit, the pump **MUST** be replaced.

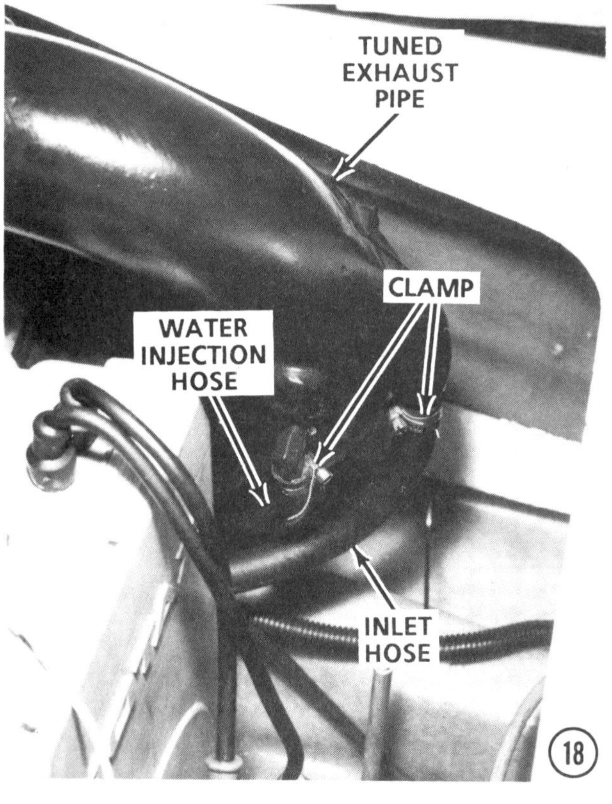

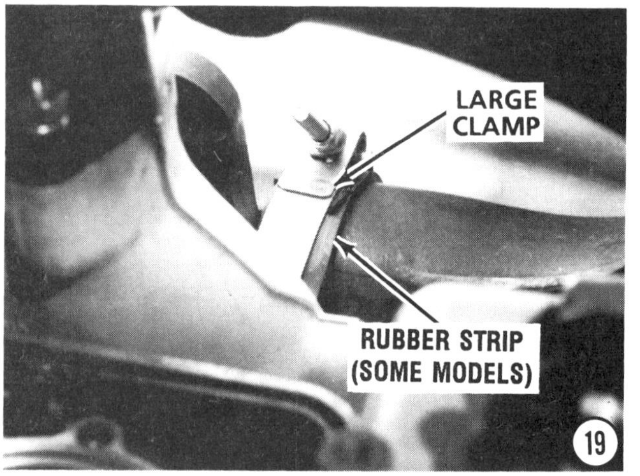

Exhaust System Removal

17- Snip the tie wrap and remove the drain tube from tuned exhaust pipe.

18- Loosen the clamps on the water injection hose and the inlet hose on the tuned exhaust pipe. Pull the hoses free of the fittings.

NOTE

On very new model engines, the tuned pipe may be in two halves.

19- Loosen the large clamp securing the tuned exhaust pipe to the engine. Remove and **SAVE** the rubber cushioning strip, if the strip was installed.

20- Loosen the two clamps on the hose connecting the tuned exhaust pipe to the muffler. This disconnect is necessary to provide some free play for the exhaust pipe to be moved forward and removed from the craft at a later time.

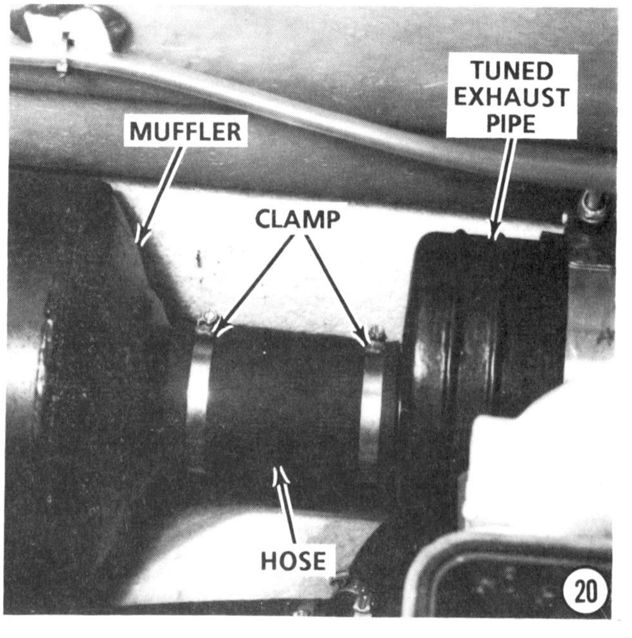

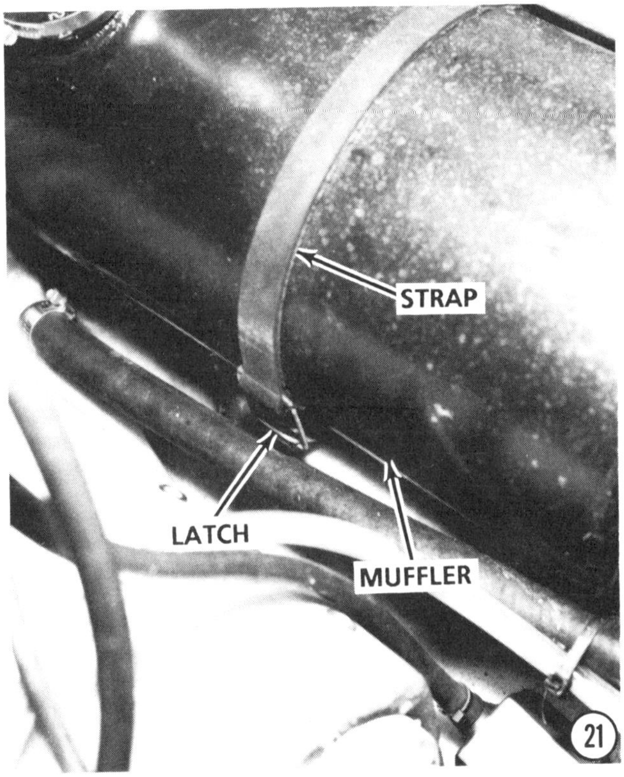

21- Unhook the muffler from the latch on the hull.

22- Remove the four nuts and washers securing the tuned exhaust pipe flange to the exhaust manifold. There is not enough room to lift the tuned exhaust pipe up over the exhaust manifold studs and out of the craft. Therefore, the exhaust pipe must be lifted up and moved aft while the studs are removed.

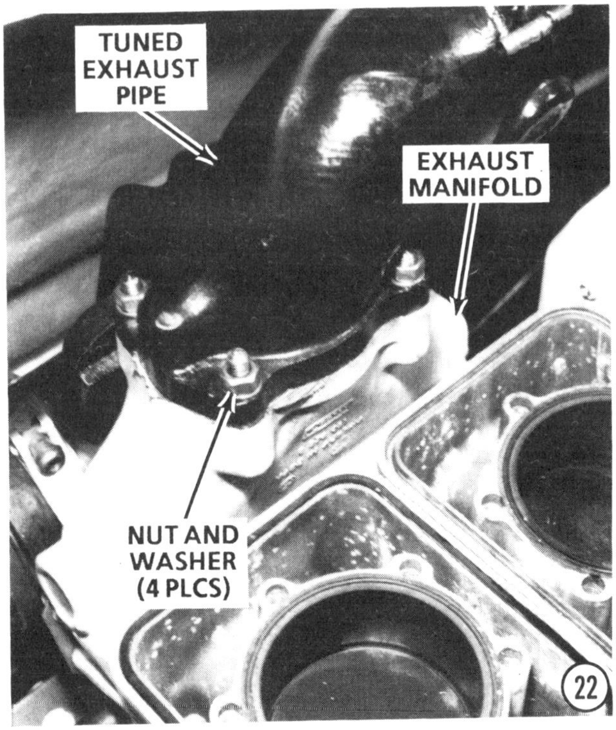

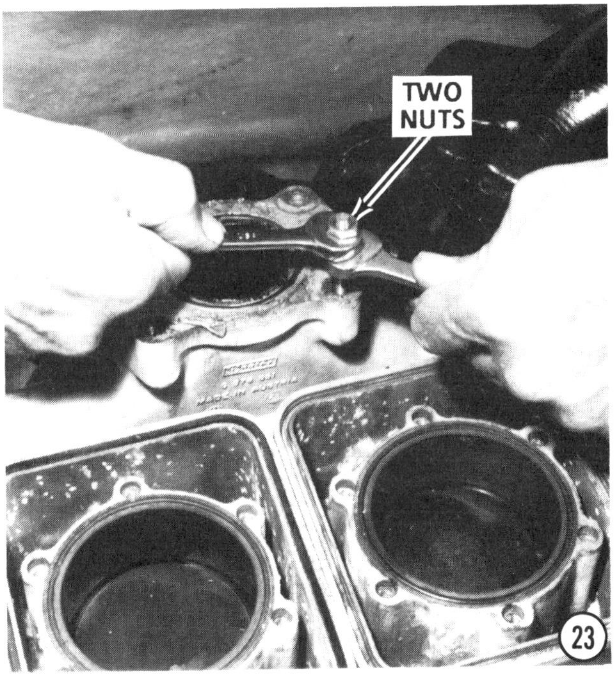

SPECIAL WORDS

Late model engines have three bolts and one stud. As a result, Steps 23 and 24 may not necessary.

23- Install two of the nuts removed in the previous steps onto one of the studs. Tighten the two nuts against each other.

24- Remove the stud by rotating the lower nut **COUNTERCLOCKWISE**. The upper nut will prevent the lower nut from threading upward on the stud. In this manner the stud may be removed from the exhaust manifold flange. Repeat this procedure and remove the other studs.

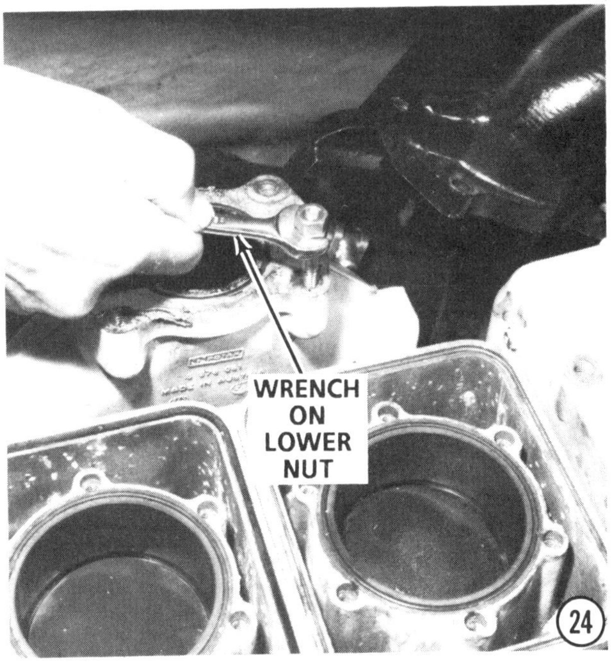

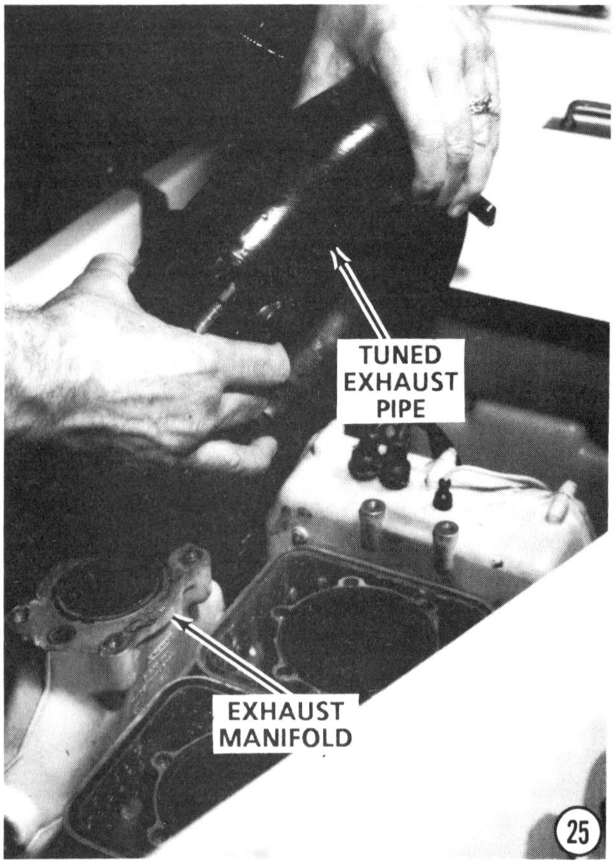

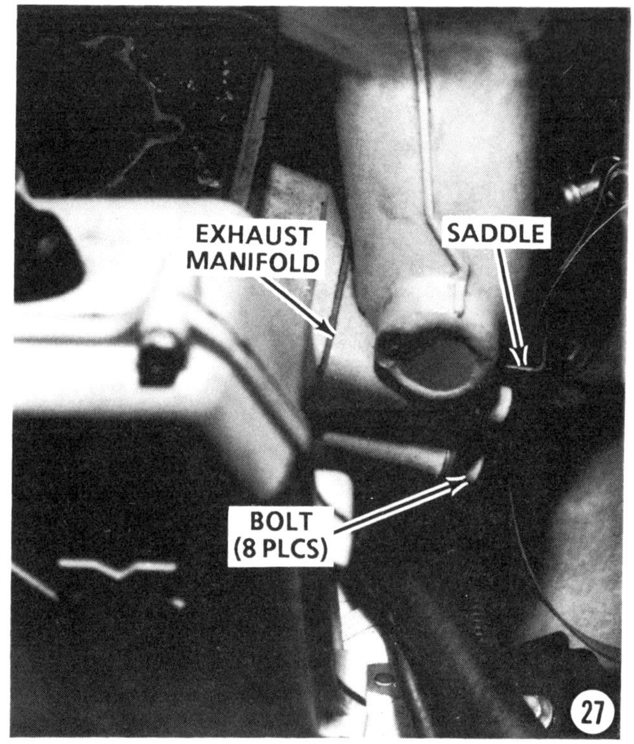

25- Ease the tuned exhaust pipe over the exhaust manifold and clear of the engine compartment.

26- Remove and discard the large exhaust passage O-ring and the two smaller water passage O-rings from the exhaust manifold flange. Remove and discard the gasket.

27- Remove the eight Allen head bolts and washers securing the exhaust manifold to both cylinders. The four forward bolts also secure a saddle, if installed, to the manifold. Remove the saddle, washer, and rubber pad. This arrangement supports the cranking motor support bracket.

28- Insert a long pointed tool between the exhaust manifold and the two cylinders, as shown. Pry the manifold away from the cylinders.

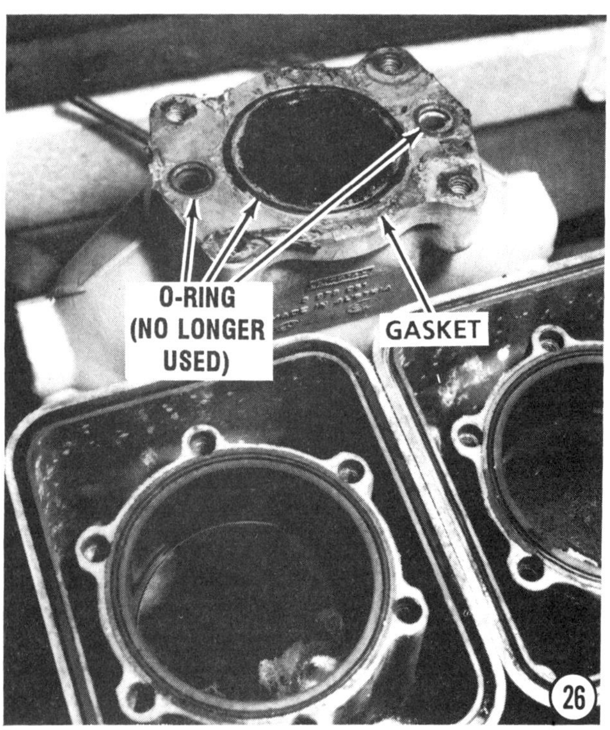

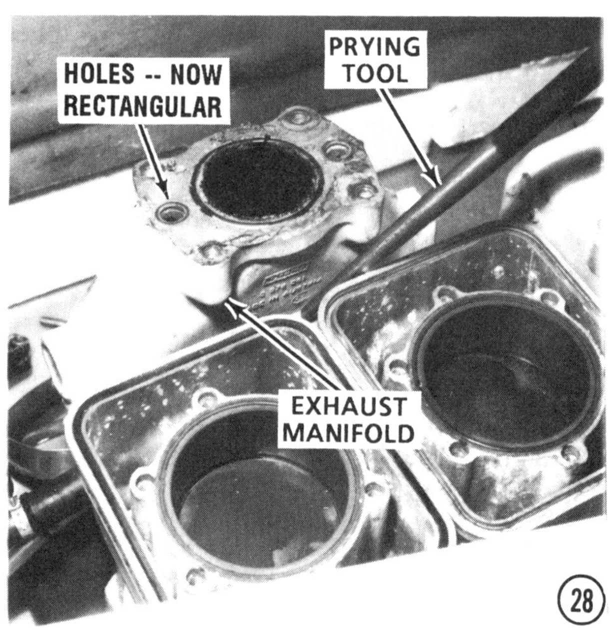

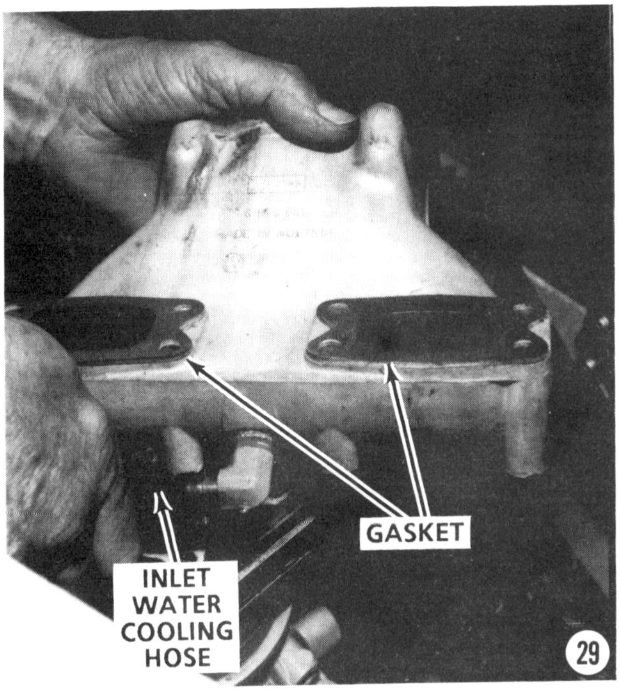

29- Lift out the exhaust manifold from the engine compartment. Remove the inlet cooling hose from the base of the exhaust manifold. Remove and discard the two gaskets. If the gasket material is stuck to the mating surfaces, take time to clean away the old gasket material. An exhaust leak will develop, if new gaskets are installed over traces of old material.

30- Remove the water hose from the outlet fitting on the port side of the engine. Exact

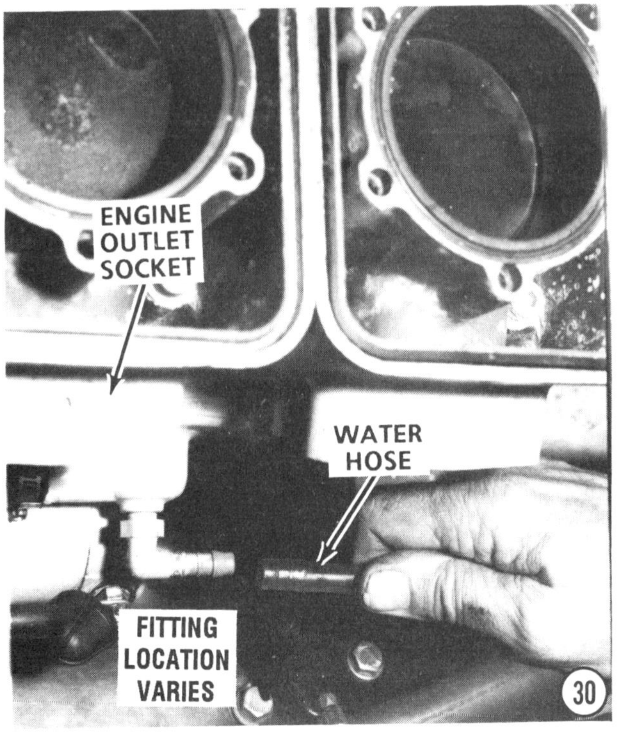

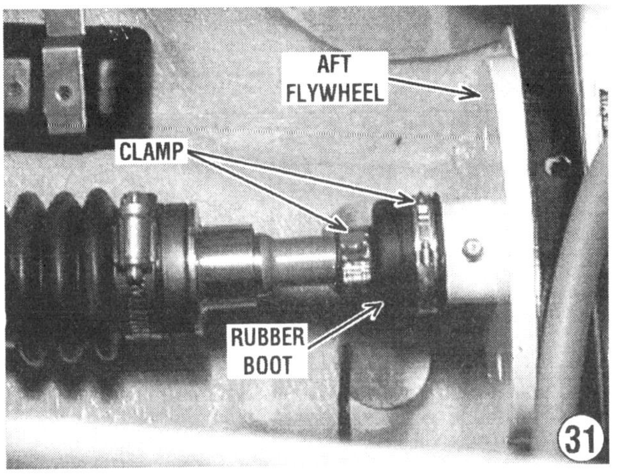

location of the socket on new model engines may vary slightly.

SPECIAL WORDS

Until 1995, the rubber boot mentioned in the next step could not be removed. The clamps are merely loosened to permit the engine to be moved forward for removal. Since 1995, the boot may be moved aft on the driveshaft to permit observing when the engine has been moved far enough forward to clear the driveshaft.

31- Prior to 1995 -- Loosen the two clamps. Since 1995 -- Loosen the two clamps and then force the rubber boot aft on the driveshaft.

Stator Plate Cover Removal

32- Remove the twelve bolts securing the cover to the stator plate.

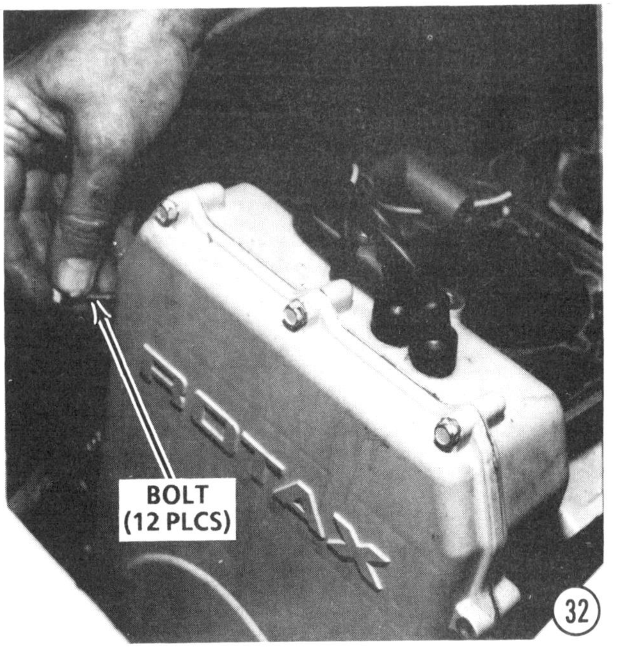

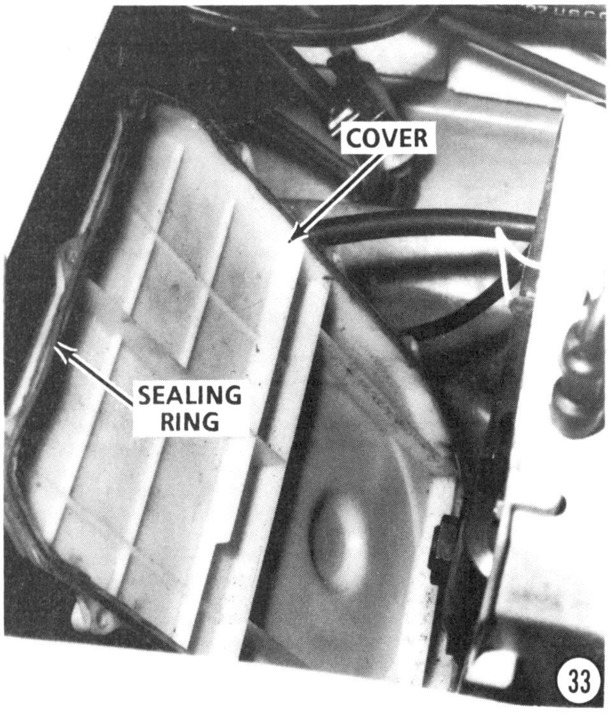

33- Lift off the cover. Remove and discard the sealing ring.

Cranking Motor Disconnect

34- Disconnect the large Red lead from the terminal on the cranking motor. Remove the upper mounting bolt and the large Black ground lead. If this is difficult to accomplish, at this time, the lead may be

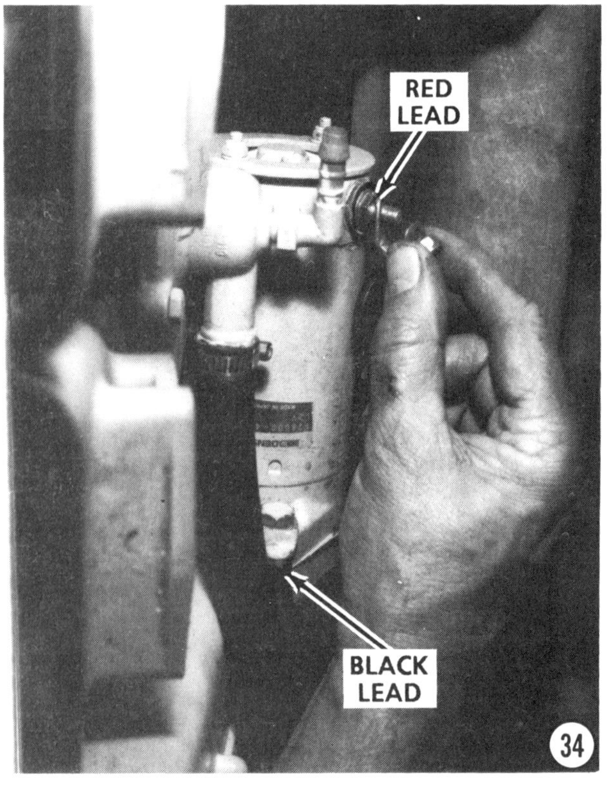

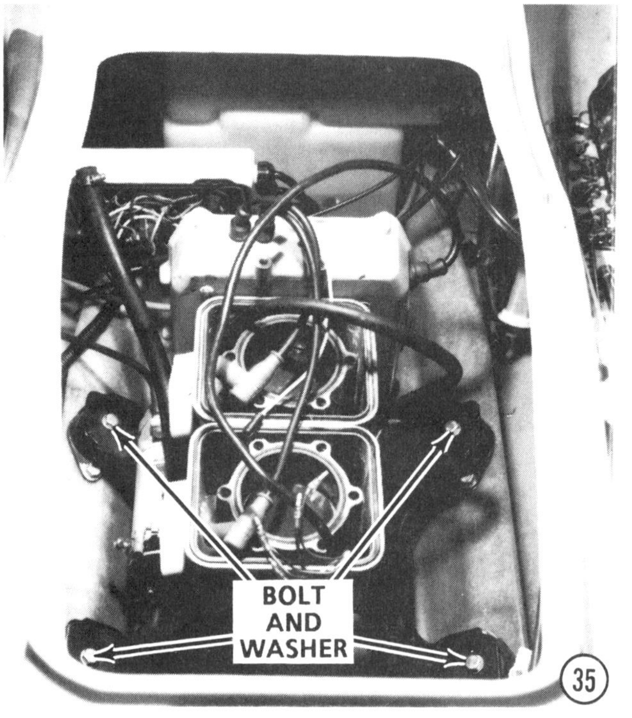

left in place until the engine is removed from the craft, **PROVIDED** the other end of the Black lead is no longer connected to the battery.

Engine Removal

35- Disconnect the overheat buzzer, tether cord switch, and the starting switch from the engine harness leads. Make one final check to ensure everything attached to the engine has been disconnected and the engine is ready to be lifted from the craft. Remove the four bolts and washers securing the engine and engine support bracket to the engine mounts.

DO NOT disturb the pieces of shim material at the engine mounts.

36- Two adults should have no difficulty lifting the engine clear of the craft. Lift

the engine and rotate the cylinders to **PORT** to permit the engine support bracket to clear the edge of the engine compartment. Watch for any hose or electrical leads still attached to the engine. Move the engine to a suitable work station.

37- DO NOT disturb the shim material or the engine mounts, unless a mount is to be replaced. Use the original amounts of shim material at each location, if new mounts are installed. Use some type of commercial cleaner and "square away" the bilge and engine compartment.

38- If there is an excessive amount of water in the bilge, the seals in the seal carrier may have failed and require replacement. Service to the seal carrier is outlined in detail in Chapter 10 -- **JET PUMP**.

ROTARY VALVE BACKLASH CHECK

The rotary valve backlash **MUST** be checked **BEFORE** any further disassembling work is performed.

Grasp the rotary valve shaft and rock it back and forth to determine the degree of movement. Rotate the flywheel to turn the crankshaft through 1/8 revolution. Check the backlash again. Repeat the procedure to check the backlash through one complete revolution of the crankshaft. The backlash **MUST** be the same at all position of the crankshaft.

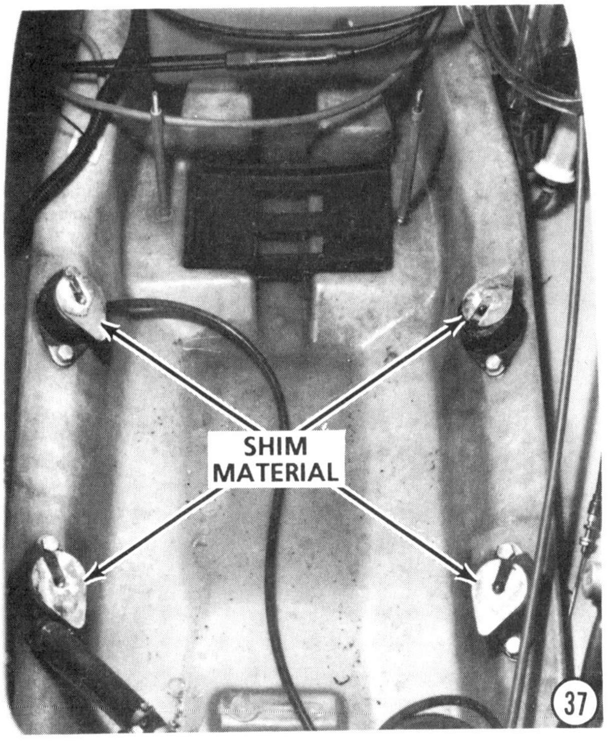

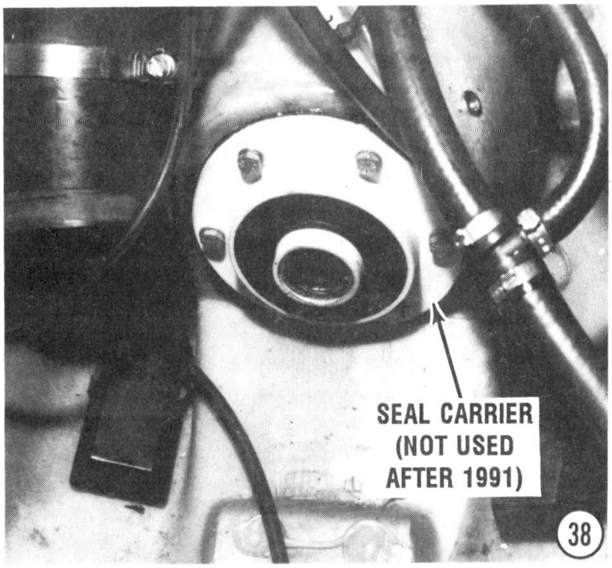

If the backlash is not consistent, the crankshaft and rotary valve shaft must be inspected to determine the cause of the binding:

The gear teeth may be stripped.
The bearing may have failed.
The rotary valve shaft may be bent.
The crankshaft may be bent.

The rotary valve shaft is serviced on Page 8-23, after the crankshaft has been removed.

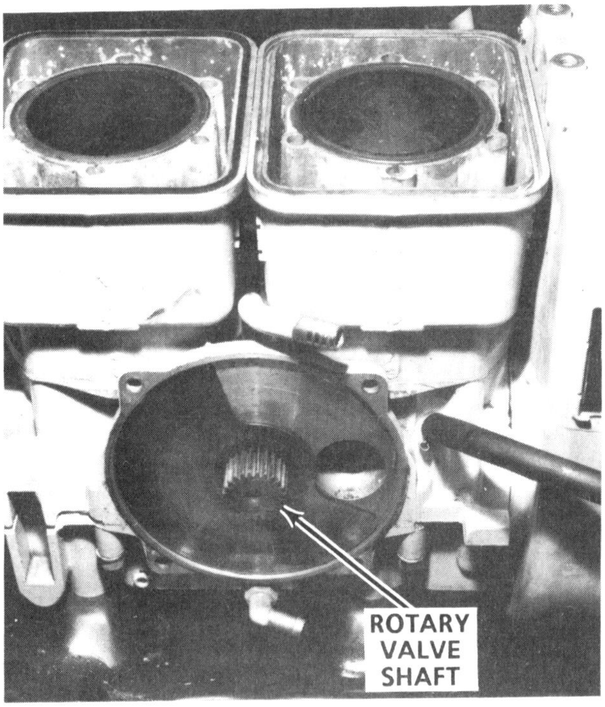

*The rotary valve backlash check **MUST** be completed **BEFORE** any further disassembling work is performed, as described in the text.*

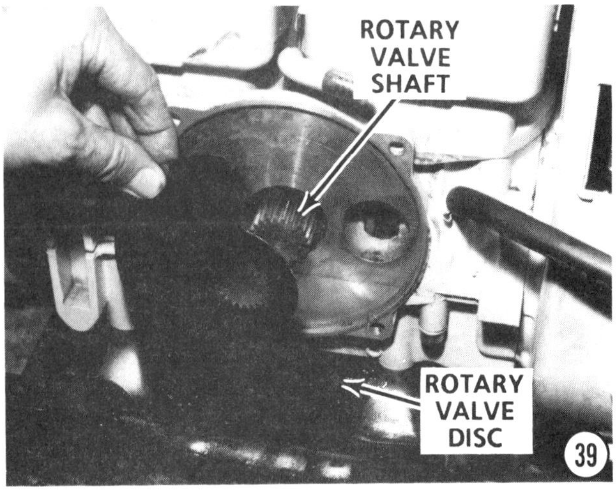

CRITICAL DISC WORDS
Before removing the rotary valve disc, establish TDC for the No. 1 cylinder, and then scribe a mark towards the top of the rotor face surface aligned with the opening edge of the rotory valve disc -- and a matching mark on the block. These marks will **ENSURE** accurate installation.

39- Slide the rotary valve disc from the shaft.

Cranking Motor Removal
40- Remove the two bolts securing the cranking motor to the stator plate -- one may have been removed earlier -- and a third bolt securing the aft end of the motor to the lower crankcase.
41- Pull the cranking motor away from the stator plate. If the motor is stuck, tap the mounting flange with a soft head mallet to jar it loose. Remove and discard the gasket.

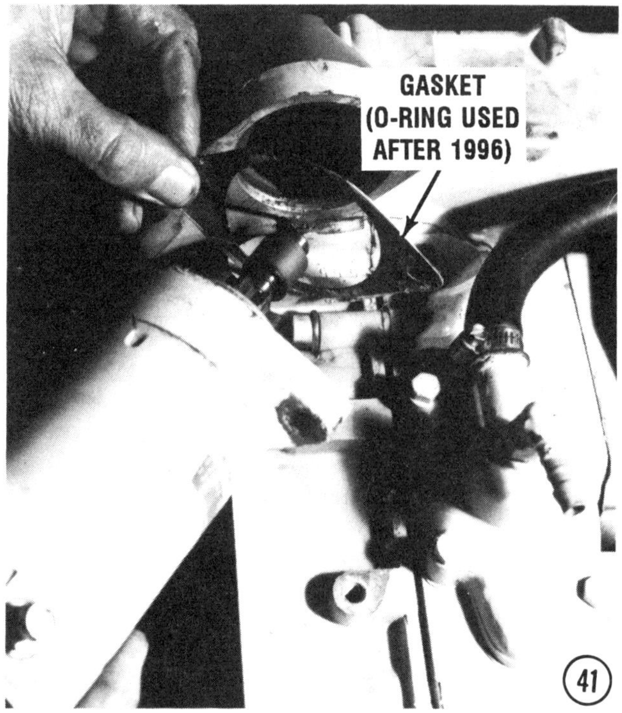

Aft Flywheel Removal

CRITICAL WORDS
NEVER attempt to lock the crankshaft by inserting any type of tool into the crankcase to wedge the crankshaft. This includes attempting to insert a rod through both small ends of the connecting rods to prevent rotation. Extensive damage to the connecting rods and/or crankshaft will surely occur.

42- Obtain Puller Plate P/N 290-876-080, Extension Handle P/N 295-000-003, and three M8x25mm bolts. Install the

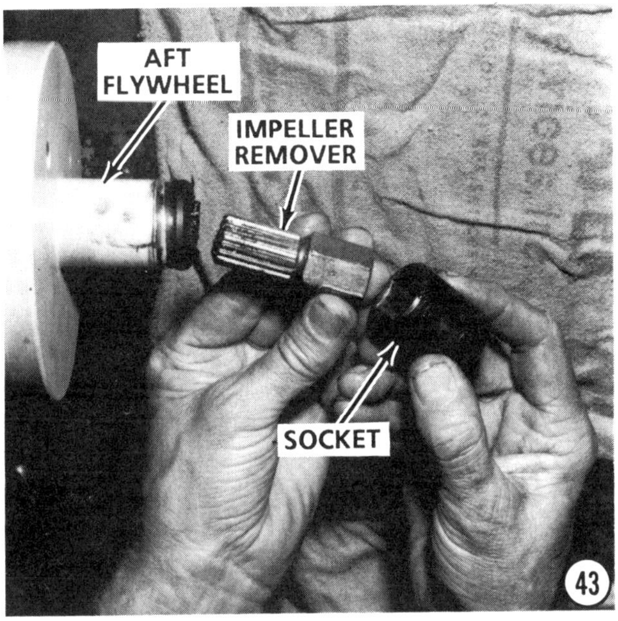

plate onto the flywheel using the three bolts. Tighten the bolts securely. This plate will be used in this step and the next four steps, therefore, leave it in place.

Install the extension handle onto the puller plate arm. Rotate the flywheel **CLOCKWISE** until the handle rests on the work bench, as an aid in loosening the flywheel.

43- Obtain Impeller Remover P/N 295-000-001. Insert the splined end of the tool into the aft flywheel. Use a suitable wrench or socket on the tool, and "break" the aft flywheel loose, by rotating the wrench or socket **COUNTERCLOCKWISE.**

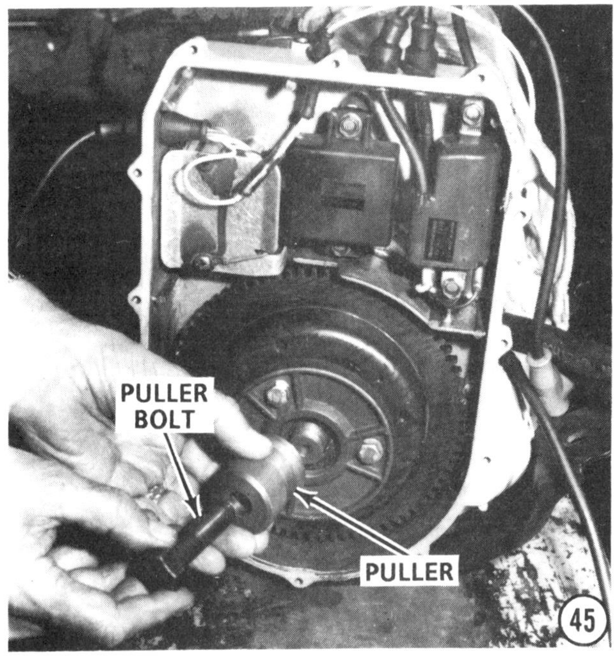

44- Continue to unthread the aft flywheel from the threaded end of the crankshaft. Remove all traces of old sealant from both sets of threads.

Forward Flywheel Removal

45- Install a 22mm thin-walled socket over the flywheel nut. The socket **MUST** fit inside the puller plate hole.

If the only available socket will not fit into the puller plate hole, grind down the outside diameter of the socket until it will fit into the hole. **DO NOT** grind on the internal threads of the puller plate.

Place a breaker bar into the socket and rotate it **COUNTERCLOCKWISE** to

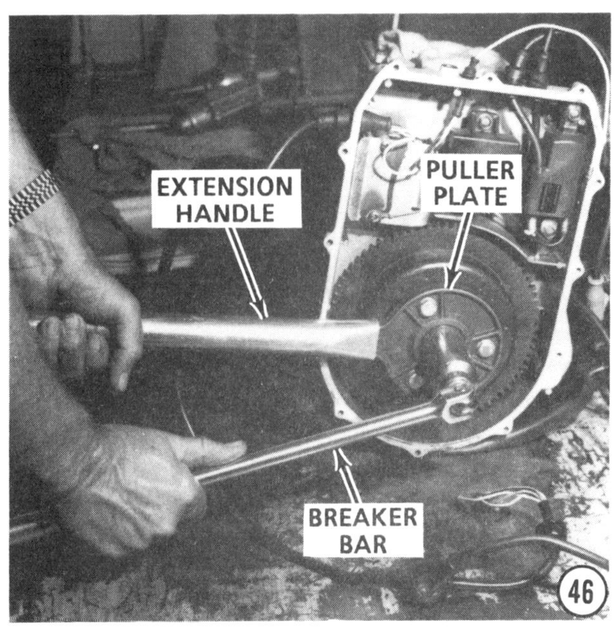

"break" the flywheel nut free. Remove the nut and washer.

46- Obtain Puller and Puller Bolt P/N 290-876-065. Thread the bolt into the puller until just a couple threads are engaged. Thread the puller into the internal threads of the puller plate. Now, Thread the bolt into the puller until the bolt makes contact with the end of the crankshaft.

47- Obtain the correct size wrench or socket and place it over the puller bolt.

Now, continue to tighten the bolt and at the same time, shock the crankshaft with a gentle to moderate tap on the end of the bolt, with a soft head mallet. This shock will assist in "breaking" the flywheel loose from the crankshaft.

DO NOT strike the bolt with a severe blow. The force of such action would be transmitted to the seals and bearings along the crankshaft and cause damage. **NEVER** strike the flywheel directly, because this type of shock would surely cause the flywheel magnets to loose their magnetism.

48- Pull the flywheel free of the crankshaft. Remove and save the Woodruff key from the crankshaft keyway.

SPECIAL WORDS ON WOODRUFF KEYS

Inspect the condition of the Woodruff key and its keyway in the crankshaft. If the key shows any sign of shearing, the cause is probably from jumping waves and the effect will be a loss of power. When the craft is clear of the water, the engine rpm approaches the limit set by the rpm limiter in an unloaded condition.

When the craft returns to the water surface, the sudden load on the impeller is transferred along the driveline. Speed of the rotating driveshaft and the crankshaft is suddenly reduced while the flywheel continues to spin rapidly, due to its mass. This condition, especially when repeated over a period of time, often shears the Woodruff key and the flywheel becomes repositioned on the crankshaft.

A very small misalignment will result in a change in ignition timing. The craft operator will notice a drastic loss of power. A worse condition would be the flywheel shifting to the degree engine start would not be possible.

Stator Plate and Magneto Assembly Removal

49- The securing screws pass through elongated slots in the magneto assembly. The assembly could be installed as much as 15° off its original position -- drastically affecting the timing characteristics of the engine. Therefore, make a set of aligned punch marks -- one on the magneto assembly -- the other on the stator plate. These marks will serve as an aid during installation to **ENSURE** correct positioning of the magneto assembly on the stator plate.

47

48

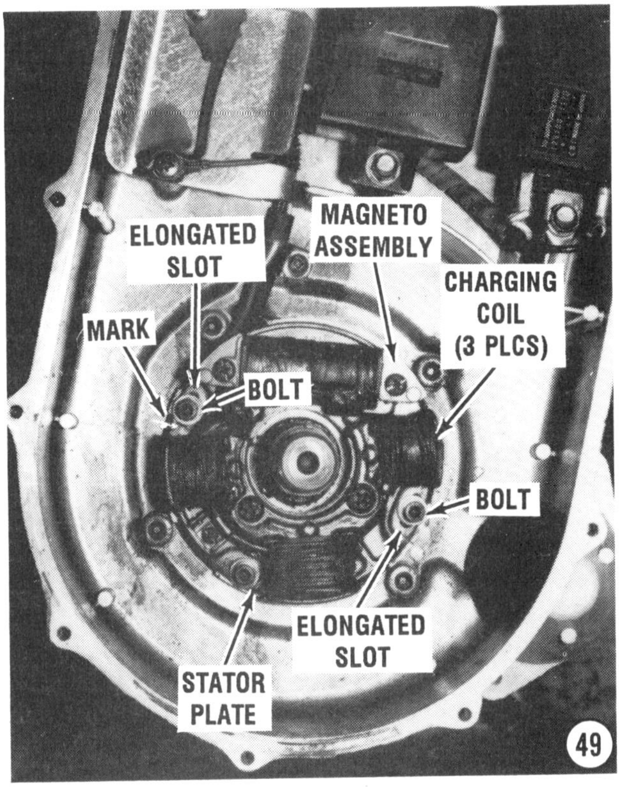

Remove the two screws securing the magneto assembly to the stator plate. These screws must be removed because they are threaded into the crankcase.

50- Remove the four bolts securing the stator plate to the crankcase. Lift off the stator plate with the magneto assembly and wiring harness attached.

Engine Support Bracket Removal

51- Remove the four nuts and washers securing the engine support bracket to the bottom of the lower crankcase. Lift the bracket off the studs.

Cylinder Removal

52- Remove the four Allen head bolts from each cylinder. These bolts are located inside the water jacket.

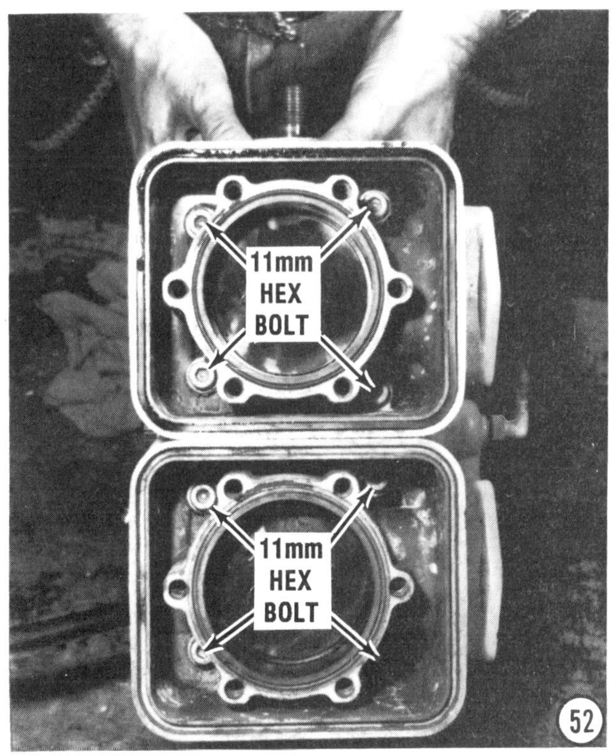

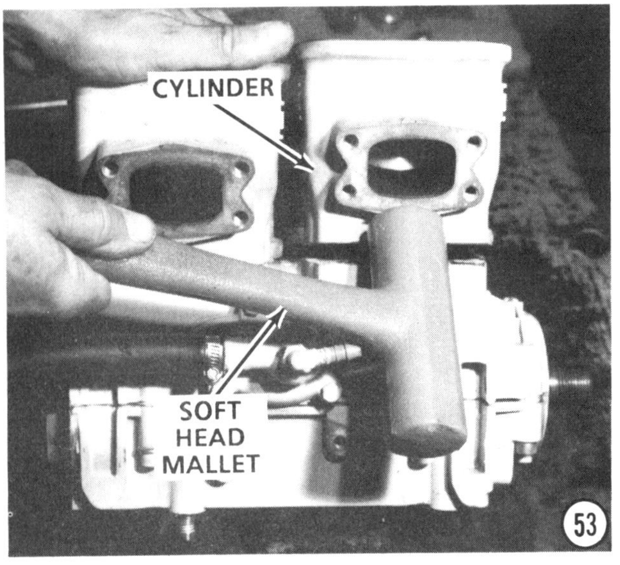

53- Using a soft head mallet, gently tap each cylinder to release it from the upper crankcase.

54- Slowly slide the cylinder up and free of the piston. **TAKE CARE** not to allow the piston to slip out of the cylinder bore and damage the piston skirt or connecting rod on the side of the upper crankcase. Remove the other cylinder in the same manner.

Piston Removal

55- Scribe a mark on the inside of the piston skirt to identify the No. 1 and No. 2 cylinders, before the piston is removed from the connecting rod, as described in the following steps.

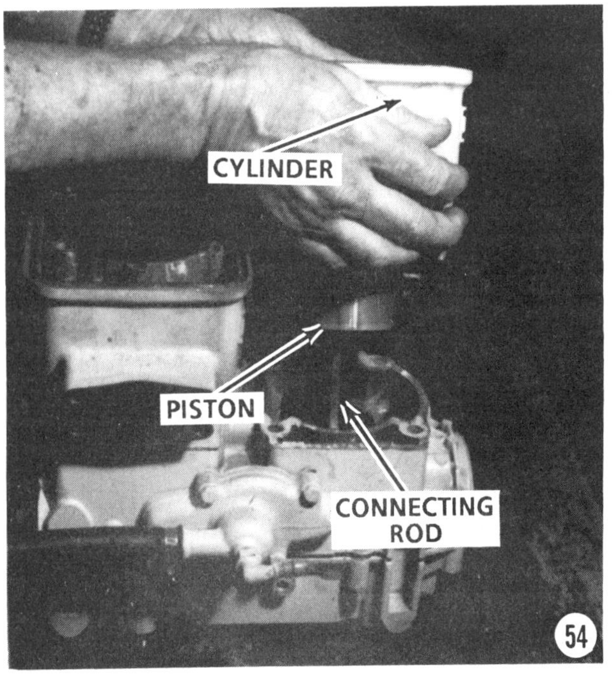

SPECIAL WORDS ON NEEDLE BEARINGS

A decal on the flame arrestor support bracket, warns there is a cageless needle bearing inside the piston. As soon as the piston is separated from the connecting rod, these needle bearings could fall into the crankcase, unless arrangements are made to catch them.

56- Raise the first piston to be disassembled up to the TDC position. Obtain Rubber Pad P/N 295-000-079. Install the

pad around the connecting rod. Temporarily install two of the bolts to prevent the pad from shifting. Stuff a shop cloth into any part of the upper crankcase which the pad does not cover. If the special pad is not available, stuff numerous shop cloths into the upper crankcase. Smooth the tops of the cloths to enable the needles to be found after they fall free.

Some mechanics place a magnet under the piston being serviced to attract the loose needles.

SAFETY WORDS

Snap rings and **C**-lockrings are made of spring steel and may pop out of the groove with considerable force. Therefore, warn others in the area and **WEAR** eye protection glasses or a shield while removing this type of ring.

57- Insert a small screwdriver or awl into the notch next to the **C**-lockring and pry the ring free. Discard the lockring. These rings stretch and distort during removal. Therefore, they can **NEVER** be used a second time.

SPECIAL WORDS ON PISTON REMOVAL

The piston may be removed using factory special tools or by using readily available shop tools.

If factory tools are available, perform Step 58 -- then proceed to Step 61.

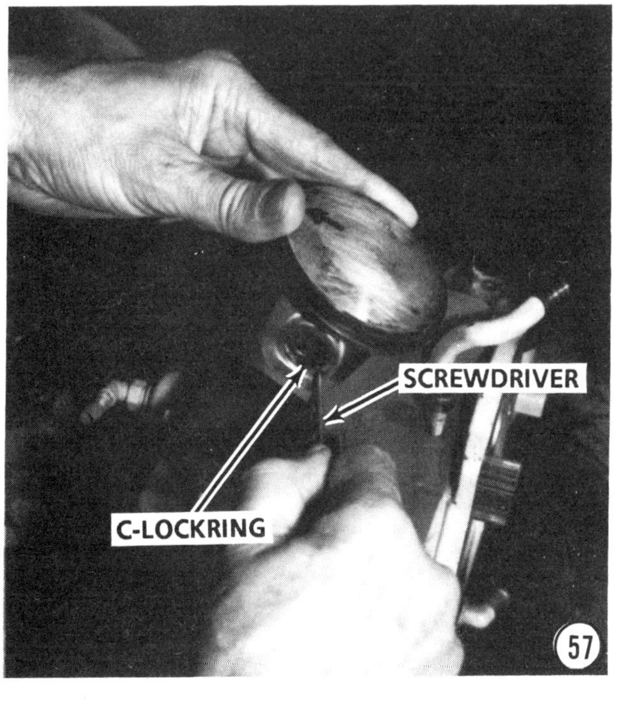

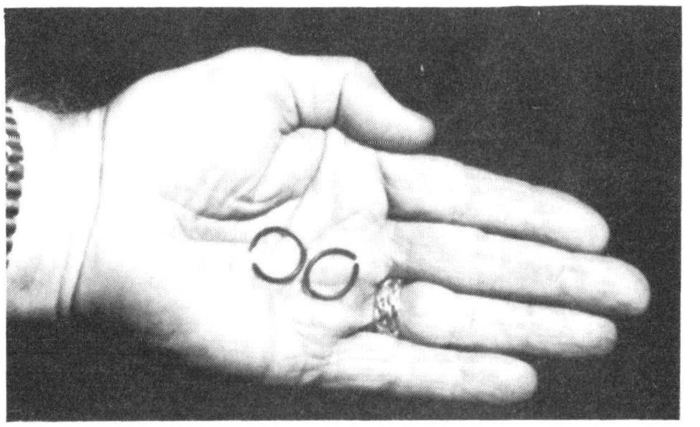

Piston pin C-Lockrings stretch and distort during removal. These rings should NEVER be used a second time.

If shop tools are to be used, perform Step 59 and on.

New needle bearings should be installed in the connecting rods, even though they appear to be in serviceable condition. New bearings will ensure lasting service after the overhaul work is completed. If it is necessary to install the used bearings, keep them separate and identified to **ENSURE** they will be installed back with the same connecting rod from which they were removed.

58- Obtain Piston Pin Puller P/N 290-877-090. Thread the puller handle in as

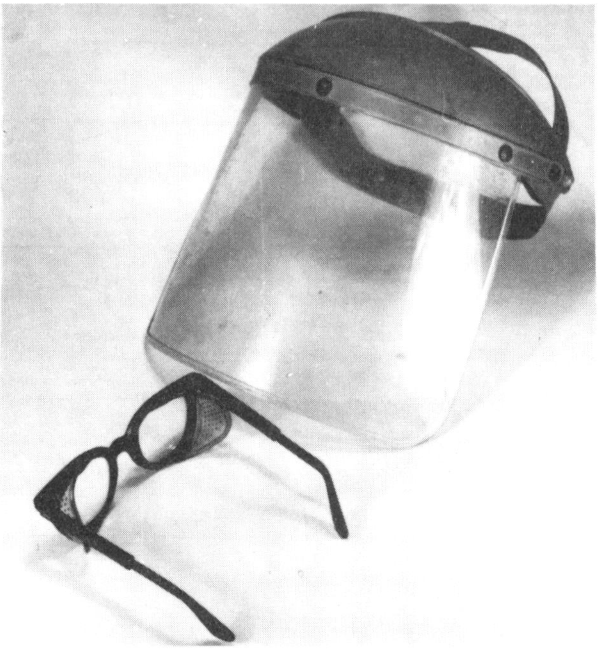

*C-lockrings are under tremendous tension during removal and installation, presenting a potential hazard. As an eye protection measure, safety glasses or a shield should **ALWAYS** be worn during work with such snap rings. Warn others in the area such work is in progress.*

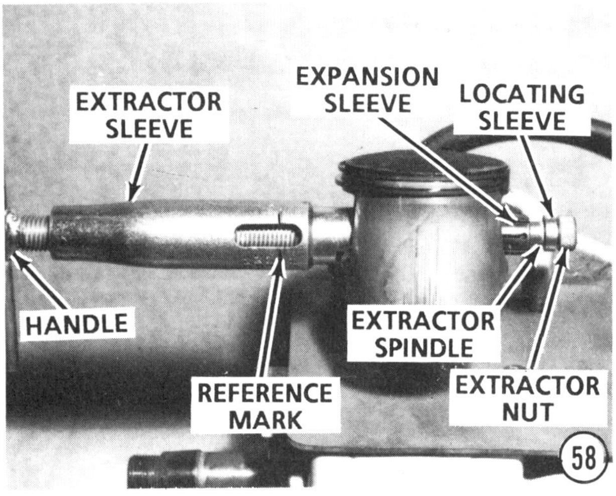

far as possible. Insert the extractor spindle into the piston pin. Slide the expansion sleeve, followed by the locating sleeve onto the spindle. Install the extractor nut onto the spindle threads.

Rotate the handle **CLOCKWISE** to pull the piston pin until the edge of the pin aligns with the reference mark on the extractor sleeve window. Now, rotate the spindle **COUNTERCLOCKWISE** until the extractor nut can be removed. Remove the spindle and extractor sleeve from the piston pin and remove the piston. The expansion sleeve will hold the loose needle bearing and thrust washer in place, preventing the needle bearing from falling apart.

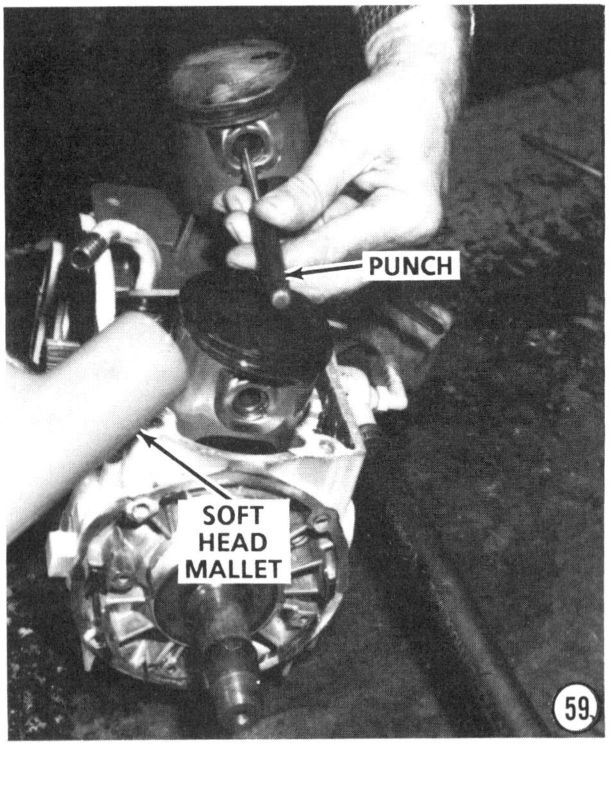

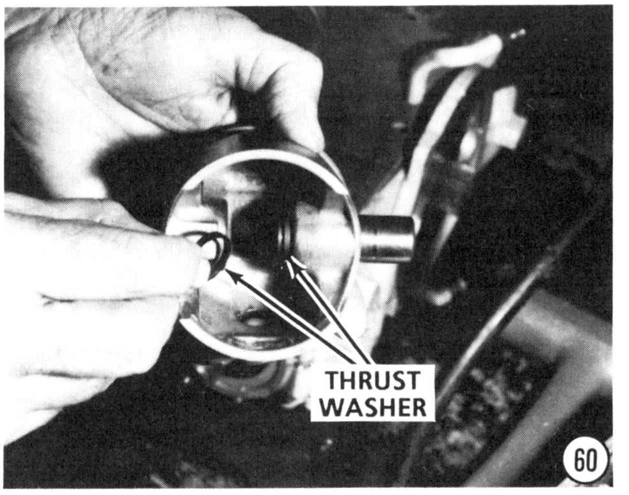

Carefully keep this assembly intact, if the needle bearings are to be used again. Use an extra expansion sleeve to disassemble the other piston using factory tools.

59- If the piston pin removal tool is not available, tap the piston pin free using a narrow punch and shoft head mallet. Be sure to tap evenly around the pin, not just on one side, because such action will cause the pin to jam.

60- Remove the two thrust washers from inside the piston. Recover all 31 needle bearings for each piston, even if they are not to be used again. In other words, all **MUST** be accounted for to ensure none have found their way into the lower crankcase.

GOOD WORDS

Good shop practice dictates to replace each piston ring during an engine overhaul. However, if the rings are to be used again, expand them **ONLY** enough to clear the piston and the groove, because used rings are brittle and break easily.

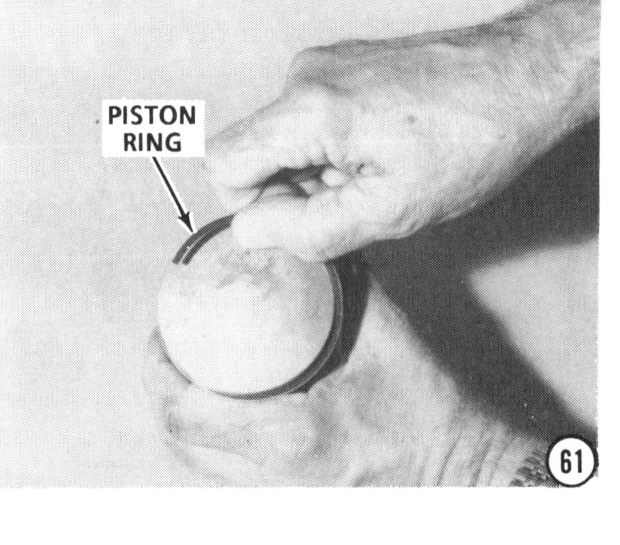

61- Gently spread the top piston ring enough to pry it out and up over the top of the piston. No special tool is required to remove the piston ring. Remove the lower ring in a similar manner. These rings are **EXTREMELY** brittle and **MUST** be handled with care if they are intended for further service.

Crankcase Separation

62- Remove the six hexagon head bolts and eight Allen head bolts securing the two halves of the crankcase together.

63- Separate the two halves of the crankcase by inserting a screwdriver between the lugs specifically provided for this purpose. **DO NOT** pry at any other location. Work around the crankcase split, prying alternately and evenly from side to side.

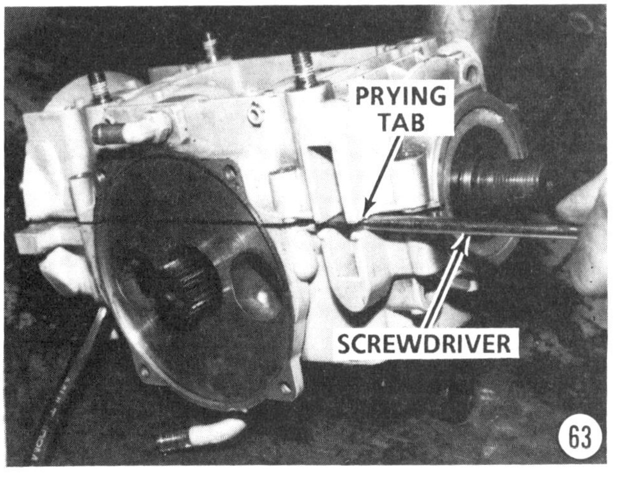

64- Carefully lift the upper crankcase half up and clear of the lower crankcase. If the two surfaces are stuck together, tap the ends of the crankshaft with a soft head mallet to jar the sealing surfaces loose. Remove all traces of sealant from both mating surfaces.

65- Lift the retaining ring free of the crankshaft worm gear, and then lift the crankshaft out of the lower crankcase.

66- From the forward -- tapered end of the crankshaft -- remove the following in

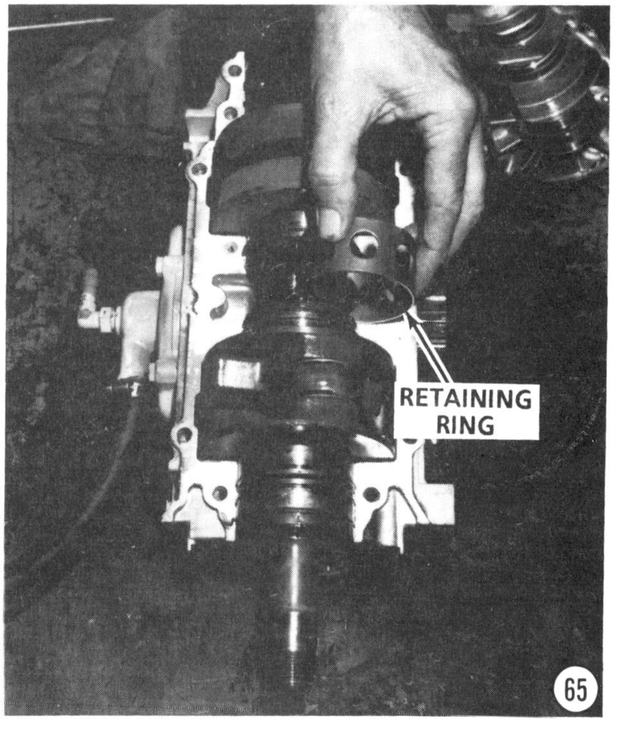

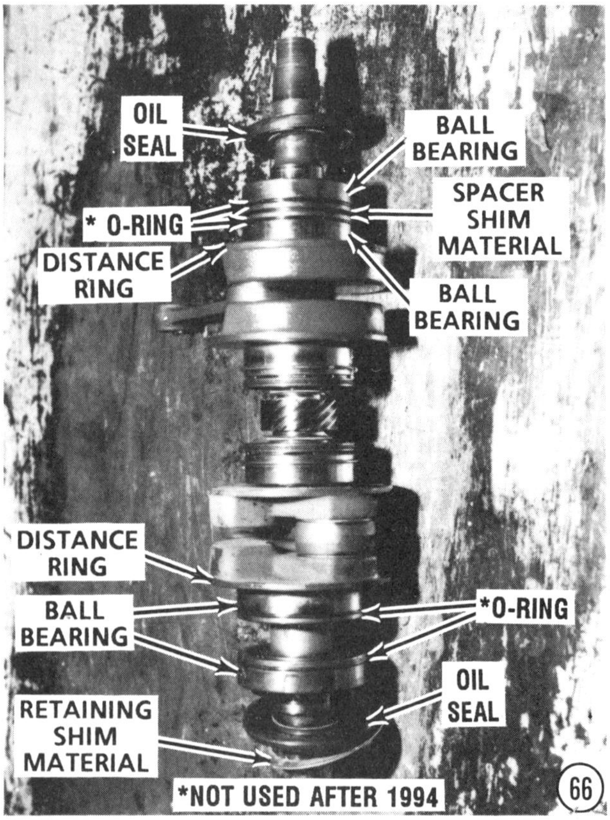

66

The rotary valve disc, attached to the rotary valve shaft, is driven directly off the engine driveshaft through a gear arrangement.

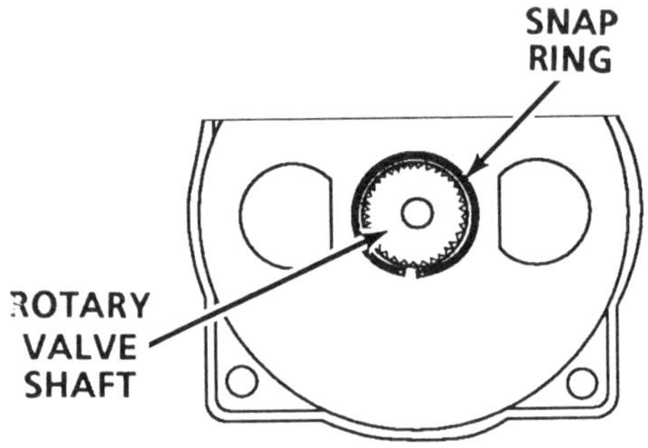

67

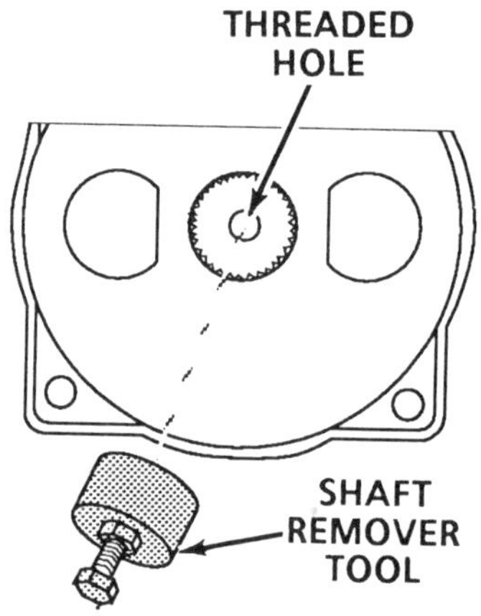

68

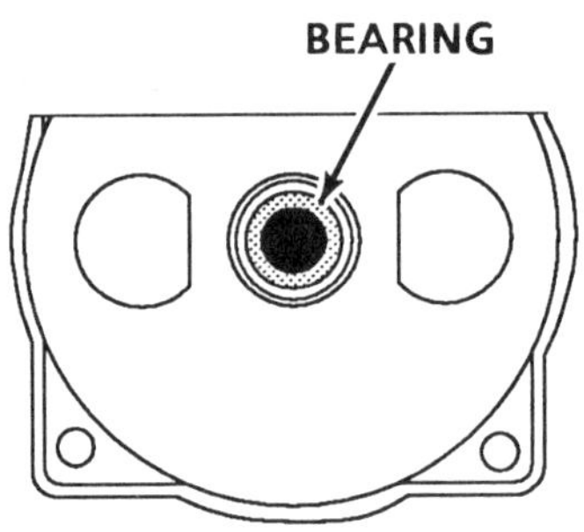

69

order: oil seal, ball bearing with an **O**-ring, spacer shim material, another ball bearing with an **O**-ring, and finally a "distance" ring. Keep these items in order on the work bench to **ENSURE** they will be installed back on the crankshaft in the proper order.

The ball bearings may or may not be a slide fit. If not, they can be pressed free with an arbor press, if they are to be replaced.

Remove the following in order from the aft end of the crankshaft: retaining shim material, oil seal, two ball bearings with **O**-rings, and a "distance" ring.

The connecting rods and rotary valve worm gear are an integral part of the crankshaft and cannot be removed. If they are no longer fit for service, the entire crankshaft **MUST** be replaced.

ROTARY VALVE SERVICE

SAFETY WORDS

Snap rings and C-lockrings are made of spring steel and may pop out of the groove with considerable force. Therefore, warn others in the area and **WEAR** eye protection glasses or a shield while removing this type ring.

67- Use a pair of external snap ring pliers and **CAREFULLY** remove the snap ring from the groove in the rotary valve shaft.

68- Obtain Rotary Valve Shaft Remover P/N 290-876-487. Slide the tool over the shaft and thread the bolt on the tool into the small threaded hole in the end of the shaft. Hold the nut welded onto the tool with the proper size wrench, and then rotate the bolt **CLOCKWISE** with another wrench until the shaft is withdrawn from the lower crankcase.

69- If the rotary valve shaft bearing on the port side of the lower crankcase need to be replaced, obtain a slide hammer with puller jaws attachment. Insert the puller jaws through the opening in the starboard side lower crankcase and hook the jaws around the back of the bearing. Pull the bearing free. **DO NOT** disturb the plug behind the bearing in the port side lower crankcase.

A quick crankshaft alignment check can be made by simply inserting a finger through each connecting rod, lifting the crankshaft to clear the work surface, and then attempting to "rotate" the crankshaft while it is suspended. If the crankshaft has the slightest amount of "twist", it will rotate in and out-of-round motion.

For all practical purposes, the engine can be considered as completely disassembled. Any further removal of items should probably done in a shop equipment with the proper tools and the shop personal possessing adequate training, experience and expertise to perform a professional job.

CLEANING AND INSPECTING

The success of the overhaul work is largely dependent on how well the cleaning and inspecting tasks are completed. If some parts are not thoroughly cleaned or if an unsatisfactory unit is allowed to be returned to service through negligent inspection, the time and expense involved in the overhaul work will not be justified with peak engine performance and long operating life.

Therefore complete, detailed, illustrated procedures for cleaning and inspecting virtually all components of the engine are presented in **Section 8-5**, beginning on **Page 8-76** The procedures presented in Section 8-5 should be followed closely and the necessary work performed with patience and attention to detail.

A "fun day" on the water is dependent on peak performance of the engine as the engineers and the manufacturer intended. The procedures and illustrations in this chapter have been formulated to give the reader an opportunity to reach the goal of optimum performance under all operating conditions. Of course, attention to good maintenance habits, as outlined in Chapter 4, certainly contribute to long engine life.

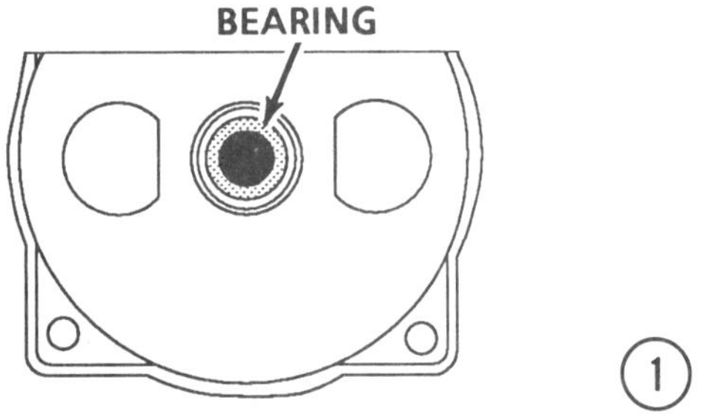

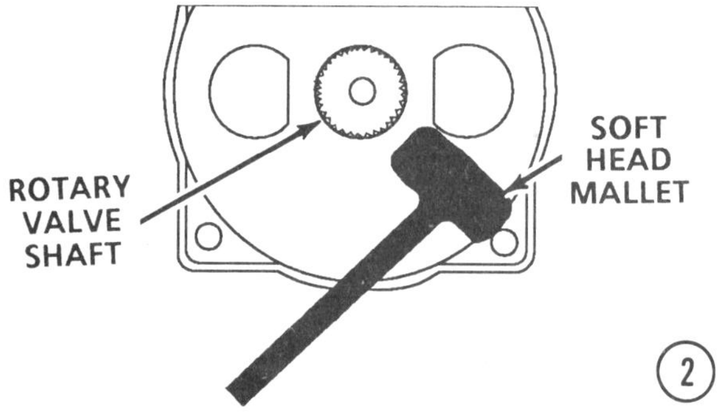

ASSEMBLING AND INSTALLATION ALL EXCEPT 787 SERIES

PLAN AHEAD

Apply a light coating of anti-sieze compound to the inner races of all four crankshaft bearings. Pack the lips of the two oil seals with Lithium lubricant. Use the same lubricant on the two O-rings. Immerse the four bearings, with the O-rings installed, into a container of oil heated to 167°F (75°C). Heating the bearings in the hot oil will expand them ever so slightly and ease the task of installation. Keep the bearings in the heated oil until the work has progressed to the point the bearings are to be installed in Step 5.

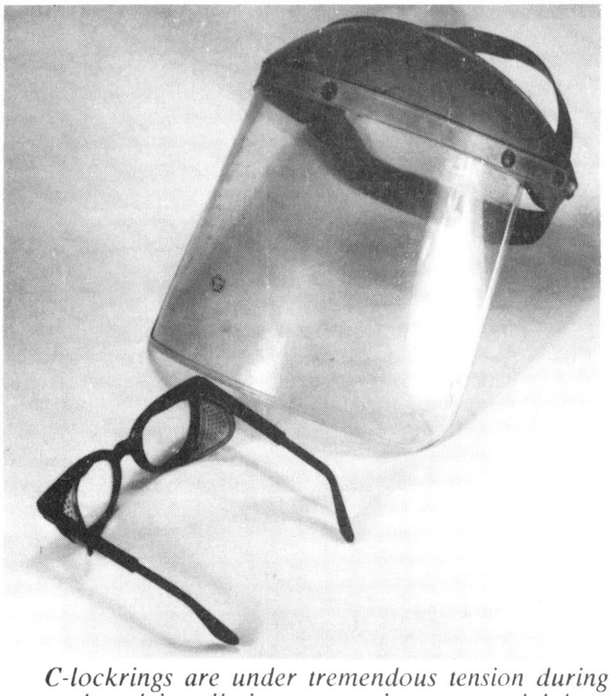

*C-lockrings are under tremendous tension during removal and installation, presenting a potential hazard. As an eye protection measure, safety glasses or a shield should **ALWAYS** be worn during work with such snap rings. Warn others in the area such work is in progress.*

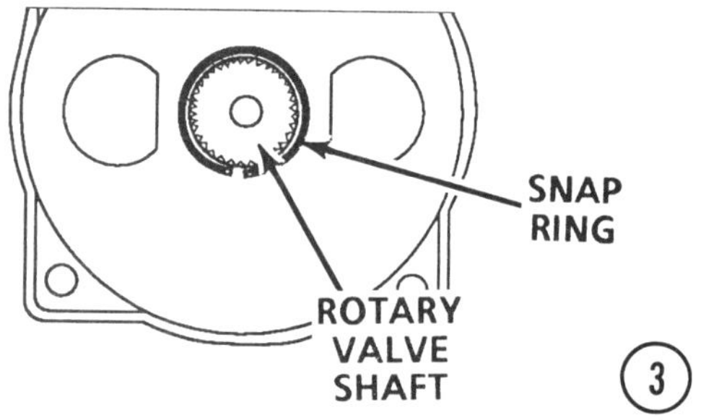

Rotary Valve Shaft Installation

1- Apply water resistant lubricant to the rotary valve shaft ball bearing. Install the bearing with "open" side facing the plug in the crankcase and the "shielded" toward the center of the crankcase. Tap the bearing into place with a suitable driver.

2- Install the rotary valve shaft into the crankcase. Use a soft head mallet and gently tap the shaft into place until it seats.

SAFETY WORDS

Snap rings and C-lockrings are made of spring steel and may pop out of the groove with considerable force. Therefore, warn others in the area and **WEAR** eye protection glasses or a shield while removing this type of ring.

3- Observe the snap ring. One face has a sharp edge -- the other edge is smooth. Using a pair of snap ring pliers, install the ring, into the groove in the rotary valve shaft with the sharp edge facing **OUTWARD.**

4- Rotate the shaft through a complete revolution to **ENSURE** there is no evidence of binding.

Crankshaft Assembling And Installation

5- Install the following items onto the forward -- tapered -- end of the crankshaft in the following order: the distance ring, a bearing with the O-ring on the side closest to the tapered end of the crankshaft, a spacer shim material piece, a second bearing with the O-ring on the side closest to the center of the crankshaft, and finally an oil seal with the lip **FACING** the bearing just installed.

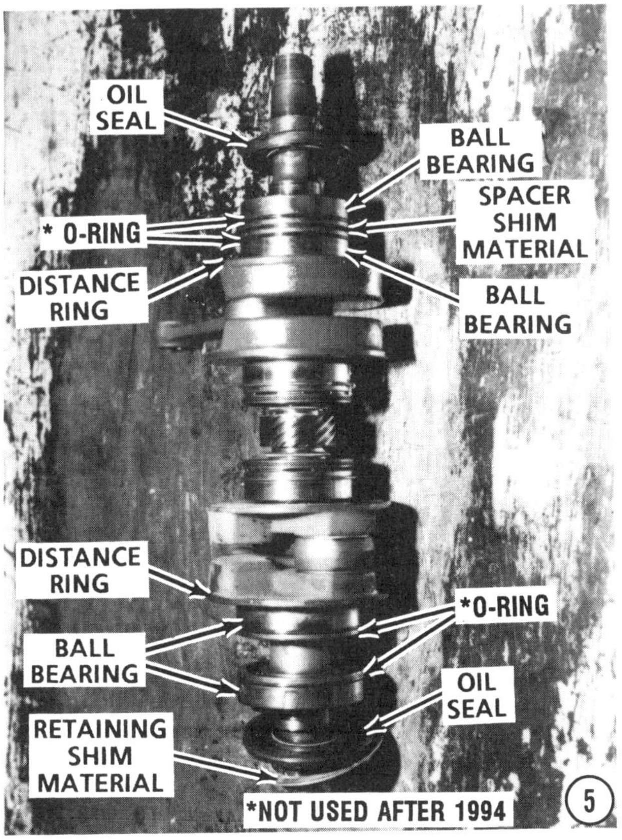

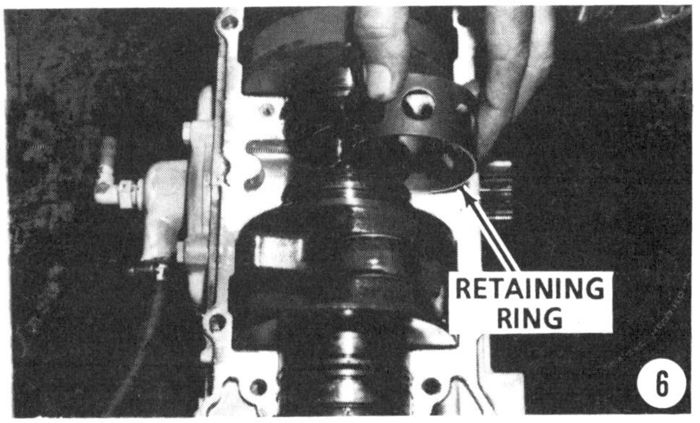

Install the following items onto the aft -- threaded end -- of the crankshaft in the following order: a distance ring, a bearing with the O-ring **FACING** the end of the crankshaft, second bearing with the O-ring on the side closest to the bearing just installed, an oil seal with the lip facing the bearing just installed, and finally the retainer shim material piece.

The two sets of O-rings should be as close to each, as possible.

The outer bearing on the aft end of the crankshaft **MUST** be spaced a specified distance from the other bearing. Obtain Distance Gauge P/N 290-876-826 and insert it between the two bearings prior to installation of the crankshaft. Remove this gauge after the crankshaft is seated.

Position the assembled crankshaft into the lower crankcase with the worm gear on the crankshaft meshing with the gear on the rotary valve shaft.

6- Install the retaining ring over the worm gear on the crankshaft.

Crankcase Assembling

7- Apply a light coating of 3 Bond #1011 gray gasket sealer onto the mating surfaces of

the two crankcase halves, working around the bolt holes. Make an honest attempt to make an even coating and to prevent the sealer from entering the bolt holes.

8- Bring the two halves of the crankcase together, with the bolt holes aligned as accurately as possible.

9- Apply a coating of Loctite 242 to the threads of the six hex bolts and the eight Allen head bolts. Install all the bolts and tighten them just **HAND TIGHT**, at this time.

Rotate the crankshaft by hand to verify it turns freely with no evidence of binding. If any binding is felt, the crankcase must be opened and the source of the binding discovered quickly -- before the Loctite sets.

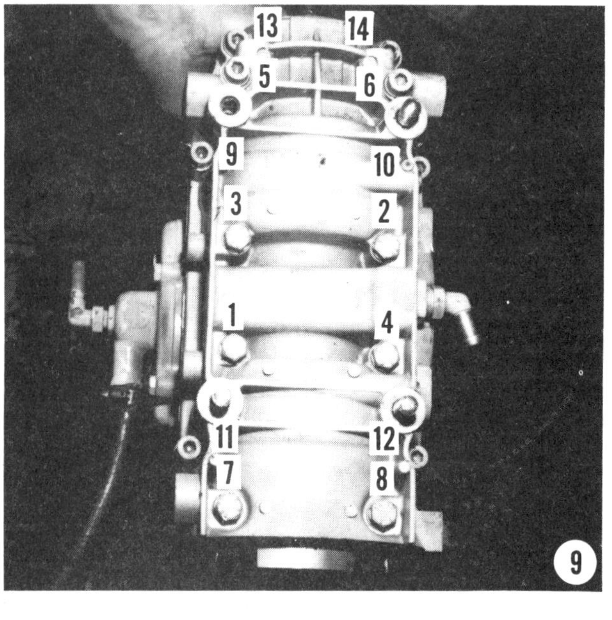

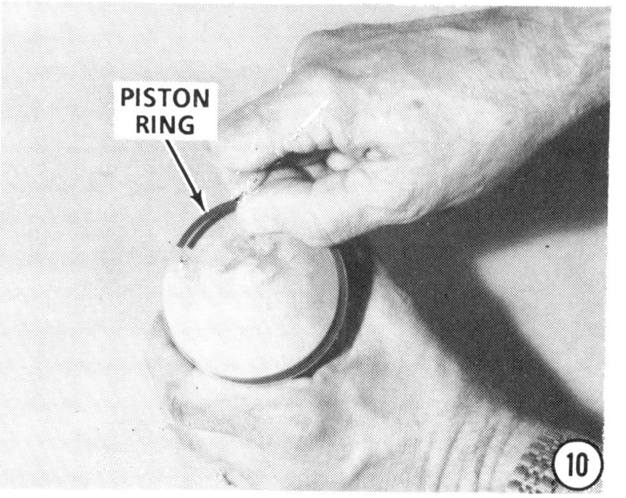

Once the determination is made the crankshaft is rotating freely, tighten the six hex head bolts to a torque value of 16 ft lb (22Nm) and the eight Allen head bolts to a torque value of 80 in lb (9Nm), all in the sequence indicated in the accompanying illustration.

Piston Installation

10- No special tool is necessary to install the piston rings. **HOWEVER**, take care to spread the ring only enough to clear the top of the piston. The rings are **EXTREMELY** brittle and will snap if spread beyond their limit.

Install a new set of piston rings onto each piston, with the rectangular ring in the lower groove. Align the ring end gap over the locating pin in each piston groove.

The purpose of the pin in the groove is to prevent the ring from rotating in the

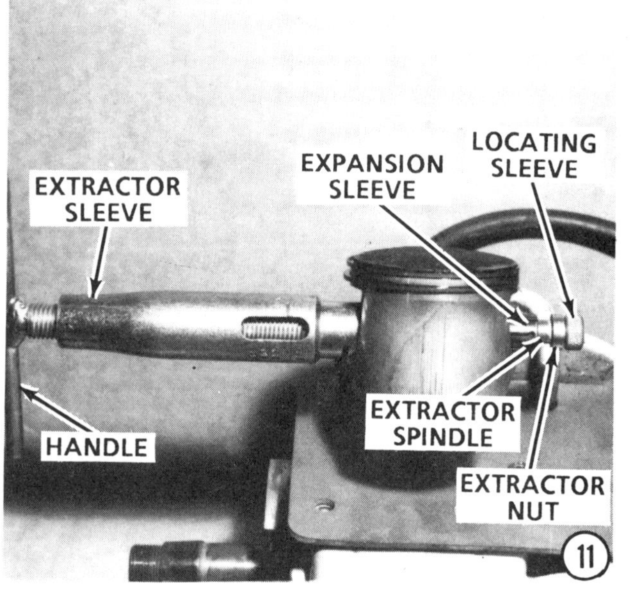

ring groove. The ring end gap must travel up and down a portion of the cylinder wall where there is no port opening. If the ring were free to rotate, an end might catch on the edge of a port opening and break.

SPECIAL PISTON INSTALLATION WORDS

Observe the intake hole in the side of the piston. The piston **MUST** be installed onto the connecting rod with the intake hole on the same side as the rotary valve housing -- the starboard side.

Two methods for installing pistons are presented. Perform Step 11, if the factory special tool is available. Perform Steps 12, 13, and 14, if the special tool is not available.

First, position the rubber mat or shop cloths over the crankcase to prevent needle bearings from being lost.

11- Dip the expansion sleeve, with needle bearings inside into a container of two-stroke engine oil. Insert the expansion sleeve between the piston pin bosses inside the piston. Place a thrust washer on each side of the sleeve. Insert the piston pin into the piston until it is flush with the inner edge of the piston pin boss. Insert the extractor spindle into the pin and install the locating sleeve and the extractor nut.

Rotate the handle **COUNTERCLOCKWISE** to pull the piston pin into the thrust washer, the needle bearings, the second

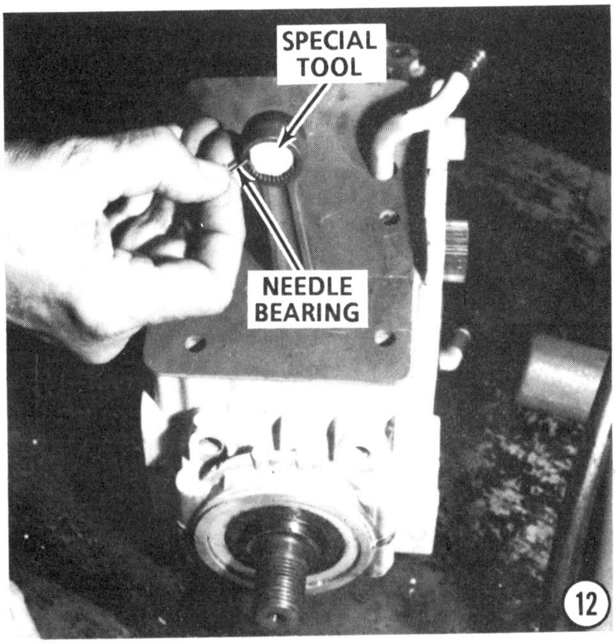

thrust washer, and the piston pin boss. The pin will drive out the expansion sleeve as the pin is pushed in.

Center the pin in the boss. The **C**-lockring grooves **MUST** be visible on both sides of the piston. Disassemble and remove the installation tools.

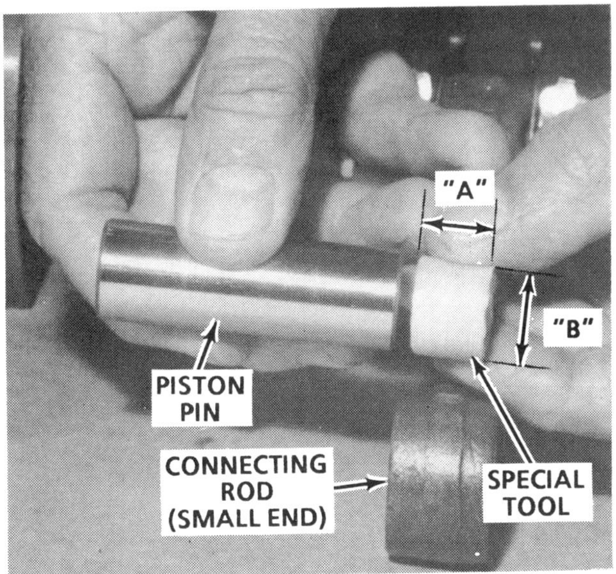

Length "A" must be just a "whisker" smaller than the length of the connecting rod small end. Diameter "B" must be just a "whisker" smaller than the diameter of the piston pin.

HOMEMADE TOOL SPECIAL WORDS

A tool must be made to keep the needle bearings in place and prevent them from falling from the perimeter of the small end of the connecting rod, while the piston pin is installed.

Find a bottle cap with straight sides which has the following dimensions: length of cap must be a "whisker" smaller than the length of the connecting rod hole. The diameter of the cap must be a "whisker" smaller than the diameter of the piston pin, as shown in the accompanying illustration.

*The piston **MUST** be installed with the intake hole on the starboard side.*

Tape, wound around the cap, can help to increase the diameter of the cap and make it the desired size.

12- Select a set of 31 needle bearings. Cover the inner circumference of the connecting rod with two-stroke engine oil. Begin to position the needle bearings -- one by one -- around the circumference. After about half the needle bearings are in place, slide the special homemade tool inside the semicircle of needle bearings. Continue inserting the needle bearings into the rod end until all 31 are in place.

13- Insert the piston pin through the piston just enough to hang a thrust washer on the end of the pin.

14- Install the piston over the end of the connecting rod. **TAKE EXTRA CARE** not to disturb the needle bearings and thrust washer. Check to be sure the intake hole in the piston is on the same side as the rotary valve housing -- the starboard side.

Insert the second thrust washer between the connecting rod end and the other boss. Slide the piston pin through the connecting rod end, thrust washer, and piston pin boss. As the piston pin slides through these parts, the pin will push the special homemade tool ahead of it. Center the piston pin by tapping it with a small punch and soft head mallet, until it is centered in the piston, with both **C**-lockring grooves visible at both ends.

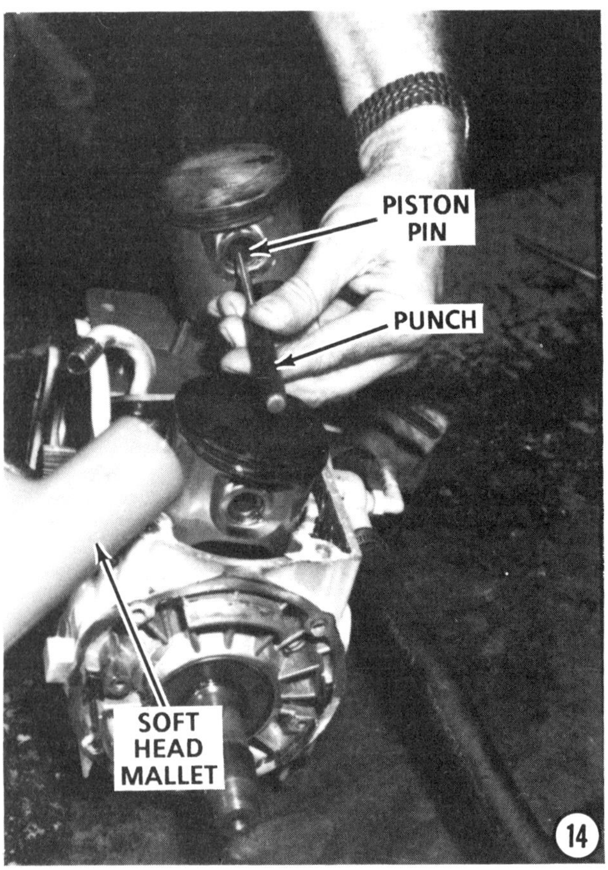

SAFETY WORDS

Snap rings and **C**-lockrings are made of spring steel and may pop out of the groove with considerable force. Therefore, warn

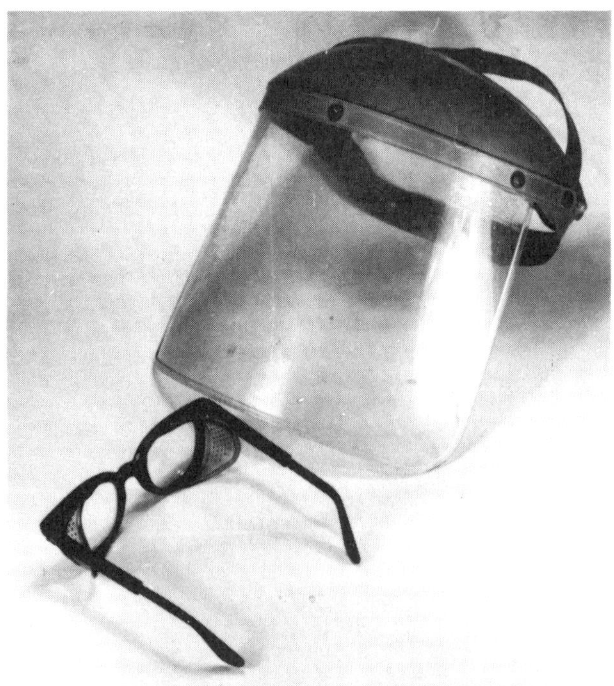

*C-lockrings are under tremendous tension during removal and installation, presenting a potential hazard. As an eye protection measure, safety glasses or a shield should **ALWAYS** be worn during work with such snap rings. Warn others in the area such work is in progress.*

others in the area and **WEAR** eye protection glasses or a shield while removing this type of ring.

15- Install a **C**-lockring in the groove at each end of the piston pin. The lockrings may be worked in **CAREFULLY,** by hand. Place the end gap of the **C**-locking ring at the bottom -- the 6 o'clock position. Remove the rubber pad and shop cloths.

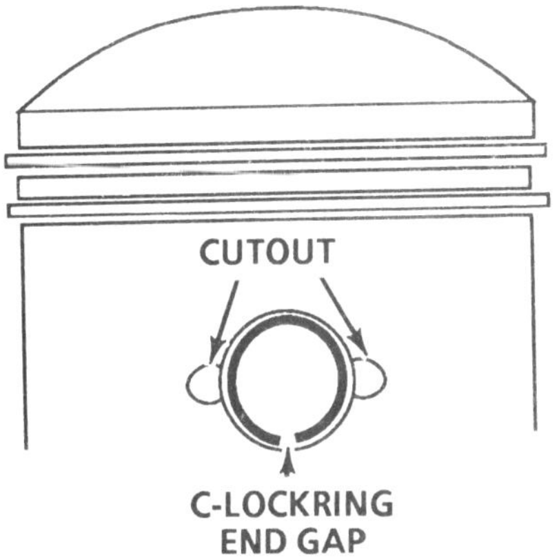

The C-lockring end gap must be positioned away from the two cutout in the piston, as indicated in this line drawing.

Cylinder Installation

16- Place a new cylinder gasket over the piston and onto the top surface of the upper crankcase. Coat the crown and upper portion of the piston sides and the rings with two-stroke engine oil. Check to be sure the ring end gap straddles the piston locating pin in each groove.

Any one of three methods may effectively be used to install the pistons into the cylinder.

a- Obtain Ring Compressor P/N 290-876-972. **UNFORTUNATELY**, this special tool cannot be used to install oversize pistons. Slide the compressor over the piston, with the tapered end going on first. Hold the piston upright, and slowly lower a cylinder down over the piston crown and rings, with the cylinder exhaust port on the opposite side to the rotary valve housing.

b- Obtain a large hose clamp and secure it over the piston rings to compress them. Hold the piston upright, and slowly lower a cylinder down over the piston crown and rings, with the cylinder exhaust port on the opposite side to the rotary valve housing.

c- Obtain the services of an assistant to compress the rings while the cylinder is lowered. Hold the piston upright, and

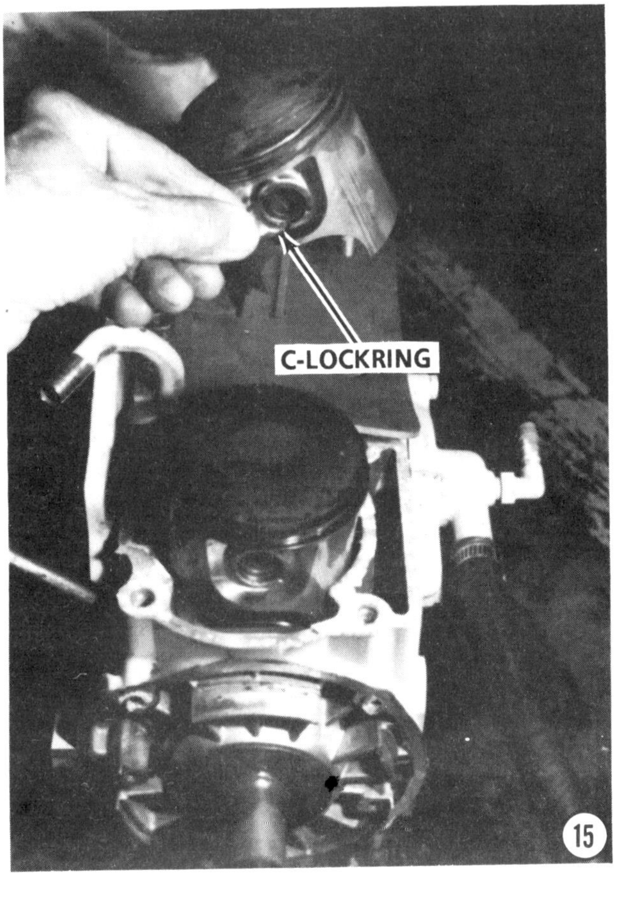

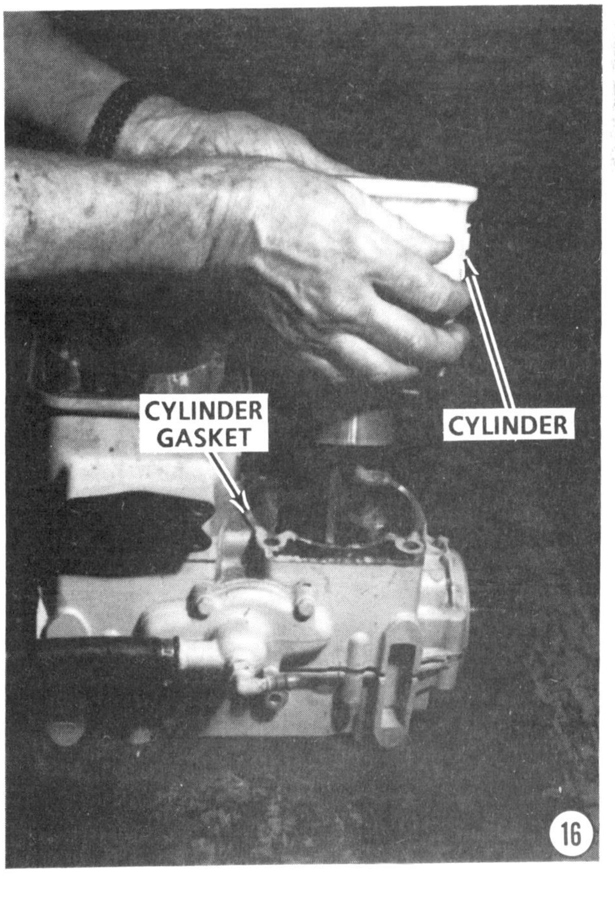

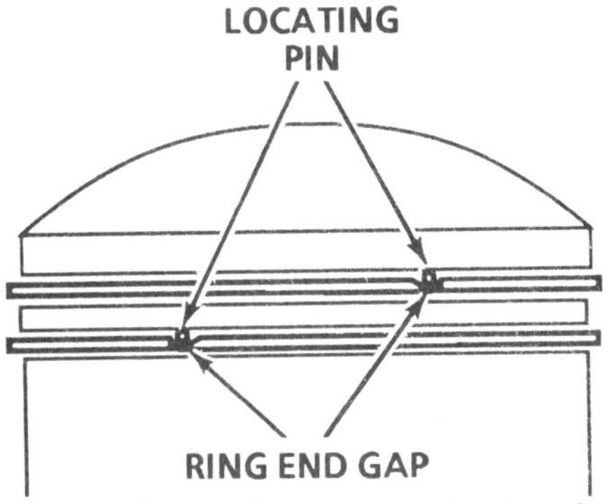

Align the ring end gap over the locating pin in each piston groove to prevent the ring from rotating in the groove. If the ring was free to rotate, an end might very well catch on the port opening and break.

slowly lower a cylinder down over the piston crown and rings, with the cylinder exhaust port on the opposite side to the rotary valve housing.

CRITICAL WORDS

If difficulty is experienced in fitting the piston into the cylinder, **DO NOT** force the piston. Such action might result in a broken piston ring. Check to be sure the ring end gap straddles the locating pin in the groove.

17- Install and tighten the four Allen head bolts securing each cylinder to the crankcase. These bolts are to be installed **DRY** -- with no sealing compound applied to the threads. Tighten the bolts alternately and evenly to a torque value of 16 ft lb (22Nm).

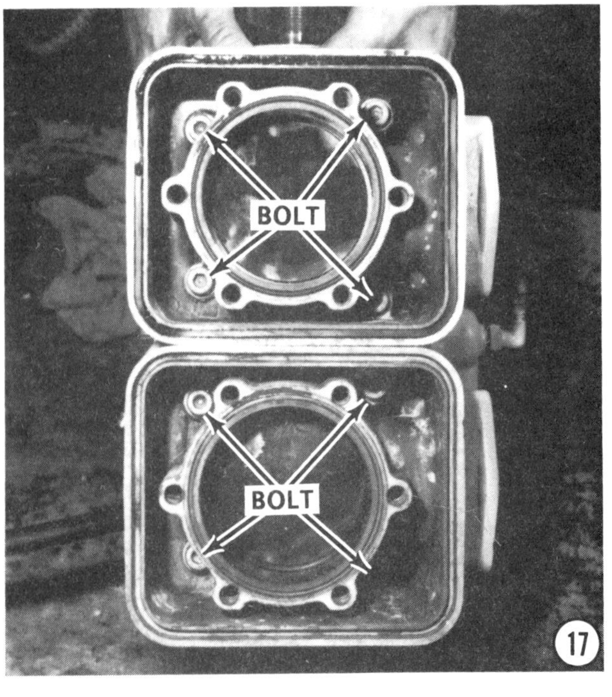

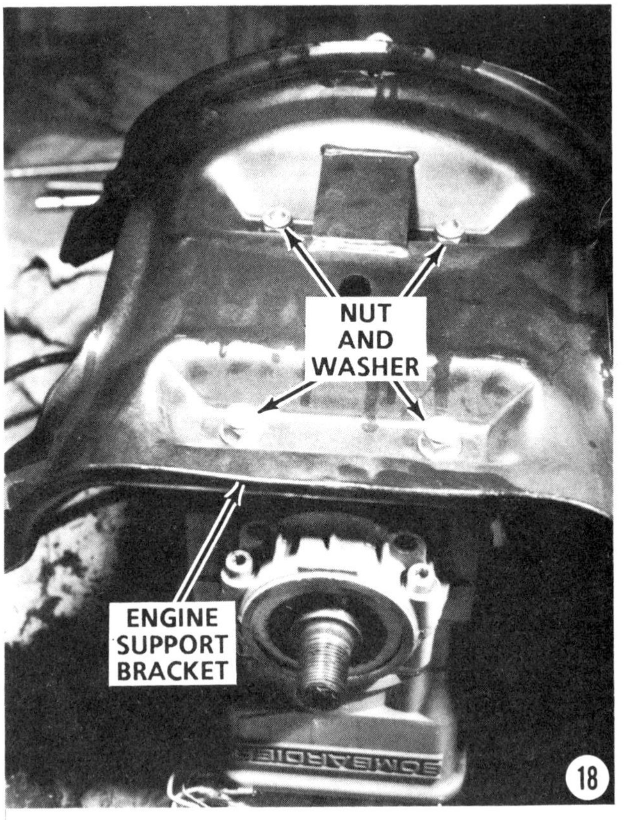

Engine Support Bracket

18- Apply a coating of Loctite 242 to the threads of the four studs on the bottom of the lower crankcase. Install the engine support bracket. Install the washers and locknuts. Tighten the nuts alternately and evenly to a torque value of 26 ft lb (35Nm).

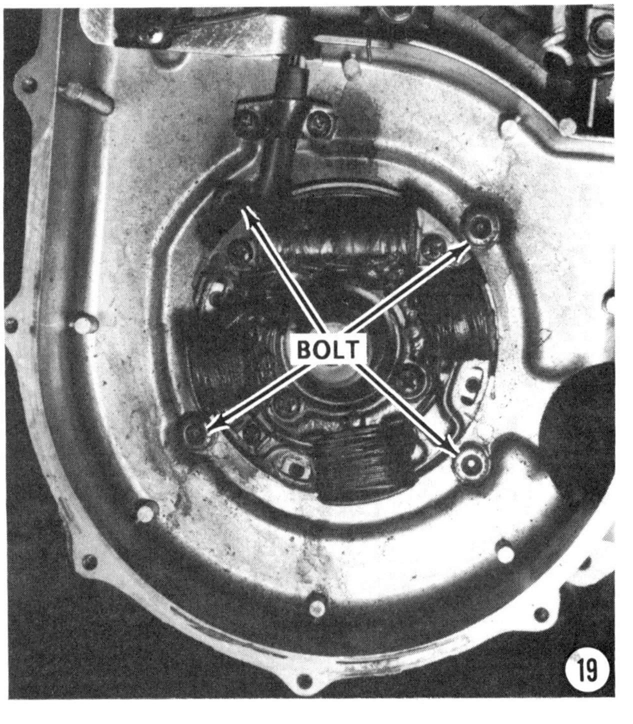

Stator Plate Installation

19- Apply a light coating of 3 Bond #1011 gray gasket sealer to the crankcase and stator plate mating surface. Make an honest attempt to make an even coating and to prevent the sealer from entering the bolt holes. Apply Loctite 242 onto the Allen head bolt threads. Install the stator plate and magneto assembly over the crankshaft. Align the bolt holes and install the bolts. Tighten the bolts alternately and evenly to a torque value of 16 ft lbs (22Nm).

20- Coat the threads of the two attaching Allen head bolts with Loctite 242. Install the magneto assembly onto the stator plate. Align the marks made during disassembling. Install the bolts, washers, and lockwashers. Tighten the bolts alternately and evenly to a torque value of 16 ft lb (22Nm). Check to be sure the two marks remained aligned while the bolts were tightened.

Flywheel Installation

21- Place a dab of thick lubricant onto the curved surface of the Woodruff key to hold it in place while the flywheel is being installed. Press the Woodruff key into place in the crankshaft recess. Wipe away any excess lubricant to prevent the flywheel from "walking" during engine operation. Check the flywheel magnets to ensure they are free of any metal particles. Double check the taper in the flywheel hub and the taper on the crankshaft to verify they are clean and free of any oil.

Now, slide the flywheel onto the crankshaft, with the Woodruff key in place. Rotate the flywheel **COUNTERCLOCKWISE** to be sure it does not contact any part of the magneto assembly or wiring on the stator plate.

22- Attach Puller Plate P/N 290-876-065 to the flywheel with three M8x25mm bolts. Install the extension handle onto the puller arm. Apply Loctite 242 onto the crankshaft taper and threads. Install the crankshaft nut and washer. Tighten the flywheel nut to a torque value of 77 ft lb (105Nm). Leave the puller plate in place over the flywheel for the next three steps.

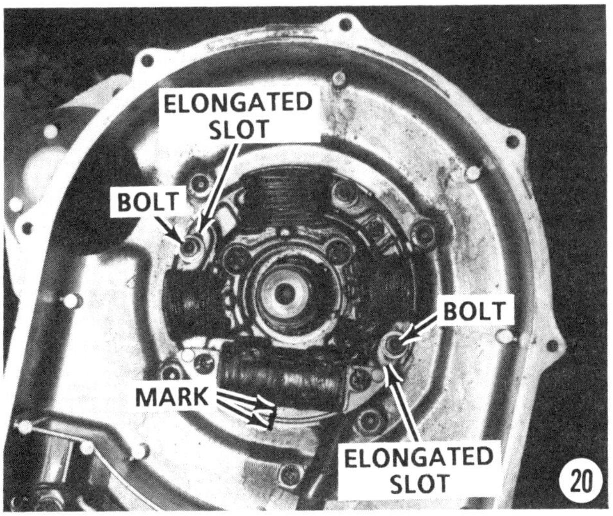

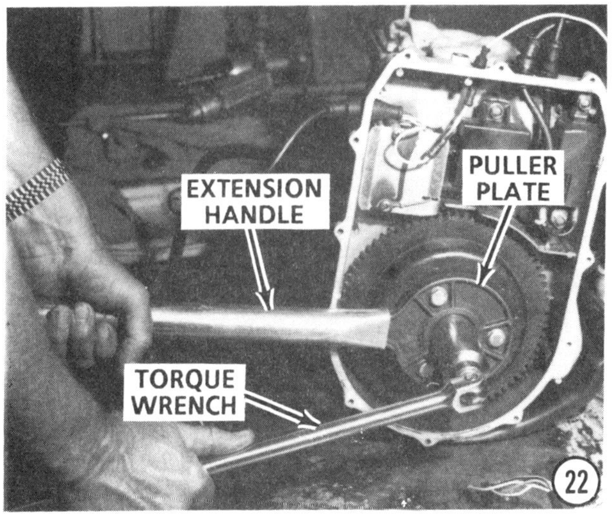

23- Apply anti-sieze compound to the threads of the aft end of the crankshaft. Thread the aft flywheel onto the crankshaft.

24- Insert Impeller Remover P/N 295-000-001 into the splines in the aft flywheel.

25- Tighten the aft flywheel to a torque value of 81 ft lb (110Nm).

Cranking Motor Installation

26- Position a new gasket over the mounting flange and insert the cranking motor into the stator plate.

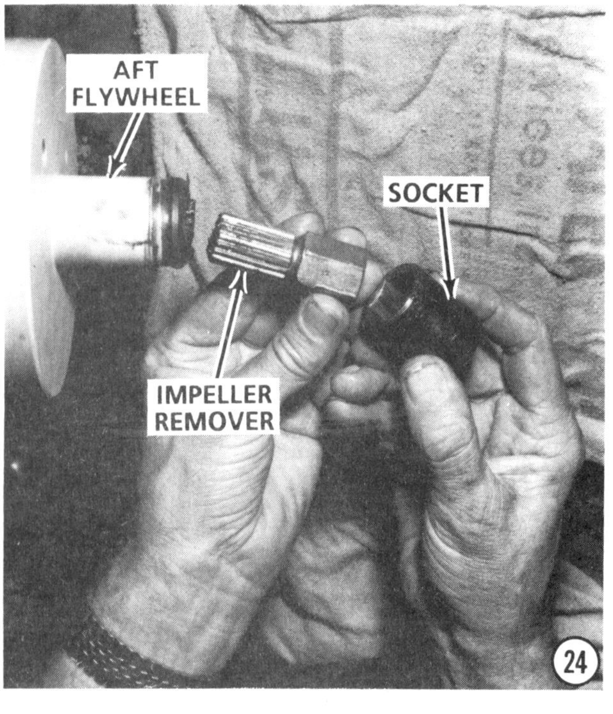

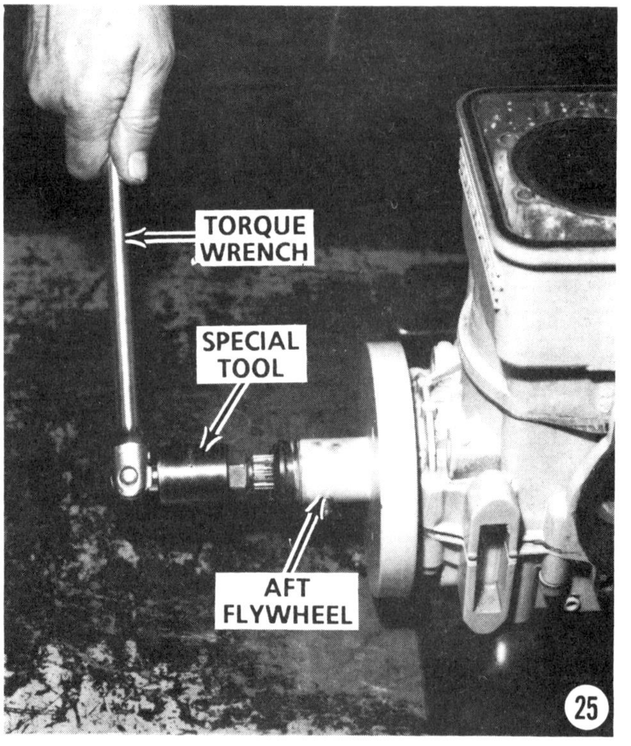

27- Apply Loctite 242 onto the threads of the three mounting bolts. Position the Black ground lead over the upper mounting bolt. Install the other two bolts, and then tighten all three to a torque value of 16 ft lb (22Nm).

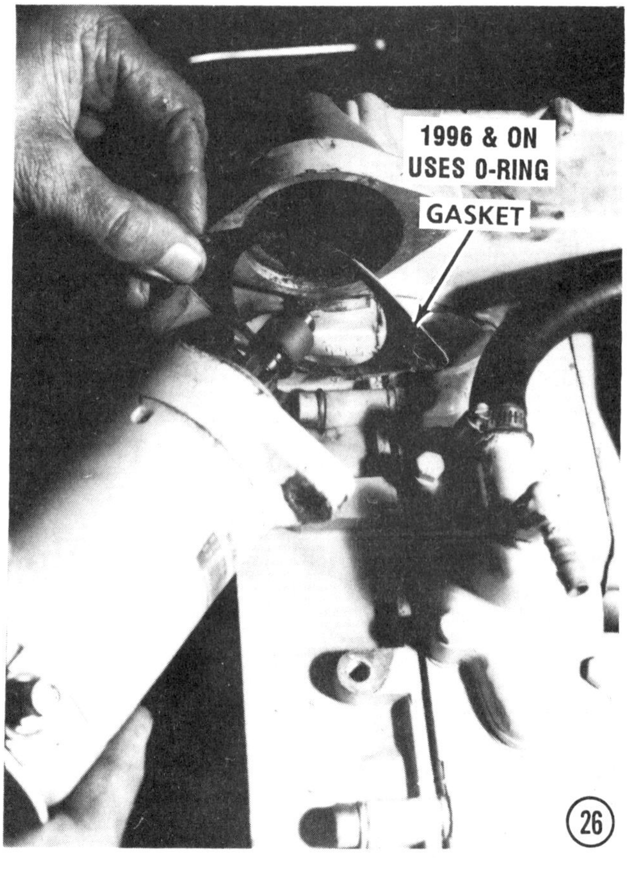

28- Make a setup with a dial indicator gauge over the forward piston. Slowly rotate thc crankshaft until the dial indicator shows the piston to be at TDC. Now, if the engine was properly timed before disassembling, the marks made during assembly -- one on the face of the rotor and the other on the block should be aligned.

Coat the rotary valve disc with Sea-Doo injection oil. Install the disc over the splines on the rotary valve shaft with the opening edge of the disc aligned with the marks made during disassembly.

If the disc edge is not perfectly aligned, remove it, turn it over, and then install it a second time. The disc is asymmetrical to allow the best fit to be made for one face, but not the other.

The edge aligned with the marks is the opening edge and the edge below the forward port is the closing edge.

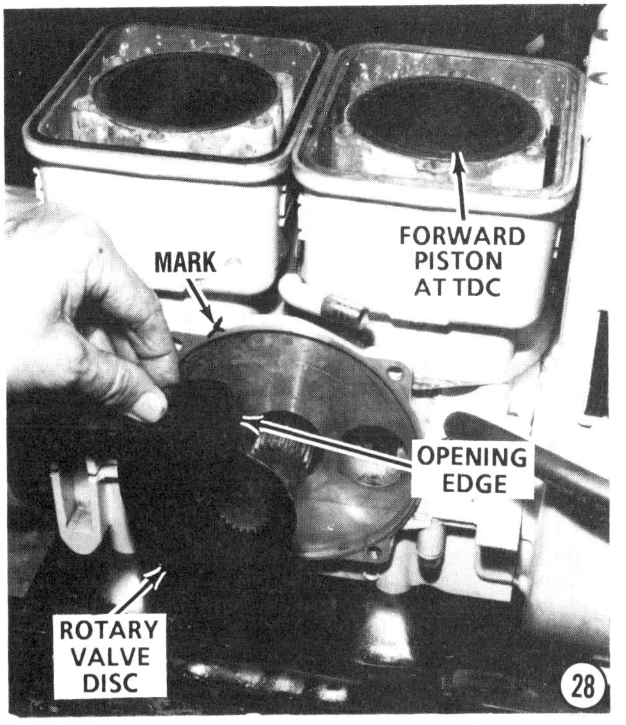

Rotary Valve Installation -- Replacement Crankcase or Verification of Timing for Original Crankcase.

CRITICAL WORDS

More than 1/2 dozen and perhaps up to a dozen degree wheels are available. Therefore, obtain the **PROPER** Degree Wheel for the engine being serviced from the Bombardier dealer. The degree wheel in the Appendix is **ONLY** a sample for reference purposes.

29- After the correct degree wheel has been obtained for the engine being serviced, make a setup with a dial indicator gauge over the forward #1 piston. Slowly rotate the crankshaft until the dial indicator indicates the piston to be at the TDC position. Place the degree wheel over the rotary valve shaft.

Align the 360° line with the lowest point of the forward inlet port. Identify the 132° line on

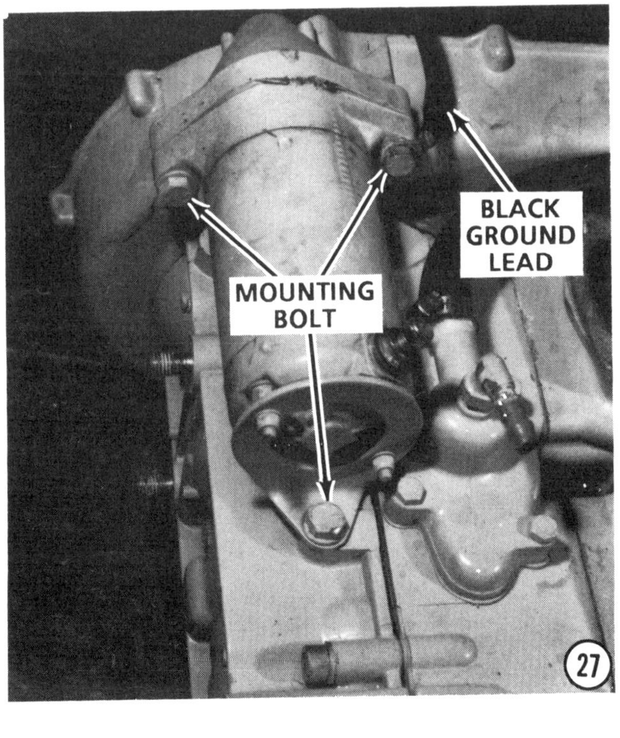

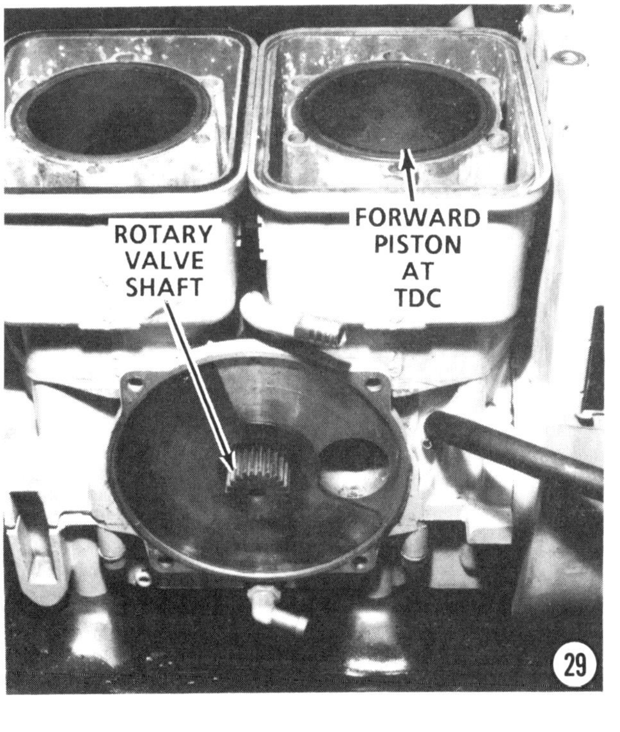

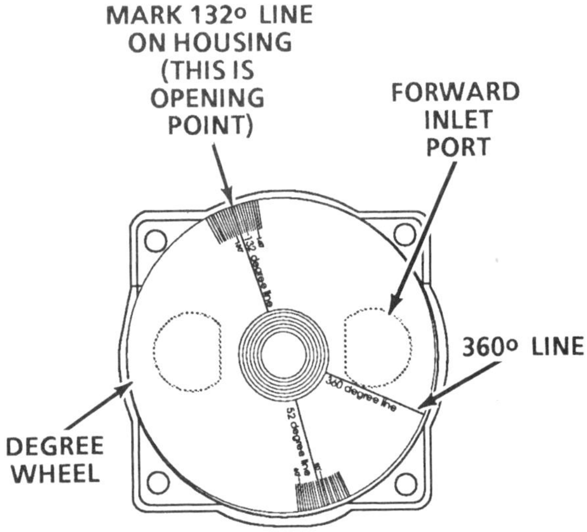

Determining the opening point of the rotary valve disc, using a degree wheel. See illustration below for closing point.

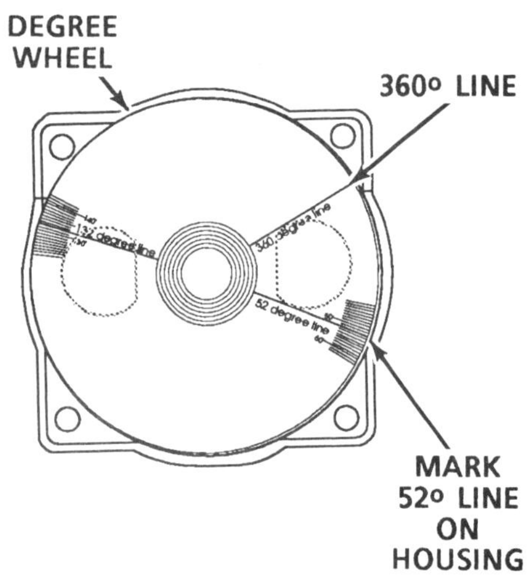

Determining the closing point of the rotary valve disc using a degree wheel, as explained in detail in the text. A couple degree wheels, suitable for cutting out and using, are included in the Appendix.

the degree wheel, and scribe a matching mark on the rotary valve housing. This mark will represent the port opening point and will be at approximately the 11 o'clock position, as depicted in the accompanying line drawing -- with the degree arrangement just described identified.

Change the position of the degree wheel on the rotary valve shaft, to place the 360° line at the highest point of the forward inlet port. Identify the 52° line and scribe a matching mark on the rotary valve housing. This mark will represent the port closing point and will be at approximately the 4 o'clock position, just below the inlet port. This arrangement is depicted in the other accompanying line drawing.

Coat both sides of the rotary valve disc with Sea-Doo injection oil. Remove the the degree wheel and install the rotary valve disc onto the shaft. Align one edge of the disc with the opening mark and the other edge with the closing mark. If the disc is not perfectly aligned with the two marks, remove the disc, turn it over and install it a second time. The disc is asymmetrical to allow the best fit to be made for one face, but not the other.

For more information on this type of rotary valve operation, see Chapter 3 -- **TUNING.**

ENGINE INSTALLATION

ENGINE MOUNT GOOD WORDS

The manufacturer states engine alignment procedures are not necessary. Proper engine alignment is performed at the factory and if the engine mounts and pieces of shim material are left **UNDISTURBED**, the engine will always remain in perfect alignment.

30- Check to be sure the shim material is still in place on top of the engine mounts. Enlist the aid of an assistant, if necessary, and lower the engine down through the opening and onto the engine mounts. Align the holes in the engine support bracket with the holes in the mounts.

31- The accompanying illustration shows a typical installation. Apply a coating of Loctite 242 onto the threads of the four mounting bolts. Install the bolts and washers. Tighten the bolts alternately and evenly to a torque value of 18 ft lbs (25Nm).

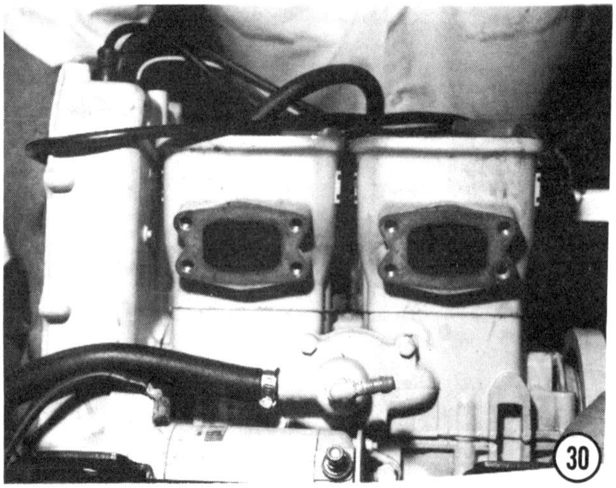

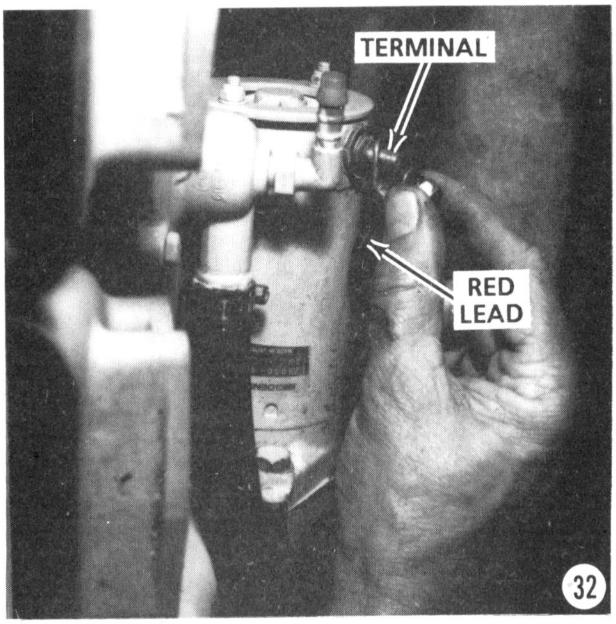

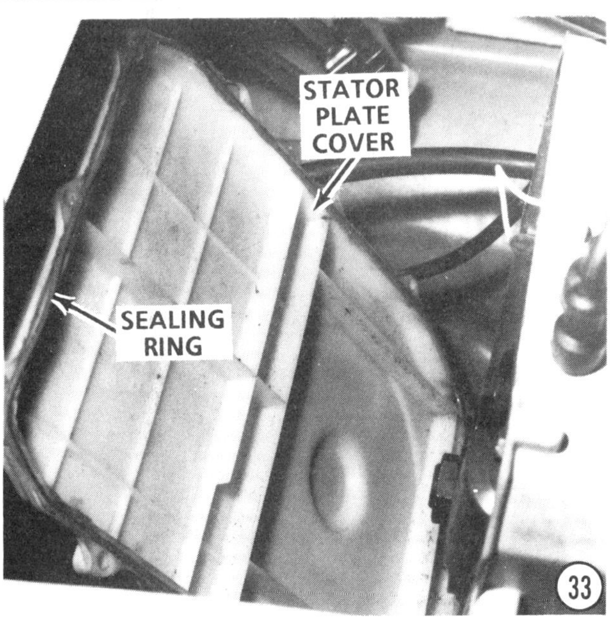

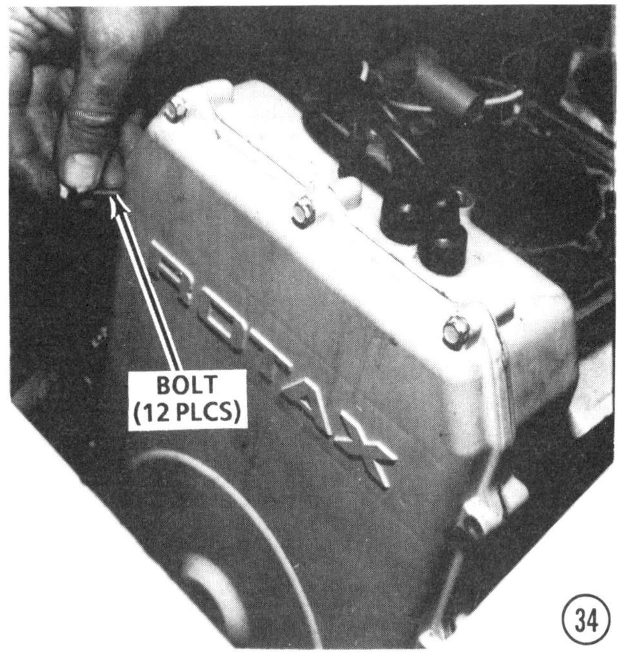

Engine Buildup

32- Connect the large Red lead onto the cranking motor terminal.

33- Apply Loctite 242 to the threads of the twelve stator plate cover securing bolts. Install a new sealing ring into the groove of the stator plate cover.

34- Install the cover and secure it in place with the attaching bolts. Tighten the bolts to a torque value of 35 in lb (4Nm).

35- Position a clamp at each end of the driveshaft boot. Install the boot between the seal carrier and the aft flywheel. Tighten both hose clamps just **SNUG**, at this

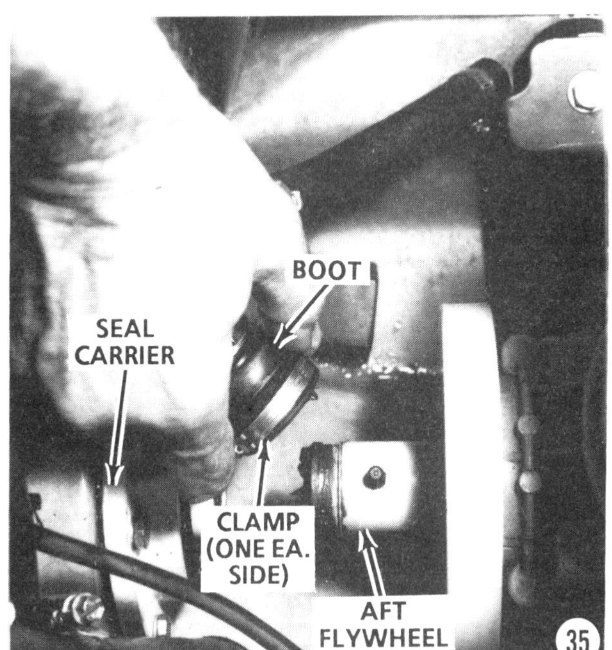

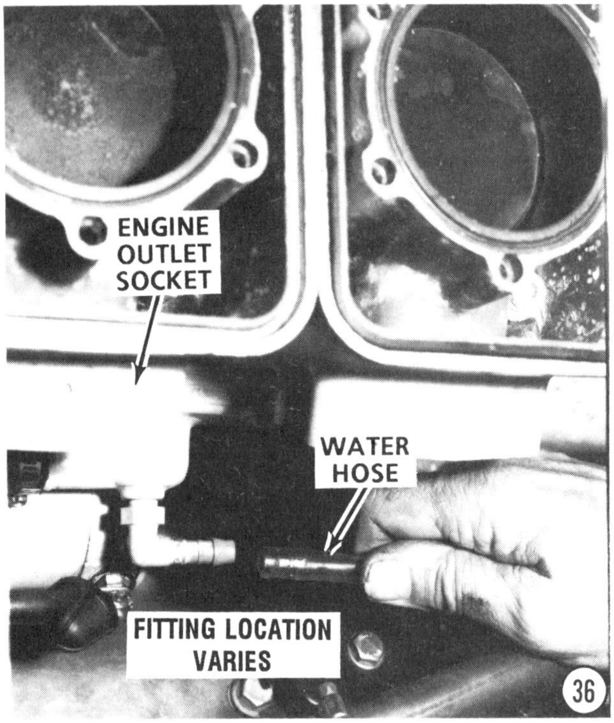

time. If the clamps are tightened securely now, it could affect engine alignment.

36- Slide a clamp over the end of the water return hose, and then slide the hose onto the fitting on the engine outlet socket. Tighten the clamp to secure the hose in place.

Exhaust Manifold Installation

37- Slide a clamp over the end of the inlet cooling water hose. Work the hose onto the fitting at the base of the exhaust manifold. Tighten the clamp to secure the hose. Position new gaskets over the exhaust ports on the intake manifold. Align the gaskets and manifold with the engine exhaust ports.

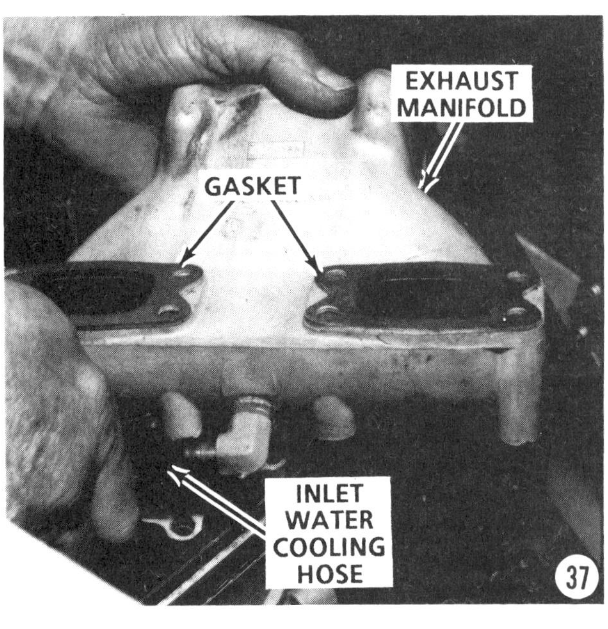

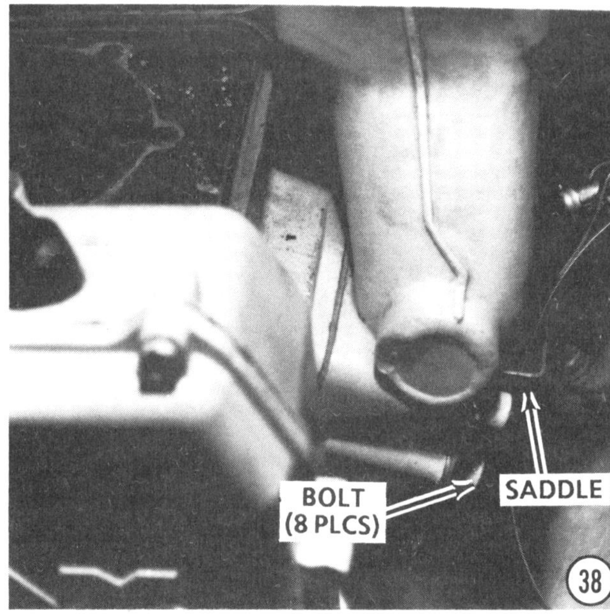

38- Apply Loctite 242 to the exhaust manifold securing bolts. **FIRST**, install and **HAND** tighten the four bolts and washers securing the exhaust manifold to the **AFT** cylinder. From 1991 and on, the saddle shown in the illustration has not been used. Install the four bolts and washers through all the parts held in place, and then **HAND** tighten the bolts. Now, tighten all bolts alternately and evenly to a torque value of 15 ft lbs (20Nm).

39- Install a new gasket onto the top of the exhaust manifold. This gasket replaces **O**-rings used on earlier model engines.

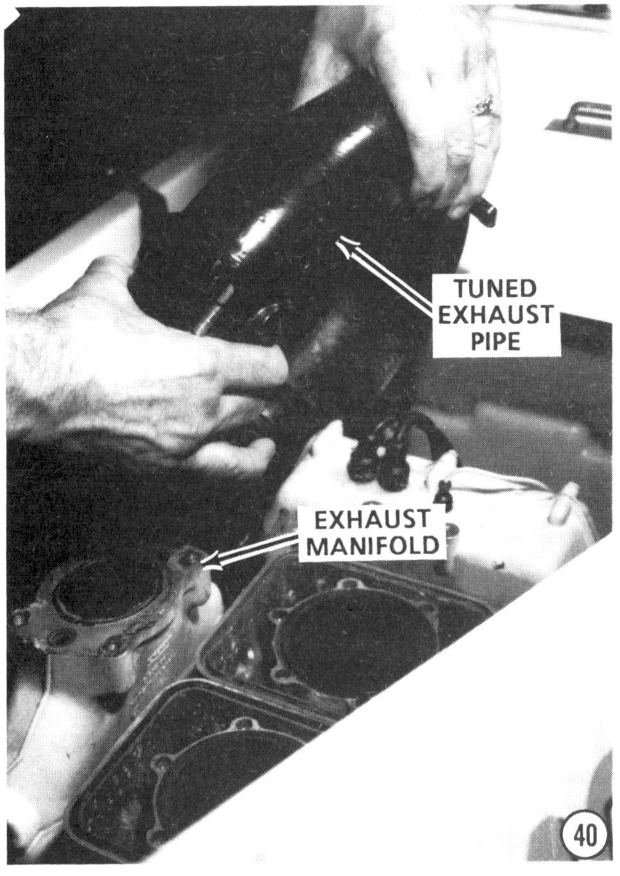

40- Ease the tuned exhaust pipe into the engine compartment and slightly forward of the exhaust manifold mounting flange.

SPECIAL STUD WORDS

Various model engines and different production years of the engine result in two arrangements of bolts and studs for the exhaust pipe. Therefore, use the same attaching hardware as was removed during disassembling.

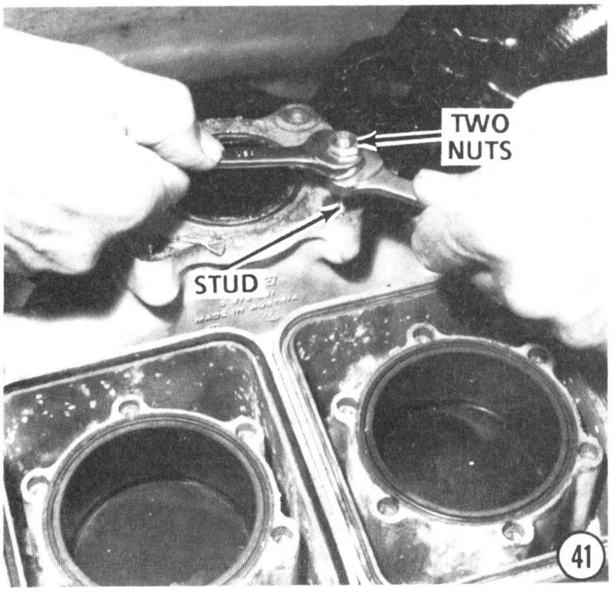

41- Apply Loctite 242 to the manifold stud threads, if it was removed. Install a stud into the exhaust manifold mounting flange. Thread two nuts onto the stud, and then tighten them against each other. Tighten the upper nut to install the stud into the exhaust manifold flange. Remove both nuts to be used on the other studs.

42- Move the tuned exhaust pipe over the studs. Install and tighten the nuts and washers to a torque value of 18 ft lbs (25Nm).

43- Tighten the two clamps on hose connecting the tuned exhaust pipe to the muffler.

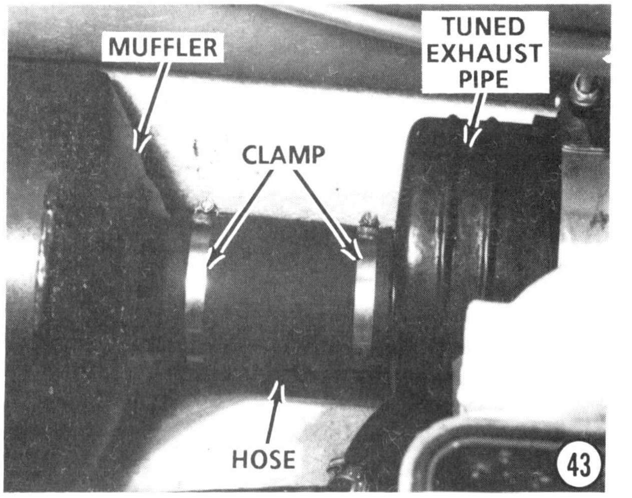

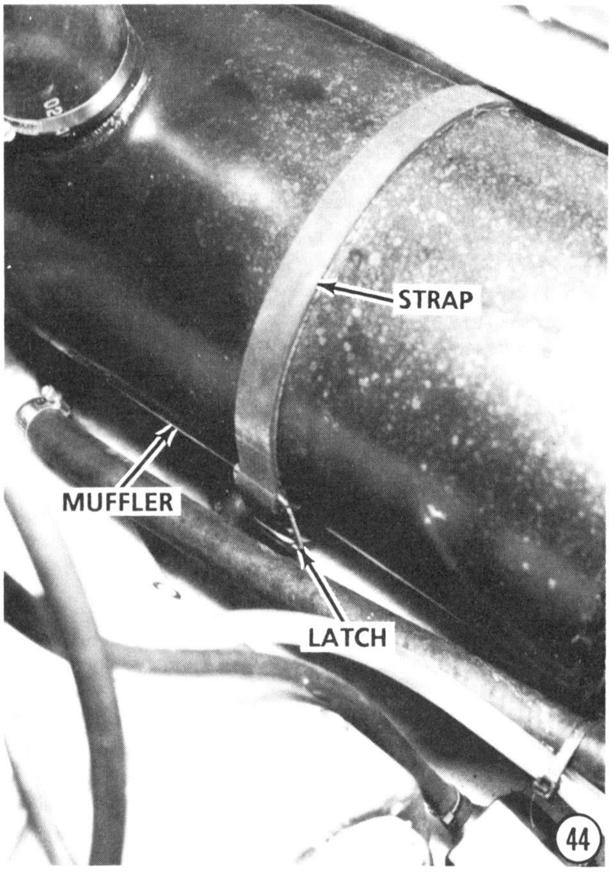

44- Hook the end of the muffler strap onto the latch in the hull.

45- Install the rubber cushioning strip, if used, between the tuned exhaust pipe and the large clamp. Tighten the clamp securely.

46- Slide the clamps over the water injection hose and the water inlet hose. Work the hoses onto the fittings on the tuned exhaust pipe. Tighten the clamps securely.

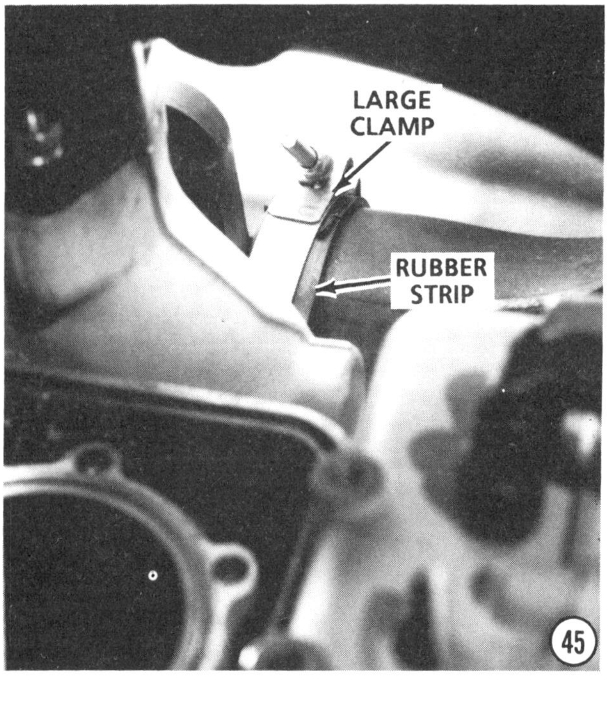

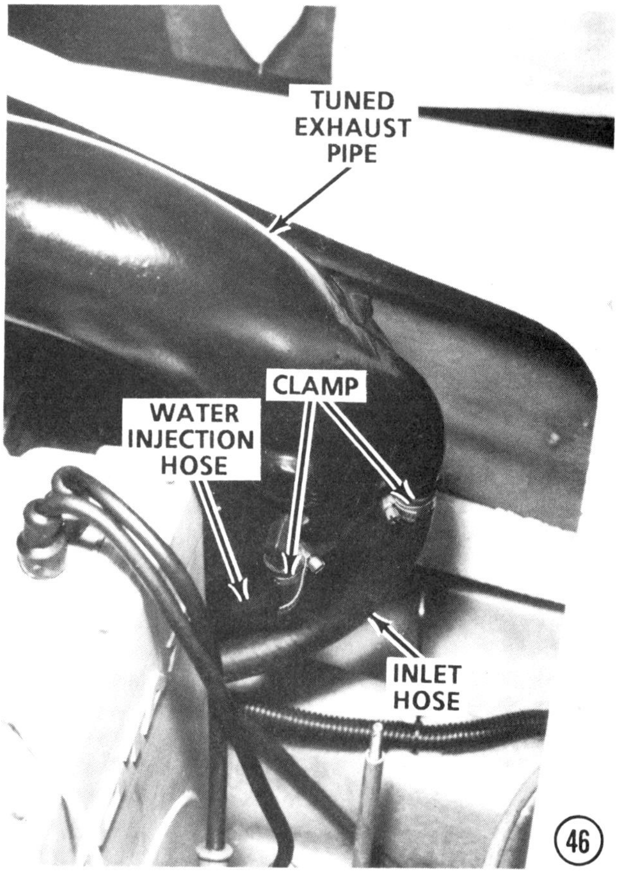

47- Connect the drain tube to the tuned exhaust pipe fitting. Use a new tie wrap over the tube to secure it in place.

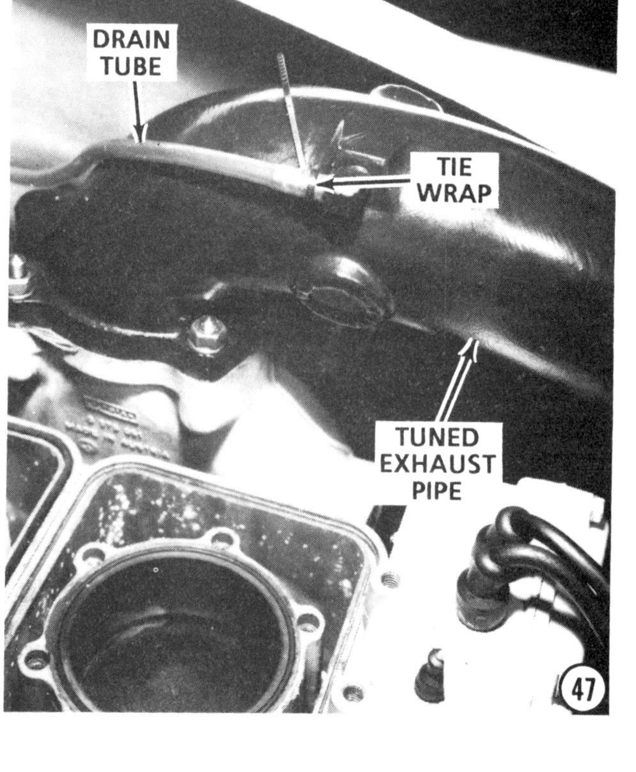

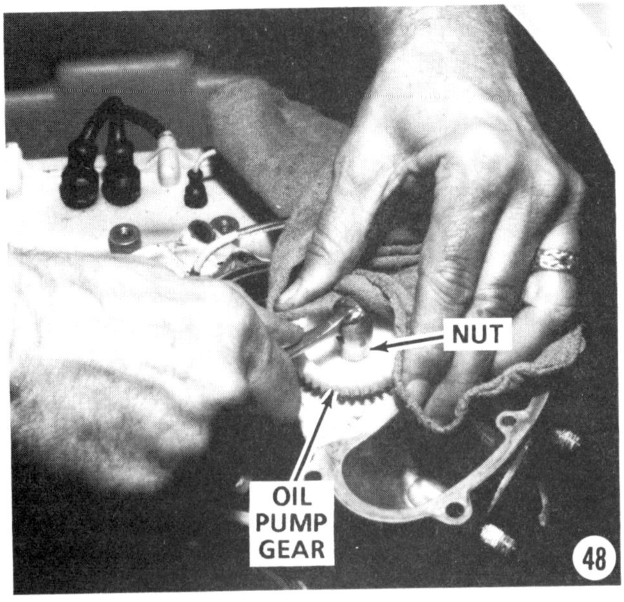

Oil Injection Pump Installation

48- Install a new O-ring over the oil pump shaft. Install the pump onto the intake manifold with the two bolts and washers. Tighten the bolts to a torque value of 53 in lb (6Nm). Install the two "banjo" bolts and fittings, with the oil lines attached, to the pump. Apply Loctite 242 to the pump shaft threads.

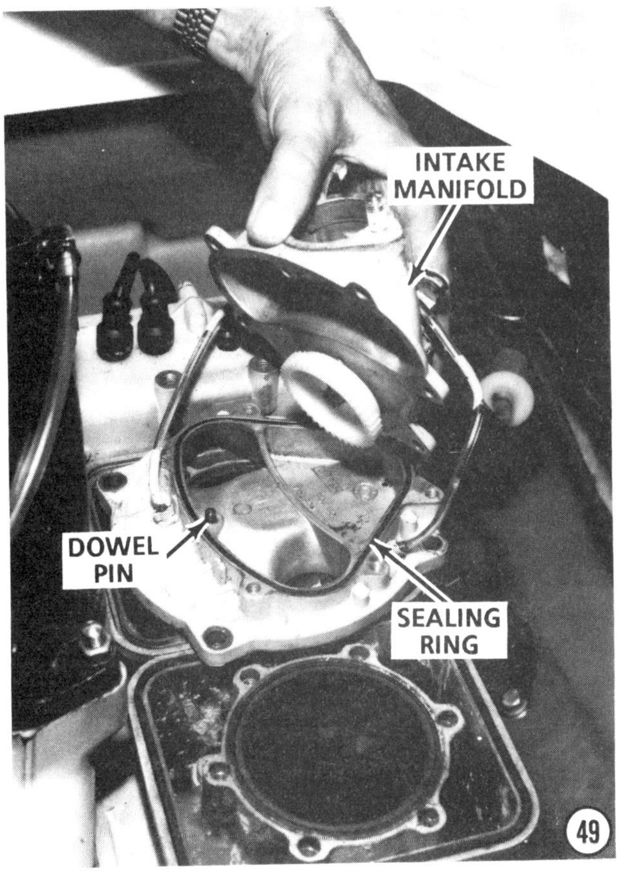

Inside the intake manifold: install the washer, gear, another washer, and then the nut onto the oil pump shaft. Obtain Gear Holder P/N 290-277-905 or shop cloth. Hold the gear firmly to prevent it from rotating while the nut is tightened to a torque value of 89 in lb (10 Nm).

49- Fit a new sealing ring into the groove of the rotary valve cover. Install the intake manifold onto the rotary valve cover, with the manifold indexed over the single dowel pin in the cover.

50- Install the six bolts and washers securing the intake manifold to the rotary valve cover. Tighten the bolts alternately and evenly to a torque value of 80 in lb (9Nm).

Throttle Cable Installation

51- Hook the end of the throttle cable into the pump lever. Push the cable into

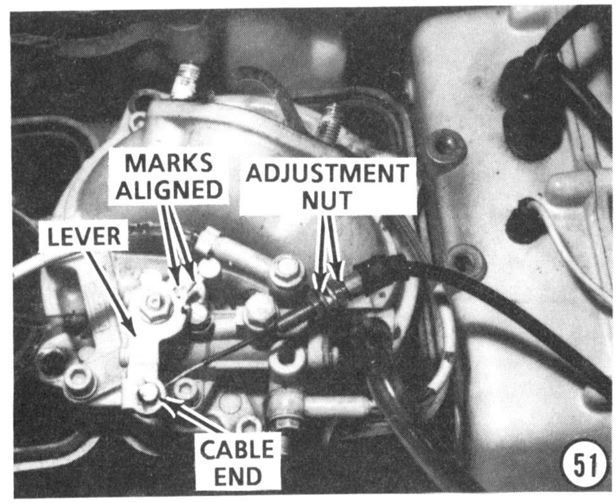

the slot with an adjustment nut on each side of the slot.

Oil Pump Synchronization

Observe the mark on the pump lever and the mark on the pump boss. These two marks **MUST** align for oil pump synchronization. Adjust the position of the two nuts to change the length of the cable, and therefore, the position of the lever until the two marks align.

Rotary Valve Cover Installation

52- Slide a clamp over the end of the oil supply hose. Remove the pincer tool (or plug in the end of the hose), and quickly work the hose onto the fitting at the base of the rotary valve. Tighten the clamp securely.

53- Fit a new sealing ring into the groove in the rotary valve housing. Install the housing against the upper and lower crankcase. Install the four securing bolts and washers. Tighten the bolts to a torque value of 15 ft lb (20Nm).

Slide a clamp over the end of the oil return hose. Remove the pincer tool (or plug in the end of the hose), and quickly work the hose onto the fitting at the top of the rotary valve. Tighten the clamp securely.

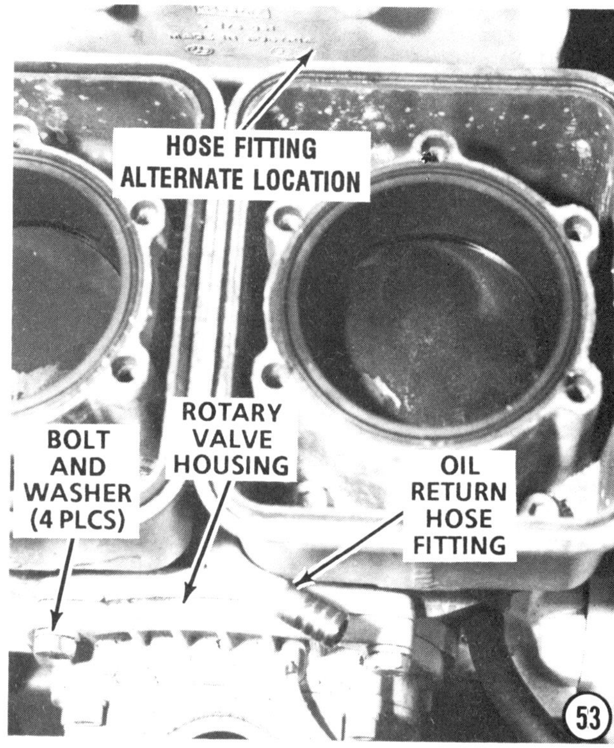

Cylinder Head Installation

54- Apply a coating of Loctite Primer N into the cylinder and water jacket grooves for each cylinder. Apply the primer into the corresponding grooves in the cylinder head. Coat both sides of the 12 cylinder head bolts and washers with the Loctite Primer. Allow about 20 minutes for the primer to dry. Apply Loctite 515 gasket eliminator over all the surfaces coated with the primer, including the grooves.

Install the sealing rings into the cylinder and water jacket groove for each cylinder.

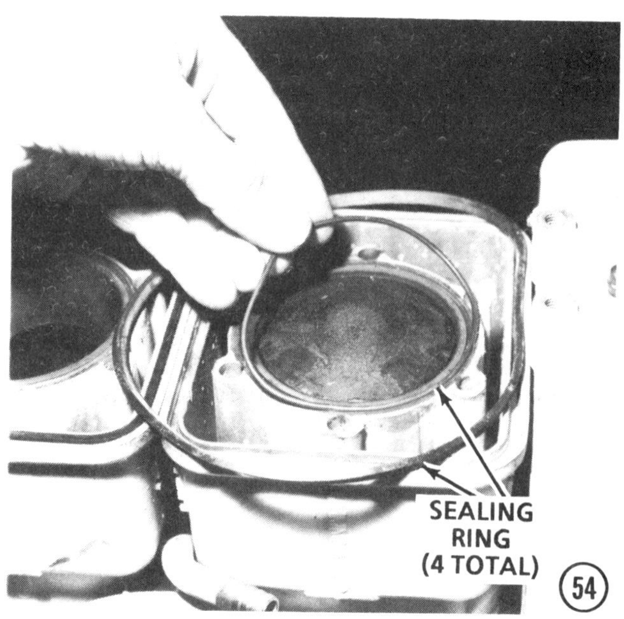

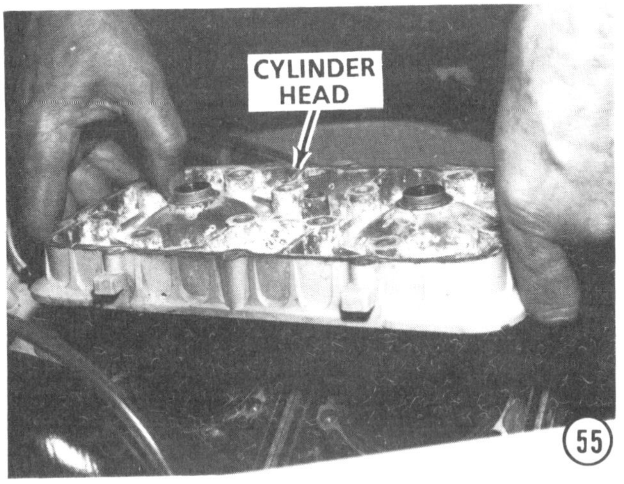

55- Lower the cylinder head into place on the two cylinders, with the bolt holes aligned. **TAKE CARE** not to deform the sealing rings.

56- Install the bolts and washers -- coated with the primer and Loctite 515. Tighten the bolts just **HAND** tight, at this time. Insert new **O**-rings into place in each spark plug opening. Place a new small **O**-ring into the temperature sensing hole, indicated in the accompanying illustration.

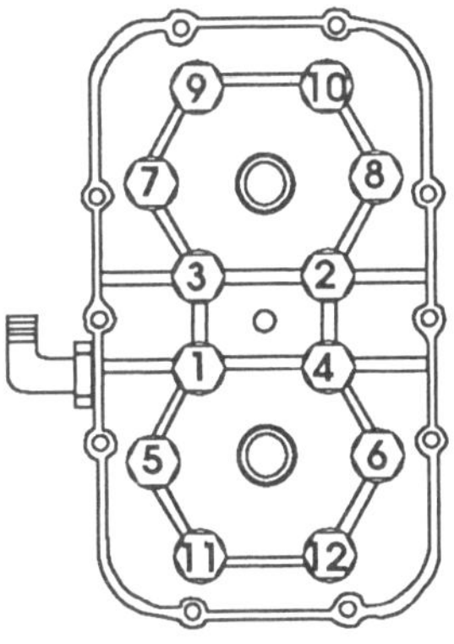

Tightening sequence for the cylinder head securing bolts. Tightening sequence for the head cover is given on Page 8-52.

57- Tighten the cylinder head bolts to a torque value of 15 ft lb (20Nm), in the sequence indicated in the accompanying captioned line drawing.

58- Install a new sealing ring into the groove in the cylinder head. Lower the cover into place on the head.

59- On some model engines, a stator support bracket is used. Therefore, position the

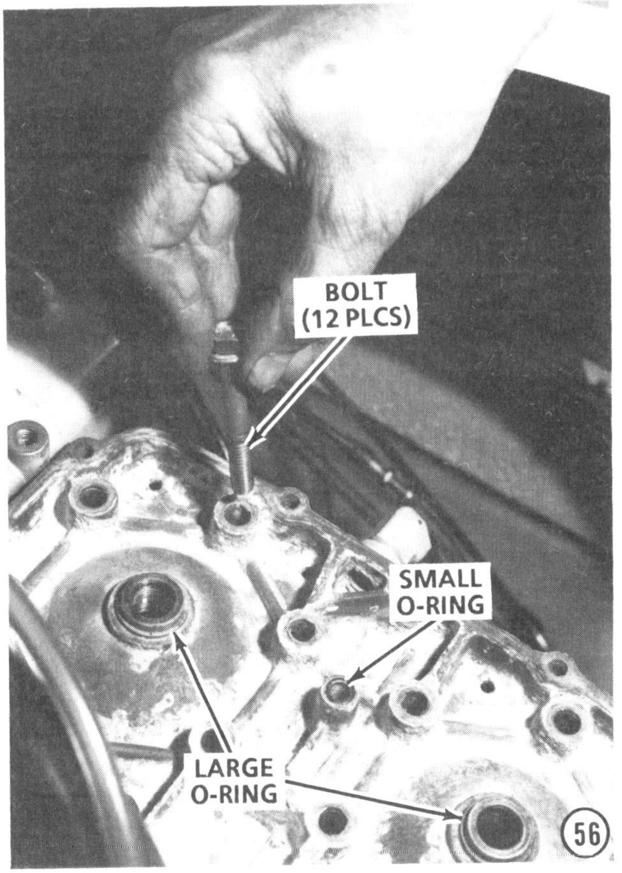

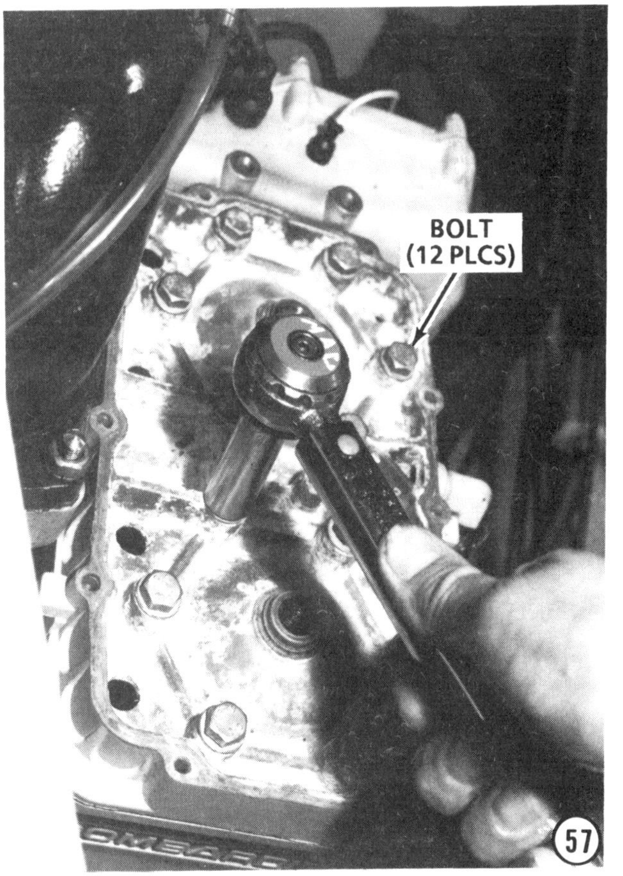

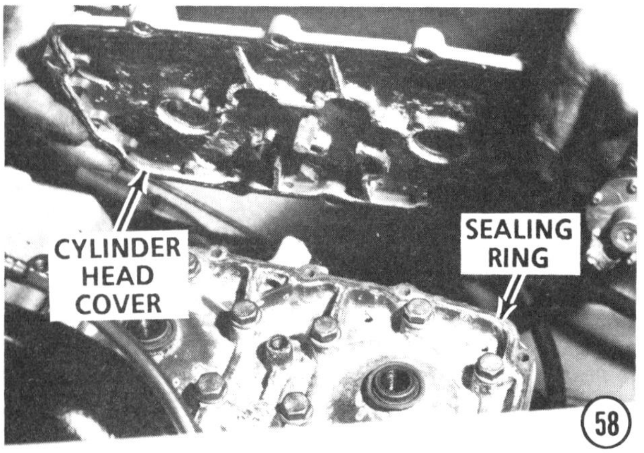

magneto plate support bracket -- if equipped -- over the two forward bolt holes and the flame arrestor support bracket over the two outer starboard bolt holes. Install the Allen head bolts and washers securing the cover to the head. Tighten the bolts to a torque value of 80 in lbs (n), in the sequence shown in the accompanying captioned line drawing on the previous page.

60- Install the two Allen head bolts securing the magneto plate support bracket to the plate. Tighten the bolts alternately and evenly to a torque value of 80 in lbs (9Nm).

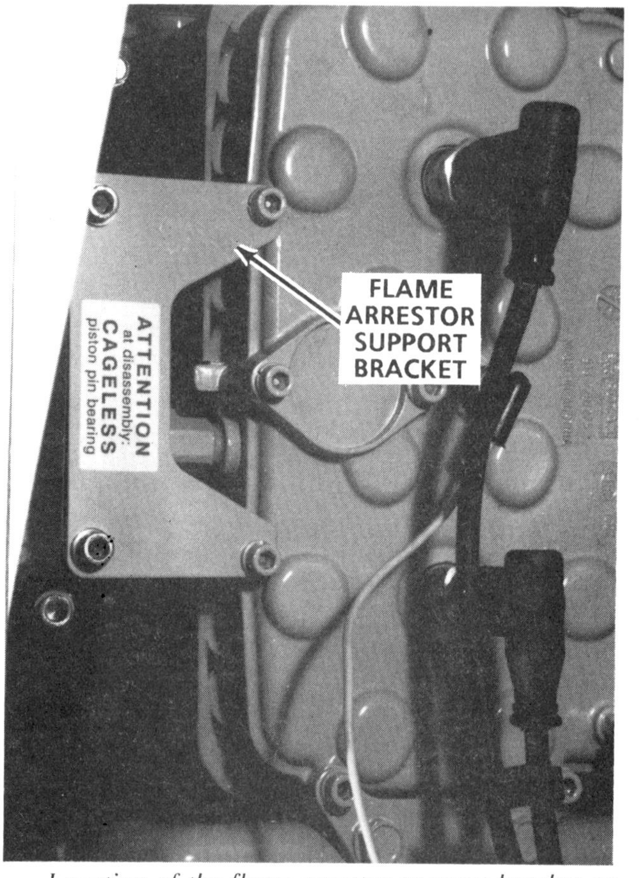

Location of the flame arrestor support bracket on the cylinder head cover.

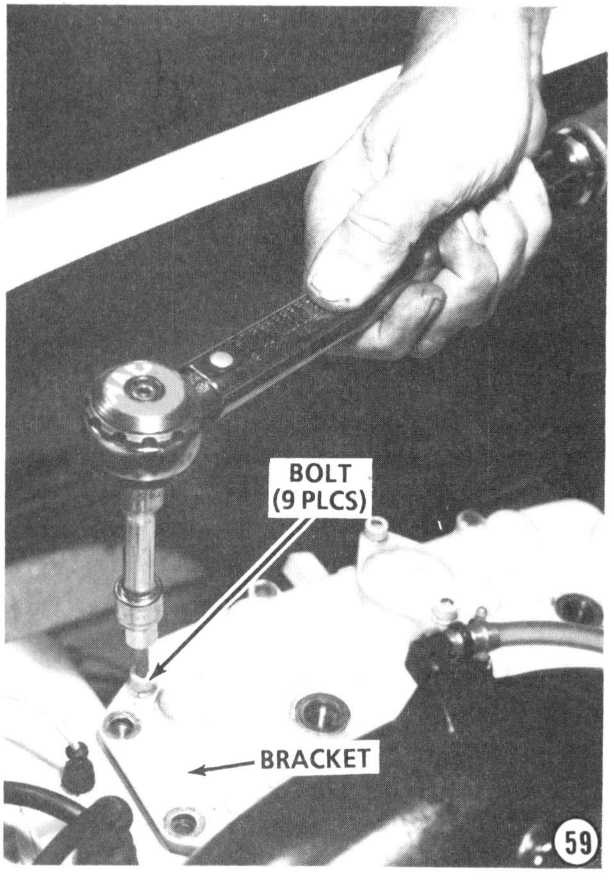

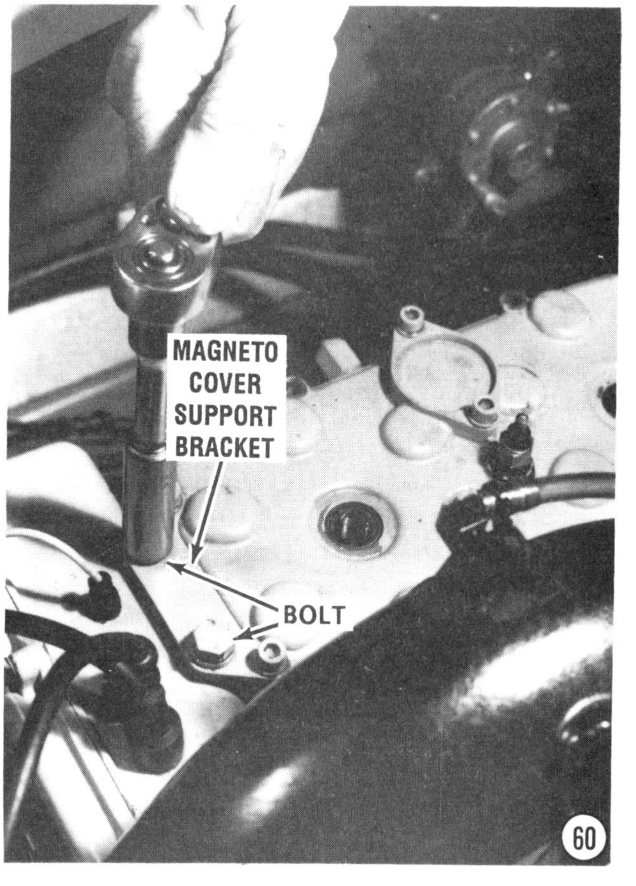

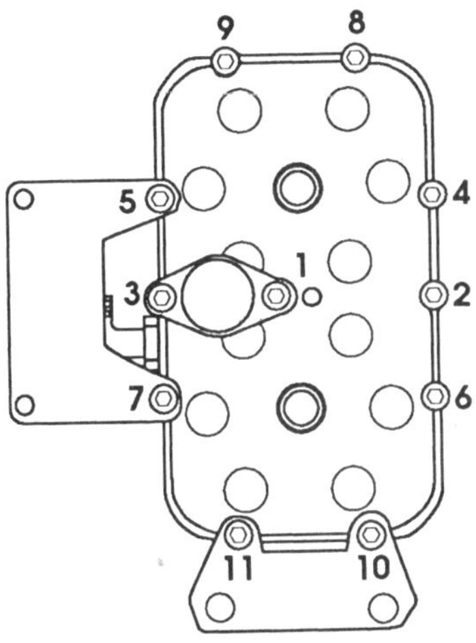

Tightening sequence for the cylinder head ***COVER*** *securing bolts.*

61- Install the spark plugs. Connect the lead to the temperature sensor.

62- Connect the spark plug leads to the spark plugs.

63- Tighten the two clamps on the boot between the aft flywheel and the seal carrier.

64- Slide a clamp over the end of engine cooling water hose. Work the hose onto the fitting on the jet pump. Tighten the clamp securely.

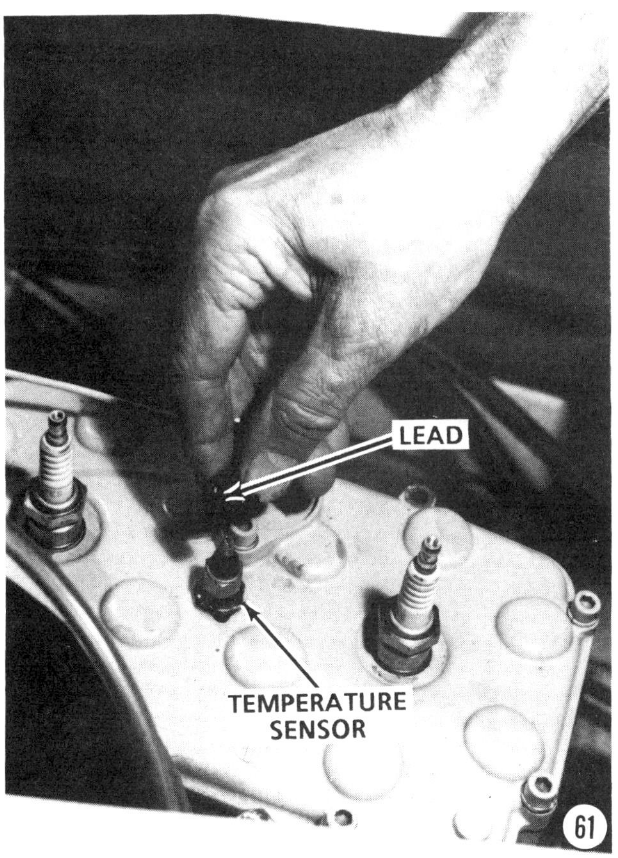

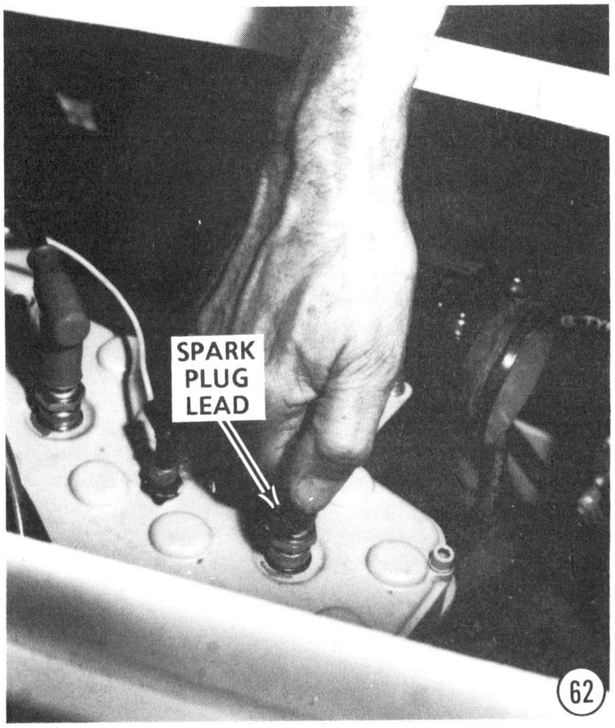

CLOSING TASKS

Install the carburetor, associated hoses, throttle and choke cables, see Chapter 6 -- **FUEL**.

Connect the stator harness lead at the electrical box.

Install he driveshaft into the jet pump, see Chapter 10 -- **JET PUMP**.

The oil injection system **MUST** be "bled" while operating the engine on a premix, as directed in "**Purging Air from Oil Injection System**", beginning on Page 6-38.

Connect the negative and positive leads to the battery.

Move the craft into a body of water or connect a garden hose to the engine flush system.

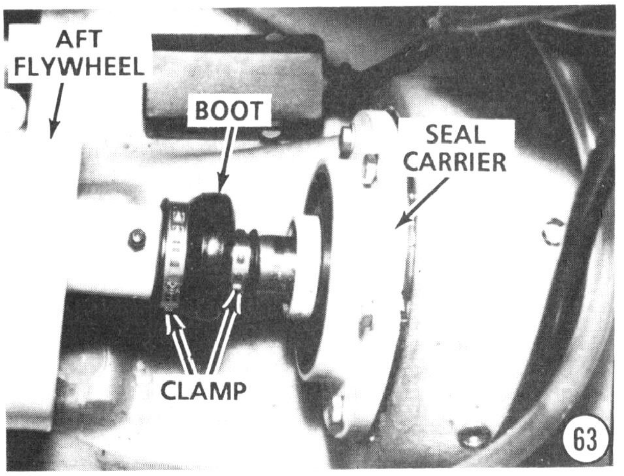

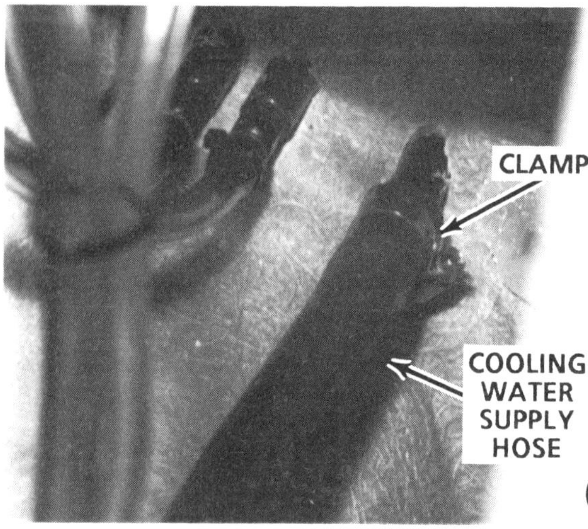

64

CAUTION

The following procedure MUST be followed whenever the engine is started with a hose attached for cooling the engine. Start the engine, and then IMMEDIATELY turn the water on. The engine MUST be started first to prevent water from backing up into the tuned exhaust pipe and into the engine. Water must circulate anytime the engine is operating to prevent overheating -- which could cause damage to moving parts and possibly cause engine seizure. NEVER allow the engine to operate without cooling water for more than 15 seconds. NEVER run the engine with a flushing attachment for more than 5 minutes. The driveline seal is not provided with cooling water if it is not immersed in water. To shut the engine down, first shut the water supply off, and then shut down the engine. If this sequence of tasks is not followed as stated, the engine could be severely damaged.

REMEMBER

Start the engine.
Turn the water on.
Turn off the water.
Shut down the engine.

Attempt to start and run the engine, without the engine compartment cover in place. This will provide the opportunity to check for water, exhaust, or fuel leaks.

After the engine is operating properly, install the engine cover. Follow the break-in procedures with a load on the jet pump.

BREAK-IN PROCEDURES

Do not operate the engine at full throttle, except for very short periods, until ten hours of operation.

a- Operate the engine at various speeds throughout the power range, but **DO NOT** maintain a steady rpm for the first two hours.

b- Operate at any speed after two hours, but not at sustained full throttle until after another eight hours of operation.

c- Mix oil to the gasoline during the break-in period -- total of 2-1/2 to 3 hours -- at a mixture of 16oz oil per a full tank of gasoline. This oil is addition to the oil supplied by the oil injection system -- for added lubrication.

d- While the engine is operating -- under jet pump load, during the initial period -- have an assistant control the craft while a check is made of the fuel, oil, exhaust, and water systems for leaks.

Make a final check, while the engine is operating, for water, fuel, and exhaust leaks, before stowing the tools, cleaning the engine compartment, and "squaring away" the work area.

8-4 COMPLETE SERVICE MODEL 787 ENGINE

Before commencing any work on the engine, an understanding of two-stroke engine operation will be most helpful. Therefore, it would be well worth the time to study the principles of two-stroke engine operation, as outlined briefly in the Section 8-2, beginning on Page 8-2.

This section is well illustrated. However, some models may have slightly different routings for electrical, fuel, oil or water lines. Therefore, a Polaroid, or equivalent instant type camera is an extremely useful item, providing the means of accurately recording the arrangement of parts and wire connections of the particular engine being serviced, before the disassembly work begins. Such a record will prove most valuable during assembling.

PRELIMINARY TASKS

Take time to identify and tag the various hoses and electrical leads which are disconnected during the disassembly process. Such a practice will ease the installation procedures. The engine ID number is located on the aft end of the engine crankcase. This number **MUST** be available when ordering replacement parts and special tools required for servicing the specific engine.

Disconnect the Negative and Positive leads from the battery terminals. Rotate the fuel selector switch to the **OFF** position.

ENGINE REMOVAL

1- Push the flame arrestor cover locking tab down, then sideways to unlock the tab. Perform this procedure for each of the cover lock tabs.

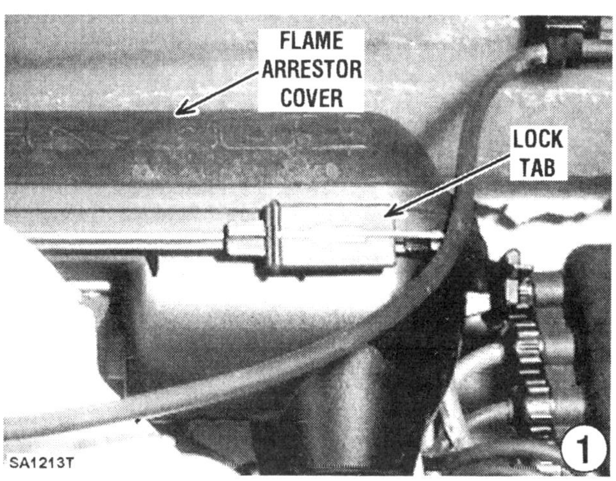

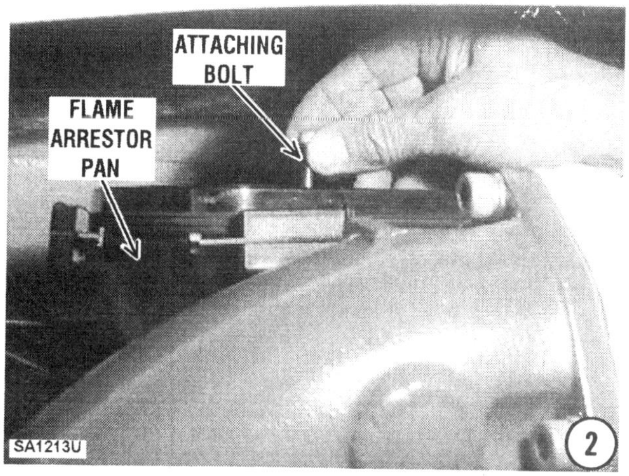

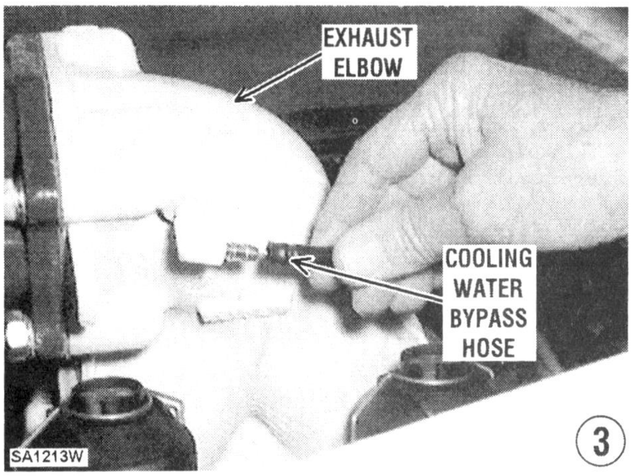

2- Remove the six 5mm Allen head bolts securing the flame arrestor screen assembly to the arrestor pan -- or base.

3- Remove the fastener securing the cooling water bypass hose to the upper portion of the exhaust elbow. Pull the hose free of the fitting. This bypass hose channels water overboard to give the operator a visual indication of coolant water circulating through the engine.

4- Rotate the flame arrestor pan 90 degrees towards the port side, then slide the arrestor pan aft and out of the engine compartment.

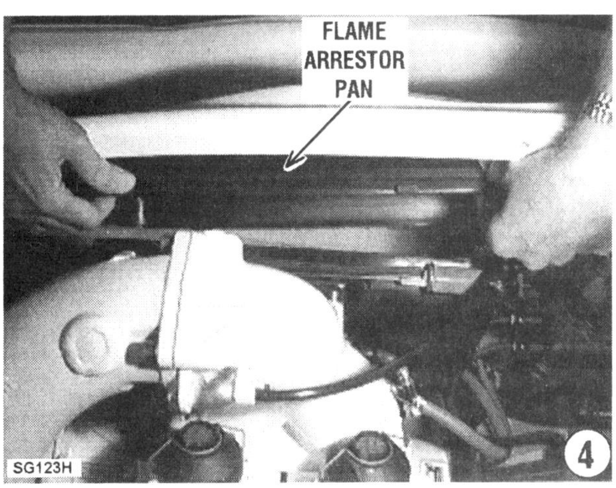

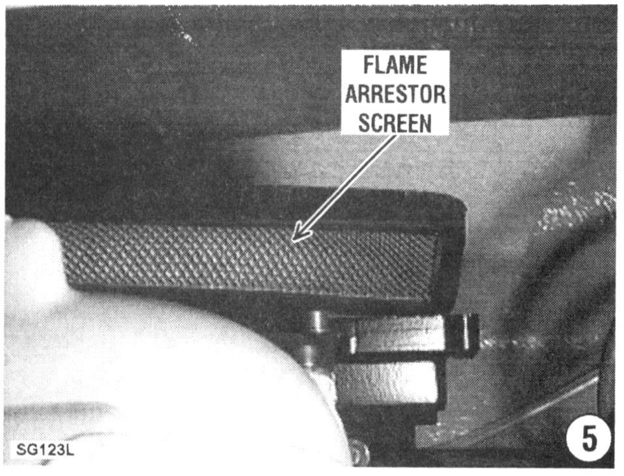

5- Pry the arrestor screen away from the bracket assembly using a sharp instrument, and then set the screen aside for further inspection.

6- Remove the four 5mm bolts securing the arrestor stainless steel bracket to the exhaust manifold.

7- Lift the arrestor base bracket free of the carburetors.

8- Loosen the 3mm Allen head locking screw securing the choke cable to the pivot arm. Temporarily remove the 10mm cable locknut and disconnect the choke cable. Install the locknut back on the pivot arm. Remove the throttle in the same fashion.

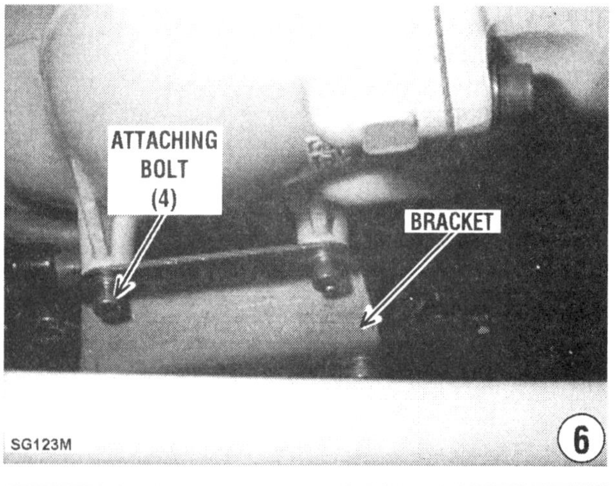

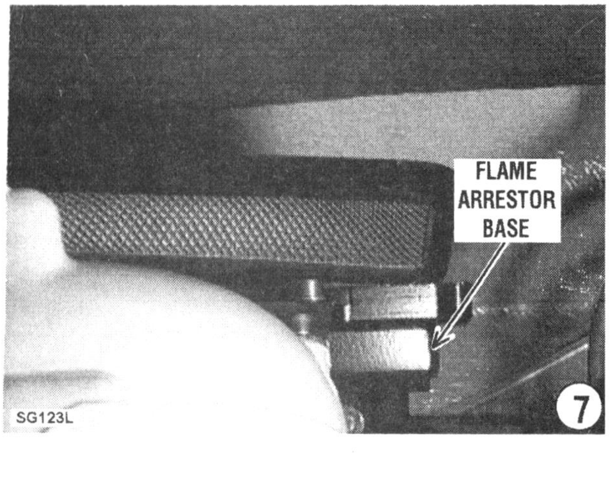

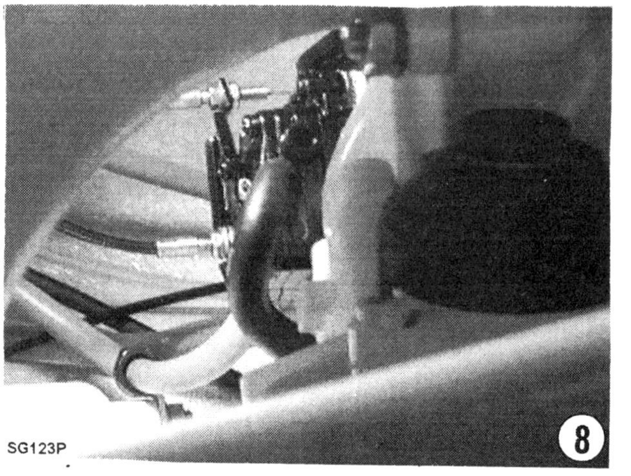

9- Obtain special tool Sea-Doo P/N 295 000 054 -- Clamp pliers. Squeeze the pliers on the tab of the fuel return hose at the aft carburetor clamp, then rotate the pliers 90 degrees to remove the clamp. These clamps may be used again.

10- Cut the pulse hose metal clamp -- **not reusable** -- then remove the pulse hose from the top of the No. 1 -- forward -- carburetor.

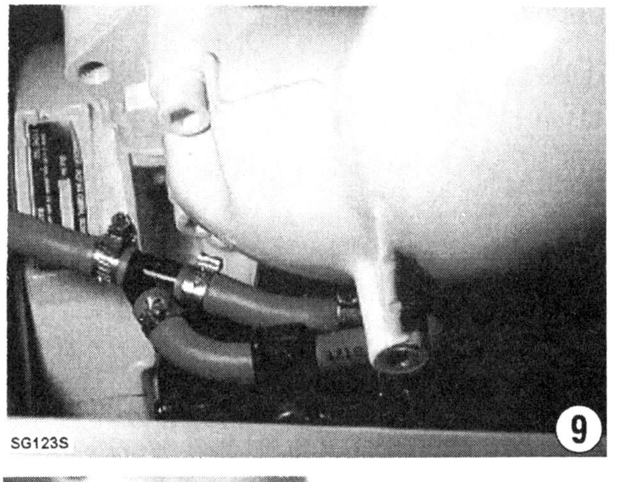

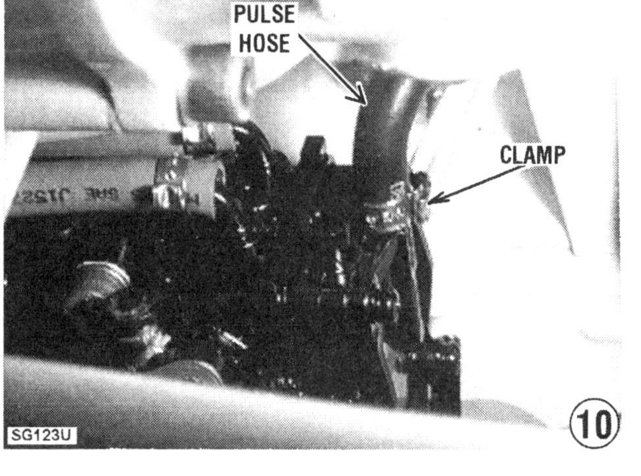

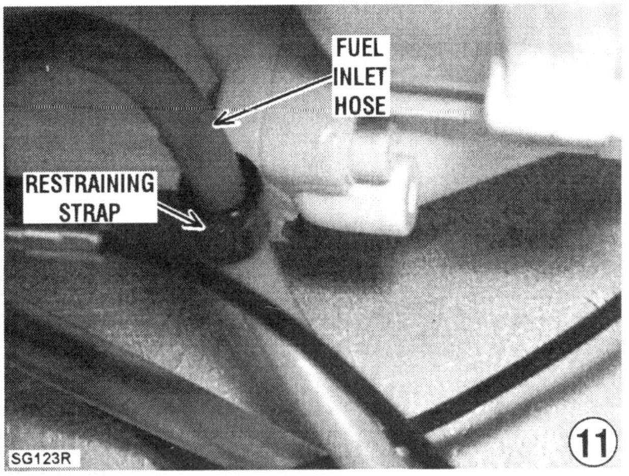

11- Remove the restraining strap securing the fuel inlet hose to the forward cranking motor shoulder.

12- Remove the four 6mm Allen head bolts securing the carburetors to the rotary valve cover.

13- Temporarily remove the locknut securing the oil pump cable to the oil injection pump located on the magneto cover. Disconnect the cable from the pump. Install the locknut after the cable is removed. Do not disconnect the

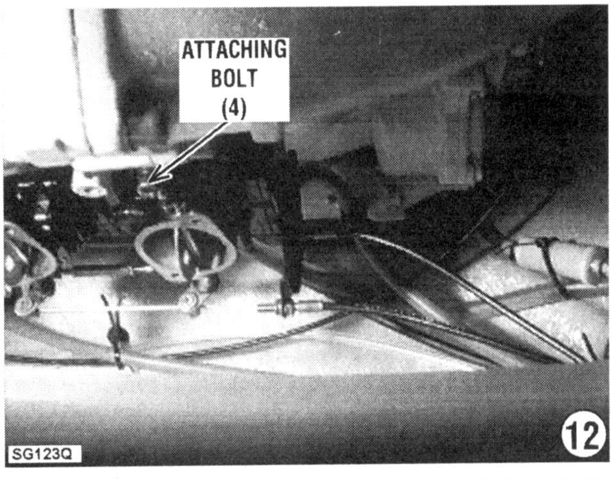

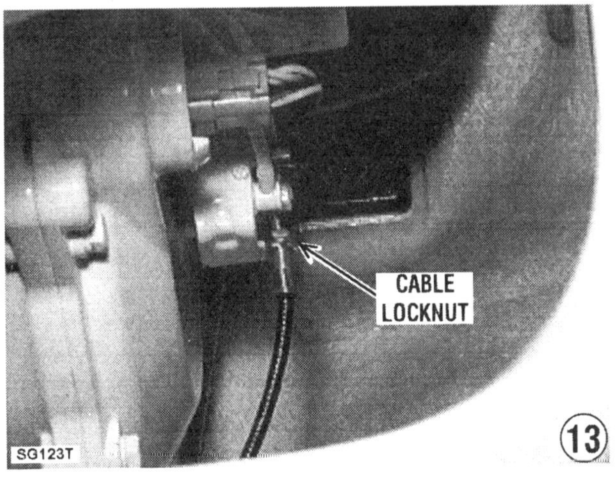

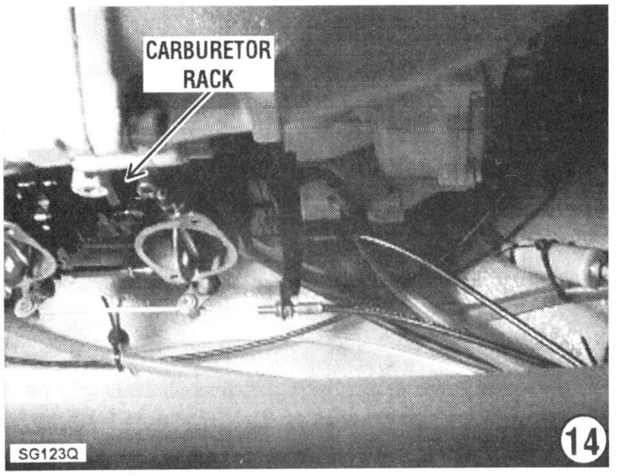

injection pump cable end connected to the carburetor assembly, at this time.

14- Lift the carburetor/s, together as an assembly, from the rotary valve and set it aside for further inspection and cleaning.

15- Remove the one-time use metal clamp securing the fuel inlet hose to the carburetor. Disconnect the inlet hose.

16- Pull the boot from the cranking motor power terminal. Remove the 10mm "nylock"

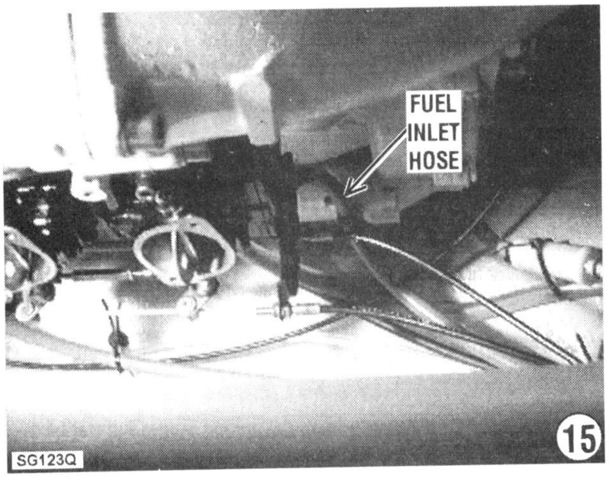

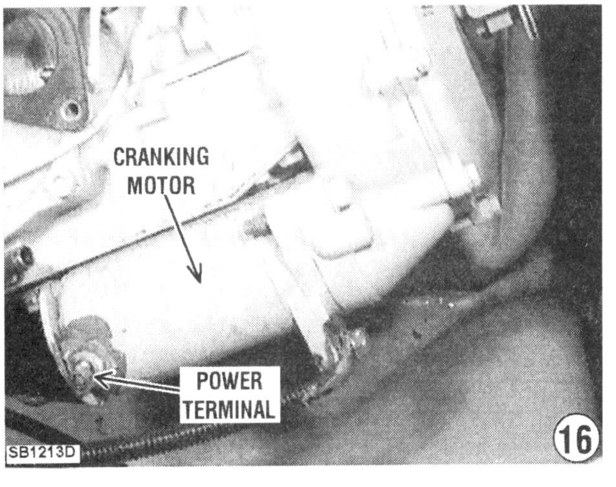

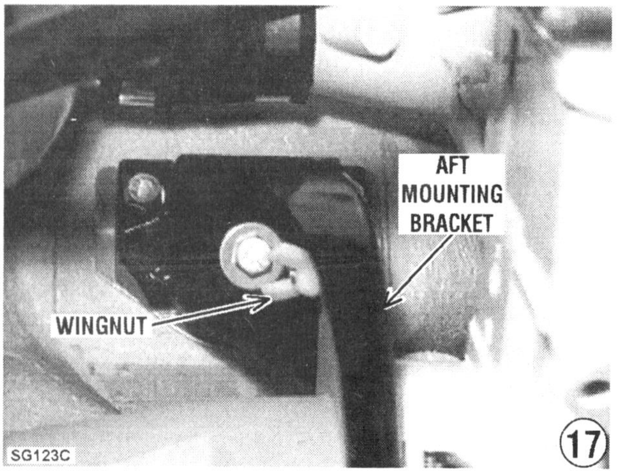

nut securing the lead to the cranking motor, and then remove the lead.

17- Remove the two plastic wing nuts and two stainless steel washers securing the aft flywheel -- PTO flywheel -- cover to the engine. Lift the cover from the flywheel and driveshaft assembly.

18- Remove the spark plug high tension leads and overheat temperature sensor lead from the cylinder head. Remove the spark plugs. Mark the plugs to identify which cylinder they came from as an inspection reference.

19- Disconnect the cooling water hoses from the aft end of the cylinder head. Take note of the embossed arrows on the cylinder head, indicating proper direction of flow for each hose.

20- Remove the hose clamps securing the cooling water hose leading from the forward end of the cylinder head to the top of the expansion chamber. Disconnect the cooling water inlet hose from the expansion chamber.

21- Remove the three 8mm Allen head bolts and one 17mm nut securing the upper expansion chamber to the exhaust elbow.

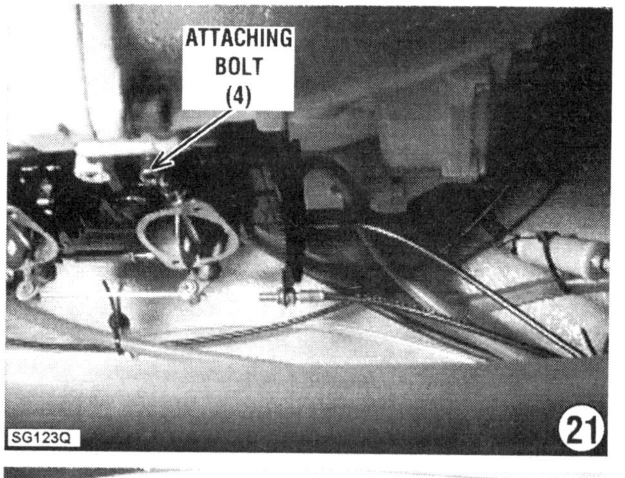

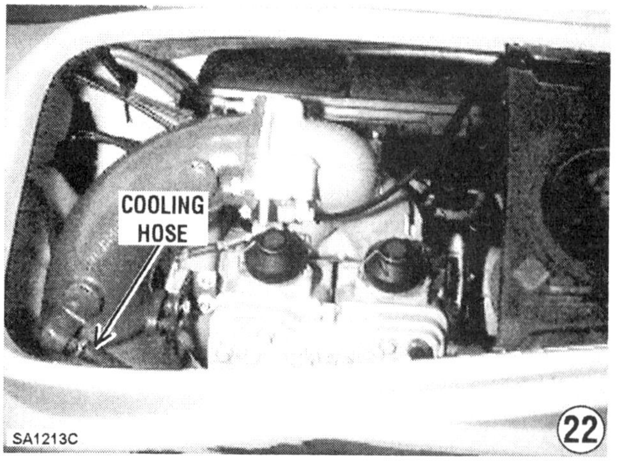

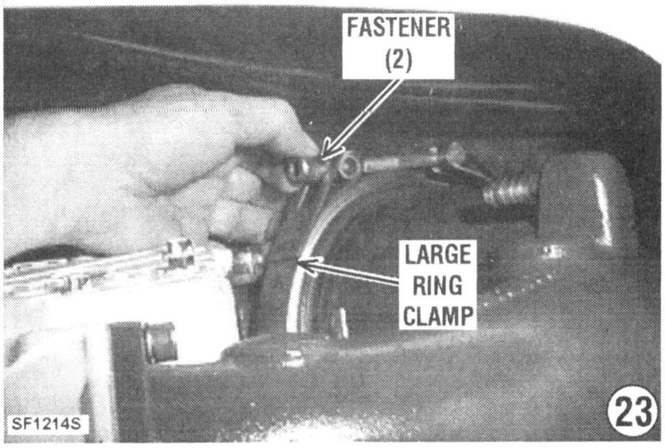

Remove the two 6mm Allen head bolts and one 13mm Allen head bolt securing the upper expansion chamber support to the magneto housing.

22- Disconnect the water inlet supply hose leading from the water regulator valve to the expansion chamber. Disconnect the hose at the expansion chamber only, at this time.

23- Remove the expansion chamber large ring clamp bolts. This clamp secures the two chamber halves together and may not be an easy task to remove. Use care and patience to remove the fasteners and pull the clamp from the expansion chamber assembly.

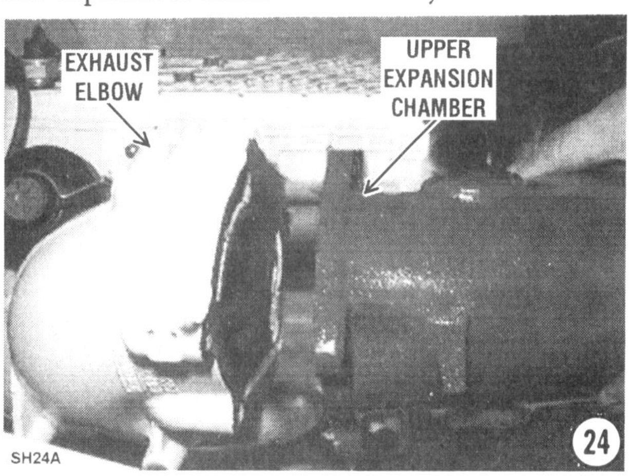

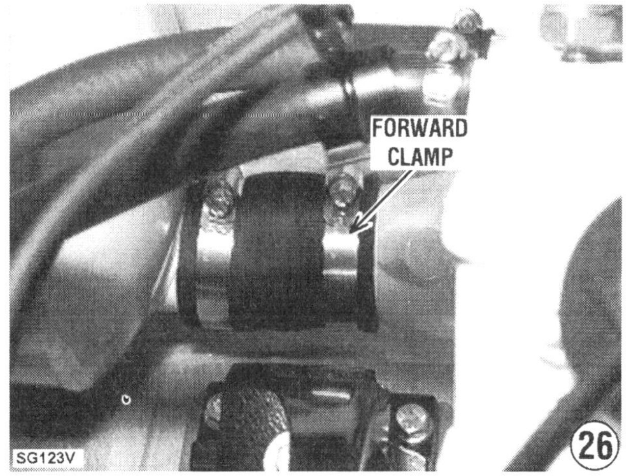

24- Using a soft rubber mallet, tap the upper expansion chamber to help free it from the exhaust elbow. Lift the expansion chamber free of the engine compartment.

25- Remove the hose clamp securing the cooling hose from the tuned exhaust muffler to the lower expansion chamber. Disconnect the cooling hose from the expansion chamber.

26- Remove the two 6mm Allen head bolts -- fore and aft -- securing the lower expansion chamber to the crankcase. Loosen and remove the forward hose clamp on the aft end of the

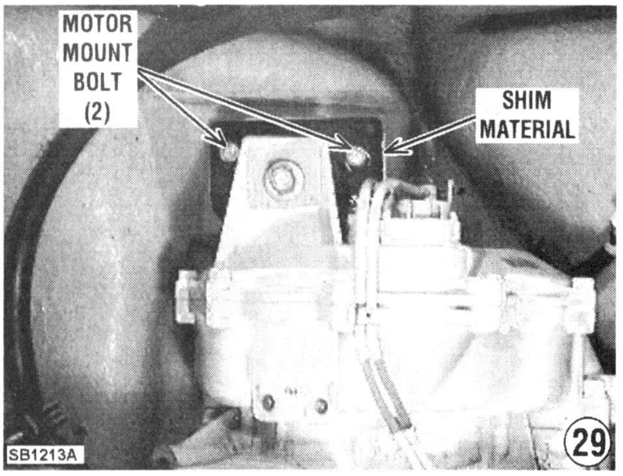

expansion chamber. The aft hose clamp on the tuned exhaust muffler remains secure, at this time.

27- Rotate the lower expansion chamber to "break" it loose from the rubber coupler attaching the chamber to the tuned exhaust muffler. Slide the expansion chamber forward, then up and out of the watercraft.

28- Disconnect the stator electrical harness quick-connect fitting.

29- Loosen, but **DO NOT** remove the forward motor mount bolts, at this time.

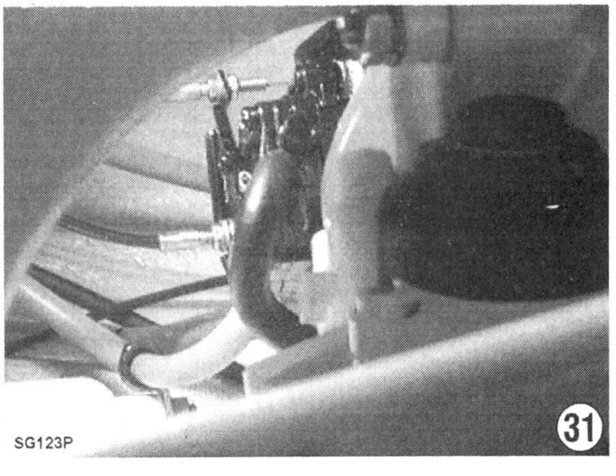

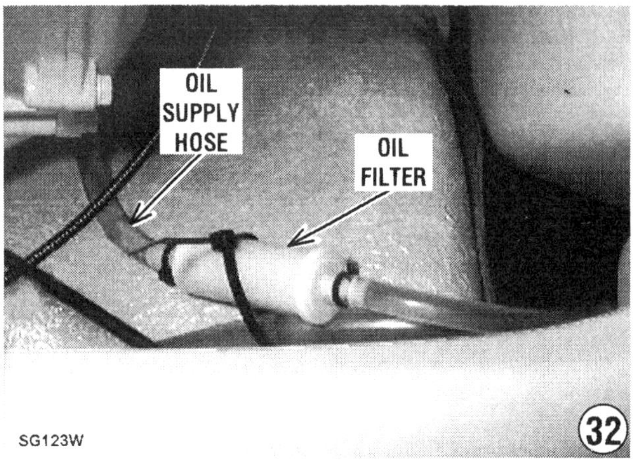

30- Remove the aft motor mount bolts.

31- Obtain hose pincher tools P/N 295 000 076, then "pinch off" the two main oil supply hoses leading from the oil tank to the lower crankcase -- port and starboard. Remove the hose only after it is pinched off.

If the clamping tool is not available, pinch the hose with fingers, disconnect the hose from the fitting, and then insert a spark plug and secure it with a zip tie.

32- Locate and pinch off the oil injection hose leading from the tank to the oil pump. This hose can be identified because it has an inline oil filter attached to it. Pinch off the hose near the oil injection pump.

33- Remove the forward engine mounting bolts. Lift up the engine slightly and remove the motor mounts **AND** shim material. Be **ABSOLUTELY** sure to indicate the location of each stack of shim material, in order to maintain proper engine-power train alignment during assembly.

SPECIAL WORDS

If the two special pliers called out in the next two steps are not available, the clamps must be cut to release them. In this case new

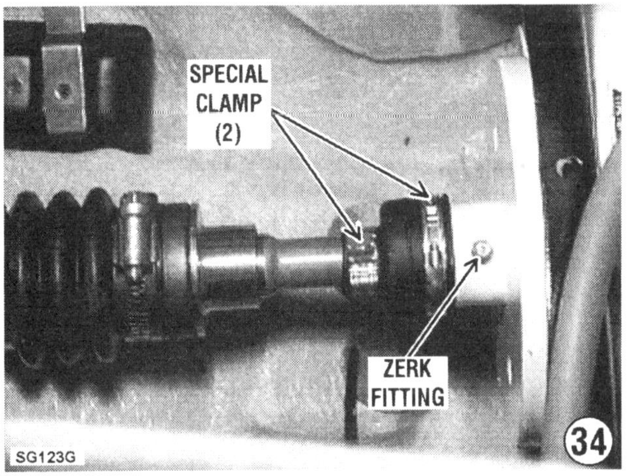

clamps **MUST** be obtained for installation. Therefore, make every effort to obtain the clamps **BEFORE** cutting the old ones.

34- Obtain special tool P/N 295 000 069 -- clamp pliers. Squeeze the hooks of the aft flywheel boot large clamp together to release the band. Remove the large clamp from the boot. Obtain special tool P/N 295 000 054 -- clamp pliers. Remove the small metal clamp from the boot by twisting the clamp to release it.

Slide the boot aft on the driveshaft. Obtain a grease gun and apply grease to the zerk fitting. This action will result in pushing the engine forward enough to release the driveshaft from the splines of the aft flywheel. The engine will now slide forward easily.

35- Move the engine forward and rotate it to port to gain access to the ground lead located on the bottom of the cranking motor.

36- Disconnect the grounding lead from the forward end of the cranking motor.

37- Disconnect the clear water "bleed" hose connected to each cylinder via a tee connector on the port side. This hose acts as a sump,

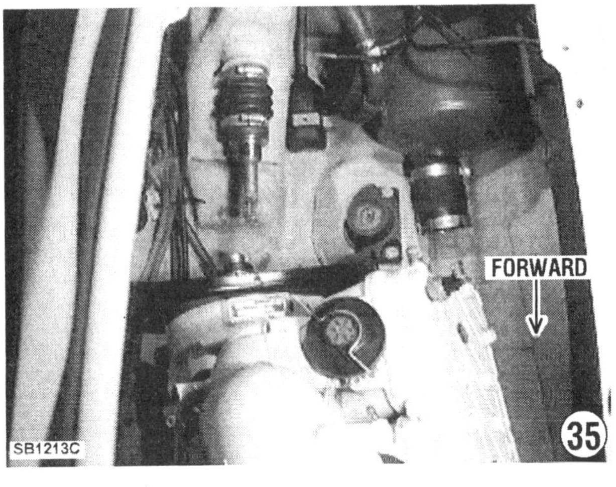

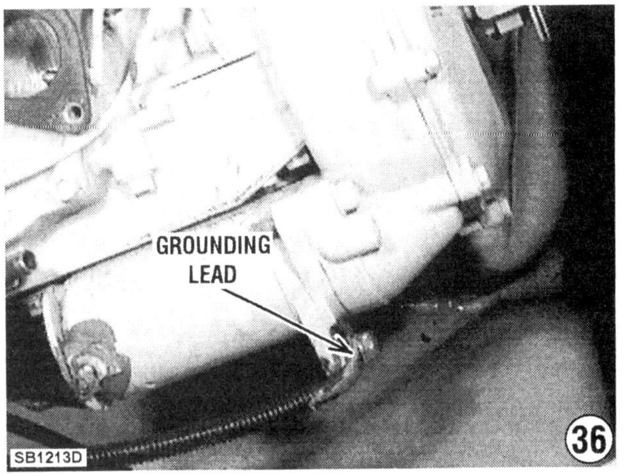

because it is the lowest part of the recirculating system. Once the engine is shut down, water drains from this connection, "bleeding" the engine dry of water.

Engine Removal
Good Words

Make one final inspection to ensure all hoses and electrical leads are either out of the way, or wrapped around the engine, in preparation to lifting the engine clear of the compartment.

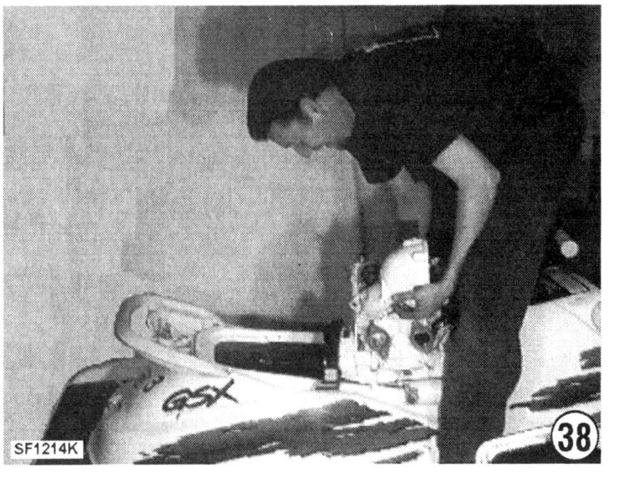

38- Obtain the services of an assistant or a suitable lifting device. If an assistant or lift is not available, the engine may be lifted out by one person, in good shape, if done carefully. Place a board -- just large enough to cover the aft portion of the watercraft. **Use the legs, not the back!** Straddle the engine compartment and move the engine slightly forward, then up and out of the craft. Lower the engine onto the board. In the accompanying illustration, the GSX model seat bracket was angled, the engine was lowered to the side of the craft.

Now, transfer the engine to a suitable work surface for disassembly.

ENGINE DISASSEMBLING MODEL 787 SERIES

This sections provides detailed illustrated procedures to disassemble virtually all components of the engine.

HOWEVER, if certain parts appear to be in satisfactory condition for further service, simply skip the steps involved in removing the particular item, and proceed with the necessary tasks to restore the engine to full operating efficiency.

The following procedures pickup the work after the engine has been removed and is on a suitable work surface.

1- Remove the two oil supply lines from the rotary valve cover. These hoses attach to the oil injection pump.

Obtain special tool P/N 295 000 069 -- Oetiker 1090 -- and remove the special metal clamp from the aft flywheel boot. Now, pull the rubber boot free of the aft end of the flywheel.

Flywheel Removal

2- Remove the bolts securing the flywheel cover to the magneto housing. Take note of

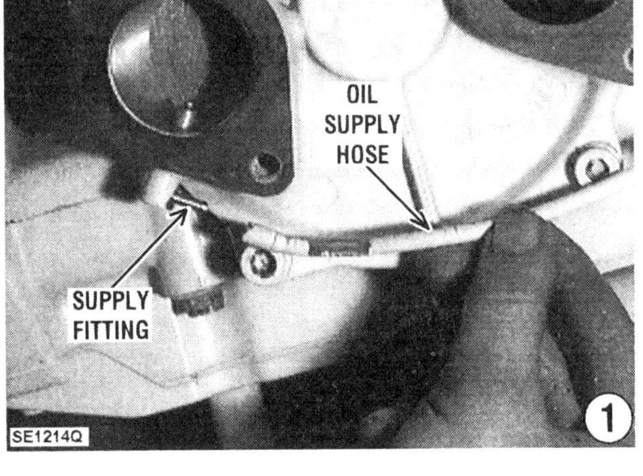

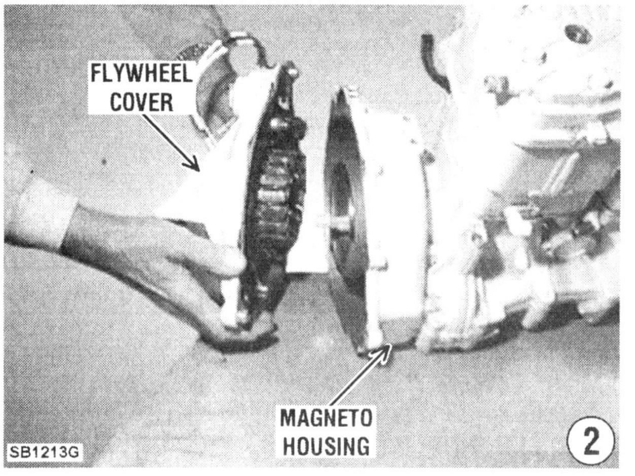

two tabs on the cover rim allowing a screwdriver or other suitable tool to be used in prying the cover from the housing. Pull the flywheel cover free of the magneto housing.

3- Be sure to locate and retain the nylon connector shaft which indexes between the oil injection pump and the crankshaft.

GOOD WORDS

The stator ring -- as shown in the accompanying illustration -- on the 787 engine is basically "permanently" attached to the flywheel. The attaching bolts are tightly fastened to the flywheel. Any attempt to remove them **WITHOUT** applying heat will almost always cause the bolt heads to shear. If the magnets are suspected to be at fault, the flywheel and stator ring should be replaced as a single unit.

As explained in the text, the stator ring on a Model 787 Series engine could almost be considered "permanently" attached to the flywheel. Therefore, it would be necessary to apply heat to the bolts in order to remove them.

Cranking Motor Removal

The cranking motor must be removed at this time, because a special flywheel holder tool will be used to prevent the crankshaft from rotating. The special tool is inserted in the back of the magneto housing -- the location of the cranking motor pinion gear.

4- Remove the bolts securing the cranking motor to the back of the magneto housing.

5-- Pull the cranking motor free of the magneto housing. The **O**-ring should stay with the cranking motor. If the **O**-ring is in satisfac-

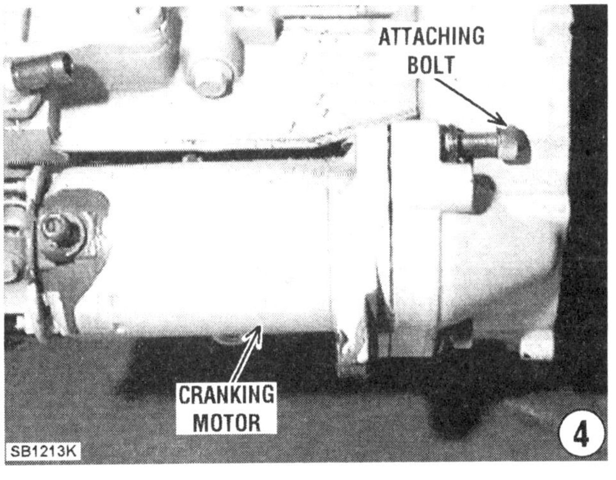

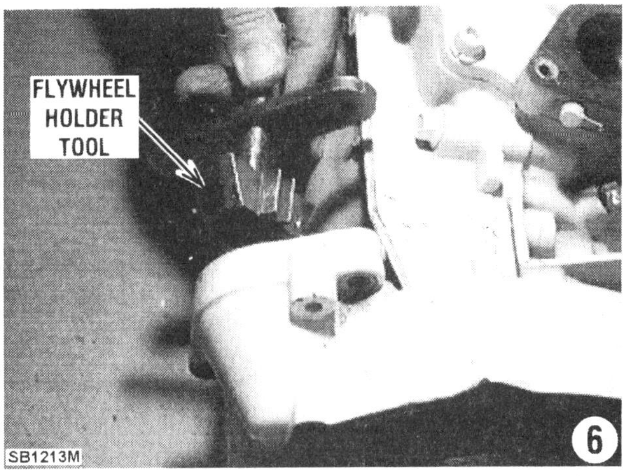

tory condition, it may be used again during assembly.

6- Obtain special tool P/N 295 000 134 -- flywheel holder. Insert the holding tool into the hole in the aft magneto housing -- location of the cranking motor mount. Be sure the teeth of the holder index properly with a tooth of the flywheel. Attach the special tool with the existing cranking motor bolts.

7- Remove the five bolts securing the aft engine mounting bracket to the aft end of the

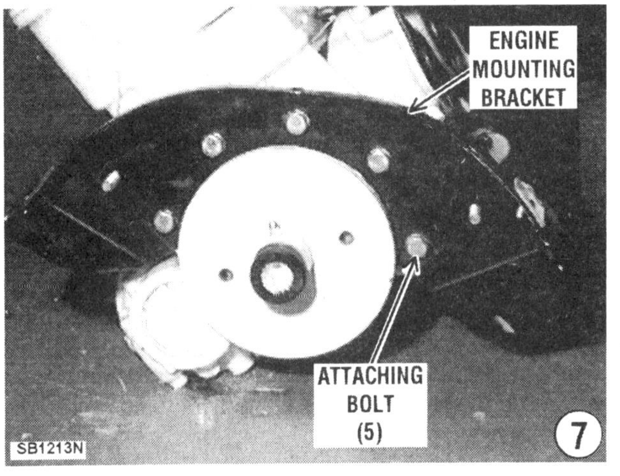

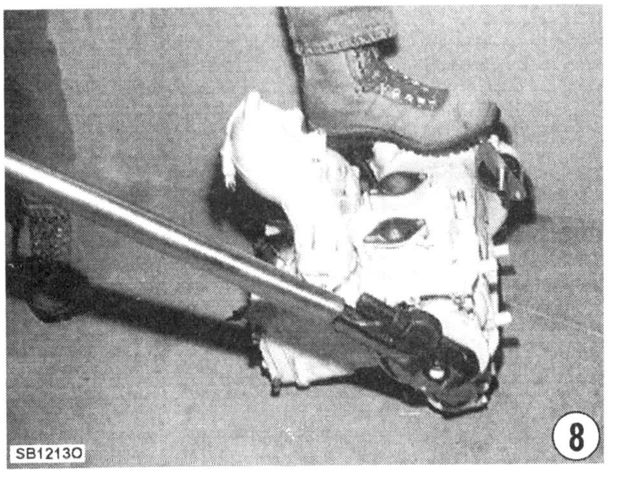

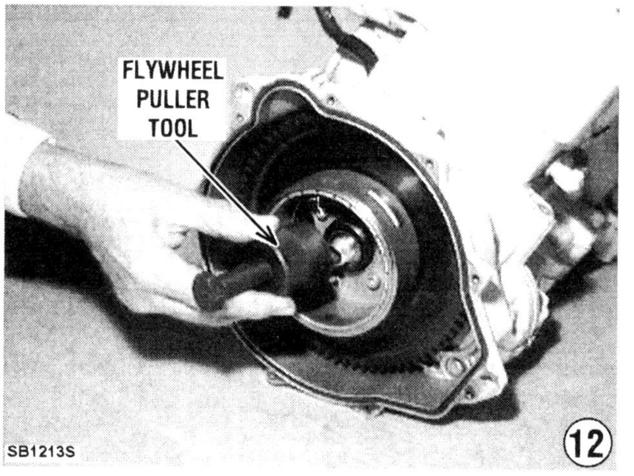

crankcase. Lift the mounting bracket from the engine and set it aside.

8- Lower the engine to the shop floor. Obtain special tool P/N 295 000 001 -- Aft Flywheel Remover tool. Index the splines of the special tool with the splines of the aft flywheel. Lightly tap the tool into position. Place a suitable wrench on the end of the tool and a "breaker bar" over the wrench, as shown. One person, holding the engine stationary with his foot, **MAY** be able to "break" the aft flywheel loose. If possible, obtain the help of an assistant to keep the engine stationary. Rotate the flywheel **COUNTERCLOCKWISE.**

The aft flywheel is fastened **SECURELY** to the crankshaft at the factory. If the flywheel fails to loosen, heat may be applied, **VERY CAREFULLY**, to accomplish the task of removal.

9- Remove the aft flywheel from the crankshaft end.

10- Obtain a suitable wrench for removal of the forward flywheel nut. The special flywheel holder tool must remain installed through the cranking motor opening. Rotate the 27mm flywheel nut **COUNTERCLOCKWISE.**

11- Remove the flywheel nut and lockwasher.

12- Obtain special tool P/N 420 976 235 -- front flywheel puller. Install the puller tool onto crankshaft end. Tap the screw of the puller with a hammer to "break loose" the flywheel.

13- Place a suitable wrench on the head of the flywheel puller tool. Rotate and "pull" the flywheel free of the crankshaft.

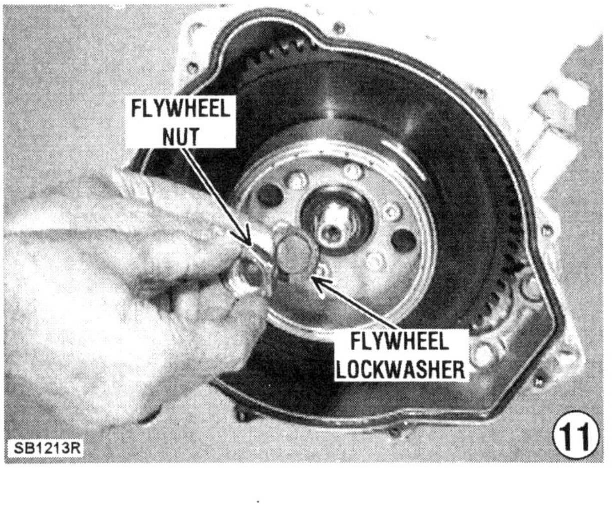

SPECIAL WORDS ON FLYWHEEL MAGNETS AND FLYWHEEL HANDLING

The outer ends of any magnet are called "poles". One end is the north pole and the other end is the south pole. The magnetic field surrounding a magnet is concentrated around these two poles.

Magnets are usually installed in pairs with the north pole of one adjacent to the south pole of its neighbor and so on. In this manner, a continuous magnetic field surrounds the inside of the flywheel. If a flywheel is accidently dropped, not only could the teeth be damaged, but the impact will weaken the magnetic strength of all the magnets housed in the flywheel.

If one or more of the magnets should break or fracture, two new magnetic poles would be created. A long magnet with two poles will become two short magnets with four poles. The new poles will possess only half the magnetic strength of an original pole. The overall magnetic field will be altered. The new field of the shorter magnets will not extend to cover the area of the flywheel.

Serious consequences apply to a CD type ignition system in the event of a flywheel magnet fracture. Pulser coils are energized by the concentrated magnetic field at the magnetic poles. If new poles are suddenly created, the trigger will receive conflicting signals from the magnets and may even attempt to fire the cylinder twice in one revolution.

All these reasons require the flywheel to be handled with **CARE.**

14- Remove and save the Woodruff key from the recess in the crankshaft.

15- Remove the ten 10mm bolts securing the magneto housing to the crankcase.

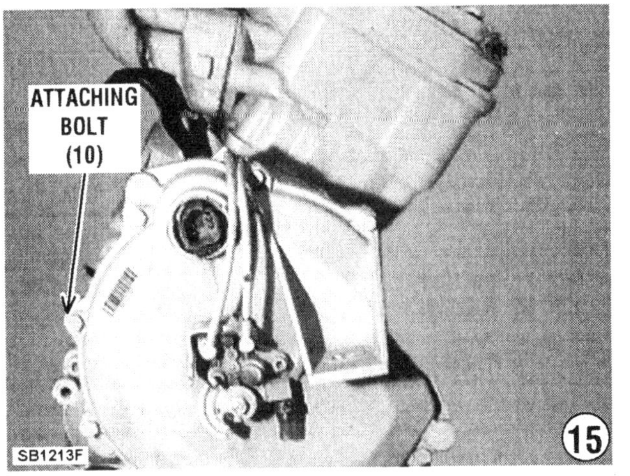

16- Tap the backside of the magneto housing with a rubber mallet to aid in separating the housing from the crankcase. Carefully pull the housing free of the crankshaft and crankcase. Check to be sure the crankshaft forward oil seal remains encased in the magneto housing. The forward crankshaft bearing is also visible.

17- Remove the 8mm Allen head bolts and lockwashers attaching the exhaust manifold to the cylinders. Tap the exhaust manifold with a rubber mallet to "jar" it loose.

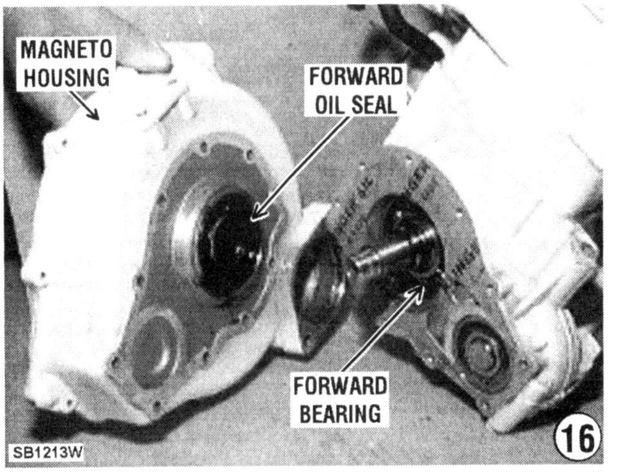

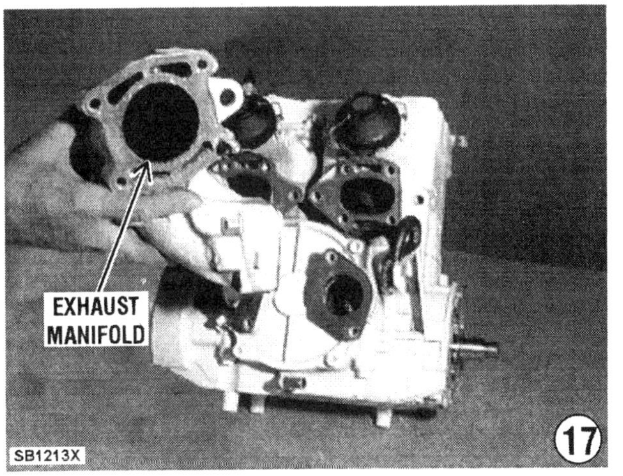

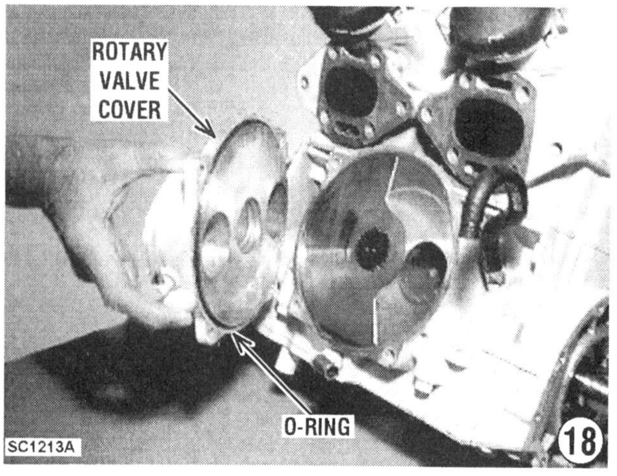

TIMING WORDS

Before removing the rotary valve cover, determine Top Dead Center -- TDC. Insert a dial indicator into the No. 1 -- forward -- spark plug opening. Observe the indicator while rotating the crankshaft. Find the exact location when the needle stops and begins to reverse direction. This position is TDC. Do not rotate the crankshaft further.

If a dial indicator is not available, use a pencil, straw, or equivalent, resting on top of the piston dome. Carefully determine the highest point to which the inserted object rises when the crankshaft is rotated. This is TDC. Do not rotate the crankshaft further.

18- Remove the four 6mm Allen head bolts and lockwasher attaching the rotary valve cover to the crankcase. Pull the valve cover free of the case. Check to be sure the large **O**-ring on the "backside" of the valve cover is intact.

GOOD WORDS ON ROTARY VALVES

During assembling, it is **CRITICAL** to install the rotary valve disc in the proper "direction" -- it may be installed reversed. **BEFORE** removal,

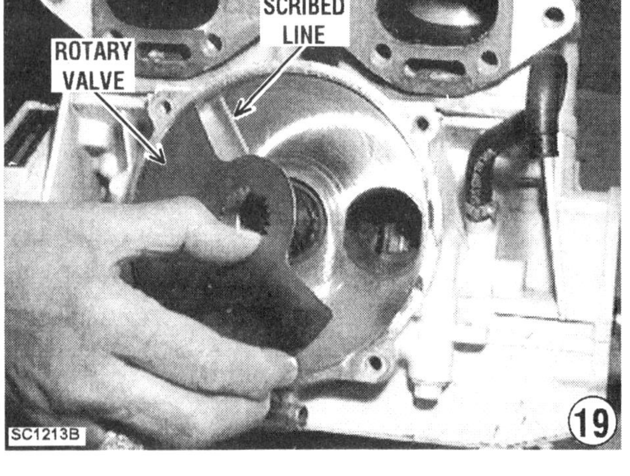

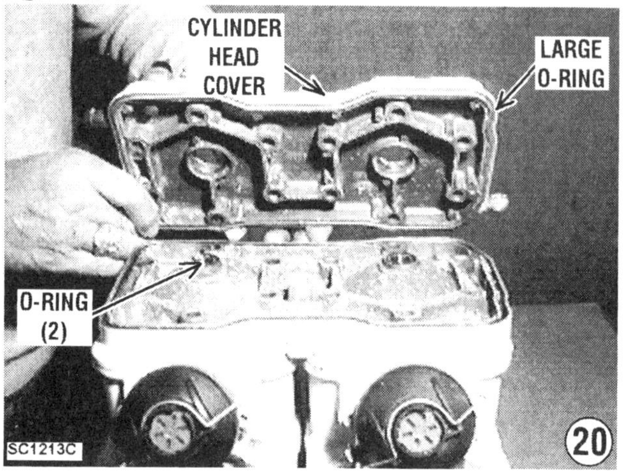

scribe a light line on the crankcase valve facing, along both edges of the rotary valve. **ONLY** in this manner can the proper positioning of the rotary valve be **ENSURED** during assembling. If these marks are not made at this time, proper installation of the rotary valve is much more involved because a degree wheel **MUST** be obtained and used. If the valve were to be installed "inversely" (backwards), the edges of the valve will be off by several degrees and will not align with the scribe marks.

19- After scribing the proper alignment of the rotary valve -- at TDC -- pull the valve free of the crankcase.

Cylinder Head Removal

20- Remove the 11mm hex bolts attaching the cylinder head cover to the cylinder head. Lift the cover free of the cylinder head and set it aside. Observe the location and condition of the mating surface and spark plug opening **O**-rings.

21- Pull the cylinder head free of the cylinders. Check the condition of each cylinder and cylinder head **O**-ring. Good shop practice dictates the **O**-ring not be used a second time.

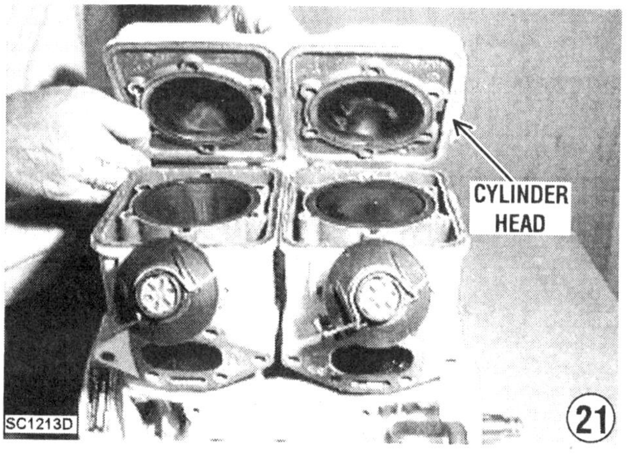

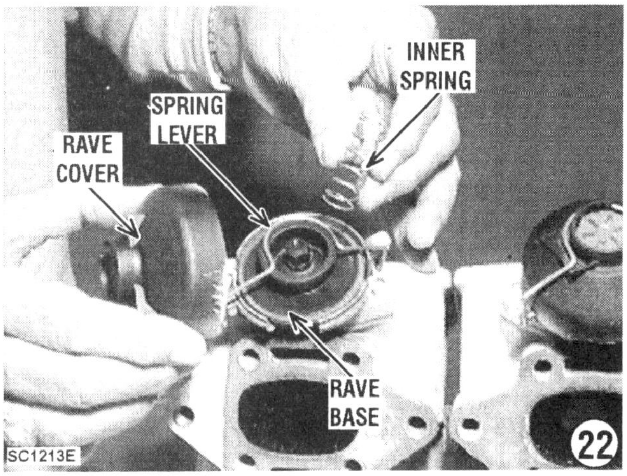

R.A.V.E. System

The 787 series engine is equipped with a RAVE system -- Rotax Adjustable Variable Exhaust -- system. This system in conjunction with the higher volume tuned pipe provides a broader power band. Adjustment and disassembly of these units is quite simple.

22- Hold the RAVE cover down and at the same time, pull back the spring lever securing the RAVE housing to the base assembly. An inner spring pushes up against the cover and will cause it to "pop off" unless it is held down while releasing the spring lever. Lift the housing from the base.

Observe the seating of the inner spring against the small red cup. These springs must be installed in the same manner -- properly seated against the red cup.

23- Rotate the black bellow cover -- it should only be fastened hand tight. Pull the black bellow cover off the base assembly. Remove the large diameter spring.

24- Remove the two 5mm Allen head bolts securing the RAVE valve housing to the cylinder. Note the location of a notch or "cut-out" in the bottom of the RAVE housing. During

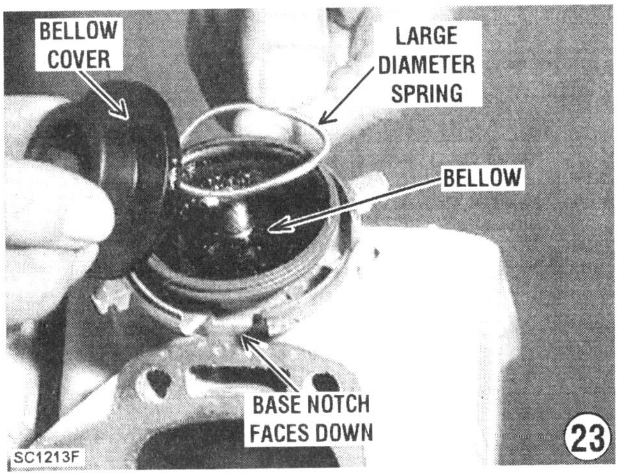

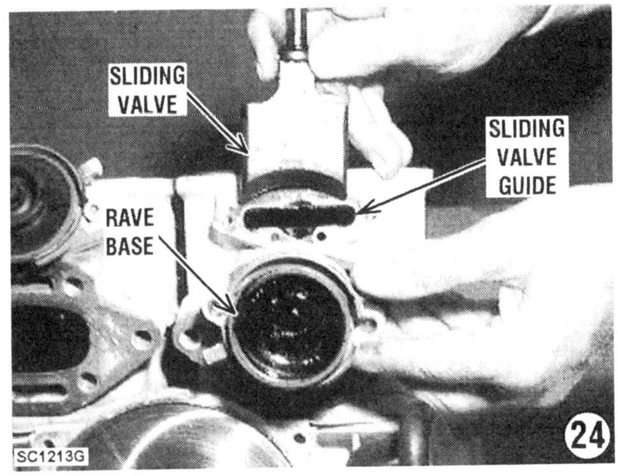

assembling be sure to install the housing with the notch facing **DOWN**, as shown in the previous illustration.

Pull the RAVE sliding valve out of the valve guide port. Observe the stamped "**TOP**" on the sliding valve. The valve must be installed with the word "**TOP**" facing up. The flat surface of the sliding valve is up, the cambered surface of the valve faces downward. If the valve is installed incorrectly, the valve will most likely catch a piston. Repeat Steps 22-24 for the other RAVE valve.

25- Remove the 13mm cylinder base bolts securing the cylinder to the crankcase. Take note of the two lengths and their appropriate locations. Gently tap the intake/exhaust side of the cylinders with a rubber mallet to "jar" them loose. Mark each cylinder -- No. 1 (forward) and No. 2 (aft). Carefully lift each cylinder straight up until clear of the piston, then set them aside.

GOOD WORDS ON CYLINDER BASE GASKET

Take note of the number of small stamped holes in the approximate middle of the cylinder

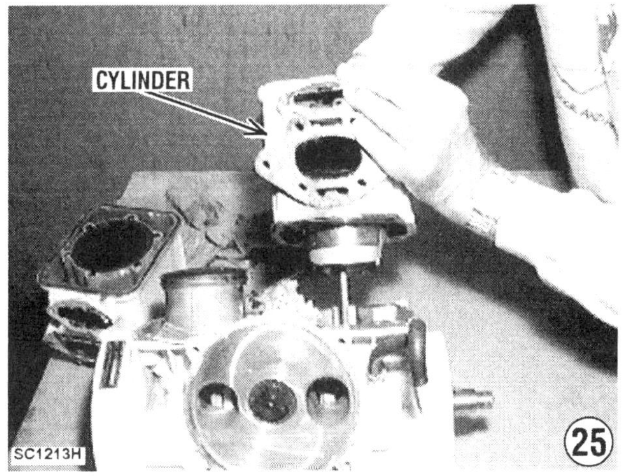

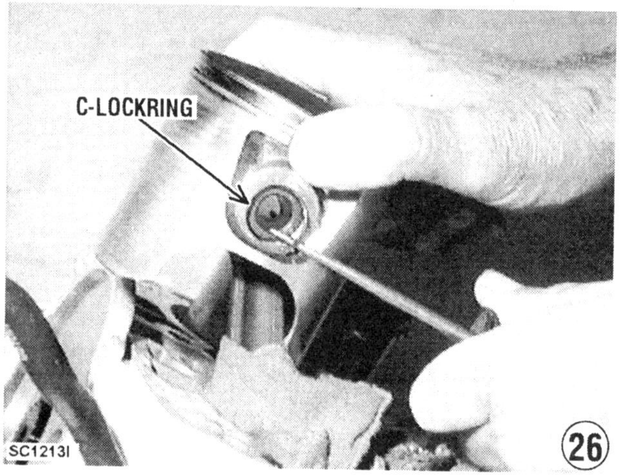

base gasket. This is an identification stamp for the proper gasket used for a specific engine model. Be **ABSOLUTELY** sure to obtain the **IDENTICAL** new gasket, otherwise leaks are likely to occur. Bombardier uses approximately a half dozen different gaskets. Therefore it is essential to count the number of holes, just described, and write the figure down!

PISTON REMOVAL

WORDS

The majority of Sea-Doo watercraft engines use **CAGELESS** needle bearings. Before proceeding, obtain some clean shop towels and cover the crankcase to prevent the loose needles from falling like rain into the crankcase. A special rubber pad is available from Sea-Doo to fit firmly over the crankcase to prevent foreign objects from entering the crankcase.

Scribe a mark on the inside of each piston skirt to identify the piston location **BEFORE** removal from the connecting rod. This mark will **ENSURE** each piston will be installed back onto the same rod and into the same cylinder from which it was removed.

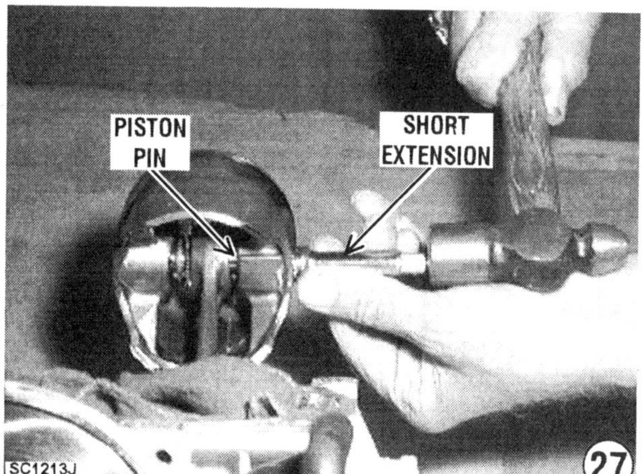

SAFETY WORDS

The piston pin C-lockrings are made of spring steel and may pop out of the groove with considerable force. Therefore, warn others in the area and **WEAR** eye protection glasses or a shield while removing the piston pin lockrings.

26- Remove the C-lockring from both ends of the piston pin using a screwdriver inserted into the small groove next to the piston pin opening. Discard the C-lockrings. These rings stretch during removal and must **NEVER** be used a second time.

27- Obtain a short socket wrench extension and hammer. Check to be sure the crankcase is completely covered and protected by the shop towels. Gently tap one end of the piston pin to "start" moving the pin out of the piston. Remove the piston pin and be sure to collect all needle bearings from the towel. Drive the piston pin from the other cylinder in a similar fashion, using care to pickup all needle bearings.

Crankcase Separation

28- Remove the twelve 13mm bolts and eight 11mm bolts securing the two halves of the

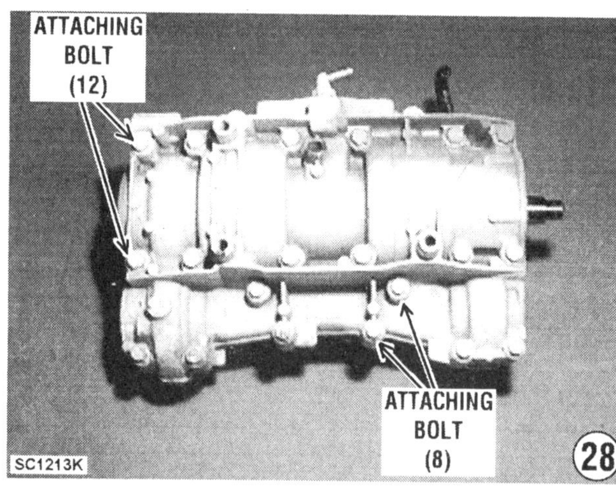

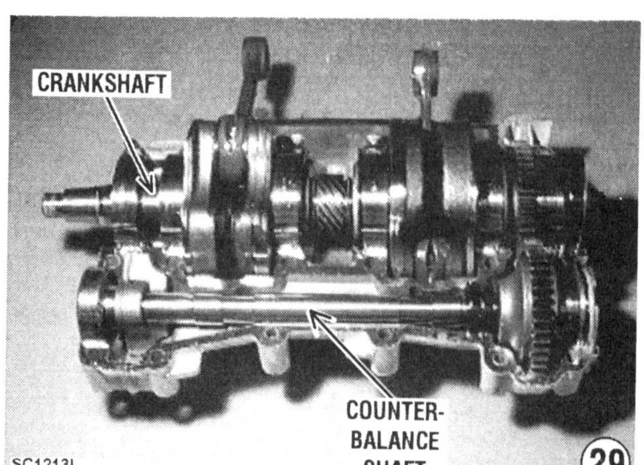

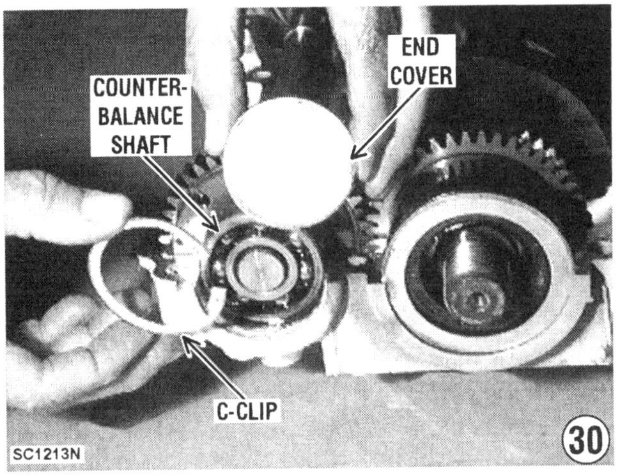

crankcase together. The 11mm bolts secure the counterbalancer shaft assembly side of the crankcase.

Using a screwdriver or other suitable tool, **CAREFULLY** pry the halves apart. Tabs are not provided, therefore proceed with care and patience. Work the screwdriver or tool around the perimeter of the case in a couple of locations to help free the halves. Do not mar the mating surfaces. Separate the upper crankcase from the lower crankcase.

29- Observe the location of the crankshaft bearing locator pins and grooves. During assembling, be sure the crankshaft is installed with these pins seated properly in their respective groove.

30- Remove the snap ring and end cover from the aft end of the counterbalancer shaft.

31- Remove the crankshaft retaining clip and aft oil seal. Take note of the small orange **O**-ring. Be sure it remains seated in the proper location.

32- Observe the alignment **GROOVE** on the aft gear of the counterbalance shaft and the alignment **MARK** on the aft gear of the crankshaft. These two index marks **MUST** align during assembly or the unit will operate out of balance.

33- Lift the crankshaft with the rods attached up and free of the upper crankcase half. Carefully place the crankshaft on a suitable work surface.

CLEANING AND INSPECTING

The success of the overhaul work is largely dependent on how well the cleaning and inspecting tasks are completed. If some parts are not thoroughly cleaned or if an unsatisfactory unit is allowed to be returned to service through negligent inspection, the time and expense involved in the overhaul work will not be justified with peak engine performance and long operating life.

Therefore complete, detailed, illustrated procedures for cleaning and inspecting virtually all components of the engine are presented in Section 8-5, beginning on Page 8-76. The procedures presented in Section 8-5 should be followed closely and the necessary work performed with patience and attention to detail.

ASSEMBLING AND INSTALLATION 787 SERIES ENGINES

Detailed procedures are given to assemble virtually all parts of the engine followed by instructions to install the engine. If certain parts were not removed or disassembled because the part was found to be fit for further service, simply skip the particular step involved and continue with the required tasks to return the engine to operating condition.

The following instructions pickup the work after all parts of the engine have been thoroughly cleaned and inspected according to the procedures outlined in Section 8-5.

The Cleaning and Inspecting section should have revealed any parts unfit for further service.

ADVICE

Before commencing the assembling tasks, check to be sure replacement parts have been obtained and are on hand.

Assembling Lower Half Series 787 Engine

PLAN AHEAD

Apply a light coating of anti-seize compound to the inner races of all four crankshaft bearings. Pack the lips of the two oil seals with Lithium lubricant. Use the same lubricant on the two **O**-rings. Immerse the four bearings, with the **O**-rings installed, into a container of oil heated to 167°F (75°C). Heating the bearings in the hot oil will expand them ever so slightly and ease the task of installation. Keep the bearing in the heated oil until the work has progressed to the point the bearings are to be installed in Step 5.

Rotary Valve Shaft Installation

1- Apply water resistant lubricant to the rotary valve shaft ball bearing. Install the bearing with the "open" side facing the plug in the crankcase and the "shielded" side toward the center of the crankcase. Tap the bearing into place with a suitable driver.

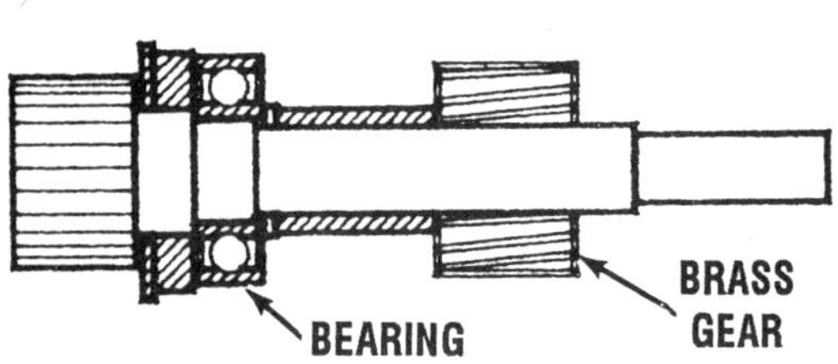

1

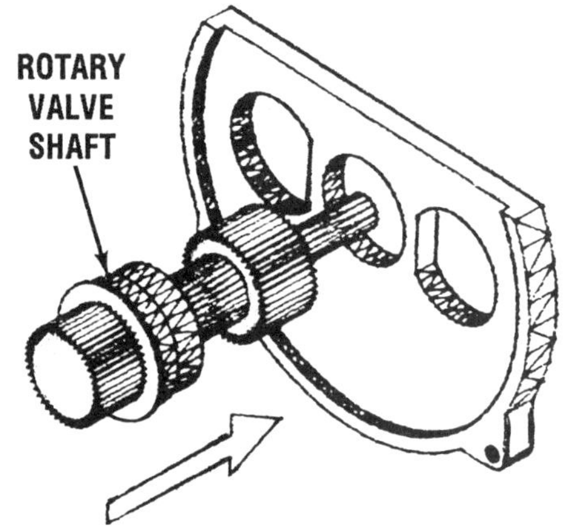

2

2- Install the rotary valve shaft into the crankcase. Use a soft head mallet and gently tap the shaft into place until it seats.

SAFETY WORDS

The C-lockrings are made of spring steel and may pop out of the groove with considerable force. Therefore, warn others in the area and **WEAR** eye protection glasses or a shield while installing the lockrings.

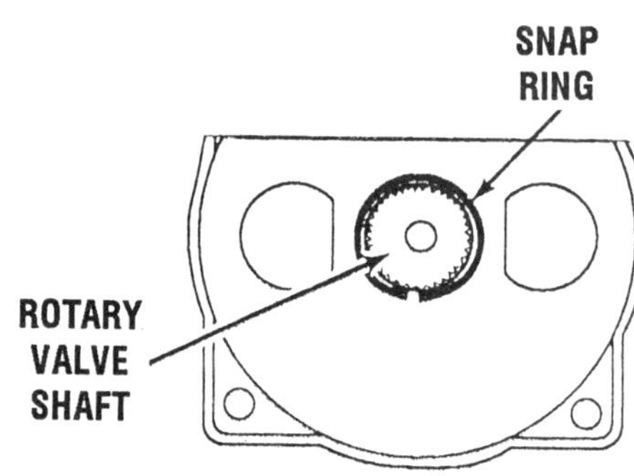

3

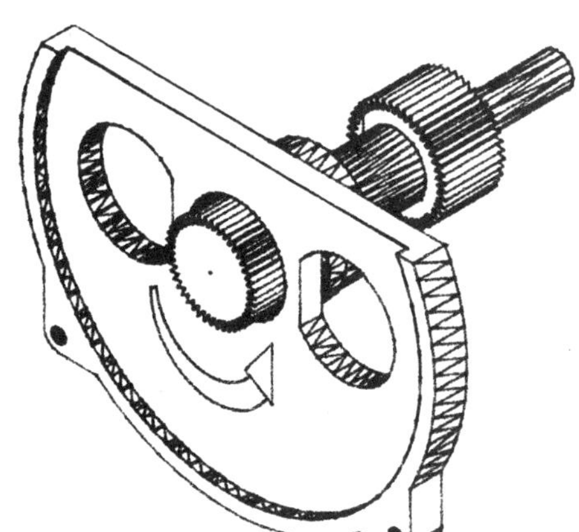

4

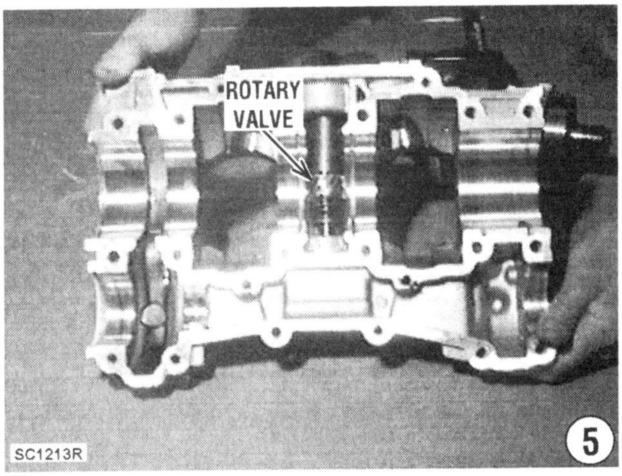

3- Observe the snap ring. One face has a sharp edge -- the other edge is smooth. Using a pair of snap ring pliers, install the ring into the groove in the rotary valve shaft with the sharp edge facing **OUTWARD**.

4- Rotate the shaft through a complete revolution to **ENSURE** there is no evidence of binding.

5- Place the upper crankcase in the inverted position on the work surface on a couple of wood blocks. The wood blocks are necessary to provide room for the connecting rods as the crankshaft is lowered into place.

Rotate the crankshaft bearing locator pins to point straight up in preparation to positioning the crankshaft in the case.

6- Seat the crankshaft into the upper crankcase. Check to be sure the index mark on the aft crankshaft ring gear is situated in an approximate nine o'clock position when facing it.

Rotate the locating pins -- lightly tap them with a punch or similar tool -- towards the counterbalance shaft. Continue to rotate each locating pin until it rests in the appropriate recess.

7- Lower the counterbalance shaft into place in the upper crankcase while observing the index groove in the aft counterbalance gear. This groove **MUST** align with the index mark observed on the aft crankshaft ring gear.

Check to be sure the two aft rubber oil seals are pushed all the way back -- aft -- on the counterbalance shaft and the seals are perpendicular to the shaft -- not crooked.

8- Slide the stainless steel retainer clip over the aft end of the crankshaft. Seat the retainer clip in position against the aft oil seal. Check to be sure the oil seal is snug against the retainer clip.

9- Install the large C-clip into the aft groove to permit the clip to seat against the counterbalance shaft. The end cap shown in the illustration will be installed once the crankcase halves are mated.

10- Obtain Gasket Eliminator sealant -- P/N 293 800 007 -- Loctite 515; or an equivalent semi-drying liquid gasket. Apply a light, even coating of this sealant to the mating surface of one crankcase half **ONLY**. Be sure to coat the

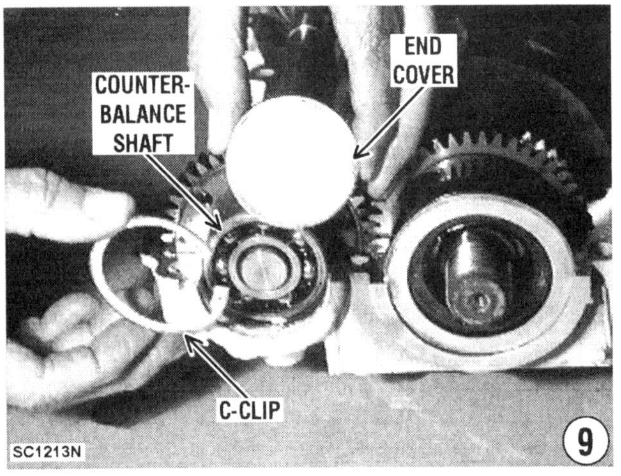

entire mating surface. **CAREFULLY** lower the upper crankcase and mate the two halves together. Check to be sure the retainer clip indexes properly in the lower half and all surfaces seat firmly. Seat the aft end plug firmly in position behind the counterbalance shaft, using a rubber mallet.

11- Apply a coating of non-permanent locking compound to the threads of all crankcase bolts. Refer to the accompanying illustration. Install the twelve 13mm crankshaft-side bolts into the lower crankcase, as shown. Bring the

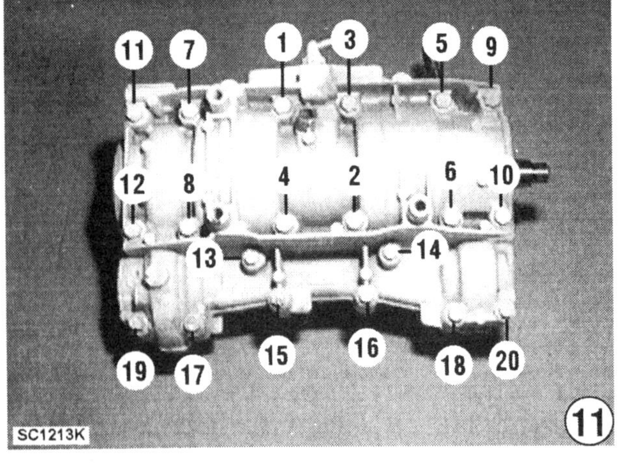

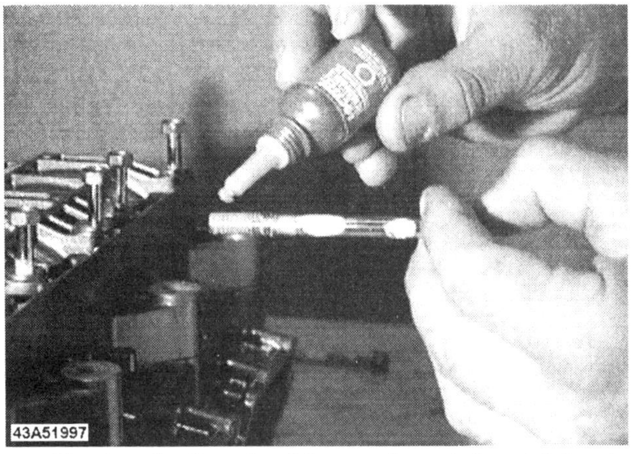

Coating the threads of the crankcase bolts with Loctite to ensure they do not work loose.

bolts up "just snug," at this time. Install the eight 11mm counterbalance shaft-side bolts and bring them up "just snug". Now, tighten all bolts in the sequence shown, to a torque value of 9 ft lbs (12Nm).

Tighten the bolts in the same sequence to a torque value of 17 ft lbs (24Nm). Finally, tighten **ONLY** the 13mm bolts to a torque value of 30 ft lbs (40Nm) in the same sequence.

The manufacturer recommends adding 1oz. (30ml) of SAE 30 oil to the counterbalance shaft gear. Remove the crankcase filler plug, add the oil, and then install the plug.

12- Obtain special aft flywheel holding tool P/N 195 000 134. Apply anti-seize compound to the threads at the aft end of the crankshaft. Thread the aft flywheel onto the aft end of the crankshaft. At this time, tighten the flywheel just "snug". It will be tightened to the required torque value later -- after the forward flywheel is installed.

13- Position the aft motor mount bracket on the aft end of the crankcase. Install the five 13mm bracket mounting bolts and tighten them securely.

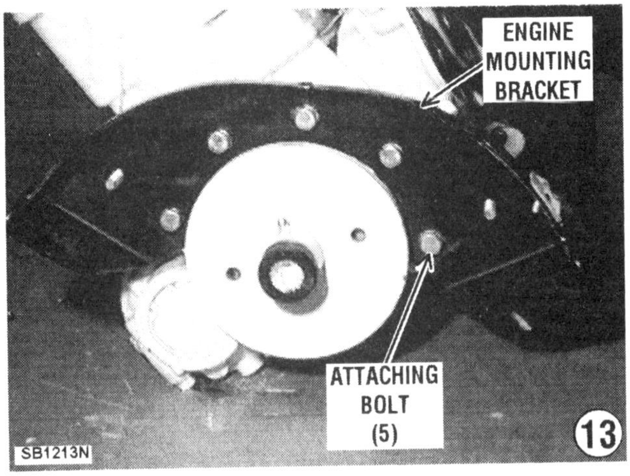

14- Inspect the mating surfaces of the crankcase halves for any excess liquid gasket compound, especially around the forward mating surfaces where the magneto housing cover is to be installed. Remove all traces of excess material. Place a **NEW** magneto housing gasket in position on the crankcase.

15- Install the magneto housing onto the crankcase. Again, check to be sure the mating surfaces seat firmly. The surfaces will not seat properly if excess gasket material is not removed.

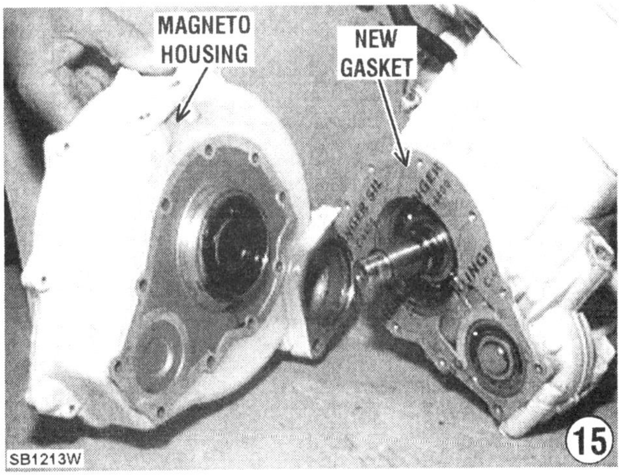

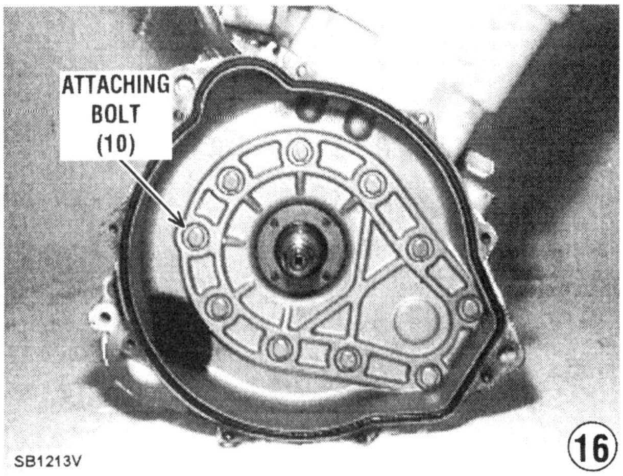

16- Apply Loctite 242 to the threads of the 10mm magneto housing bolts. Install the bolts securing the magneto housing onto the crankcase. Tighten the bolts in a criss-cross sequential pattern to a torque value of 7 ft lbs (9Nm).

17- Press the Woodruff key into the slot in the crankshaft. Slide the flywheel onto the crankshaft with the notch in the flywheel aligned over the Woodruff key. Move the flywheel onto the crankshaft as far as possible. Be **ABSOLUTELY** sure the notch aligns with

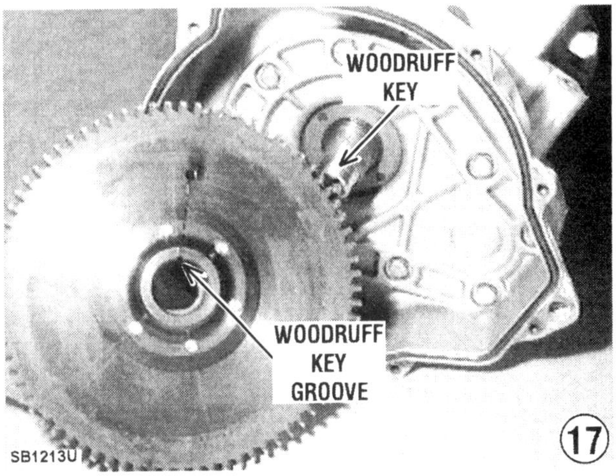

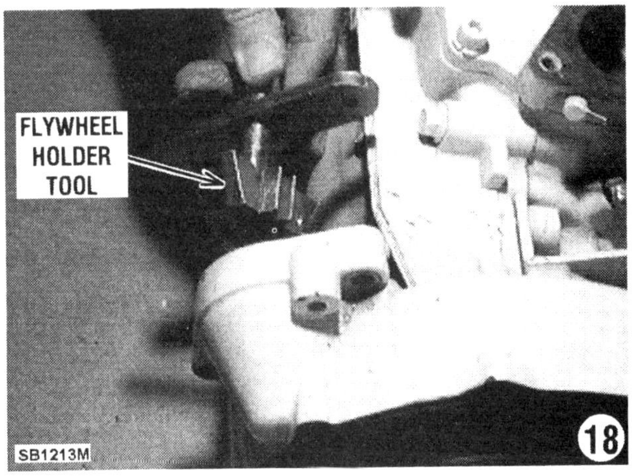

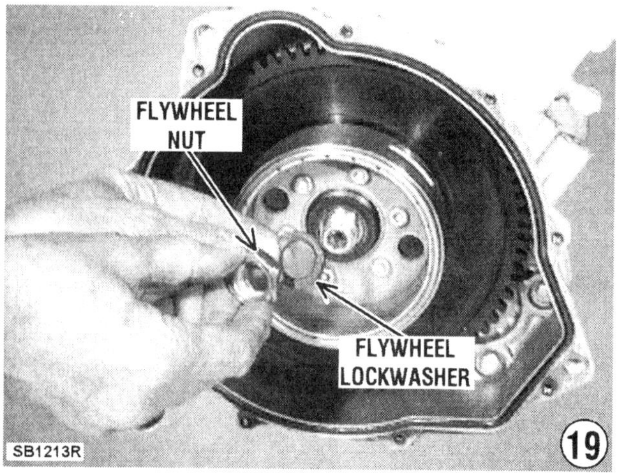

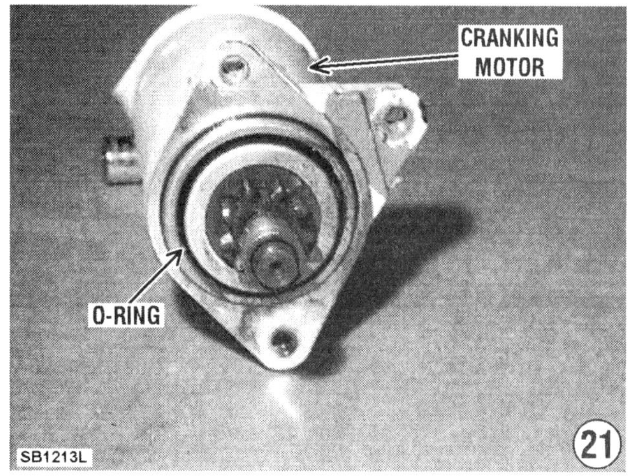

the Woodruff key. If not, the key may "fold over", damaging the flywheel **AND** crankshaft. Rotate the aft flywheel back-and-forth to verify the forward flywheel rotates, indicating the flywheel notch is properly positioned over the Woodruff key.

18- Apply Loctite 271, or equivalent, to the forward threaded end of the crankshaft. Insert the flywheel holder tool in the cranking motor opening in the back of the magneto housing. Secure the special tool in place with the attaching bolts.

19- Position the lockwasher, then the flywheel nut, in place on the end of the crankshaft. Install the flywheel nut and tighten it to a torque value of 105 ft lbs (142Nm).

With the forward flywheel holding tool still in place through the cranking motor opening, tighten the aft flywheel to a torque value of 81 ft lbs (110Nm). Slide the rubber boot over the end of the aft flywheel. Secure the boot with the reusable clamp removed during disassembly, using the special tool.

20- Insert the nylon oil injection connecting rod into the slot in the forward end of the crankshaft. Rotate the aft flywheel and at the same time, position the forward flywheel cover against the magneto housing mating surface. Continue rotating the aft flywheel until the connecting rod indexes with the oil injection pump slot. Once the nylon rod indexes into the slot, the flywheel cover will seat firmly against the magneto housing.

Cranking Motor Installation

21- Remove the bolts attaching the flywheel holder tool to the magneto housing. Pull the special tool free of the flywheel and set it aside. Check to be sure the cranking motor **O**-ring is in satisfactory condition.

22- Apply Loctite 242, or equivalent, to the cranking motor attaching bolt threads. Position the cranking motor into the back of the magneto housing. Install the three lockwashers, then the attaching bolts securing the cranking motor to the housing. The longest bolt is positioned at the top of the cranking motor. Tighten the bolts to a torque value of 18 ft lbs (25Nm).

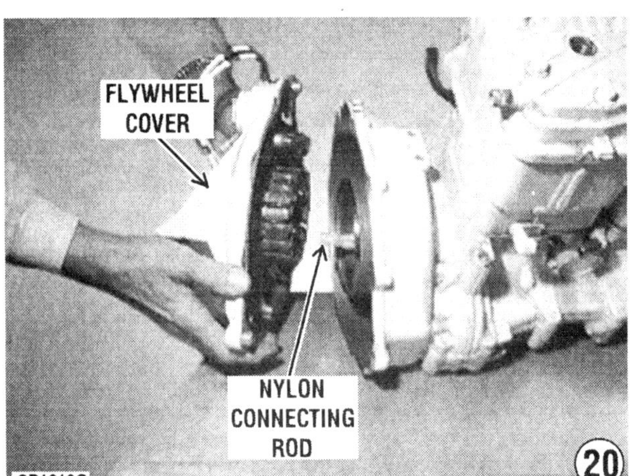

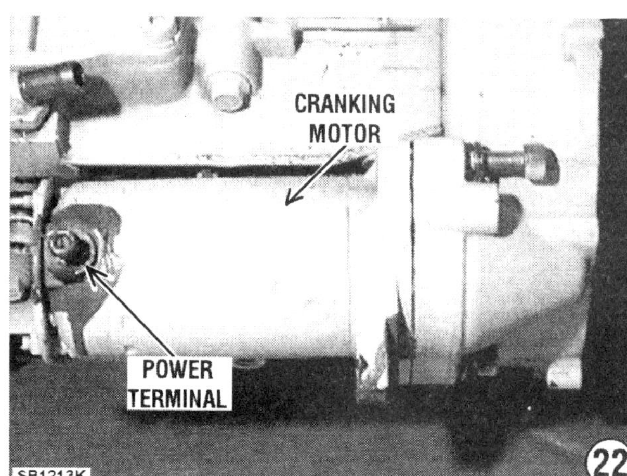

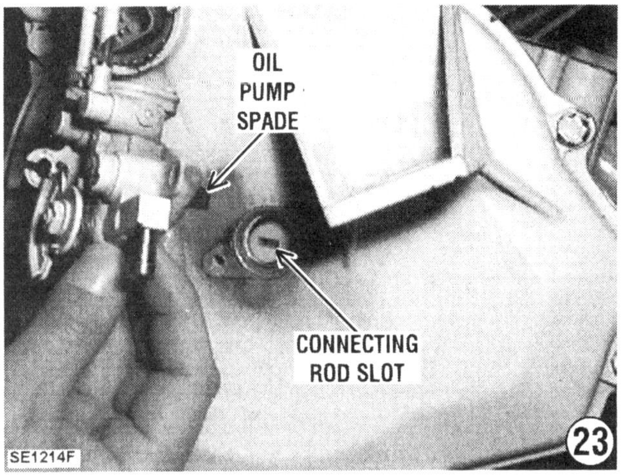

Oil Injection Pump Installation

23- Position the oil injection pump in place on the flywheel cover -- if it was removed during disassembling. Rotate the aft flywheel and at the same time, press the oil pump spade against the connecting rod. Once the spade indexes with the slot in the nylon connecting rod, install and tighten the two attaching bolts "just snug".

Be sure to bleed the oil injection system before operating the engine. It is very likely some air will enter the system while performing lubrication system tasks. The air will prevent the oil from circulating properly, causing certain damage to the engine.

Bleed the pump by unscrewing the 7mm bleeder bolt **SLIGHTLY** until oil begins to flow. Once air bubbles are no longer present, tighten the bleeder bolt.

Piston Installation

24- Lower a **NEW** cylinder base gasket into position on the upper crankcase "deck". Check to be absolutely sure it is the correct gasket for the engine being serviced -- identified by stamped holes in the gasket between the cylinder openings.

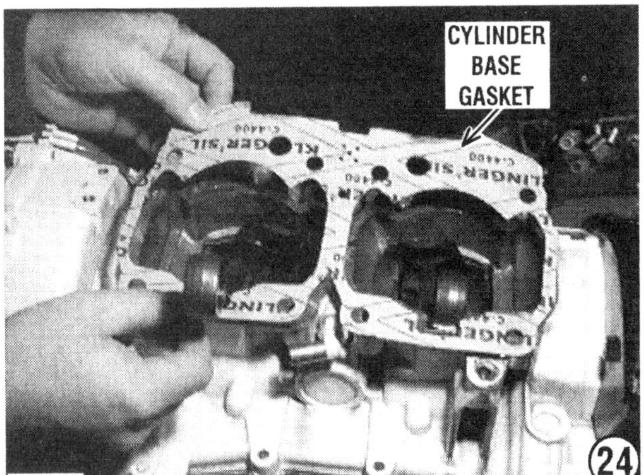

NEEDLE BEARING WORDS

Good shop practice dictates cageless needle bearings not be used a second time. New needles are encased in a nylon holder which greatly eases the task of installing them.

25- As a precaution, place a clean shop rag over the crankcase deck. Install a set of **NEW** cageless needle bearings into the connecting rod small end by first pushing the bearings out slightly from the nylon holder. Slide the exposed bearings into the connecting rod small end.

26- Press the needle bearings into position, then discard the nylon holder. Install new cageless bearings into the other connecting rod in the same manner.

27- Observe the notch in the middle of the gasket of each "cylinder" in the accompanying illustration. The side port or opening of each piston **MUST** face the notch in the gasket -- the

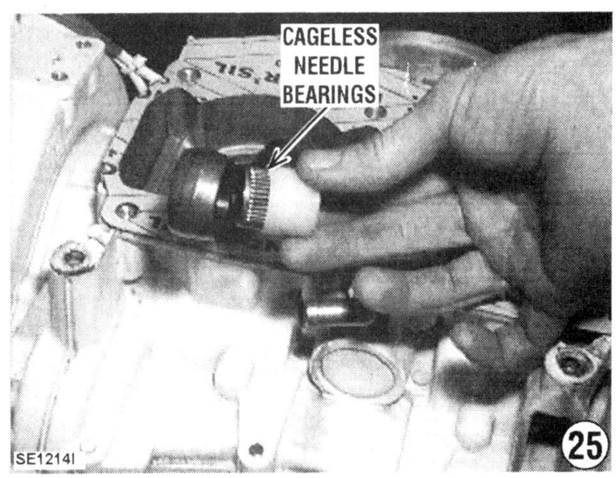

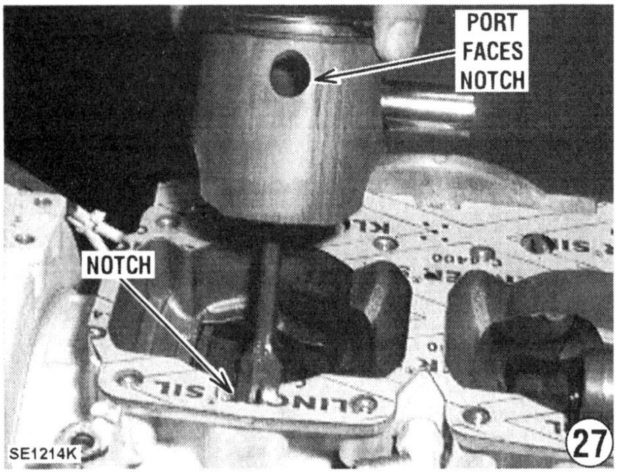

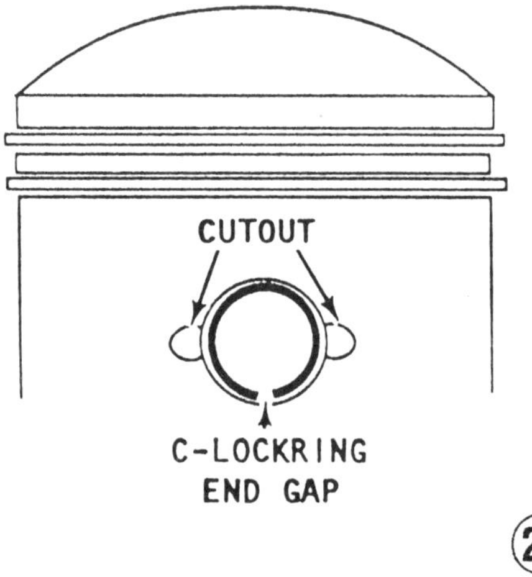

piston opening points to the port side -- opposite the rotary valve cover. Position the piston over the connecting rod with the opening facing port. Insert the piston pin into the piston and needle bearings.

28- Center the pin in the piston, using a small ballpeen hammer and gently press the pin into place. The inner needle bearing holder will be driven out the other side of the piston.

SAFETY WORDS

The piston pin C-lockrings are made of spring steel and may pop out of the groove with considerable force. Therefore, warn others in the area and **WEAR** eye protection glasses or a shield while inserting the piston pin lockrings.

29- Install a C-lockring at each end of the piston pin. Press the lockring into position using special tool P/N 290 877 016 -- model 787 engines only. Insert the lockring into the special tool, then seat the lockring against the pin and tap the tool with a hammer. The C-lockring will be properly seated.

If this tool is not available, the lockrings may be worked in by hand. Position the ring end gap away from the cutout in the piston -- place the end gap straight up or straight down. Use care to prevent deforming the lockring during installation. Snap the three remaining C-lockrings for both pistons in the same fashion. Check to be sure the piston pin lockrings do not move freely after installation and the end gap of the lockring is away from the piston cut out. (Installation of piston rings is covered in the Cleaning and Inspection section of this chapter.) Check the piston ring end gaps to be sure they "straddle" the locating pin in each piston ring groove.

30- Compress the piston rings and carefully lower the No. 1 -- forward cylinder into place over the piston. Seat the cylinder on the "deck" of the crankcase.

Compress the piston rings and carefully lower the No. 2 -- aft cylinder -- into place over the piston. Seat the cylinder against the crankcase, and verify the base gasket is intact, not "rolled", or "pinched over".

Apply Loctite 515, or equivalent, to the threads of the cylinder bolts. Apply just a "dab" of synthetic grease under each bolt head. Install the cylinder bolts just hand tight, at this

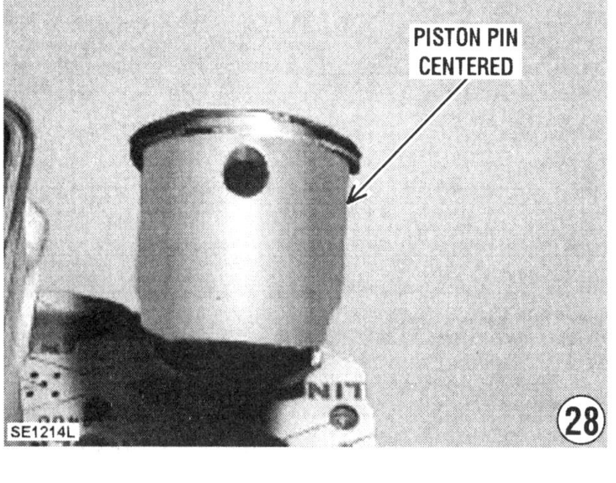

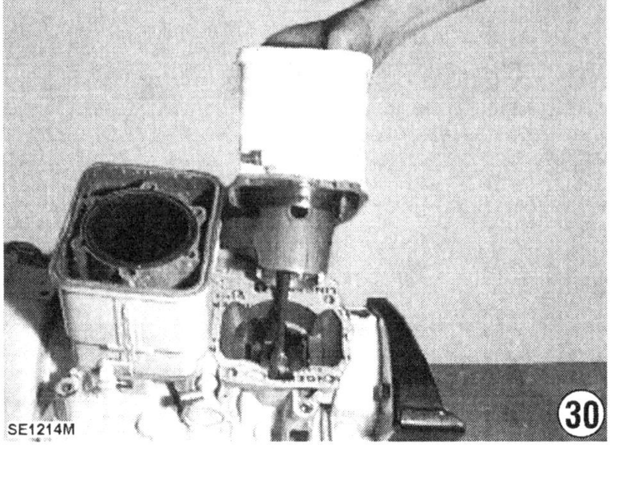

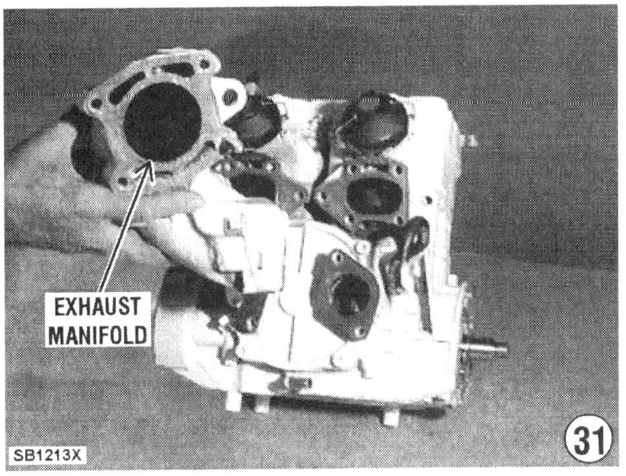

time. Because the cylinders are separate, the exhaust manifold must be installed before the cylinder bolts are tightened, to permit slight cylinder movement to achieve proper exhaust manifold installation. The exhaust manifold must be installed in the manner described in the next step and at this time, because it is not possible to accomplish the task if the cylinders are secure **OR** if the engine is installed in the craft.

Exhaust Manifold Installation

31- Apply a "dab" of synthetic grease to the threads of the four bolts to be installed at the lowest location of the manifold. Apply the grease to the aft bolt of the forward cylinder and to the threads of the aft bolt of the aft cylinder location. Apply Loctite 515, or equivalent to the threads of the bolts for the two remaining locations. Place a **NEW** exhaust manifold gasket in position against the cylinders. Note the gasket can only be properly indexed one way. Position the exhaust manifold against the gasket. Install the manifold attaching bolts and tighten them to a torque value of 30 ft lbs (40Nm).

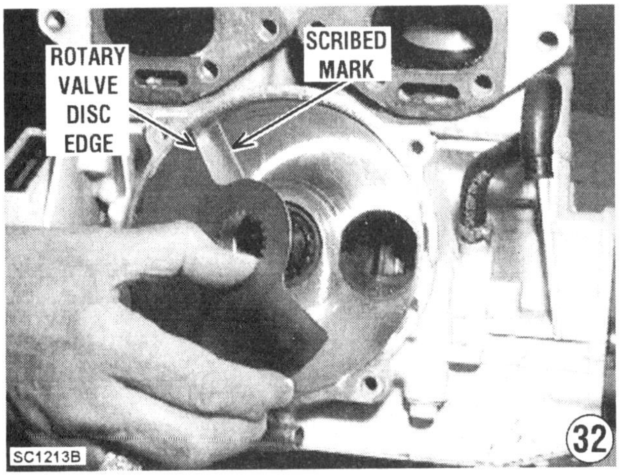

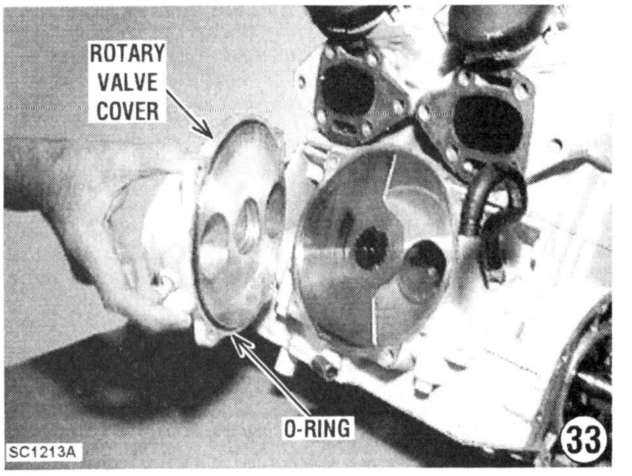

After the exhaust manifold is installed, tighten the cylinder bolts to a torque value of 30 ft lbs (40Nm).

Rotary Valve Installation

32- Slide the rotary valve onto the rotary shaft. Seat the valve disc on the crankcase, aligned with the scribed marks made during the disassembling procedure.

CRITICAL ROTARY VALVE WORDS

If the scribe marks were not made, as instructed, during disassembling, proper positioning the rotary valve is much more involved and can only be accomplished using a "Degree Wheel". Therefore, obtain the degree wheel and follow the directions included with the wheel.

33- Install the rotary valve large **O**-ring by seating one end, then pulling the ring around and letting it gently "snap" into the groove. If the **O**-ring is "rolled" into place, as is usually done with an **O**-ring, the ring will simply "pop"

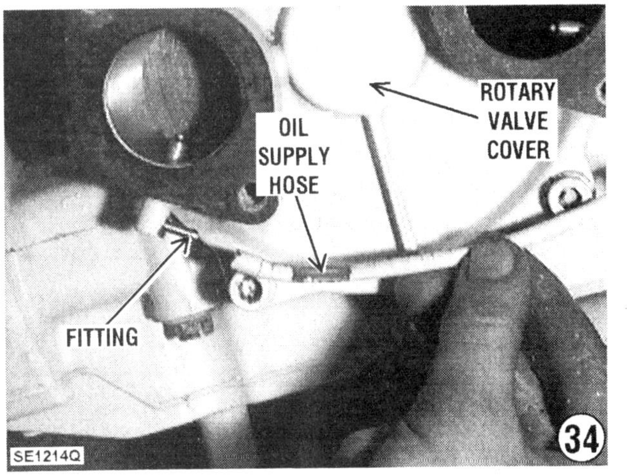

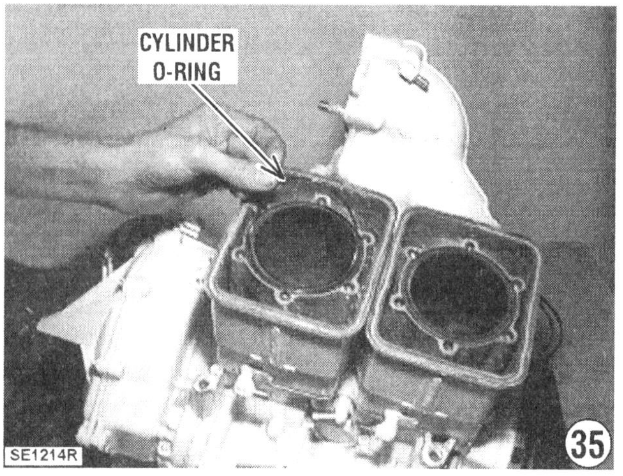

right back out of the groove.

Position the rotary valve cover over the valve. Insert the four 6mm attaching bolts and tighten them securely, in an alternating pattern.

34- Attach the two oil delivery hoses from the injection pump to the intake port fittings.

Cylinder Head
Installation

Before setting the **O**-rings in place, inspect the cylinder and water jacket grooves of each cylinder, to be sure they are clean and free of debris.

35- Install a **NEW** cylinder -- smaller diameter O-ring into the groove at the top of the cylinder.

36- Set a **NEW** water jacket **O**-ring -- large rectangular one -- into the groove of the cylinder "wall". Check to be sure the ring is completely seated around the entire perimeter of the cylinder.

Repeat steps 35 and 36 for the other cylinder.

37- Verify the cylinder head spark plug openings are clean and free of debris. Install a **NEW** spark plug **O**-ring in each opening.

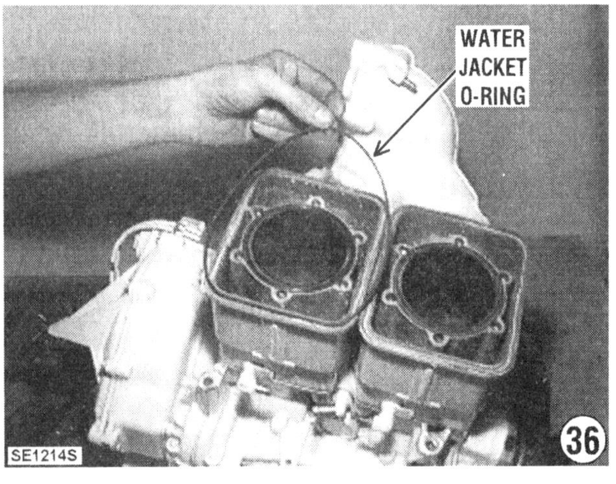

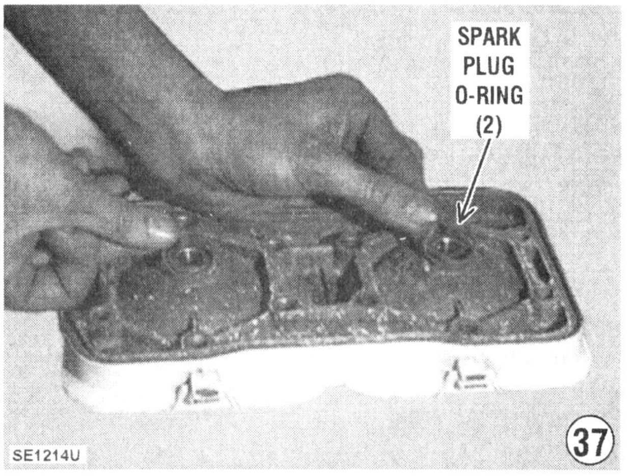

38- Install the large cylinder head **O**-ring into the groove of the head. Check to be sure the ring is seated uniformly around the head.

39- Observe the two tabs -- one per cylinder -- on the side of the cylinder head. This side of the cylinder head faces **PORT** -- left. The other side of the cylinder head has only one tab, which faces the exhaust manifold.

Lower the cylinder head **EVENLY** and squarely into position over the cylinders. Be sure to seat the head firmly on the cylinders, without repositioning the **O**-rings.

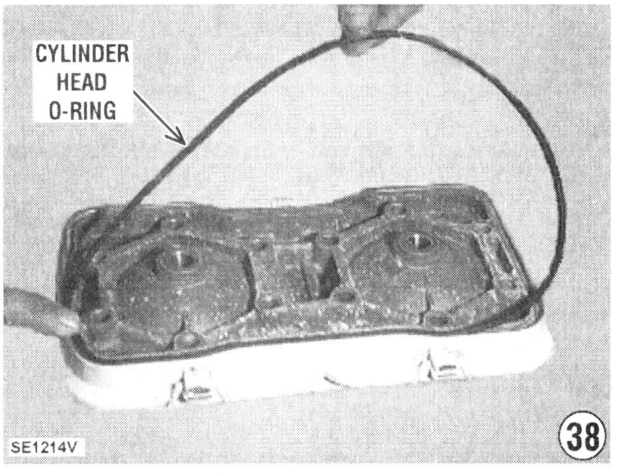

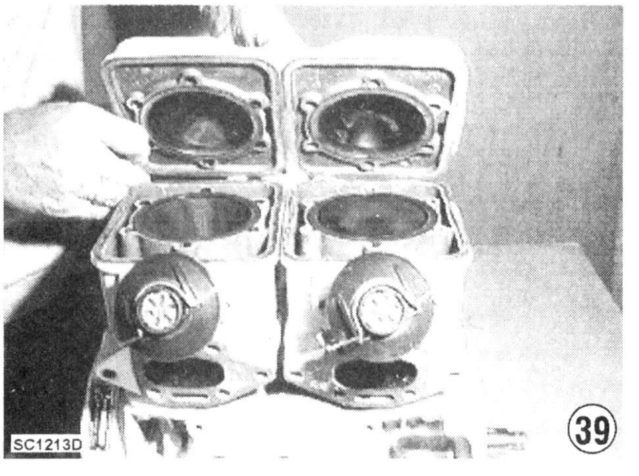

After the head is in position, the head cover is lowered onto the head. The text explains critical positioning.

Lower the head cover into position over the cylinder head. Be sure the cylinder head **O**-ring remains seated in the groove. The head cover should be positioned with the embossed arrows indicating water flow direction is **AFT**. The head cover was tilted in the accompanying illustration for photographic clarity.

40- Tighten the head cover bolts in the sequence shown to a torque value of 17 ft lbs (24Nm).

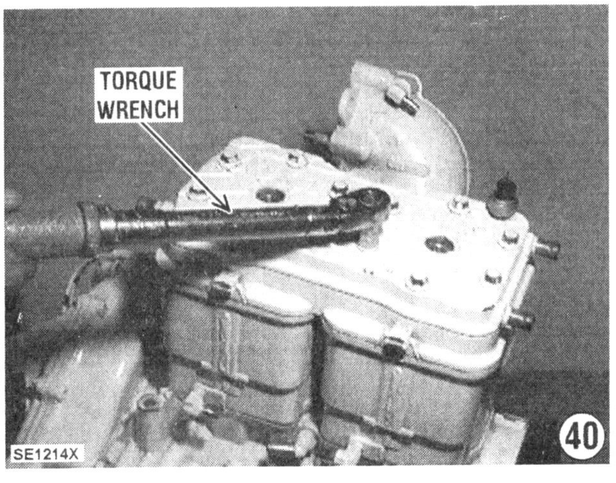

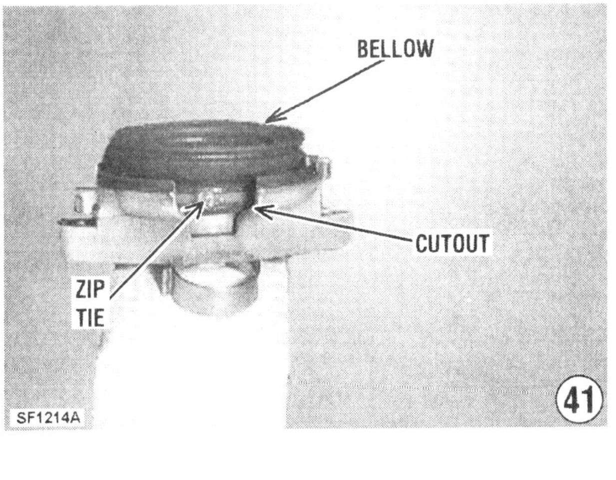

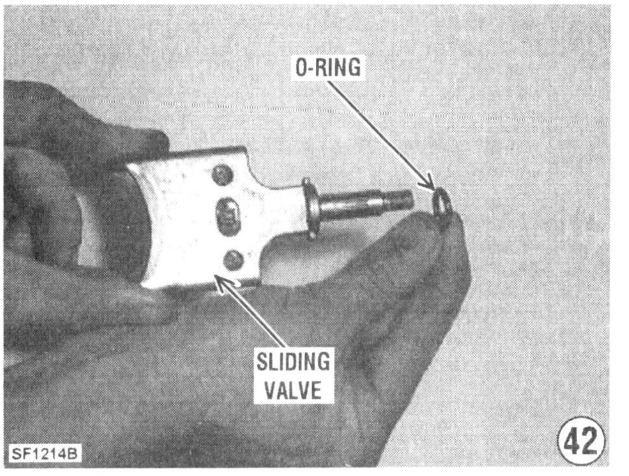

RAVE valve
Installation

The RAVE valve housing may be cleaned with contact cleaner or WD-40. The housing should **NEVER** be cleaned with a carburetor cleaner or solvent. Such action would surely destroy the rubber materials within the assembly.

41- If the bellow has been removed from the rotary valve housing, lower the bellow into the valve housing. Secure the base of the bellow with a zip tie. Take note of the "cutout" in the RAVE valve housing. The housing is installed with the cutout facing **DOWNWARD**.

42- Place a new **O**-ring over the shaft of each sliding valve.

43- Observe the word **"TOP"** embossed on one side of the sliding valve. This mark **MUST** face **UP**, towards the cylinder head. Slide the valve, with **"TOP"** facing **UP**, into the valve guide.

44- Place a **NEW** gasket over the shaft of the sliding valve and into position on the cylinder port. Lower the housing into position on

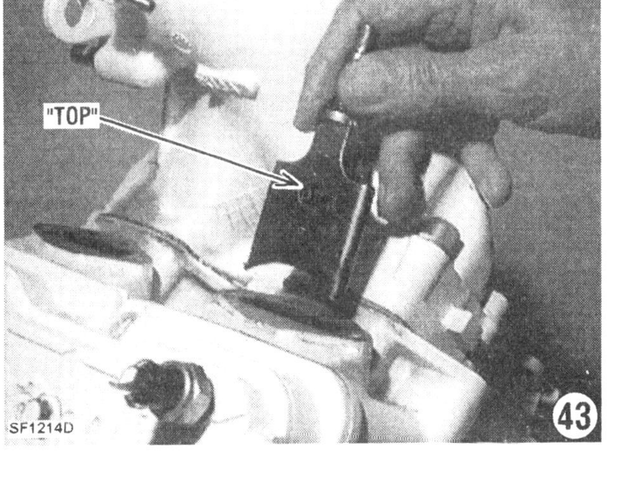

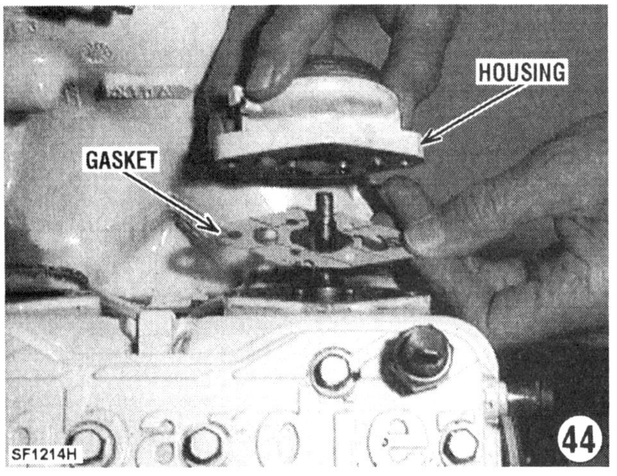

the cylinder with the "cutout" facing **DOWNWARD**, toward the exhaust manifold. This cutout is a vacuum port, allowing the bellow to operate. Therefore, if the housing is installed incorrectly, the bellow will not operate properly.

45- Apply Loctite 242 to the valve housing attaching bolts. Install the 5mm bolts and tighten them securely. Lower the large diameter spring "keeper" into position over the bellows cap. Seat the spring into the groove on one side, then pull the spring around and seat it into position.

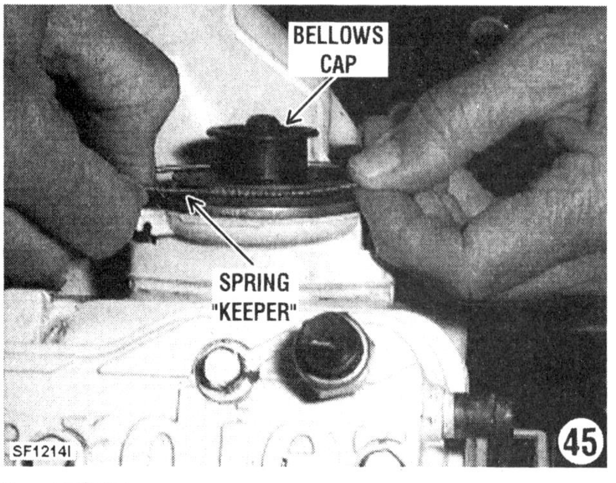

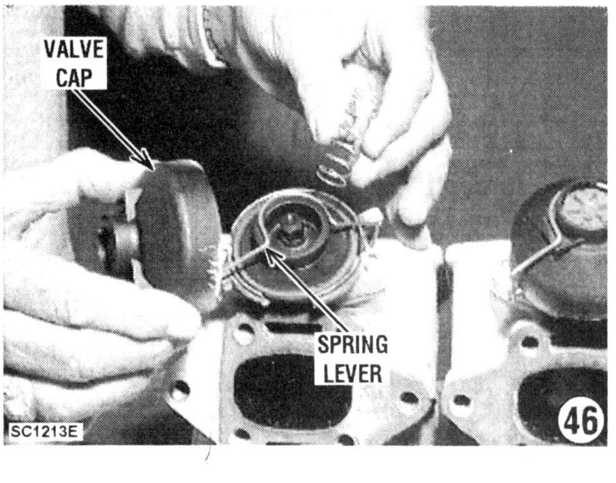

46- Place the black cap over the bellow and secure it in position by sliding the spring lever up over the cap. The red adjustment knob should be turned in or out, until the top of the knob rests flush with the black cap. This knob adjusts the speed at which the RAVE valve opens, due to spring pressure. This is the optimum setting for the RAVE valve.

MODEL 787 ENGINE INSTALLATION

Check to be sure any attached hoses or wires are securely wrapped around the engine, or tucked away to prevent entanglement while lowering the engine into position.

1- Enlist the aid of an assistant or straddle the engine compartment in the same manner used during engine removal. Move the engine into the craft slightly forward of the engine mounts. Carefully shift the engine aft, aligning the splines of the aft flywheel with the splines of the driveshaft. Gently rock the engine back-and-forth to index the splines, and then continue to move the engine aft. The shaft is properly indexed once the motor mount holes align with the engine.

2- Install the clear water drain hose to the "T" fitting on the port side crankcase fitting.

3- If the engine mounting holes do not properly align with the engine, use a long screwdriver to provide leverage against the engine lug to ease the engine into the proper position. Apply Loctite 242 to the threads of the aft motor mount bolts. Install a flat washer, followed by a lockwasher, and then install the two mounting bolts. Tighten the aft bolts **ONLY** hand-tight, at this time.

4- Attach the cranking motor power lead to the terminal on the aft end of the cranking

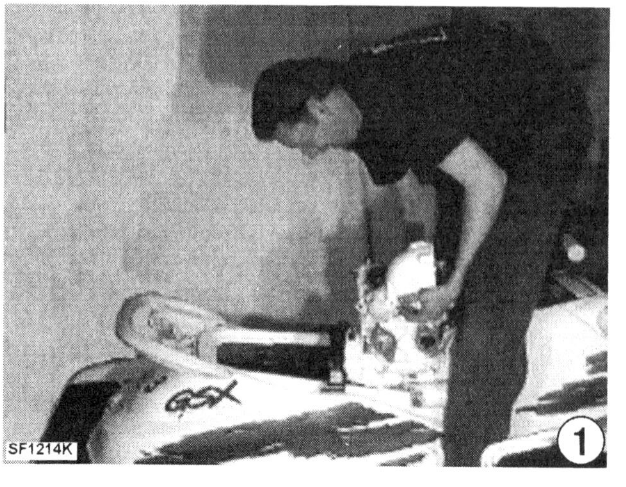

motor. Tighten the bolt securely. Coat the terminal with dielectric grease, and then cover it with the boot.

Advice On Grounding

The ground lead is typically attached to the lowest cranking motor lug. In this position, it is not easily accessible and may be susceptible to corrosion at this lower location due to standing water in the compartment. The next step provides an alternative method.

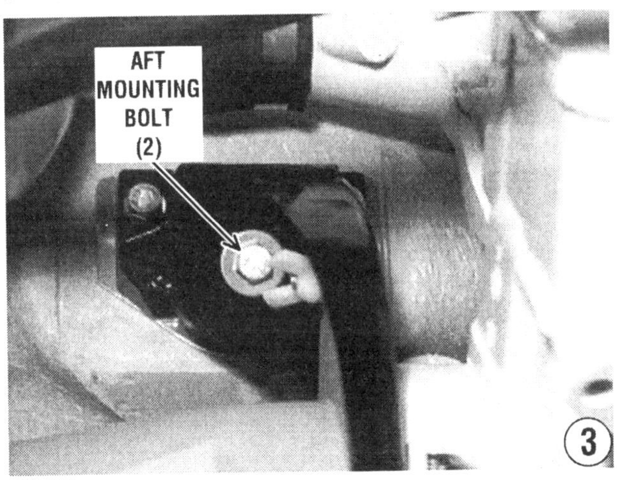

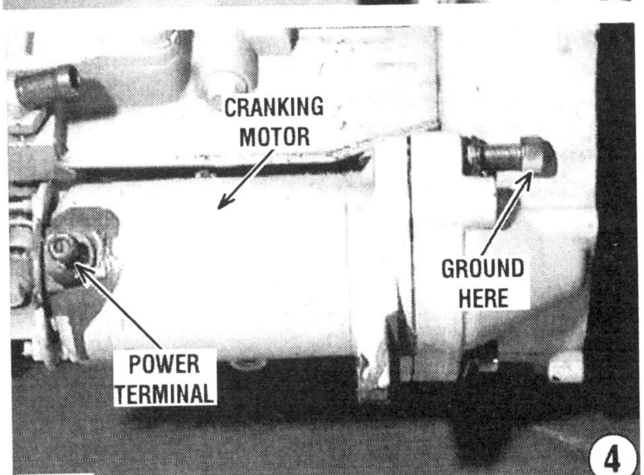

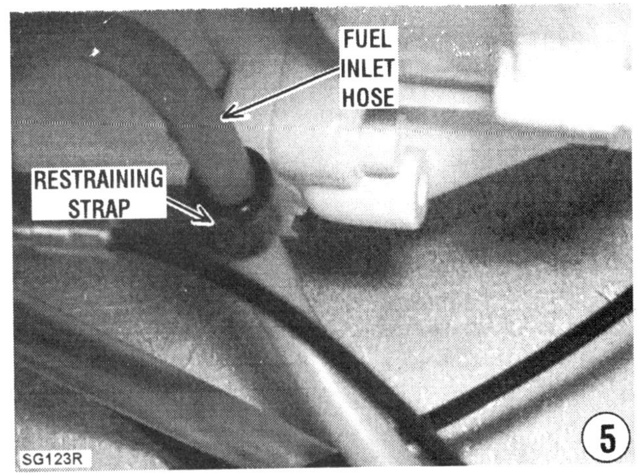

5- Attach the ground lead to the top cranking motor shoulder -- be sure to use the long bolt in this location. Install the inlet fuel hose clamp on the top cranking motor bolt, at this time. Tighten the bolt securely.

6- Install the front motor mounts. Take time to verify the shim material being installed is positioned back in the same location from which it was removed. Raise the engine slightly to permit sliding the motor mounts and shim material into place. Check to be sure the cylindrical portion of each motor mount faces **AFT**.

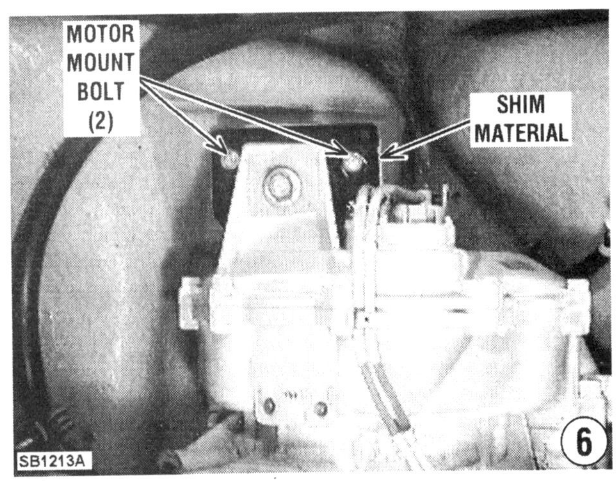

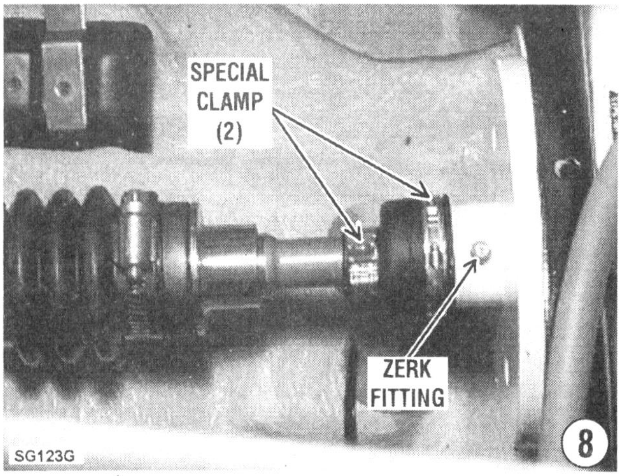

If the forward holes are not aligned properly, use a suitable screwdriver as leverage to ease the engine into the proper position. Apply Loctite 242 to the forward motor mount bolts. Install the bolts and tighten them to a torque value of 18 ft lbs (25Nm).

7- Tighten the aft motor mount bolts to a torque value of 18 ft lbs (25Nm).

8- Slide the rubber boot over the end of the aft flywheel. Secure the boot in position with the metal clamp and the same special tool used during removal.

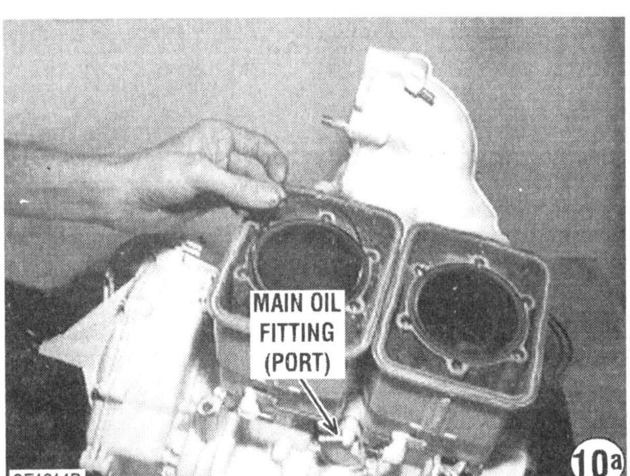

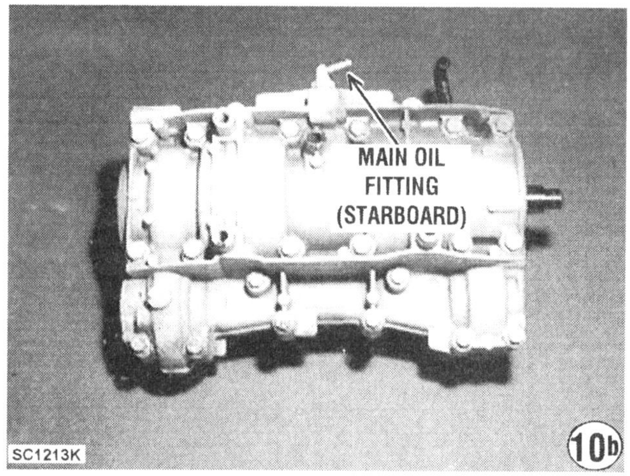

Rotate the aft flywheel until the zerk fitting on the flywheel points upward. Using a grease gun, apply grease to the fitting, and at the same time, observe the driveshaft. Applying grease should push the driveshaft aft slightly.

9- Attach the front oil supply hose to the oil pump. Secure the hose to the injection pump with a zip tie or compatible clamp. Obtain a small container to catch oil ejected from the loosened bleed screw. Bleed the pump by opening the bleed screw -- just a turn or so -- to allow oil to escape. A small thin washer is affixed to the back side of the bleed screw and would be difficult to find if it fell into the bilge. Once the air has been removed from the pump, and a steady stream of oil without any bubbles is observed, tighten the bleed screw securely.

10a-- Attach the motor main oil hoses to the lower fitting on each side of the engine upper crankcase.

10b- The starboard fitting is the inlet; the port fitting is the return for the oil system. Tighten the clamps securely, then remove the "pinch-off" tools used at disassembly.

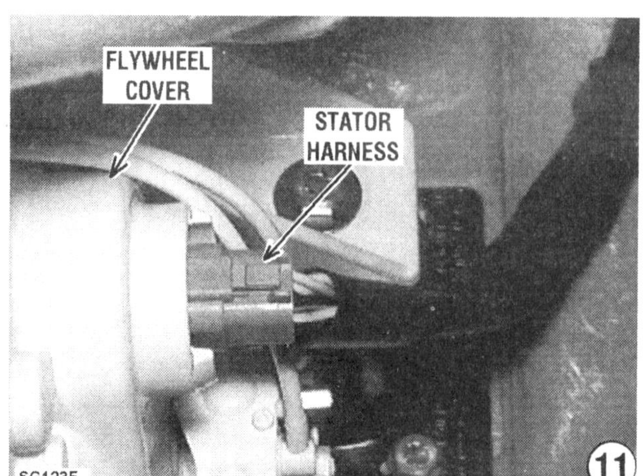

Note the small plastic base on the hull near the expansion chamber. The return hose must be positioned on this base and secured with a zip tie to ensure the oil return hose does not rest against the hot exhaust pipe. Bring the oil supply hose -- with filter -- and the supply oil hose from the pump together and fasten them with a zip tie.

11- Connect the stator electrical harness to the quick disconnect coupler located on the upper portion of the flywheel cover.

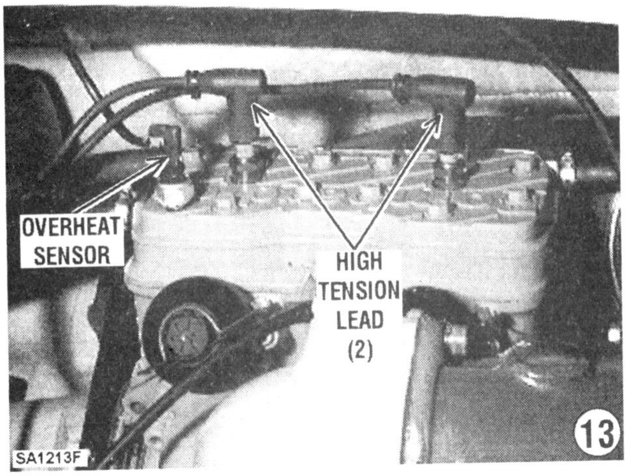

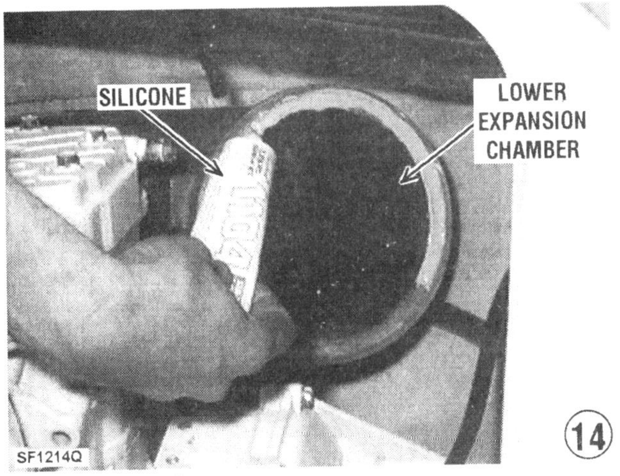

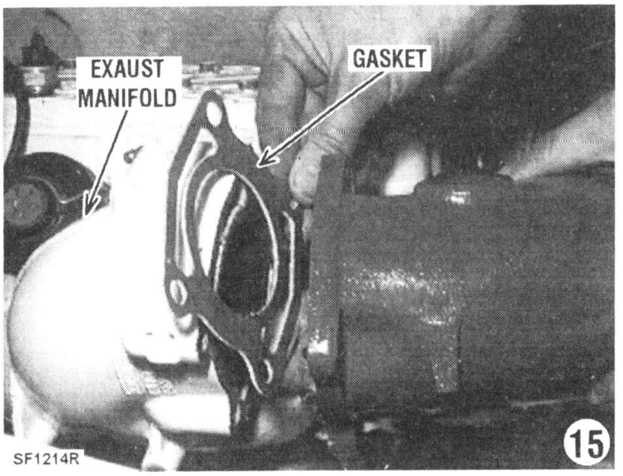

12- Lower the cover for the driveshaft and aft flywheel into position over the flywheel. Place a stainless steel washer on each stud followed by a wingnut. Tighten the wingnuts securely.

13- Attach the cooling hoses to their respective fittings on the cylinder head cover and exhaust pipes. The hose from the water regulator module -- located on the tuned exhaust muffler -- supplies water to the engine. This hose attaches to the cylinder head cover fitting with the arrow pointing forward. During disassembly. instructions were given to tag the hoses indicating their position.

ENGINE BUILD-UP

14- Move the lower expansion chamber into the engine compartment in front of the engine, and then work it around to the port side. Shift the chamber aft into position against the crankcase. Install and tighten the two attaching bolts securely.

If the brass sealing ring was removed from the forward end of the lower expansion chamber, install it now and secure it in place with a bead of silicone.

15- Place a NEW gasket between the exhaust manifold and upper expansion chamber. Install the 17mm nuts securing the expansion chamber to the manifold, and then tighten them just hand tight, at this time.

16- Install the stainless steel clamp securing the expansion chamber halves together. Be sure to position the ring to permit the clamp bolts to be tightened. Tighten these attaching bolts securely. Install the clear plastic thread protectors over the threads of the ring bolts. Now, tighten the bolts securing the upper expansion chamber to the exhaust manifold to a torque value of 30 ft lbs (40Nm).

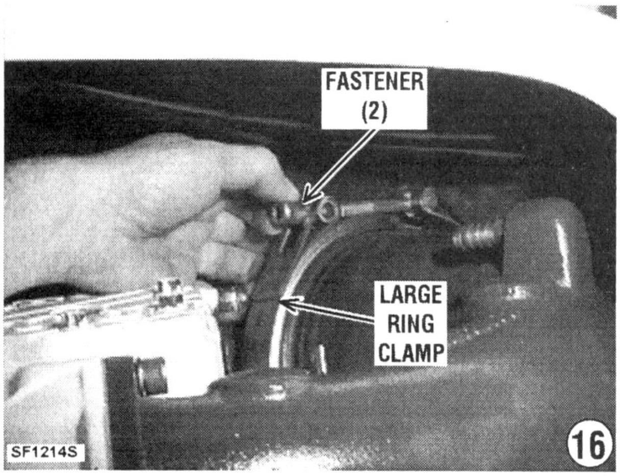

17- Install the bolt securing the upper expansion chamber bracket to the magneto housing. Tighten the bolt to a torque value of 30 ft lbs (40Nm).

18- Connect the small cooling inlet hose leading from the water regulator valve to the lower end of the upper expansion chamber. Secure the inlet hose to the fitting with a zip tie. Connect the large cooling water inlet hose leading from the cylinder head cover to the upper expansion chamber. Tighten the hose clamp

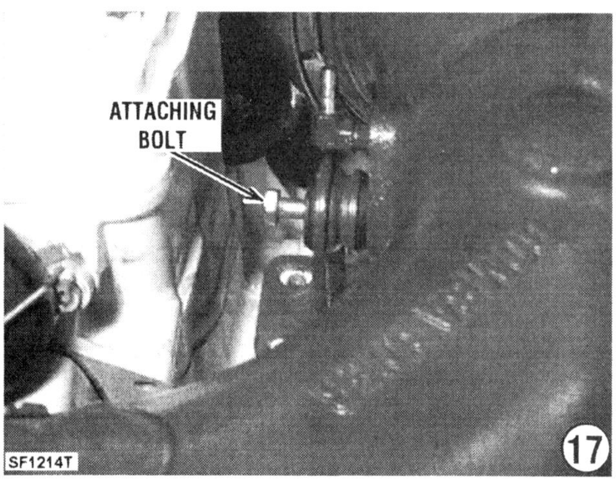

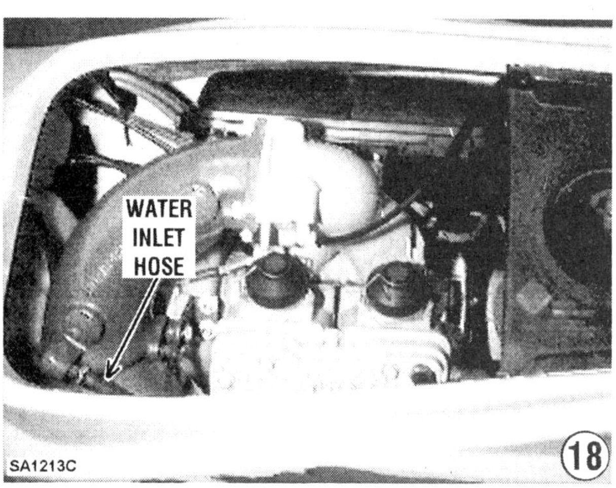

securely. Check to be sure all larger hoses are securely fastened with hose clamps.

19- Attach the temperature sensor lead to the sensor. Install the spark plugs and tighten them to a torque value of 17 ft lbs (24Nm). Connect the high tension lead to each spark plug, and then work the rubber boot down over the connection. Check to be sure the boot is firmly seated.

20- Install a new rubber carburetor base gasket on each rotary valve port in the cover. Lower the carburetor rack into position on the rotary valve cover. Install the four attaching bolts and tighten them securely. Cover the carburetors with a clean shop towel to prevent debris from contaminating the carburetors.

Connect the fuel inlet hose to the No. 1 lower end of the -- forward -- carburetor. Connect the pulse hose to the top side of the No. 1 carburetor. Connect the fuel return hose to the aft end of the No. 2 -- aft -- carburetor. Secure each hose with a zip tie -- except the fuel return hose. This hose uses a special clamp. Observe the hose bracket, located on the forward engine mount bracket. Wrap a zip tie

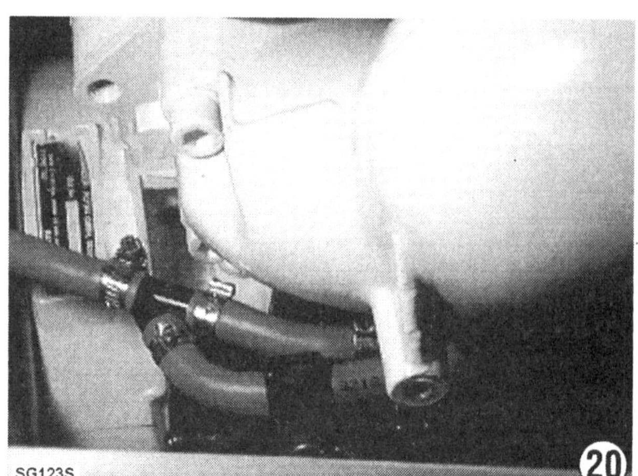

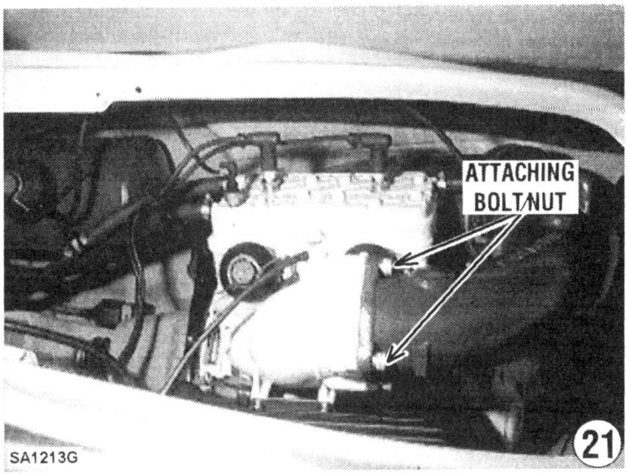

around the fuel return hose and the oil injection cable and secure them to this bracket.

21- Connect the throttle and choke cables to the No. 1 carburetor. Due to the angle of the engine, care and patience will go a long way when connecting these cables. While the choke locknuts are loose, slide the cable in place in the lever arm, then tighten the tension on the choke cable. Next, tighten the locknuts on each side. Tighten the 3mm Allen head set screw on the No. 1 choke lever arm first, then tighten the

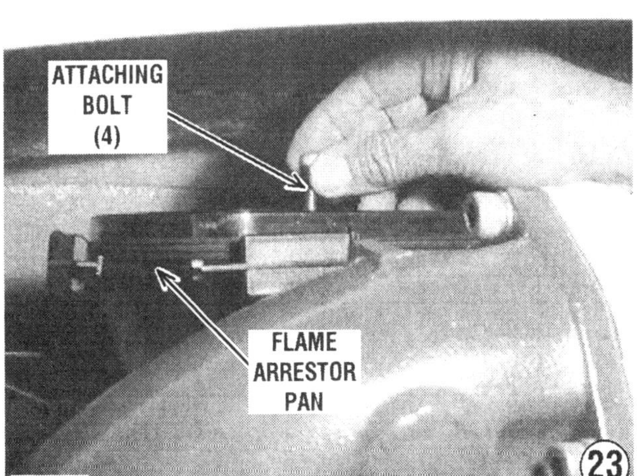

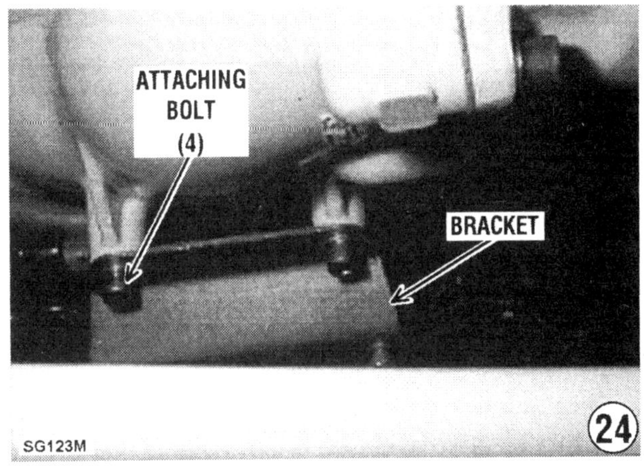

No. 2 cable set screw. Test the throttle and choke levers to be sure the cable moves freely without binding.

22- Connect the oil pump cable to the oil pump. Rotate the oil pump lever **COUNTER-CLOCKWISE** until it almost makes contact with the bracket -- say with 1/4" clearance. When the lever is rotated it should be possible to verify there is spring pressure against the lever. Without spring pressure, the lever arm will not actuate. Operate the throttle to be sure the oil injection pump lever functions properly.

23- Remove the shop towels placed over the carburetors. Lower the flame arrestor base into position on the carburetor rack. Install the four 5mm attaching bolts. Tighten the bolts securely. Place the shop towel over the arrestor base, to prevent any foreign object from falling into the carburetor throat.

24- Position the stainless steel bracket securing the flame arrestor pan to the exhaust manifold. Install the four lockwashers, then the 5mm attaching bolts and tighten them securely. Remove the shop towel from the arrestor base. Lower the stainless steel flame arrestor screen into place. Be sure the cambered top -- bowed portion of screen -- faces **UP**.

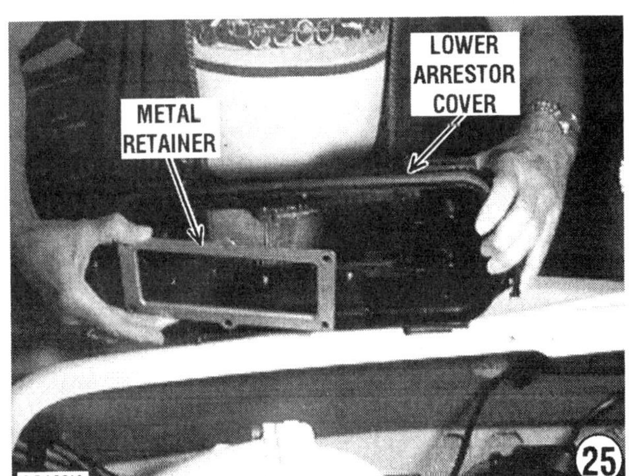

25- Move the large lower arrestor cover, with the metal retainer, into position on the arrestor base -- over the arrestor screen. Secure the cover in place with six 5mm Allen head bolts. Tighten the bolts securely.

Install the flame arrestor top cover on the lower cover. Slide the tabs in the opposite direction to lock the cover into place.

26-- Attach the cooling water bleed hose from the aft end of the craft to the top of the exhaust manifold.

CLOSING TASKS

Connect the negative and positive leads to the battery.

Move the craft into a body of water or connect a garden hose to the engine hose fitting at the aft end of the cylinder head.

CAUTION

The following procedure MUST be followed whenever the engine is started with a hose attached for cooling the engine. Start the engine, and then turn the water on. DO NOT turn the water on first. The engine MUST be started first to prevent water from backing up into the tuned exhaust pipe and into the engine. Water must circulate anytime the engine is operating to prevent overheating -- which could cause damage to moving parts and possibly cause engine seizure. NEVER allow the engine to operate without cooling water for more than 15 seconds. NEVER operate the engine with a flush attachment for more the 5 minutes total time. The driveline seal is not provided with cooling water if it is not immersed in water. To shut the engine down, first shut the water supply off, and then shut down the engine. If this sequence of tasks is not followed, as stated, the engine could be severely damaged.

REMEMBER

Start the engine.
Turn on the water.
Turn off the water.
Shut down the engine.

Attempt to start and operate the engine, without the engine compartment cover in place. This will provide the opportunity to check for water, exhaust, oil, or fuel leaks.

After the engine is operating properly, install the engine cover. Follow the "Break-in Procedures" with a load on the jet pump.

BREAK-IN PROCEDURES

Do not operate the engine at full throttle, except for very short periods, until ten hours of operation.

a- Operate the engine at various speeds throughout the power range, but **DO NOT** maintain a steady rpm for the first two hours.

b- Operate at any speed after two hours, but not at sustained full throttle until after another eight hours of operation.

c- Mix oil to the gasoline during the break-in period -- total of 2-1/2 to 3 hours -- at a mixture of 16oz oil per a full tank of gasoline. This oil is addition to the oil supplied by the oil injection system -- for added lubrication.

d- While the engine is operating -- under jet pump load, during the initial period -- have an assistant control the craft while a check is made of the fuel, oil, exhaust, and water systems for leaks.

8-5 CLEANING AND INSPECTING ALL ENGINES

PHOTOGRAPHIC WORDS

The photographs used in this section are intended to depict a "typical" setup or procedure for any small two-stroke engine. Therefore, do not be concerned if the component being serviced does not exactly match the one in the illustration. Bear-in-mind, the procedure and the end result are the most important purposes of this entire section.

The success of the overhaul work is largely dependent on how well the cleaning and inspecting tasks are completed. If some parts are not thoroughly cleaned, or if an unsatisfactory unit is allowed to be returned to service through negligent inspection, the time and expense involved in the work will not be justified with peak engine performance and long operating life. Therefore, the procedures in the following sections should be followed closely and the necessary work performed with patience and attention to detail.

Crankshaft Service

Clean the crankshaft with solvent and wipe it dry with a lint free cloth. Inspect the connecting rod for cracks, or scratches. Inspect the crankshaft oil seal surface for nicks, sharp edges or burrs which might damage the oil seal during installation or might cause premature seal wear. **ALWAYS** handle the crankshaft carefully to avoid damaging the highly finished bearing surfaces. Blow out all oil passages with compressed air. The oil passageway leads from the rod to the main bearing journal. **TAKE CARE** not to blow dirt into the main bearing journal bore.

A quick check for crankshaft alignment can be made by inserting finger into each connecting rod and attempting to "rotate" the crankshaft in the air. A twisted crankshaft will twist and turn in an out-of-round motion.

Inspect the threads at both ends for signs of abnormal wear.

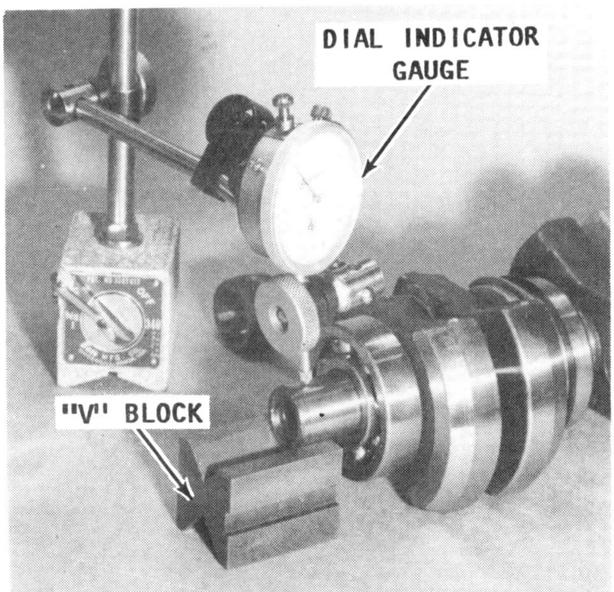

Typical setup of a crankshaft with "V" blocks and dial indicator to measure main bearing journals on an assembled two-stroke crankshaft. The bearings and connecting rod will not affect the reading.

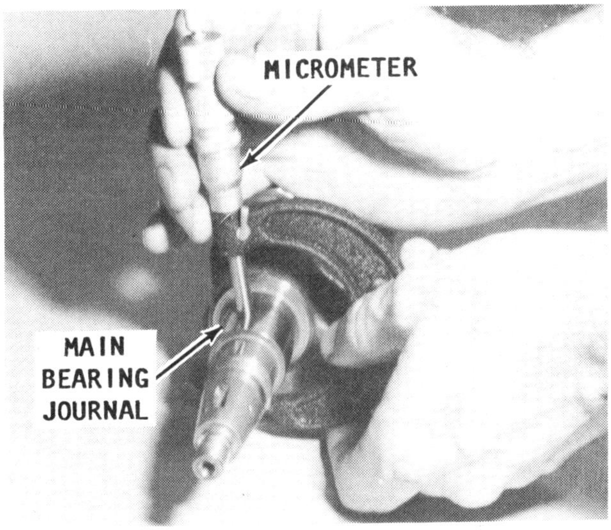

Method of using a micrometer to take the first measurement on a crankshaft main journal for out-of-round and taper.

Install a dial indicator gauge above the main bearing journals. Rotate the crankshaft and measure the runout (or the out-of-round) and the taper at both ends and in the center journal on all two-cylinder models.

The limits for all engines manufactured -- 1992, 1993 and 1994 for both ends in 0.0031 in (0.08mm). For 1995, 1996 and 1997 engines -- limit for forward end is 0.002 in. (0.05mm) and for the aft end 0.001 (0.030mm).

If "V" blocks or a dial indicator are not available, a micrometer may be used to measure the diameter of the journal. Make a second measurement at right angles to the first. Check the difference between the first and second measurement for our-of-round condition. If the journals are tapered, ridged, or out-of-round by more than the specification allows, the journals should be reground, or the crankshaft replaced.

Any out-of-round or taper shortens bearing life. Good shop practice dictates new main bearings be installed with a new or reground crankshaft.

The connecting rod axial "play" is checked at the piston end to determine the amount of wear at the crankshaft end of the connecting rod. Checking the "play" is described in the following paragraph.

Set up the crankshaft in the "V" blocks. Set up a dial indicator to make contact with the flat surface of the piston end of the rod. Now, hold the crankshaft steady in the "V" blocks and at the same time, rock the piston end of the rod along the same axis as the crankshaft. If the dial indicator needle moves through more then 0.047" (1.2mm), the "play" is considered excessive. The connecting rods are an integral part of the crankshaft and **CANNOT** be removed. If they are no longer fit for service, the entire crankshaft must be replaced.

Inspect the crankshaft oil seal surfaces to be sure they are not grooved, pitted, or scratched. Replace the crankshaft if it is severely damaged or worn. Check all crankshaft bearing surfaces for rust, water marks, chatter marks, uneven wear or overheating. Clean the crankshaft surfaces with 320-grit carborundum cloth. **NEVER** spin-dry a crankshaft ball bearing with compressed air.

Clean the crankshaft and crankshaft ball bearing with solvent. Dry the parts, but not the ball bearing, with compressed air. Check the crankshaft surfaces a second time. Replace the crankshaft if the surfaces cannot be cleaned properly for satisfactory service. If the crankshaft is to be installed for service, lubricate the surfaces with light oil. **DO NOT** lubricate the crankshaft ball bearing, at this time.

After the crankshaft has been cleaned, grasp the outer race of the crankshaft ball bearing installed on each end of the crankshaft, and attempt to work the race back-and-forth. There should not be excessive "play". A very slight amount of side "play" is acceptable because there is only about 0.001" (0.025mm) clearance in the bearing.

Lubricate the ball bearing with light oil. Check the action of the bearing by rotating the outer bearing race. The bearing should have a smooth action and no rust stains. If the ball bearing sounds or feels rough or catches, the bearing should be removed and discarded.

Connecting Rod Service

Inspect the connecting rod bearings for rust or signs of bearing failure. **NEVER** intermix new and used bearings. If even one bearing in a set needs to be replaced, all bearings at that location **MUST** be replaced.

Clean the inside diameter of the piston pin end of the connecting rod with crocus cloth.

Clean the connecting rod **ONLY** enough to remove marks. **DO NOT** continue once the marks have disappeared.

Assemble the piston end of the connecting rod with loose needle bearings. Insert the piston pin and check for vertical "play". The piston pin should have **NO** noticeable vertical "play".

If the pin is loose or there is vertical "play", check for and replace the worn part/s.

Inspect the piston pin and matching rod end for signs of heat discoloration. Overheating is identified as a bluish bearing surface color and is caused by inadequate lubrication or by operating the engine at excessive high rpm.

Piston Service

Inspect each piston for evidence of scoring, cranks, metal damage, cracked piston pin boss, or worn pin boss. Be especially critical during inspection if the craft and engine has been submerged. If the piston pin is bent, the piston and the pin **MUST** be replaced as a set for two reasons. First, a bent pin will damage the boss when it is removed. Secondly, a piston pin is not sold as a separate item.

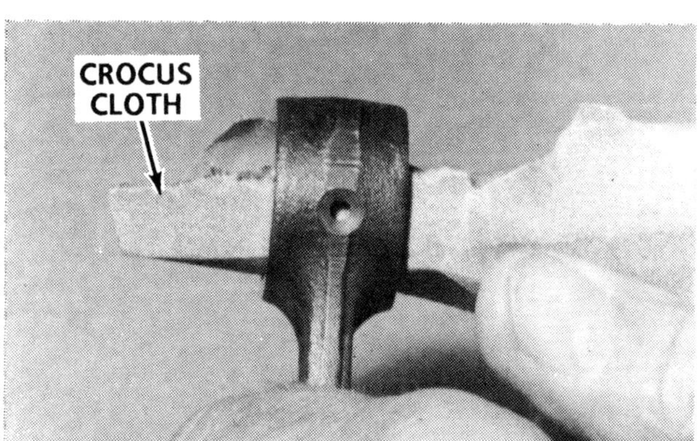

Cleaning the piston end of a connecting rod with crocus cloth.

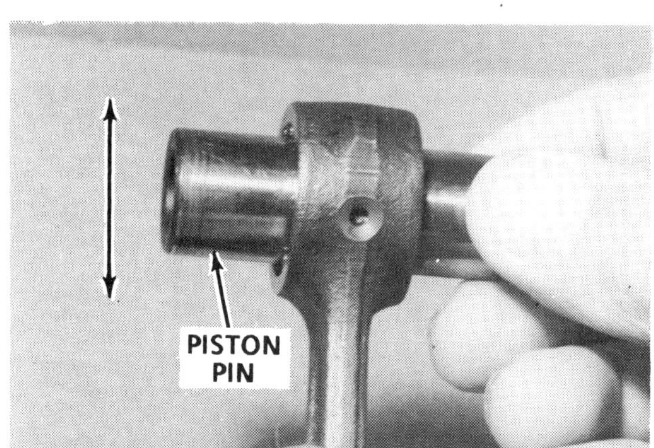

Checking the piston end of a connecting rod for vertical free "play" using a piston pin.

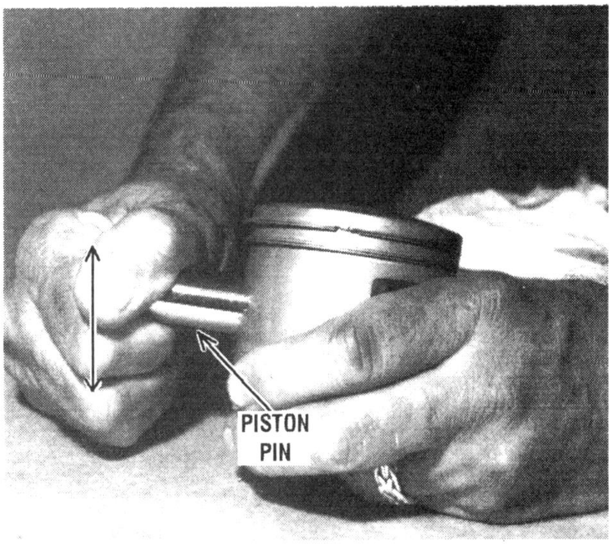

*Checking free "play" between the piston pin and the piston boss. There should be **NO** "play".*

Check the piston ring grooves for wear, burns, distortion, or loose locating pins. During an overhaul, the rings should be replaced to **ENSURE** lasting repair and proper engine performance after the work is completed. Clean the piston dome, ring grooves and the piston skirt. Clean carbon deposits from the ring grooves using the recessed end of a broken piston ring.

NEVER use a rectangular ring to clean the groove for a tapered ring, or use a tapered ring to clean the groove for a rectangular ring.

NEVER use an automotive type ring groove cleaner, because such a tool may loosen the piston ring locating pins.

Clean carbon deposits from the top of the piston using a soft wire brush, carbon removal solution or by sand blasting. If a

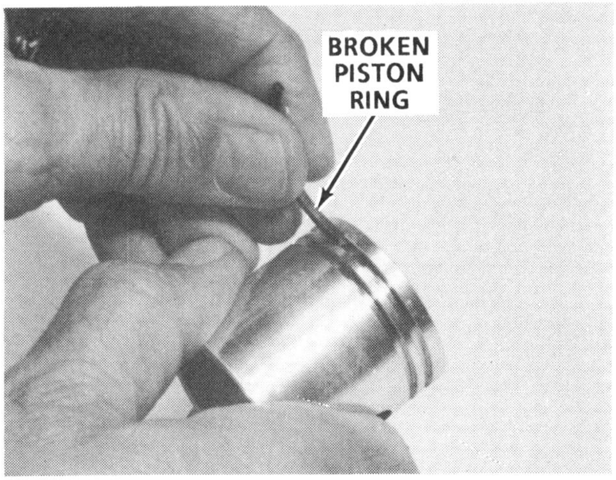

*Using part of a broken piston ring to clean the ring groove. Exercise **CARE** not to disturb the locating pin in each groove.*

Locating pins prevent the rings from rotating and catching on a port in the cylinder wall.

wire brush is used, **TAKE CARE** not to burr or round machined edges. Clean the piston skirt with crocus cloth.

Install the piston pin through the first boss only. Check for vertical free "play". There should be **NO** vertical free "play". The presence of "play" is an indication the piston boss is worn. The piston is manufactured from a softer material than the piston pin. Therefore, the piston boss will wear more quickly than the pin.

Excessive piston skirt wear **CANNOT** be visually detected. Therefore, good shop practice dictates the piston skirt diameter be measured with a micrometer.

Piston diameters are measured at a definite distance up from the bottom of the skirt at right angles to the piston pin axis.

It is believed, this crown seized with the cylinder wall when the unit was operated at high rpm and the timing was not adjusted properly. At the same instant, the rod apparently pulled the lower part of the piston downward, severing it from the crown.

The pitted damage to this piston crown was probably caused by a broken piston ring working its way into the combustion chamber. The little "hills" then became "hot" spots on the crown, contributing to "dieseling" after the engine was shut down.

Size limits are as follows:

Std. - New	2.990"	(75.94mm)
1st Oversize	3.000"	(76.19mm)
2nd Oversize	3.020"	(76.44mm)

Ring End Gap

Before the piston rings are installed onto the piston, the ring end gap clearance

The rings on this piston were broken, possibly during installation, and then caused extensive damage to the piston. Ring parts found their way into the combustion chamber and caused damage to the piston crown.

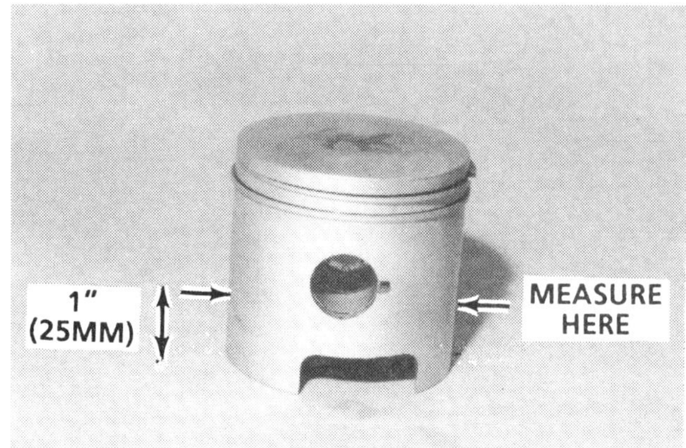

Measure the piston diameter at the specified distance above the bottom of the piston skirt.

for each ring must be determined. The purpose of the piston rings is to prevent the blowby of gases in the combustion chamber. This cannot be achieved unless the correct oil film thickness is left on the cylinder wall. This thin coating of oil acts as a seal between the cylinder wall and the face of the piston ring. An excessive end gap will allow blowby and the cylinder will lose compression. An inadequate end gap will scrape too much oil from the cylinder wall and limit lubrication. Lack of adequate lubrication will cause excessive heat and wear.

IDEALLY, the ring end gap measurement should be taken **AFTER** the cylinder bore has been measured for wear and taper **AND** after any corrective work, such as boring or honing, has been completed.

The rings on this piston became stuck due to lack of adequate lubrication, incorrect timing, or overheating.

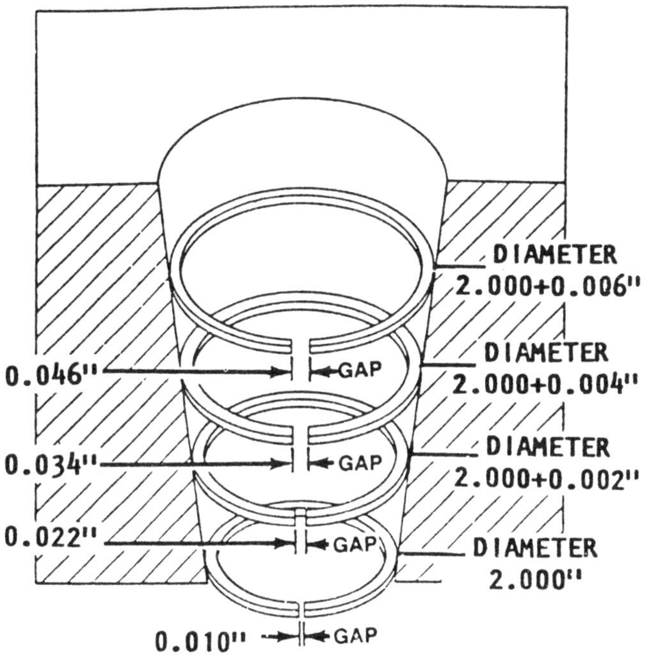

The cylinder taper drastically affects ring and end gap, as shown in this cross-section line drawing.

If the ring end gap is measured with a taper to the cylinder wall, the diameter at the lower limit of the ring travel will be smaller than the diameter at the top of the cylinder.

If the ring is fitted to the upper part of a cylinder with a taper, the ring end gap will not be great enough at the lower limit of ring travel. Such a condition could result in a broken ring and/or damage to the cylinder wall and/or damage to the piston and/or damage to the cylinder head.

If the cylinder is to be only honed, not bored, **OR** if only cleaned, not honed, the ring end gap should be measured at the lower limit of ring travel.

To perform the end gap measurement, the piston crown is used to insert the ring into the cylinder just above the ports, as indicated in the accompanying illustration.

Be sure the ring is parallel to the surface of the block, then measure the ring end gap with a feeler gauge.

If the end gap is greater than the amount listed, replace the entire ring set.

If the end gap is less than 0.010-0.016" (0.25-0.40mm) for both top and bottom rings, carefully file the ends of the ring -- just a little at a time -- until the correct end gap is obtained. If the gap is greater than 0.39" (1.00mm), the ring must be replaced. Remember, if one ring requires replacement, the set **MUST** be replaced.

Inspect the piston ring locating pins to be sure they are tight. There is one locating pin in each ring groove. If the locating pins are loose, the piston **MUST** be replaced.

Ring-To-Groove Clearance

This clearance only applies to the lower rectangular piston ring. Measure the clearance between the bottom of the ring and the groove. The clearance must not exceed 0.008" (0.2mm). If the clearance is greater than specified, the piston **MUST** be replaced.

The piston is manufactured from softer material than the ring. Therefore, the wear will occur at the piston rather than the ring. As a general rule of good shop practice, if the piston is replaced, a new set of rings are installed on the new piston.

Oversized Pistons and Rings

Scored cylinder blocks can be saved for further service by reboring and installing oversize pistons and piston rings. **HOWEVER**, if the scoring is over 0.0075" (0.13mm) deep, the block cannot be effectively rebored for continued use.

Check with the local dealer for oversize piston availability

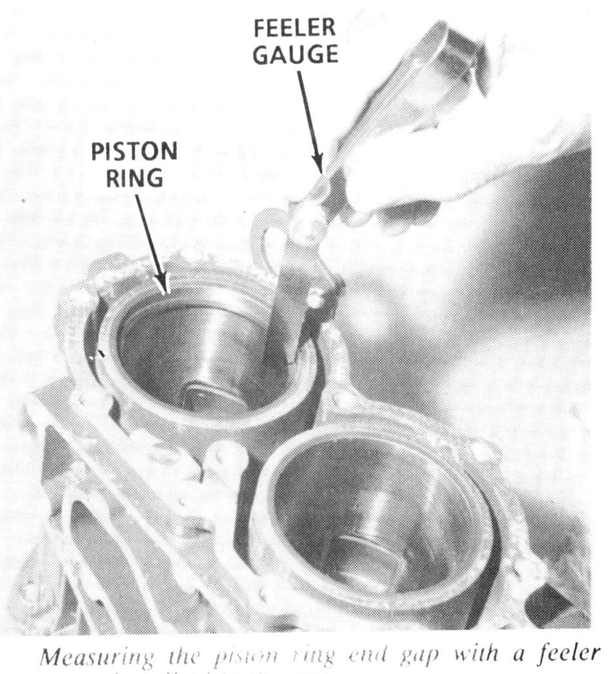

Measuring the piston ring end gap with a feeler gauge, as described in the text.

If oversize pistons are not readily available, the local marine shop may have the facilities to "knurl" the piston, making it larger.

Cylinder Block Service

Inspect the cylinder block and cylinder bores for cracks or other damage. Remove carbon with a fine wire brush on a shaft attached to an electric drill or use a carbon remover solution.

Use an inside micrometer or telescopic gauge and micrometer to check the cylinders for wear. Check the bore for out-of-round and/or oversize bore. If the bore is tapered, out-of-round, or worn more than the wear limit specified by the manufacturer, the cylinder should be rebored -- provided oversize pistons and rings are available.

Check with the local dealer **PRIOR** to reboring. If oversize pistons and matching ring sets are not available, the block **MUST** be replaced or re-sleeved.

Four inside cylinder bore measurements must be taken for each cylinder to determine an out-of-round condition: the maximum taper, and the maximum bore diameter.

In the accompanying illustration, measurements **D1** and **D2** are diameters measured at the top of the cylinder at right angles to each other. Measurements **D3** and **D4** are diameters measured from the top of the cylinder at right angles to each other.

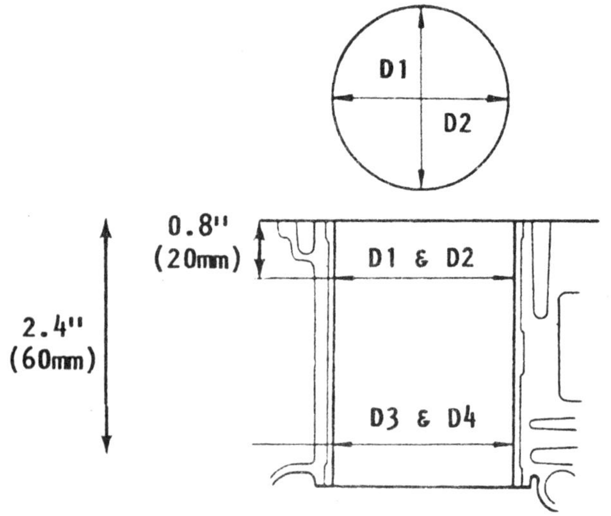

Top view diagram and cross-section of a typical cylinder to indicate where measurements are to be taken for wear limit, taper, and out-of-round limits.

Out-of-Round

Measure the cylinder diameter at **D1** and **D2**. The manufacturer requires the difference between the two measurements should be less than 0.002" (0.050mm).

Maximum Taper

Measure the cylinder diameter at **D1, D2, D3,** and **D4.** Take the largest of the **D1** and **D2** measurements and subtract the smallest measurement at **D3** or **D4**. The answer to the subtraction -- the cylinder taper -- should be less than 0.003" (0.08mm).

Replacement Cylinder Sleeves

Replacement cylinder sleeves are available in 0.25mm and 0.5mm oversizes. Check with your local dealer for installation.

Piston Clearance

Piston clearance is the difference between a maximum piston diameter and a minimum cylinder bore diameter. If this clearance is excessive, the engine will de-

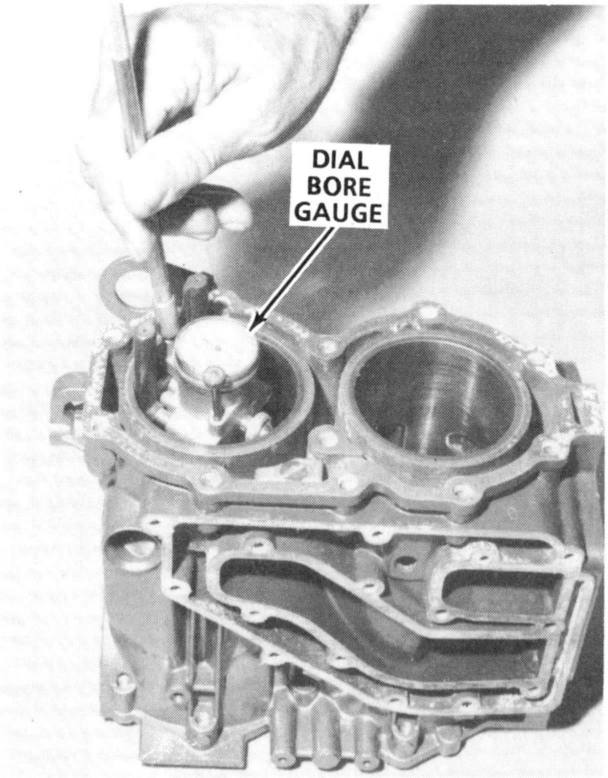

Effective method of checking the cylinder for taper using an inside micrometer. One measurement should be taken near the top and another near the bottom of the cylinder. The difference between the two measurements is the amount of cylinder taper.

velop the same symptoms as for excessive cylinder bore wear -- namely loss of ring sealing ability, loss of power, increased engine noise, unnecessary vibration, and excessive oil consumption.

Maximum piston diameter was described earlier in this section. Minimum cylinder bore diameter is usually determined by measurement **D3** or **D4,** also described earlier in this section.

If the piston clearance exceeds 0.008" (0.2mm), either the piston or the cylinder block **MUST** be replaced.

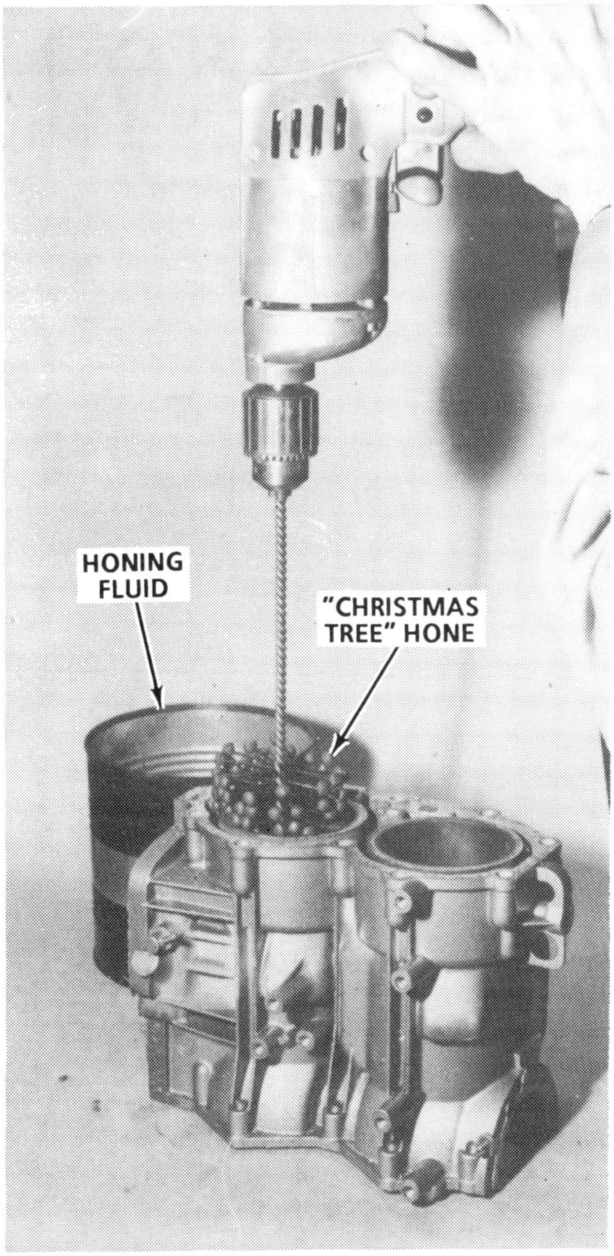

*Excellent example of refinishing a cylinder wall using an electric drill and a "Christmas Tree" hone. **ALWAYS** keep the tool moving in long even strokes the entire cylinder depth. Use a continuous **LIBERAL** amount of honing fluid.*

HONING CYLINDER WALLS

Hone the cylinder walls lightly to seat the new piston rings, as outlined in this section. If the cylinders have been scored, but are not out-of-round, or the bore is rough, clean the surface of the cylinder with a cylinder hone as described in the following paragraphs.

SPECIAL WORDS

If overheating has occurred, check and resurface the spark plug end of the cylinder block, if necessary. This can be accomplished with 240-grit sandpaper and a small flat block of wood.

To ensure satisfactory engine performance and long life following the overhaul work, the honing work should be performed with patience, skill, and in the following sequence.

a- Follow the hone manufacturer's recommendations for use of the hone and for cleaning and lubricating during the honing operation. A "Christmas Tree" hone may also be used.

b- Pump a continuous flow of honing oil into the work area. If pumping is not practical, use an oil can. Apply the oil generously and frequently on both the stones and the work surface.

c- Begin the stroking at the smallest diameter. Maintain a firm stone pressure against the cylinder wall to assure fast stock removal and accurate results.

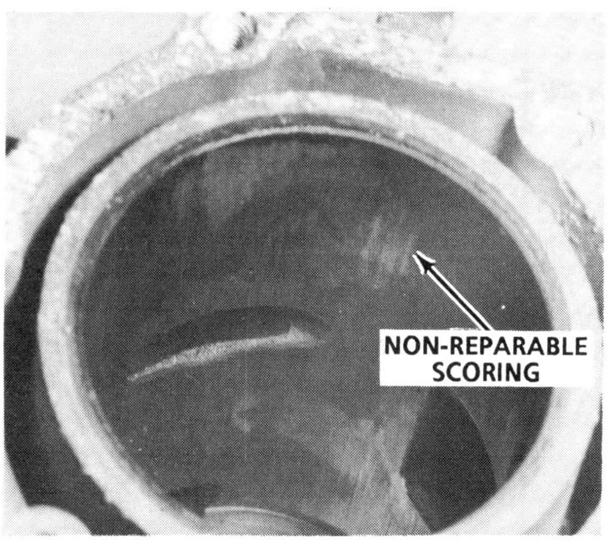

The walls of this cylinder were damaged beyond repair when a piston ring broke and worked its way into the combustion chamber.

d- Expand the stones as necessary to compensate for stock removal and stone wear. The best cross-hatch pattern is obtained using a stroke rate of 30 complete cycles per minute. Again, use the honing oil generously.

e- Hone the cylinder walls **ONLY** enough to de-glaze the walls.

f- After the honing operation has been completed, clean the cylinder bores with hot water and detergent. Scrub the walls with a stiff bristle brush and rinse thoroughly with hot water. The cylinders **MUST** be thoroughly cleaned to prevent any abrasive material from remaining in the cylinder bore. Such material will cause rapid wear of new piston rings, the cylinder bore, and the bearings.

g- After cleaning, swab the bores several times with engine oil and a clean cloth, and then wipe them dry with a clean dry cloth. **NEVER** use kerosene or gasoline to clean the cylinders.

h- Clean the remainder of the cylinder block to remove any excess material spread during the honing operation.

SPECIAL CRANKCASE WORDS

The upper and lower crankcase halves are matched at the factory. They are not interchangeable and are not available as separate items -- only as a matched set.

This cylinder head was removed from the same engine as the pitted piston shown on Page 8-80. In order to return this head to a serviceable condition, the combustion chamber surface must be refinished to a smooth surface using a stone. If left in its present condition, the sharp corners of the gouges will become "hot" spots resulting in pre-ignition and possible "dieseling" after the engine is shut down.

Rotary Valve

Inspect the teeth on the rotary valve shaft gear. Any wear pattern -- there probably will be one -- should be even on all teeth.

Check for the presence of brass filings in the bottom of the lower crankcase. Install the bearing (if it was removed), and the rotary valve shaft into the lower crankshaft. Make a setup with a dial indicator with the probe making contact on the shaft, as shown in the accompanying illustration. Now, rotate the shaft and measure the maximum deflection of the shaft. The maximum deflection should not exceed 0.003" (0.08mm). If a greater deflection is indicated, replace the shaft.

Hold the rotary valve housing -- without its sealing ring in place -- against the lower crankcase. Measure the clearance between the housing and the crankcase, with a feeler gauge. The measured clearance should be 0.012-0.020" (0.3-0.5mm). Take a clearance measurement with the upper crankcase. If the measured clearance is not within the limits given, replace the rotary valve housing.

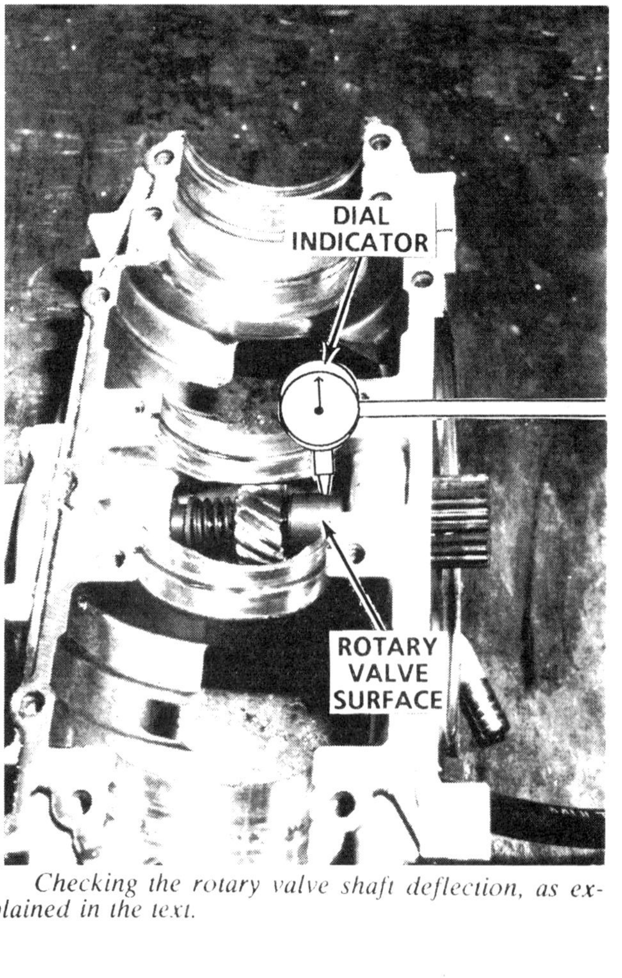

Checking the rotary valve shaft deflection, as explained in the text.

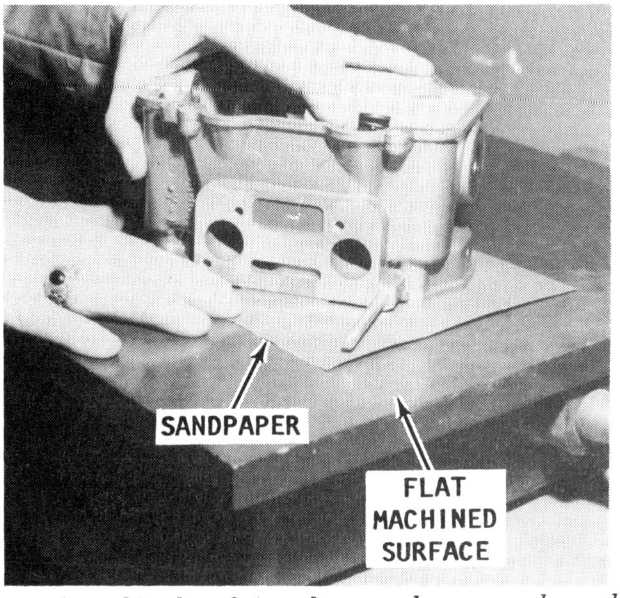

A machined surface and wet sandpaper may be used to resurface a cylinder head, as explained in the text. A piece of glass or mirror may also be used, but do ***NOT*** *attempt this task on a workbench.*

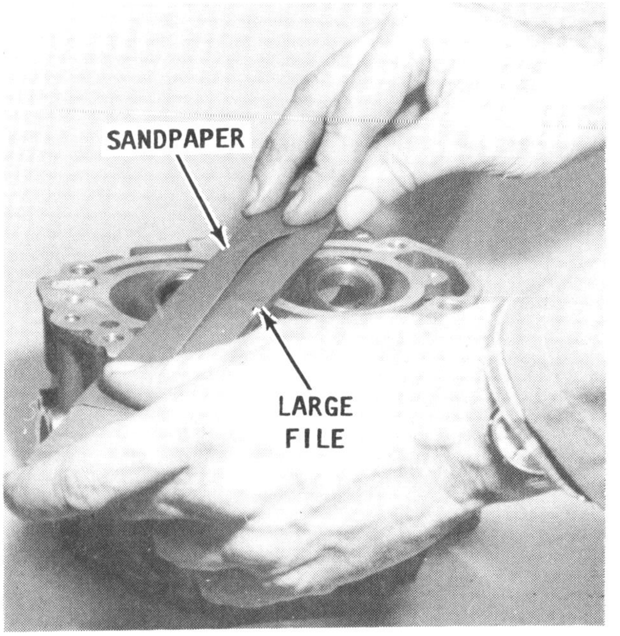

A ***LARGE*** *file and piece of wet sandpaper may be used to resurface a head or cylinder block when a suitable flat surface is not available.*

BLOCK AND CYLINDER HEAD WARPAGE

First, check to be sure all old gasket material has been removed from the contact surfaces of the block and the cylinder head. Clean both surfaces down to shiny metal, to ensure a true measurement.

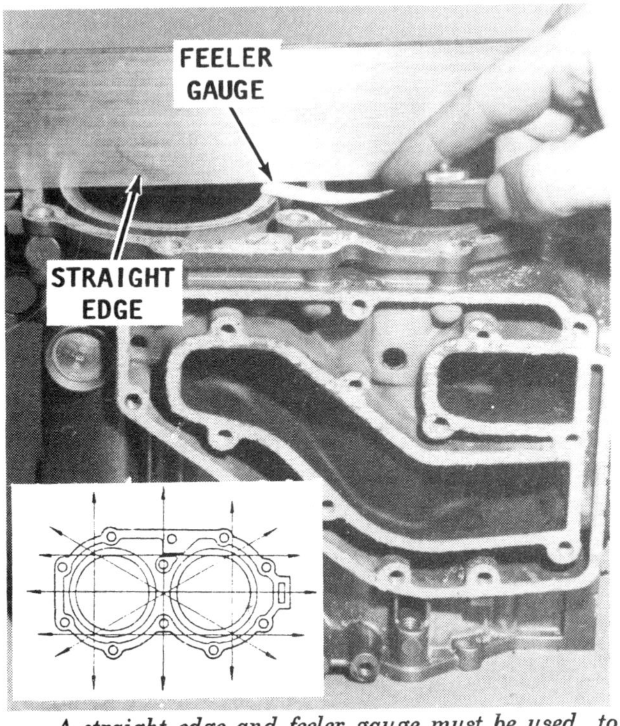

A straight edge and feeler gauge must be used to check engine-to-head mating surface on any block for warpage. Move the straight edge to at least eight different positions, as indicated in the superimposed drawing.

Next, place a straight edge across the gasket surface. Check under the straight edge with a 0.004" (0.m) feeler gauge. Move the straight edge to at least eight different locations. If the feeler gauge can pass under the straight edge -- anywhere contact with the other is made -- the surface will have to be resurfaced.

The block or the cylinder head may be resurfaced by placing the warped surface on 400-600 grit **WET** sandpaper, with the sandpaper resting on a **FLAT MACHINED** surface. If a machined surface is not available, a large piece of glass or mirror may be used. **DO NOT** attempt to use a workbench or similar surface for this task. A workbench is never perfectly flat and the block or cylinder head will pickup the imperfections of the surface and the warpage will be made worse.

Sand -- work -- the warped surface on the wet sandpaper using large figure "**8**" motions. Rotate the block, or head, through 180 degrees (turn it end-for-end) and spend an equal amount of time in each position to avoid removing too much material from one side.

If a suitable flat surface is not available, the next best method is to wrap 400-600 grit wet sandpaper around a **LARGE** file. Draw the file as evenly as possible in one sweep across the surface. Do not file in one place. Draw the file in many, many directions to get as even a finish as possible.

As the work moves along, check the progress with the straight edge and feeler gauge. Once the 0.004" (0.m) feeler gauge will no longer slide under the straight edge, consider the work completed.

If the warpage cannot be reduced using one of the described methods, the block or cylinder head should be replaced.

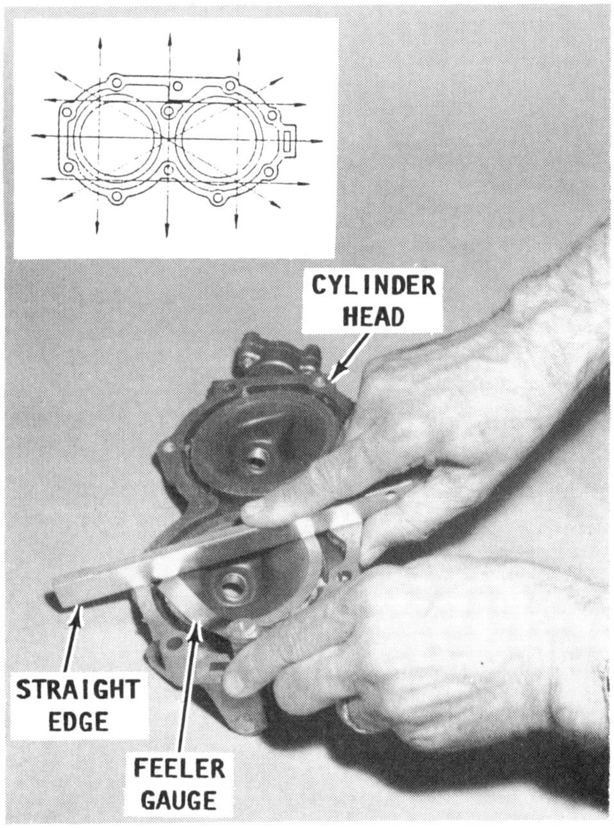

Any cylinder head surface may be checked for warpage using a straight edge and feeler gauge. Move the straight edge to at least eight different positions, as indicated in the superimposed line drawing.

8-6 SEALANTS, ADHESIVES, LUBRICANT, AND FUEL STABILIZERS

Bombardier chemical engineers have developed products to be used with their watercraft and snowmobiles. Throughout this manual, the authors recommend the application of the manufacturer's products as a first choice. Some products listed are alternatives of equal value, and may be used with confidence, if the manufacturer's line is not available.

Four sealants are recommended and they are **NOT** interchangeable. Each is designed to perform under a different set of conditions.

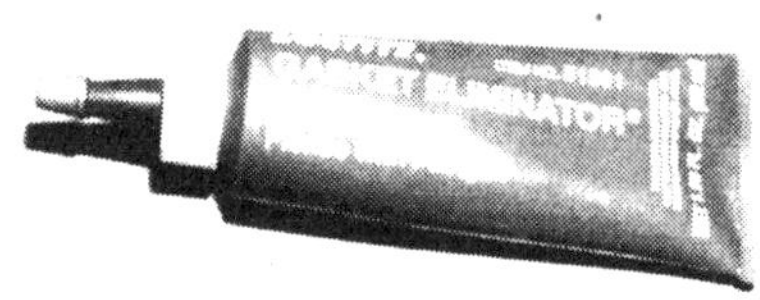

The manufacturer recommends, and the authors instruct, the use of numerous Loctite products during engine assembly, in fact, throughout this manual.

VHT[r] (very high temperature) Quick Gasket compound is a silicone rubber based adhesive recommended for use where rubber gaskets are required next to a metal surface at high temperature. This sealant is used on the exhaust manifold cover gaskets. This substance has a milky white color.

Fuel Stabilizer

"Sta-Bil" Fuel Conditioner and Stabilizer is recommended during engine operation and during the storage period. This fluid absorbs water in the fuel system and protects against corrosion.

If used during operation, this fuel additive will prevent the formation of gum and varnish deposits and greatly extend the period between required carburetor overhauls.

When added to the fuel during storage, the additive will prevent the fuel from "souring" for up to twelve full months.

Bombardier chemical engineers have developed their own line of lubricants and sealants for use with Bombardier watercraft and snowmobiles.

9
ELECTRICAL

9-1 INTRODUCTION

The battery, cranking system, trim system, temperature warning system, and the fuel and oil level sensors are all considered subsystems of the electrical system. Each of these areas will be covered in this chapter beginning with the battery. The charging system is considered a subsystem of ignition and is covered in detail in Chapter 7 -- **IGNITION**. A brief description of the charging system will be included in this chapter for complete coverage of the electrical system.

9-2 BATTERIES

The battery is one of the most important parts of the electrical system. Because of its job and the consequences -- failure to perform when needed -- the best advice is to purchase a well-known brand, with an extended warranty period, from a reputable dealer. Many owners are opting for the newer "maintenance-free" -- sealed -- batteries, which don't require the careful monitoring known to "conventional" batteries.

Personal Watercraft Batteries

The battery developed for use in a personal watercraft is the same kind used for many years on motorcycles. Personal watercraft batteries are required to perform under the most rigorous conditions.

Personal watercraft batteries have a much heavier exterior case than the usual automobile battery to withstand the violent pounding and shocks imposed on it as the craft moves through rough water. The battery case must also be vibration and seepage resistant. Therefore, a personal watercraft battery should always be the best the owner can afford.

The caps of marine batteries are "spill proof" to prevent acid from spilling into the bilges when the craft is subjected to violent maneuvers. These special caps are designed to vent gases regardless of the battery position, even when upside down.

Because of these features, the personal watercraft battery will recover from a low charge condition and give satisfactory service over a much longer period of time.

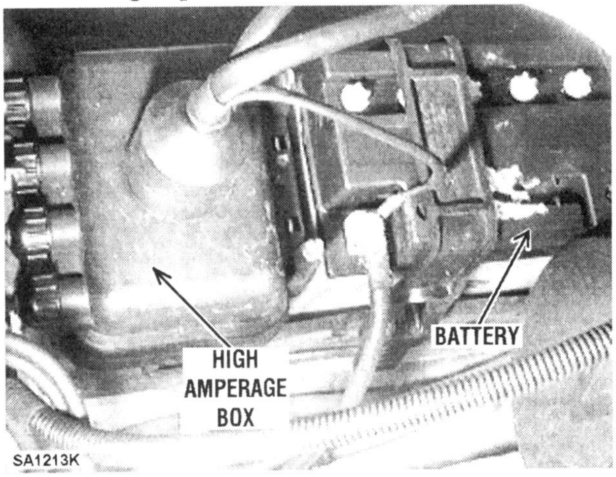

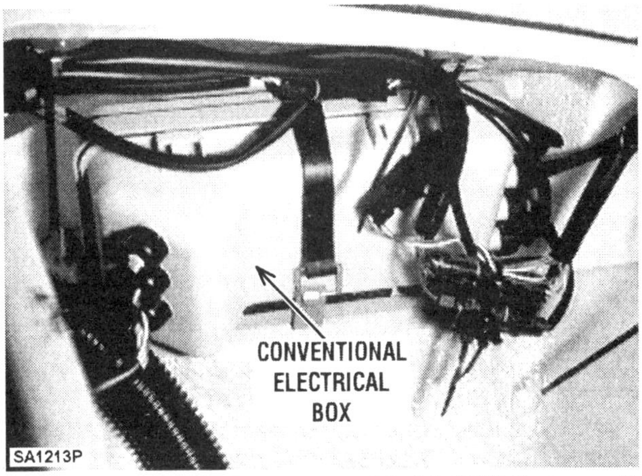

Location of the battery on a Model GSX Series craft. The high amperage electrical box is clearly visible next to the battery. The conventional electrical box is located in a forward compartment.

Battery Construction

A battery consists of a number of positive and negative plates immersed in a solution of diluted sulfuric acid. The plates contain dissimilar active materials and are kept apart by separators. The plates are grouped into what are termed elements. Plate straps on top of each element connect all of the positive plates and all of the negative plates into groups.

The battery is divided into cells which hold a number of the elements apart from the others. The entire arrangement is contained within a hard plastic case. The top is a one-piece cover and contains the filler caps for each cell. The terminal posts protrude through the top where the battery connections for the craft are made. Each of the cells is connected to its neighbor in a positive-to-negative manner with a heavy strap called the cell connector.

Battery Ratings

Four different methods are used to measure and indicate battery electrical capacity:

After the protective boot is moved back on the battery cable, the cable can be removed using a slotted screwdriver or a socket.

1- Ampere-hour rating
2- Cold cranking performance
3- Reserve capacity
4- Watt-hour rating

The **ampere-hour** rating of a battery refers to the battery's ability to provide a set amount of amperes for a given amount of time under test conditions at a constant temperature of 80° (27°C). Amperes x hours equals ampere-hour rating. Therefore, if the battery is capable of supplying 3 amperes of current for 5 consecutive hours, the battery is rated as a 15 ampere-hour battery.

The ampere-hour rating is useful for some service operations, such as slow charging or battery testing.

The recommended ampere-hour rating for a personal watercraft battery is 15-19 ampere-hours.

Cold cranking performance is measured by cooling a fully charged battery to 0°F (-18°C) and then testing it for 5 seconds to determine the maximum current flow. In this manner the cold cranking amperes rating is the number of amperes available to be drawn from the battery before the voltage drops below 9.8 volts. A typical cold cranking ampere rating for a personal watercraft battery is 240 amps.

Reserve capacity of a battery is considered the length of time -- in minutes -- at 80°F (27°C) a 25 ampere current can be maintained before the voltage drops below 10.5 volts. This test is intended to provide an approximation of how long the engine, including electrical accessories such as bilge pump, could operate satisfactorily if the stator assembly or charging coil did not produce sufficient current. A typical rating is 25 minutes.

Watt-hour is a very useful rating of battery power. It is determined by multiplying the number of ampere hours times the voltage. Therefore, a 12-volt battery rated at 16 ampere-hours would be rated at 192 watt-hours (16 x 12 = 192).

If possible, the new battery should have a power rating equal to or higher than the unit it is replacing.

Battery Installation

Every battery installed in a personal watercraft must be secured in a well-protected ventilated area. If the battery area lacks adequate ventilation, hydrogen gas which is given off during charging could become very explosive. This is especially true if the gas is concentrated and confined.

Battery Service

The battery requires periodic servicing and a definite maintenance program will ensure extended life. If the battery should test satisfactorily, but still fails to perform properly, one of five problems could be the cause.

1- Slow speed engine operation for long periods of time resulting in an undercharged condition.

2- A defect in the charging system. A faulty stator assembly or charging coil, defective rectifier, or high resistance somewhere in the system could cause the battery to become undercharged.

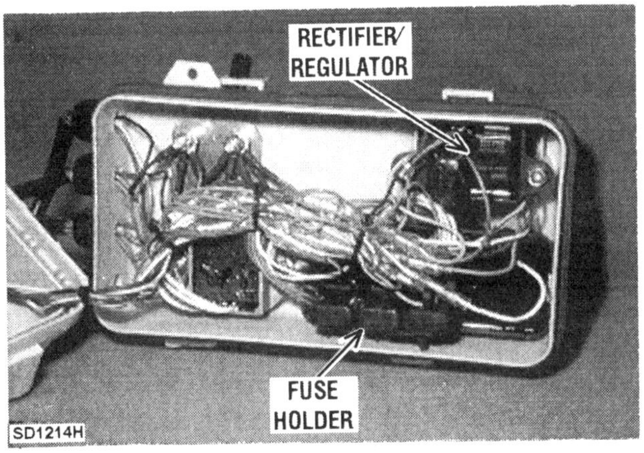

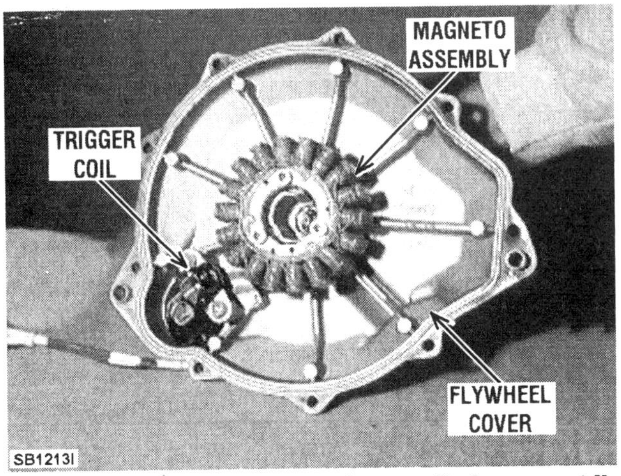

If the battery tends to weaken or discharge rapidly, the rectifier/regulator -- top -- or the stator may be at fault -- bottom.

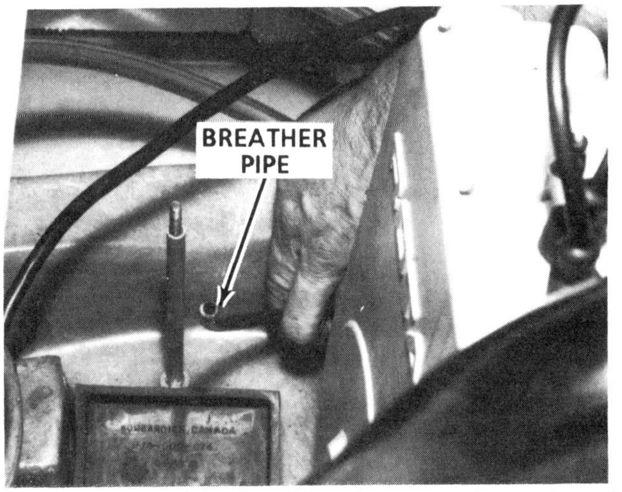

Location of the breather pipe -- on the starboard side of the battery.

3- Failure to maintain the battery in good order. This might include a low level of electrolyte in the cells; loose or dirty cable connections at the battery terminals; or possibly an excessively dirty battery top.

Remove the battery from the engine compartment for service. Inspect and service the battery, cables and connections. Check for signs of corrosion. Inspect the battery case for cracks or bulges, dirt, acid, and electrolyte leakage.

Make sure the breather pipe is attached to the battery and not pinched by any part of the engine compartment.

Cleaning

Dirt and corrosion should be cleaned from the battery just as soon as it is discovered. Any accumulation of acid film or dirt will permit current to flow between the terminals. Such a current flow will drain the battery over a period of time.

The battery top should be kept clean. A water and baking soda solution applied with a brush will neutralize battery electrolyte and clean the top nicely.

Clean the exterior of the battery with a solution of diluted ammonia or a soda solution to neutralize any acid which may be present. Flush the cleaning solution off with clean water. **TAKE CARE** to prevent any of the neutralizing solution from entering the cells, by keeping the caps tight.

A poor contact at the terminals will add resistance to the charging circuit. This resistance will cause the voltage regulator to register a fully charged battery, and thus cut down on the stator assembly or charging coil output adding to the low battery charge problem.

Scrape the battery posts clean with a suitable tool or with a stiff wire brush. Clean the inside of the cable clamps to be sure they do not cause any resistance in the circuit.

Battery Sulfation

Battery sulfation can be identified by the formation of white crystals forming on top of exposed battery plates. This usually occurs when a battery has been stored in a discharged condition over a period of time and the electrolyte has evaporated leaving the tops of the plates exposed to the air. This condition may also occur if the battery is not serviced at regular intervals during the season and the electrolyte not replenished when low.

Once white crystals form, the plates are permanently damaged and the battery will never hold a charge. It must be replaced.

Filling the Battery

Remove each filler cap using a pair of pliers. Fill each cell to the proper level with distilled water. The level of electrolyte should be between the upper and lower level marks during operation. When filling the battery, **WITHOUT CHARGING,** fill to the upper line and allow the battery to stand for twenty minutes. Check the level again and replenish, as necessary. If filling the battery **PRIOR** to charging, fill to the lower lever only, because the electrolyte level will rise during charging.

During hot weather and periods of heavy use, the electrolyte level should be checked more often than during normal operation. Add distilled water to bring the level of electrolyte in each cell to the proper level. **TAKE CARE** not to overfill, because adding an excessive amount of water will cause loss of electrolyte and any loss will result in poor performance, short battery life, and will contribute quickly to corrosion. **NEVER** add electrolyte from another battery. **NEVER** add sulphuric acid because such acid would make the electrolyte too stong and permanently damage the battery. **NEVER** add ordinary tap water. The minerals in tap water will shorten battery life.

Battery Testing

A hydrometer is a device to measure the percentage of sulphuric acid in the battery electrolyte in terms of specific gravity. When the condition of the battery drops from fully charged to discharged, the acid leaves the solution and enters the plates,

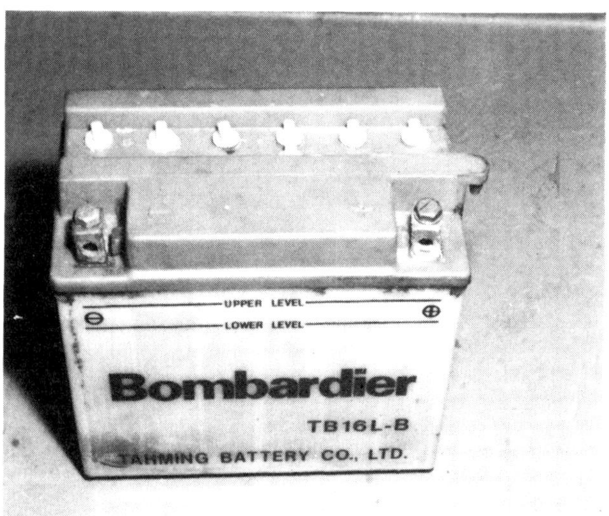

The battery MUST be filled correctly. The proper procedure for this task is explained in the text.

Pay attention to the "Poison/Danger" warning decal on the back of the battery.

causing the specific gravity of the electrolyte to drop.

Use a temperature corrected hydrometer to test the specific gravity of the electrolyte. At 20°C (68°F) the hydrometer reading should be 1.20-1.26 on the scale.

It may not be common knowledge, but hydrometer floats are calibrated for use at 80°F (27°C). If the hydrometer is used at any other temperature, hotter or colder, a correction factor must be applied. Remember, a liquid will expand if it is heated and will contract if cooled. Such expansion and contraction will cause a definite change in the specific gravity of the liquid, in this case the electrolyte.

A quality hydrometer will have a thermometer/temperature correction table in the lower portion, as shown in the accompanying illustration. By knowing the air temperature around the battery and from the table, a correction factor may be applied to the specific gravity reading of the hydrometer float. In this manner, an accurate determination may be made as to the condition of the battery.

The following six points should be observed when using a hydrometer.

1- **NEVER** attempt to take a reading immediately after adding water to the battery. Allow at least 1/4 hour of charging at the recommended rate to thoroughly mix the electrolyte with the new distilled water. This time will also allow for the necessary gases to be created.

2- **ALWAYS** be sure the hydrometer is clean inside and out as a precaution against contaminating the electrolyte.

3- If a thermometer is an integral part of the hydrometer, draw liquid into it several times to ensure the correct temperature before taking a reading.

4- **BE SURE** to hold the hydrometer vertically and suck up liquid only until the float is free and floating.

5- **ALWAYS** hold the hydrometer at eye level and take the reading at the surface of the liquid with the float free and floating.

Disregard the light curvature appearing where the liquid rises against the float stem. This phenomenon is due to surface tension.

6- **DO NOT** drop any of the battery fluid on the boat or on your clothing, because it is extremely caustic. Use water and baking soda to neutralize any battery liquid that does accidentally drop.

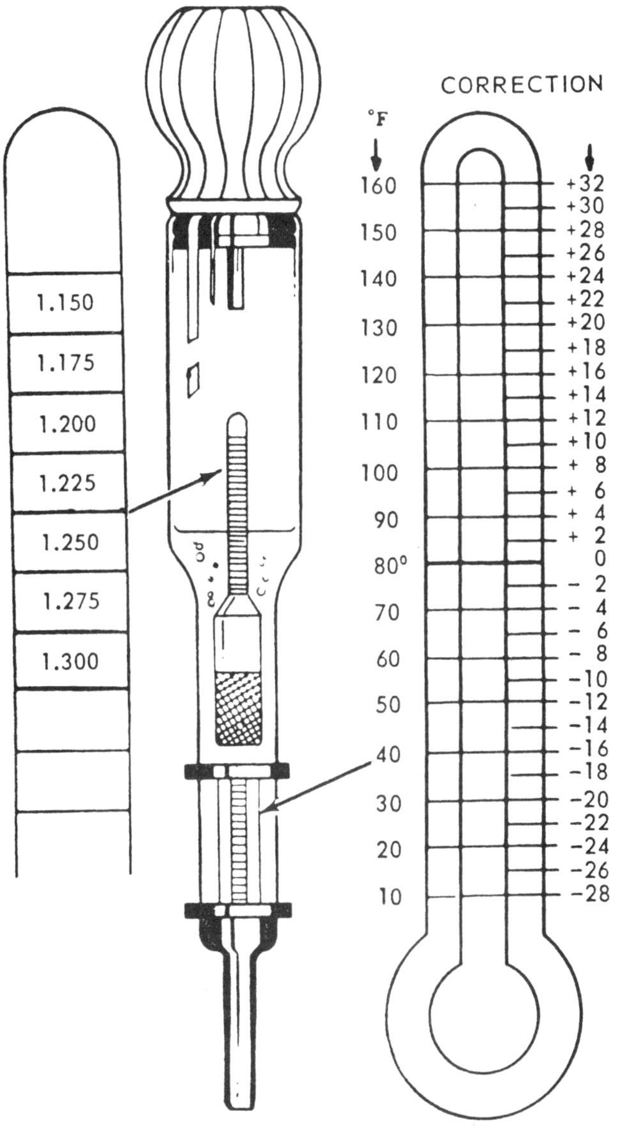

*A check of the electrolyte in the battery should be on the maintenance schedule for any craft. A hydrometer reading of 1.300, or in the Green band, indicates the battery is in satisfactory condition. If the reading is 1.150, or in the Red band, the battery **MUST** be charged. Observe the six safety points listed in the text, when using a hydrometer.*

After withdrawing electrolyte from the battery cell until the float is barely free, note the level of the liquid inside the hydrometer.

Reading the Hydrometer

Different manufacturers of battery hydrometers use different methods of displaying specific gravity readings. Three methods are used:

Red, white, and green bands
Small floating balls
Graduated scale

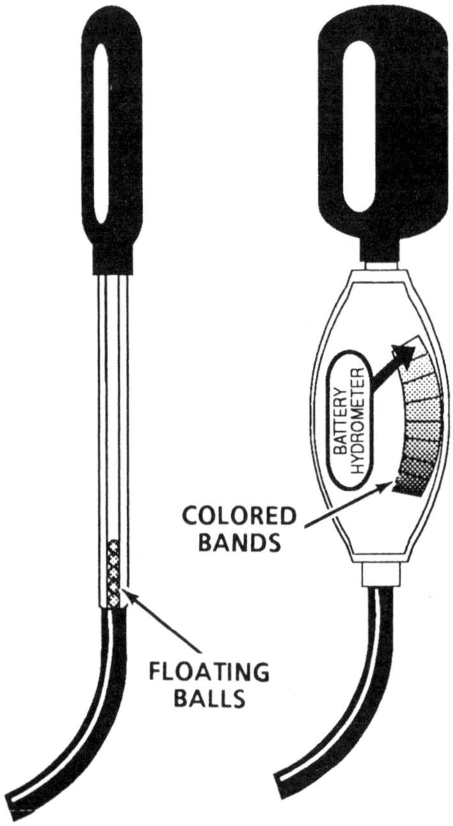

Two types of battery hydrometers, both quite different from the "old style". Their use and accurate reading techniques are explained in the text.

Colored Band Hydrometer

When using a battery hydrometer with colored bands of red, white, and green, the electrolyte level is interpreted as follows:

Draw electrolyte from the battery cell until the float is barely free, note the level of the liquid inside the hydrometer.

If the level is within the green band range for all cells, the condition of the battery is satisfactory. If the level is within the white band for all cells, the battery is in fair condition.

If the level is within the green or white band for all cells except one, which registers in the red, the cell is shorted internally. No amount of charging will bring the battery back to satisfactory condition.

If the level in all cells is about the same, even if it falls in the red band, the battery may be recharged and returned to service. If the level fails to rise above the red band after charging, the only solution is to replace the battery.

Small Float Ball Hydrometer

If using a battery hydrometer with small floating balls, the electrolyte level is interpreted as follows:

If four or five balls are floating on top of the electrolyte drawn into the hydrometer, the battery is 100% charged. If three balls are floating, the battery is 75% charged. If two balls are floating, the battery is 50% charged. If only one ball is floating the battery is 25% charged. If no balls float, the battery is 0% charged.

If the same number of balls float for each cell, the condition of the battery is satisfactory.

If all the cells but one float balls, the cell is shorted internally. No amount of charging will bring the battery back to satisfactory condition.

If the number of floated balls for all cells is about the same, even if it is only one or two, the battery may be recharged and returned to service. If the number of balls fails to increase after charging, the only solution is to replace the battery.

Graduated Scale Hydrometer

If using a battery hydrometer with a graduated scale, note the level of the liquid inside the hydrometer. If the level is between the 1.280 and 1.265 mark, the battery is 100% charged. If the level is closer to the 1.210 mark, the battery is 75% charged. If the level is closer to the 1.160 mark, the battery is 50% charged. If it is closer to the 1.120 mark the battery is only 25% charged. If the level is below 1.100 the battery is at 0% charge.

If the level is the same and above 1.160 for all cells, the condition of the battery is satisfactory. If the level is the same and above 1.120 for all cells, the battery is in fair condition.

If the level is the same for all cells except one, which registers low on the scale, the cell is shorted internally. No amount of charging will bring the battery back to satisfactory condition.

If the level in all cells is about the same, even if it registers low on the scale, the battery may be recharged and returned to service. If the level fails to rise above 1.210 after charging, the only solution is to replace the battery.

Charging the Battery

If it is necessary to charge the battery, leave the filler caps lightly resting on the cell openings to allow the gases to escape.

The charging current should not exceed 1.9 amps for 10 hours.

NEVER use a fast charger, such as those found at automotive service stations, to charge a personal watercraft battery. These chargers operate at too high an output, ideal for automotive batteries, but very damaging to personal watercraft batteries. If such a battery is fast charged, excessive heat will be generated which will warp the plates and cause internal shorting. The plates will also shed active material which will accumulate at the bottom of the battery and cause internal shorting.

If, under the recommended charging procedures, the temperature of the electrolyte rises above 115°F (45°C) during charging, reduce the charging rate and increase the charging time proportionately. If possible, stop, allow the battery to cool and then resume charging for the recommended length of time.

After charging, push each filler cap into place and wipe the top of the battery clean before installation.

Installing the Battery

Check to be sure the battery is fastened securely in position. The battery case and cover **MUST** be fastened securely to the hull to prevent any movement of the battery in the holder.

Clean the battery posts and battery cable ends with a wire brush to ensure good, clean connections. If the battery posts or cable terminals are corroded, the cables should be cleaned separately with a baking soda solution and a wire brush. Apply a thin coating of Multi-purpose Lubricant to the posts and cable clamps before making the connections. The lubricant will help to prevent corrosion.

Identify the "POS" or "+" and "NEG" or "-" embossed symbols. Correctly connect the battery cables to the battery terminals. If the cables are connected incorrectly, the ignition system **WILL** be destroyed the first time the engine is started.

Check to be sure the breather pipe is attached to the battery and not pinched by any part of the engine compartment.

Jumper Cables

If a booster battery is used for starting an engine, the jumper cables must be connected correctly and in the proper sequence to prevent damage to either battery, or to the diodes in the regulator/rectifier.

ALWAYS connect a cable from the positive terminal of the dead battery to the positive terminal of the good battery **FIRST. NEXT,** connect one end of the other cable to the negative terminal of the good battery and the other end to a good ground on the engine. **DO NOT** connect the negative jumper from the good battery to the negative terminal of the low battery. Such action will almost always cause a spark which could ignite gases escaping through the vent holes in the battery filler caps. Igniting the gases may result in an explosion destroying the battery and causing severe personal **INJURY.**

By making the negative (ground) connection on the engine, if an arc is created, it will not be near the battery.

DISCONNECT the battery ground cable before replacing a stator assembly or charging coil, or before connecting any type of meter to the ignition system.

If it is necessary to use a trickle charger on a dead battery, **ALWAYS** disconnect one

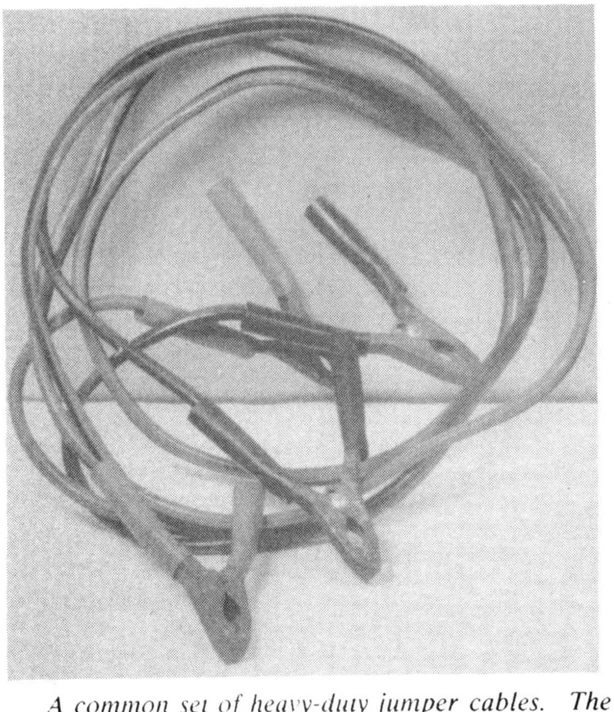

A common set of heavy-duty jumper cables. The booster battery must be connected correctly and in the proper sequence, as outlined in the text. Proper procedure will prevent damage to the battery or the diodes in the rectifier/regulator.

of the cables from the battery **FIRST,** to prevent burning out the diodes in the regulator/rectifier.

NEVER use a trickle charger as a booster to start the engine because the diodes will be **DAMAGED.**

GOOD WORDS

If the craft must be jump started from a vehicle battery, **DO NOT** perform the procedure while the automobile is running. Chances are, the car battery will be receiving a charge from the alternator. This extra electrical current will be delivered to the receiver battery. This condition would **SURELY** overload the ignition system, causing extensive damage to the components.

Storage

If the craft is to be laid up for the winter or for more than a few weeks, special attention must be given to the battery to prevent complete discharge or possible damage to the terminals and wiring. Before putting the craft in storage, disconnect and remove the battery. Clean the battery thoroughly of any dirt or corrosion, and then charge it to full specific gravity reading. After the battery is fully charged, store the battery in a clean cool dry place where it will not be damaged or knocked over, preferably on a couple blocks of wood. Storing the battery up off the deck, will permit air to circulate freely around and under the battery and will help to prevent condensation.

NEVER store the battery with anything on top of it or cover the battery in such a manner as to prevent air from circulating around the filler caps. All batteries, both new and old, will discharge during periods of storage, more so if they are hot than if they remain cool. Therefore, the electrolyte level and the specific gravity should be checked at regular intervals. A drop in the specific gravity reading is cause to charge the battery back to a full reading.

One of the largest personal watercraft battery manufacturers recommends that a battery in storage should be checked every two weeks and charged if necessary. The electrolyte level can also be replenished at this time to prevent sulfation occurring.

In cold climates, care should be exercised in selecting the battery storage area. A fully charged battery will freeze at about 60° below zero. A discharged battery, almost dead, will have ice forming at about 19° above zero.

9-3 TACHOMETER

An accurate tachometer can be connected to any engine. Such an instrument provides an indication of engine speed in revolutions per minute (rpm). This is accomplished by measuring the number of electrical pulses per minute generated in the primary circuit of the ignition system.

The meter readings range from 0 to 6,000 rpm, in increments of 100. Tachometers have solid-state electronic circuits which eliminates the need for relays or batteries and contributes to their accuracy. The electronic parts of the tachometer susceptible to moisture are coated to prolong their life.

The manufacturer recommends the use of a Bombardier Electronic Tachometer. Instructions on how to connect the leads are provided with the meter. However, as a general rule, on a two stroke engine with CDI type ignition, the tachometer leads are usually connected across the charging coil leads, refer to Chapter 7 for charging coil lead color identification.

9-4 ELECTRICAL SYSTEM GENERAL INFORMATION

The electrical system consists of three circuits:

- Charging circuit
- Cranking motor circuit
- Ignition circuit

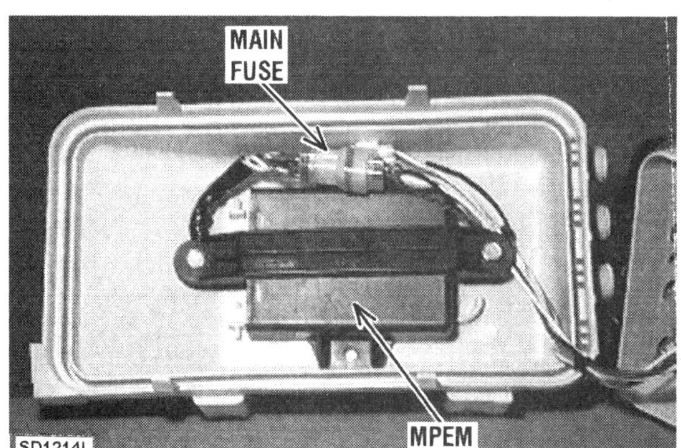

The conventional electrical box houses the rectifier/regulator -- in the other half of the box, not visible -- various fuses, and the MPEM unit on a GSX Series.

Charging Circuit

Components of the charging circuit include:

Permanent magnets attached to the inside perimeter of the flywheel.

Battery charging coils installed on the magneto assembly -- a cluster of three coils.

Rectifier/regulator located in an electrical box.

External battery.

Necessary wiring to connect it altogether.

The negative side of the rectifier/regulator is grounded. The positive side of the rectifier/regulator is connected to the battery. The negative side of the battery is connected to a good ground on the engine.

The alternating current generated in the stator windings passes to the rectifier/regulator. This module changes the alternating current (AC) to direct current (DC) to charge the 12-volt battery.

On all engines **EXCEPT** the Model 787 Series, the magneto assembly is located behind and protected by, the flywheel. The magneto assembly is mounted on the inside of the flywheel cover. Therefore, the assembly, including the battery charging coil, seldom causes problems in the charging circuit. Most problems in the charging circuit can usually be traced to the rectifier/regulator, or to the battery. If either the magneto assembly or the rectifier/regulator fails the troubleshooting tests, the defective unit cannot be repaired, it **MUST** be replaced.

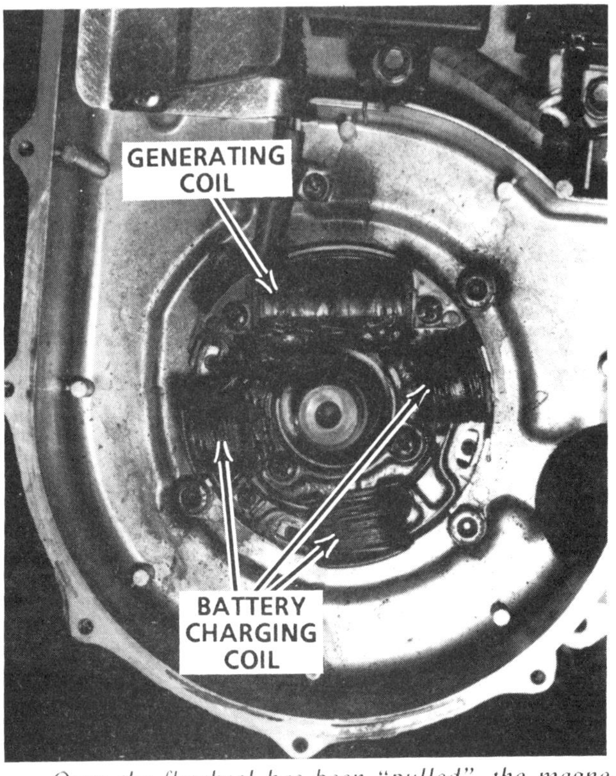

Once the flywheel has been "pulled", the magneto assembly, with battery charging coils, and the generating coils are exposed.

See Chapter 7 -- **IGNITION** to service and test components in the charging system.

Cranking Motor Circuit

The cranking motor circuit consists of a cranking motor and a starter-engaging mechanism. A starter solenoid is used as a heavy-duty switch to carry the heavy current from the battery to the cranking motor. The starter solenoid is actuated by either depressing the **START** button.

Ignition Circuit

The ignition circuit is covered extensively in Chapter 7.

9-5 CRANKING MOTOR CIRCUIT SERVICE

DESCRIPTION

As the name implies, the sole purpose of the cranking motor circuit is to control operation of the cranking motor to crank the engine until the air/fuel mixture ignites and the flywheel develops enough torque to continue under its own steam. The circuit includes a solenoid or magnetic switch to connect or disconnect the motor from the battery. The operator controls the switch with a push button.

The cranking motor is a series wound electric motor which draws a heavy current from the battery. It is designed to be used only for short periods of time to crank the engine for starting. To prevent overheating the motor, cranking should not be continued for more than 30-seconds without allowing the motor to cool for at least three minutes. Actually, this time can be spent in making preliminary checks to determine why the engine fails to start.

Theory of Operation

Power is transmitted from the cranking motor to the engine flywheel through meshed gears or clutch mechanisms. All Bombardier models are equipped with a Bendix drive on the end of the armature shaft. When the motor is operated, the pinion gear moves forward and meshes with the teeth on the flywheel ring gear.

On a cranking motor equipped with a Bendix drive: When the engine starts, the pinion gear is driven faster than the armature shaft, and as a result, it screws out of mesh with the flywheel. A rubber cushion is built into the Bendix drive to absorb the shock when the pinion meshes with the flywheel ring gear. The parts of the drive **MUST** be properly assembled for efficient operation. If the drive is removed for cleaning, **TAKE CARE** to assemble the parts in the correct order. If the screw shaft assembly is reversed, it will strike the splines and the rubber cushion will not absorb the shock.

Cranking Motor Noises

The sound of the motor during cranking is a good indication of whether the cranking motor is operating properly or not. Naturally, temperature conditions will affect the speed at which the cranking motor is able to crank the engine. The speed of cranking a cold engine will be much slower than when cranking a warm engine. An experienced operator will learn to recognize the favorable sounds of the engine cranking under various conditions.

Faulty Symptoms

If the cranking motor spins, but fails to crank the engine, the cause is usually a corroded or gummy Bendix drive. The drive should be removed, cleaned, and given an inspection.

If the cranking motor cranks the engine too slowly, the following are possible causes and the corrective actions that may be taken:

a- Battery charge is low. Charge the battery to full capacity.

b- High resistance connections at the battery, solenoid, or motor. Clean and tighten all connections.

c- Undersize battery cables. Replace cables with sufficient size.

Maintenance

The cranking motor does not require periodic maintenance or lubrication, except for lubrication of the pinion gear. If the motor fails to perform properly, the checks outlined in the previous paragraph should be performed.

Naturally, the motor will have to be removed if the corrective actions outlined under **Faulty Symptoms** above, does not restore the motor to satisfactory operation.

Otherwise, the cranking motor is not normally removed from the engine until it fails.

CRANKING MOTOR TROUBLESHOOTING

Before wasting too much time troubleshooting the cranking motor circuit, the following checks should be made. Many times, the problem will be corrected.

a- Battery fully charged.

b- Main fuse is "good" (not blown).

c- All electrical connections clean and tight.

d- Wiring in good condition, insulation not worn or frayed.

The following troubleshooting procedures are presented in a logical sequence.

Do not operate the cranking motor for more than 15 seconds. Prolonged cranking motor operation will cause overheating and damage the motor.

After each test allow the cranking motor to cool for a minute or so.

Never press the start button to activate the cranking motor while the engine is operating. Such action will damage the pinion and/or flywheel gears.

Cranking Circuit Tests

1- Remove both spark plug leads from the plugs and ground the ends to the engine to prevent accidental engine start. Keep the leads grounded for the following tests.

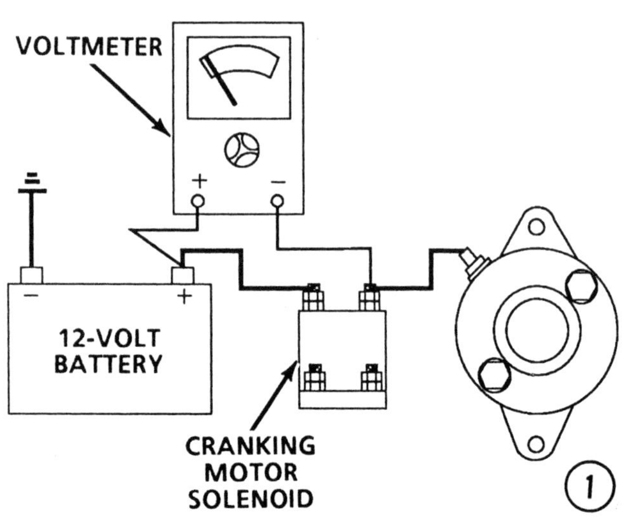

Select the Vx1 DC scale on the meter. Observe the starter solenoid. Both large terminals have Red leads connected to them. One terminal is connected to the battery and the other terminal is conected to the cranking motor. Make contact with the negative Black meter lead to the terminal on the starter solenoid which has a Red lead from the battery connected to it.

Make contact with the positive Red meter lead to the positive battery terminal. The voltmeter should register less then 0.25V DC. If the reading exceeds 0.25V DC, clean the connections at both ends of the positive battery cable and repeat the test. If the reading does not change, replace the cable.

2- Keep the meter on the same scale. Make contact with the negative Black voltmeter lead to the negative terminal on the battery. Make contact with the positive Red voltmeter lead to a good engine ground. The voltmeter should register less then 0.25V DC. If the reading exceeds 0.25V DC, clean the connections at both ends of the negative battery cable and repeat the test. If the reading does not change, replace the cable.

CRITICAL WORDS

During the following tests -- Step 3 and Step 4, both meter leads must be placed on the solenoid terminals **WHILE** the engine is being cranked. Both meter leads must be removed from the solenoid terminals **WHILE** the engine is being cranked and **BEFORE** the engine has stopped cranking. If these precautions are not followed, the voltmeter may be damaged.

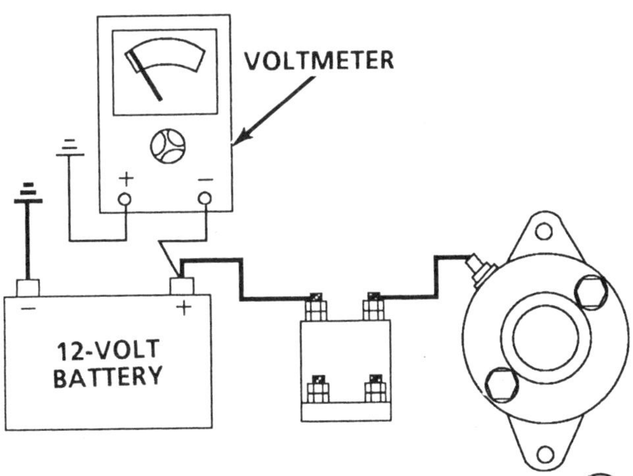

2

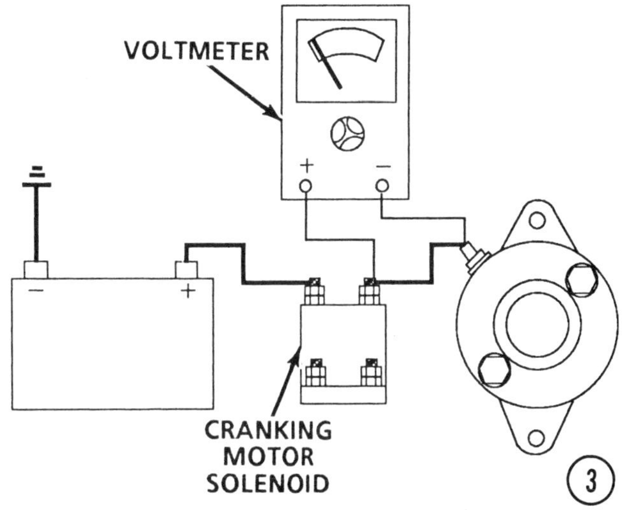

3

3- Keep the meter on the same scale. Obtain the services of an assistant. Have the assistant crank the engine, while the negative Black voltmeter lead makes contact with the solenoid terminal which is also connected to the cranking motor. Then, make contact with the positive Red voltmeter lead to the solenoid terminal which is also connected to the battery.

If the meter reading exceeds 0.25V DC, replace the cranking motor solenoid.

4- Keep the meter on the same scale and the assistant at the **START** button. Observe the same precautions as before -- in Step 3. With the engine cranking, make contact with the negative Black voltmeter lead to terminal on the cranking motor. Make contact with the positive Red voltmeter lead to the solenoid terminal which also has the Red lead from the cranking motor attached to it.

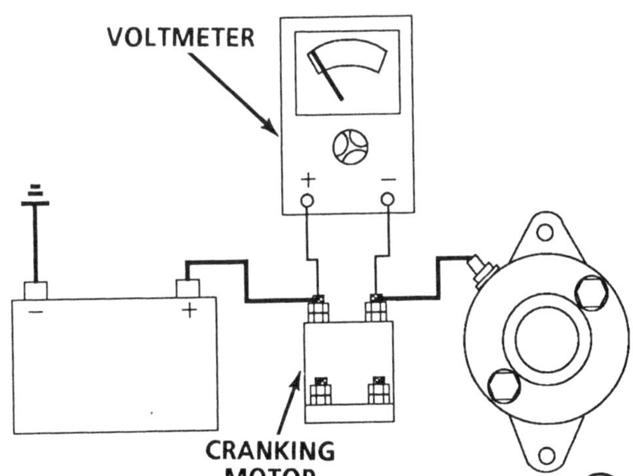

4

If the meter reading exceeds 0.25V DC, clean the connections at both ends of the solenoid-to-cranking motor cable and repeat the test. If the reading does not change, replace the cable.

9-6 CRANKING MOTOR SOLENOID REMOVAL FOR TESTING

Location

The cranking solenoid is housed inside the high amperage or "conventional" electrical box. The cover must be removed to gain access to the solenoid. The cranking motor solenoid is actually a switch between the battery and the engine. The switch cannot be serviced. Therefore, if testing indicates the switch to be faulty, it **MUST** be replaced. The solenoid must be removed from the electrical box for testing purposes.

Preliminary Tasks

Remove the electrical box cover.

Disconnect the single large Red lead from the upper left terminal.

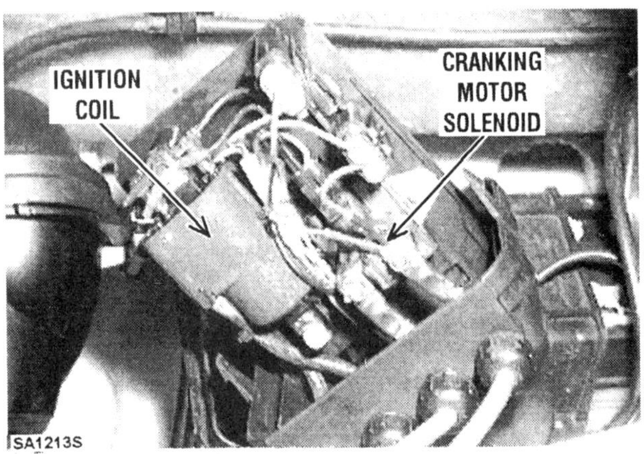

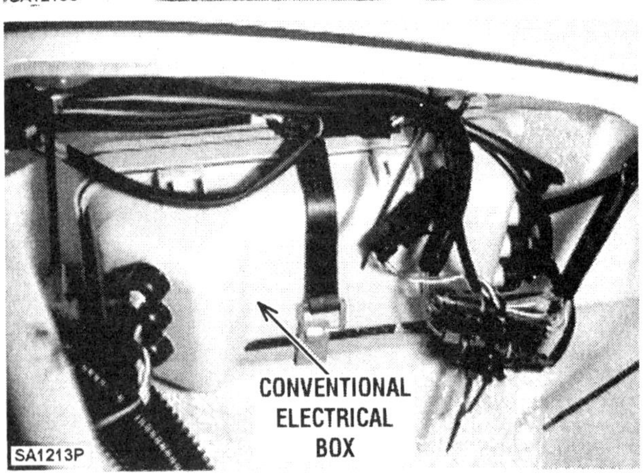

On some models, the ignition coil and cranking motor solenoid are located in a dedicated high amperage electrical box -- top. The rectifier/regulator remains in the conventional electrical box -- bottom.

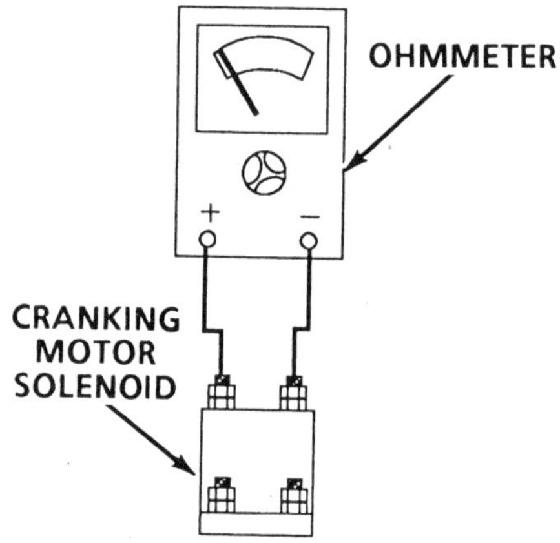

1

Disconnect the two to three -- depending on model being serviced -- Red leads from the upper right terminal.

Disconnect the two Black/White leads from the lower left terminal.

Disconnect the Yellow/Red lead from the lower right terminal.

Cranking Solenoid Testing

1- Select the Rx1 ohm scale on the ohmmeter. Make contact with the two meter leads across the two upper solenoid terminals. The meter should read an infinite resistance. If the meter indicates a resistance of less than infinity, the solenoid is defective and must be replaced.

2- Keep the ohmmeter on the same scale. Obtain a 12V battery and a pair of jumper leads. Using the set of jumper leads, connect the battery across the two lower terminals. Make contact with the two meter leads across the two upper solenoid

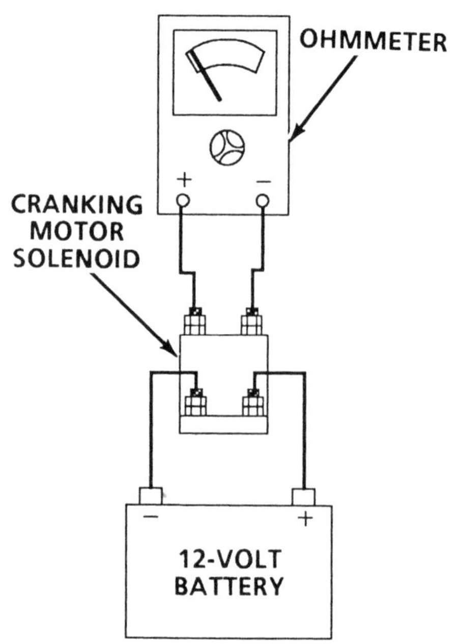

2

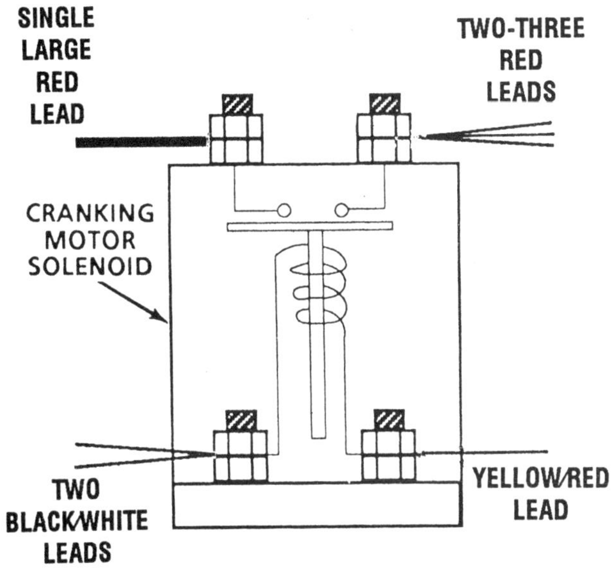

Simple diagram to indicate the lead connections on the cranking motor solenoid housed inside the conventional or high amperage electrical box.

terminals. If the solenoid emits an audible "click" and the ohmmeter registers zero ohms, the solenoid is good. If no "click" is heard, or if the meter registers a high or infinite resistance, the solenoid is defective and must be replaced.

If the cranking motor solenoid failed the test, it must be replaced. Typically, two bolts secure the solenoid to the electrical box. When installing the replacement solenoid, apply Loctite 242 to the threads of the bolts and install the new solenoid in place. Tighten both bolts securely.

Connect the Yellow/Red lead to the lower right terminal.

Connect the two Black/White lead to the lower left terminal.

Connect the 2-3 Red leads -- depending on Model -- to the upper right terminal.

Connect the single large Red lead to the upper left terminal.

Place cover over the electrical box and fasten it securely.

9-7 CRANKING MOTOR SERVICE

Description

Marine cranking motors are very similar in construction and operation to the units used in the automotive industry.

NEVER operate a cranking motor for more than 30-seconds without allowing it to cool for at least three minutes. Continuous operation without the cooling period can cause serious damage to the cranking motor.

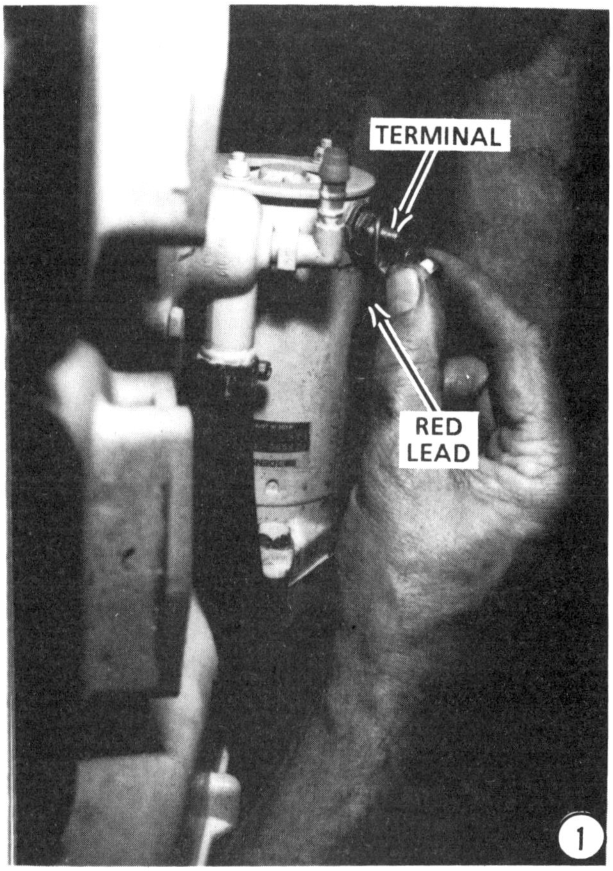

Cranking Motor Removal

1- Disconnect the large Red lead from the terminal on the cranking motor.

2- Remove the upper mounting bolt and the large Black ground lead.

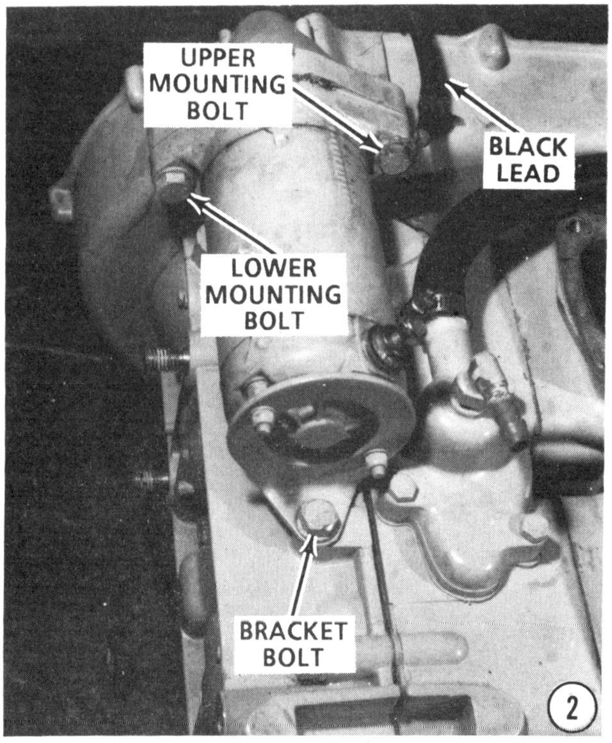

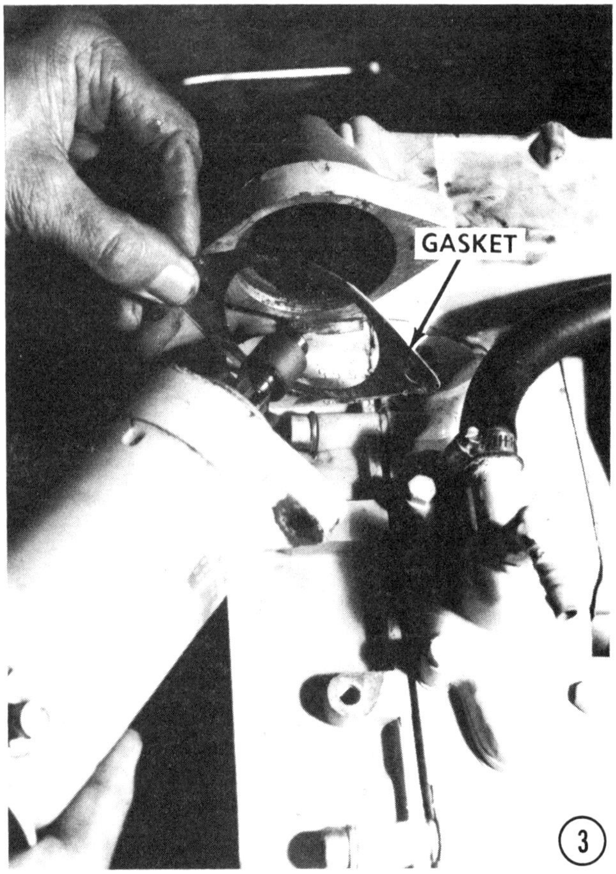

Remove the lower mounting bolt and a third bolt securing the aft end of the motor to the lower crankcase.

3- Pull the cranking motor away from the aft end of the magneto housing. If the motor is stuck, tap the mounting flange with a soft head mallet to jar it loose. Remove and discard the gasket.

SPECIAL WORDS

Using a six-inch extension with a 13mm shallow socket, it is possible to remove the cranking motor without removing the tuned exhaust pipe, on many models.

Cranking Motor Disassembling

1- Remove the two nuts and washers from the thru-bolts, and then lift off the support bracket. Remove another set of washers from the thru-bolts.

2- Make a mark across the mating surface of the clutch housing and the motor frame. Make another mark across the mating surface of the motor frame and the end cap. These marks will be most valuable later during assembling.

Without some type of alignment marks, the task of inserting the thru-bolts and

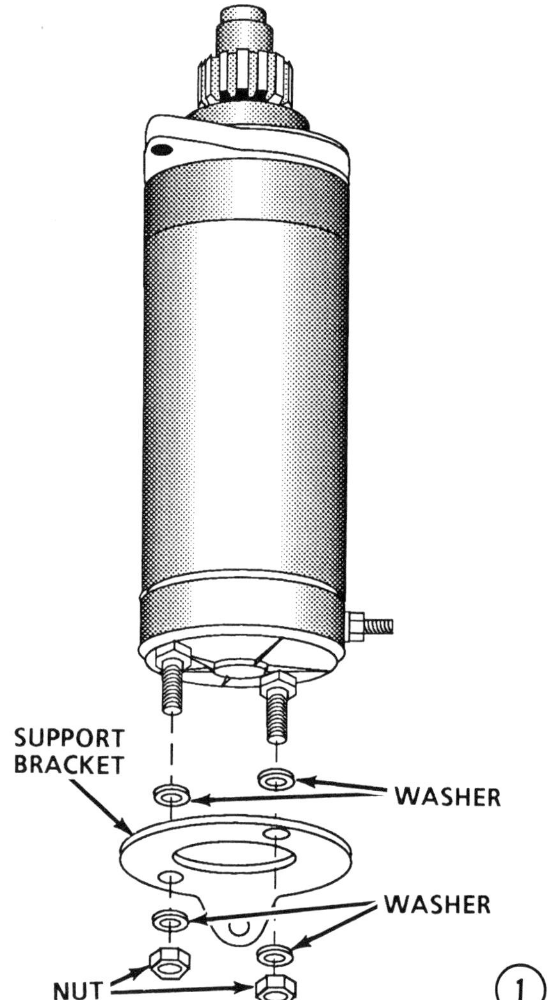

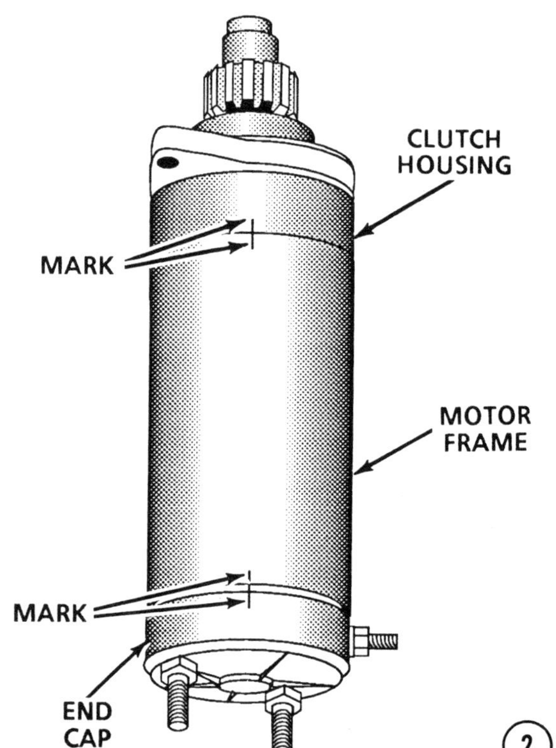

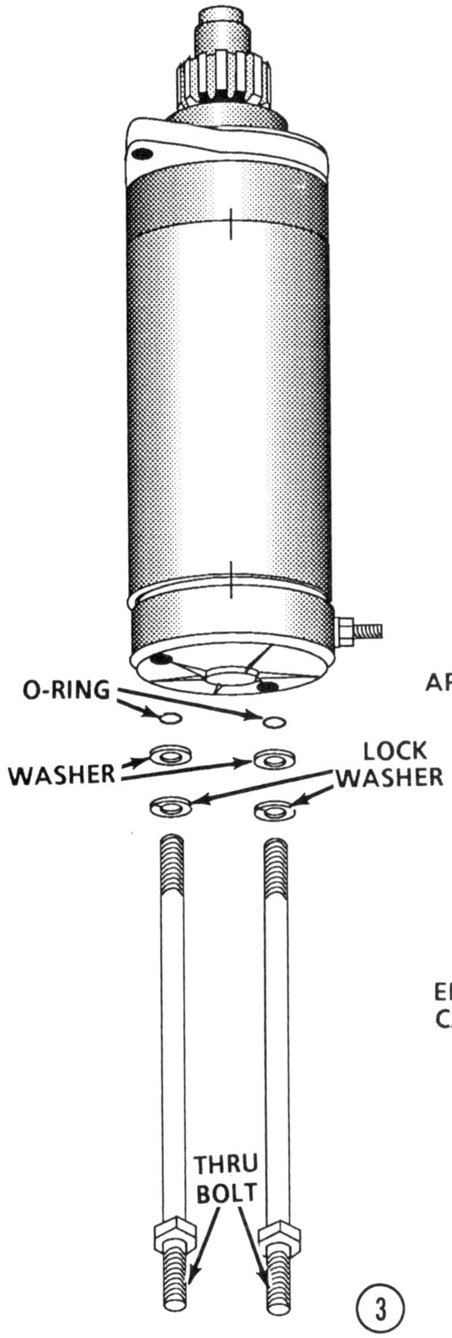

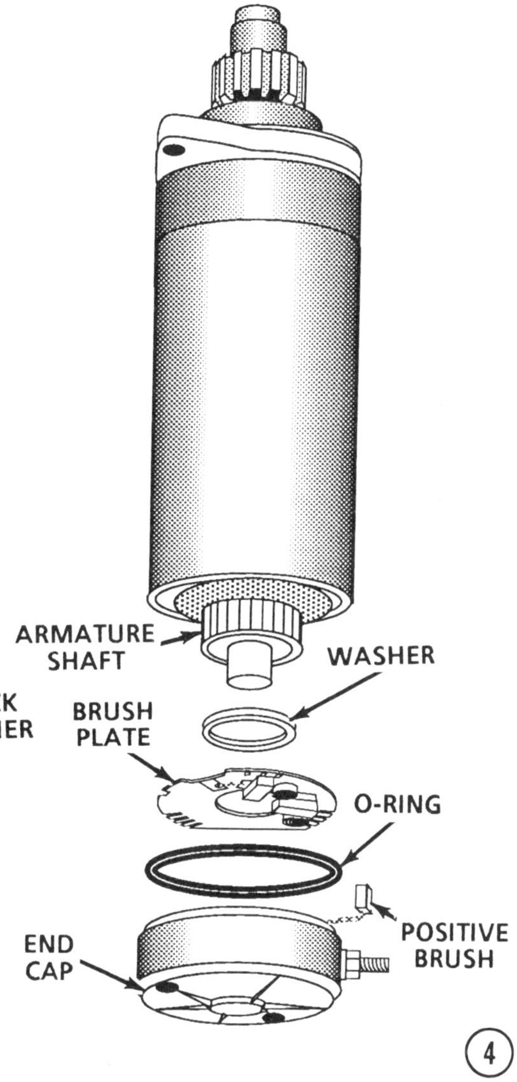

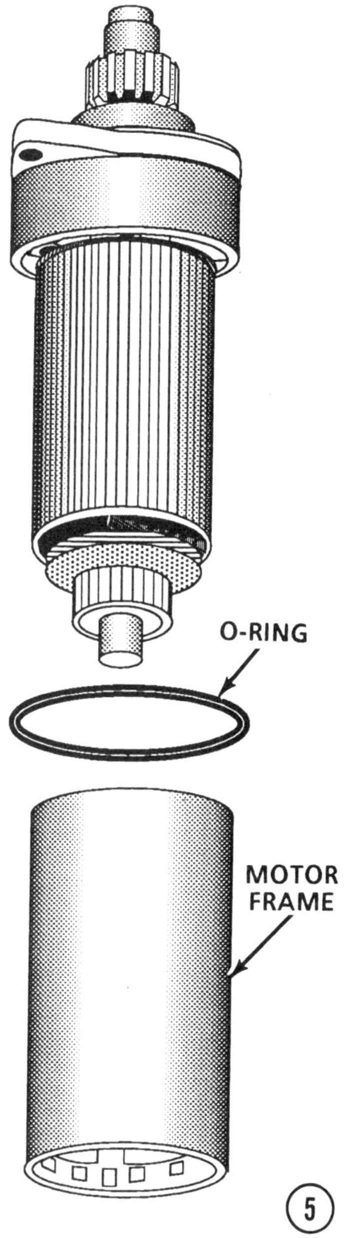

starting the threads into the end cap is almost impossible.

3- Remove the two long thru-bolts, lockwashers, plain washers, and the O-rings.

4- Slide the end cap free of the armature shaft, followed by an O-ring, brush plate, and a washer. Remove the positive brush from the brush holder on the plate to separate the brush holder from the end cap. The negative brush will remain in the holder on the plate.

5- Pull the motor frame from the armature shaft. Remove and discard the O-ring.

SAFETY WORDS

Snap rings and C-lockrings are made of spring steel and may pop out of the groove with considerable force. Therefore, warn others in the area and **WEAR** eye protection glasses or a shield while removing this type ring.

6- Push the collar down the shaft until a screwdriver can be inserted under the snap ring. Pry the snap ring out of the groove.

7- Slide the collar and spring off the shaft.

8- Rotate the pinion gear assembly **COUNTERCLOCKWISE** to remove the assembly from the splines on the shaft. Re-

C-lockrings are under tremendous tension during removal and installation, presenting a potential hazard. As an eye protection measure, safety glasses or a shield should ***ALWAYS*** *be worn during work with C-lockrings or snap rings. Warn others in the area such work is in progress.*

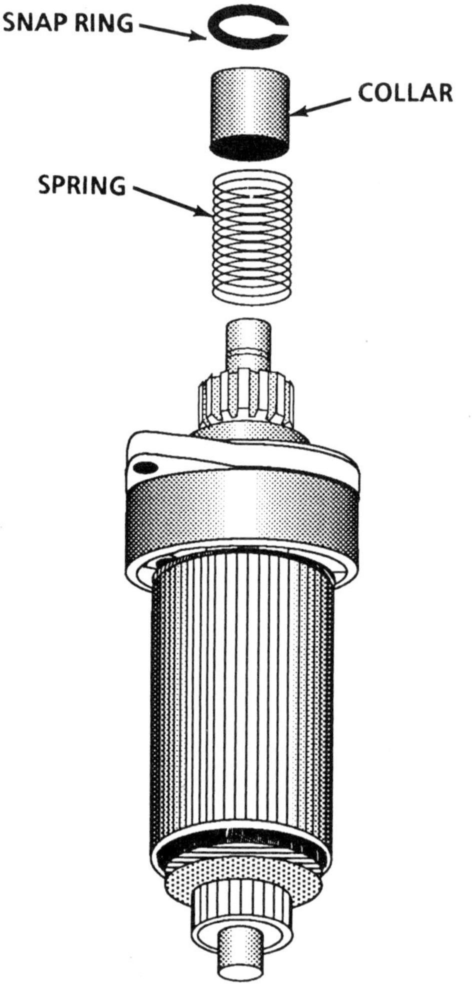

7

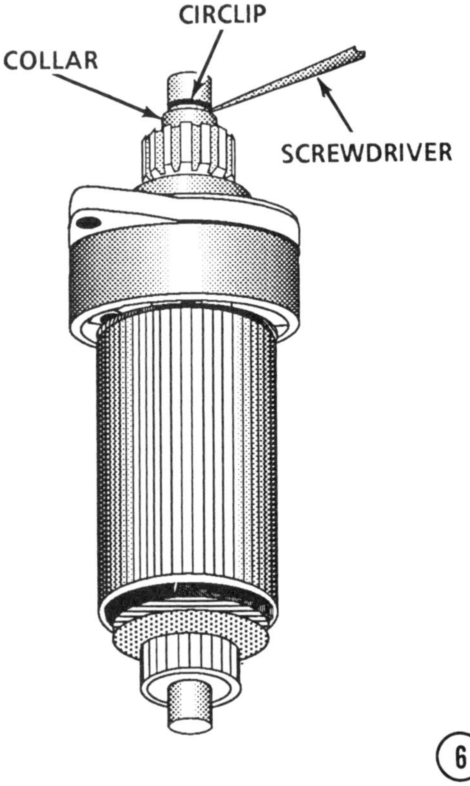

6

move the washer and clutch housing from the shaft.

SPECIAL WORDS

Make a note as to which washer was located at which end of the shaft. Place the washers in order on the workbench in the order of removal. Keeping the washers in order will be most valuable later during the assembling work. Incorrect placement of these washers will cause the armature to be misaligned with the brushes and the field magnets during assembling.

CLEANING AND INSPECTING

Inspect the pinion gear teeth for chips, cracks, or a broken tooth. Check the splines inside the pinion gear for burrs and to be sure the gear moves freely on the armature shaft.

BAD NEWS

This type of cranking motor pinion gear assembly cannot be repaired if the unit is

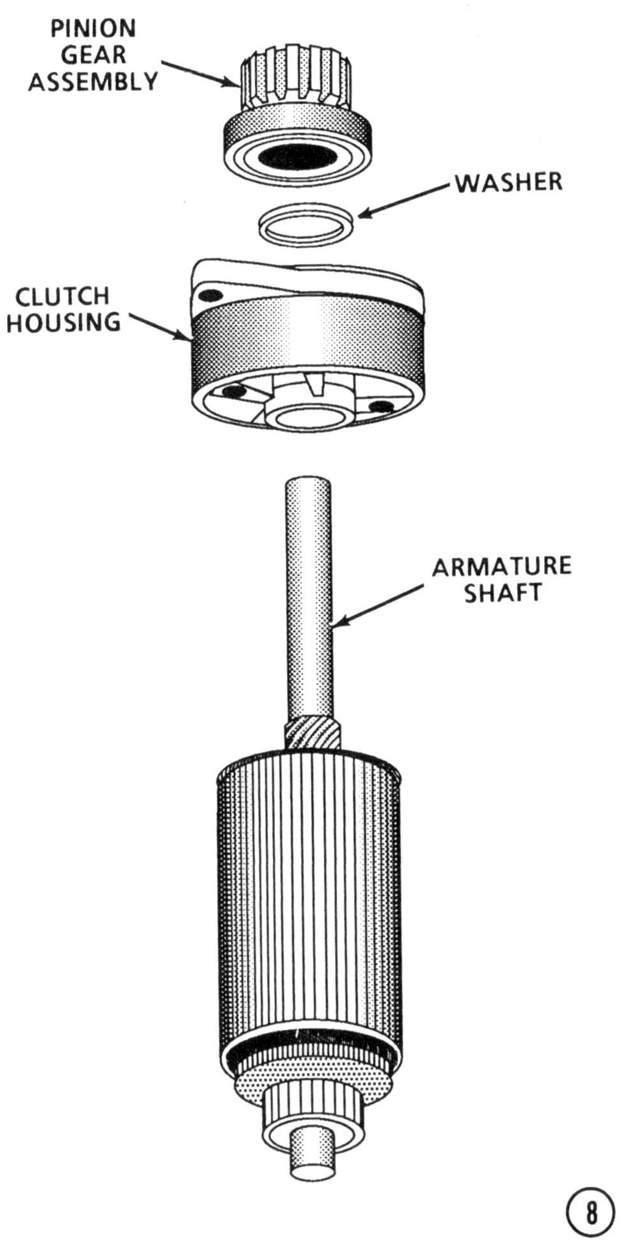

defective. Replacement of the pinion gear assembly as a unit is the only answer.

Check to be sure the return spring is flexible and has not become distorted. Check both ends of the spring for signs of damage.

Clean the armature shaft and check to be sure the shaft is free of any burrs. If burrs are discovered, they may be removed with crocus cloth.

Clean the commutator on the armature shaft and check to be sure the shaft is free of any burrs. If burrs are discovered, they may be removed with crocus cloth.

Clean the field coils, armature, commutator, armature shaft, brush assembly, and end caps with a brush or compressed air.

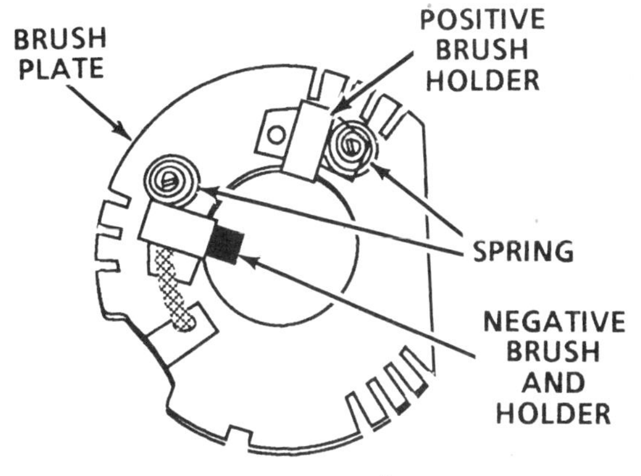

Replace both brushes if either brush has worn down below the wear limit given in the text.

Wash all other parts in solvent and blow them dry with compressed air.

Inspect the insulation and the unsoldered connections of the armature windings for breaks or burns.

Check the commutator for runout. Inspect the armature shaft and both bearings for scoring.

Turn the commutator in a lathe if it is out-of-round by more than 0.005" (0.13mm).

Measure the diameter of the commutator. If the measurement is less than 1.209" (30.7mm), the armature shaft **MUST** be replaced.

Check the springs in the brush holder to be sure none are broken. Check the insulated brush holder for a short to ground.

The armature, fields, and brush holders must be checked before assembling the cranking motor. See Section 9-8, beginning on next page, for detailed procedures to test cranking motor parts.

The negative brush is mounted on the brush plate. The positive brush is attached to the positive terminal. When both the brushes are installed onto the plate, each brush is held in place with wound spring

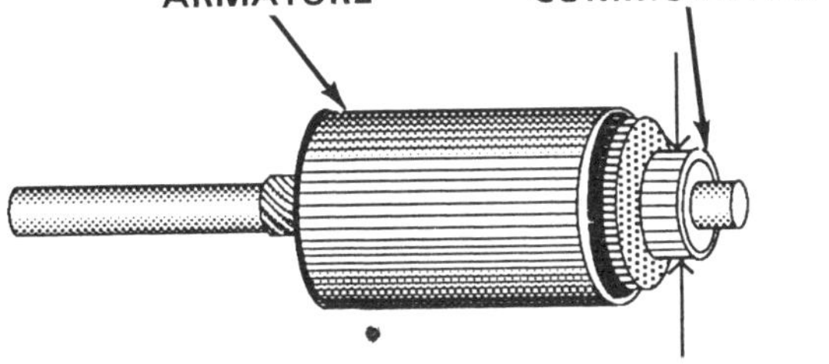

If the diameter of the commutator is less than 1.209" (30.7mm), the armature MUST be replaced.

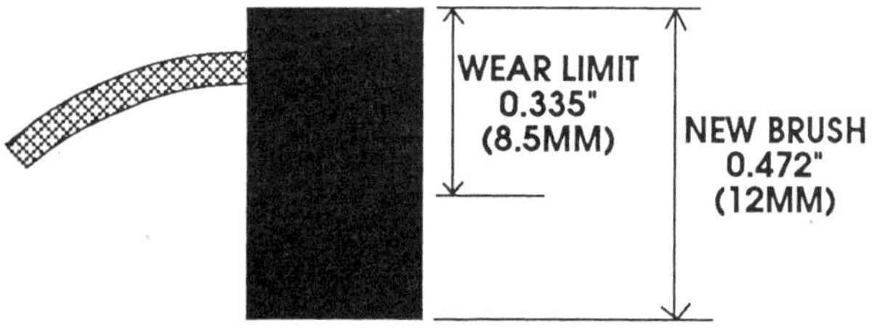

Replace both brushes if either of them has worn down below the wear limit.

steel coils mounted horizontally to provide the tensioning force against the commutator. These coils are permanently riveted to the brush plate and are not removable. If they are broken or missing, a new plate must be purchased.

Measure the length of each brush. A new brush measures 0.472" (12mm). If a brush is less than 0.335" (8.5mm) in length, it must be replaced.

To replace the positive brush, the positive terminal must be removed from the end cap. Remove the following parts and place them in order on the workbench so they may be installed in exactly the same order: a nut, a spring washer, a large insulator, and finally an **O-ring** -- all installed on the terminal on the outside of the end cap. Inside the end cap, the brush lead is soldered to the terminal.

Apply a coating of G.E. Versilube G341M lubricant on the armature shaft splines, pinion gear splines, and in the clutch housing bore. Apply 2-stroke engine oil onto all metal bushings.

9-8 TESTING CRANKING MOTOR PARTS

SPECIAL WORDS

Most marine shops and all electrical motor rebuild shops will test an armature for a modest charge. If the armature has a short, it **MUST** be replaced.

Check the armature for a short circuit by placing it on a growler and holding a hack saw blade over the armature core while the armature is rotated. If the saw blade vibrates, the armature is shorted. Clean between the armature bars, and then check again on the growler. If the saw blade still vibrates, the armature must be replaced. Occasionally carbon dust from the brushes will short the armature. Therefore, blow the slots in the armature clean with compressed air.

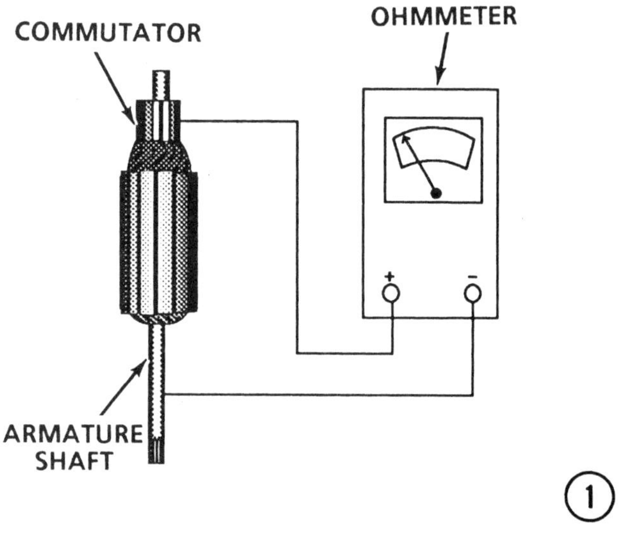

1- Make contact with one probe of the test light or ohmmeter on the armature core or shaft. Make contact with the other probe on the commutator. If the light comes on, or the meter indicates continuity, the armature is grounded and must be replaced.

Turning the Commutator

2- True the commutator, if necessary, in a lathe. **NEVER** undercut the mica because the brushes are harder than the insulation. Undercut the insulation between the commutator bars 0.008" (0.20mm), to the full width of the insulation and flat at the bottom. A triangular groove is not satisfactory. After the undercutting work is completed, clean out the slots carefully to remove dirt and copper dust. Sand the commutator lightly with No. 500 or No. 600 sandpaper to remove any burrs left from the undercutting.

3- Test light probes, placed on any two commutator bars, should light and indicate continuity.

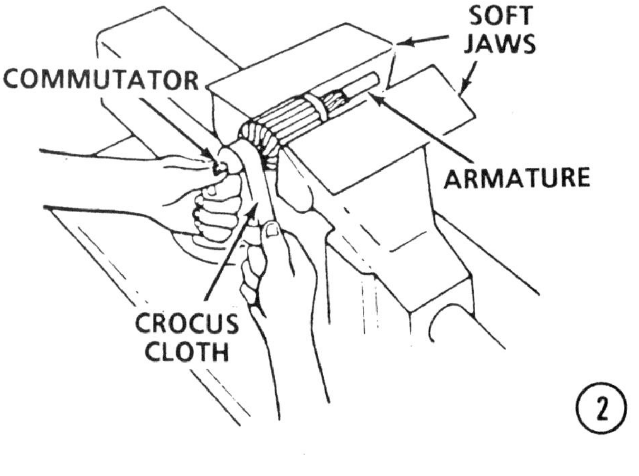

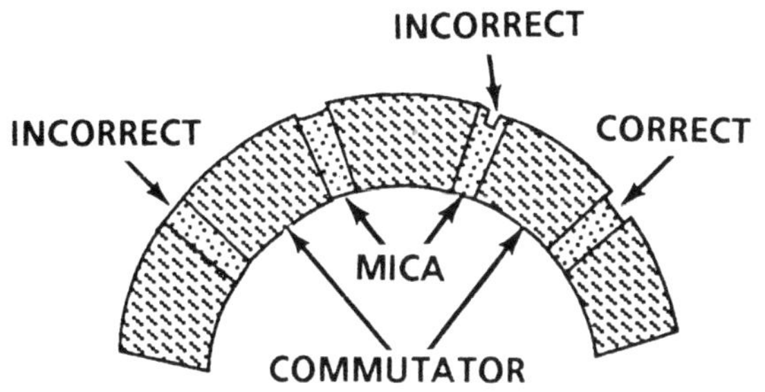

Simple line drawing to illustrate correct and incorrect mica cutting on a commutator.

4- Check the armature a second time on the growler for possible short circuits.

Positive Brush

5- Make contact with one ohmmeter lead to the positive brush. Make contact with the other ohmmeter lead to the positive terminal on the end cap. The meter should register continuity. If the meter registers no continuity, the positive brush has an open in the circuit and must be replaced.

Negative Brush

6- Make contact with one lead of the ohmmeter to the negative brush. Make contact with the other test lead to the brush plate. The ohmmeter should indicate continuity. If the meter indicates no continuity, there is an open in the negative lead or the lead is not grounded properly to the brush plate.

7- Make contact with the ohmmeter leads across the positive and negative brush holders on the brush plate. The meter should register no continuity. If the meter registers continuity, the positive brush holder is shorted to the plate and the plate must be replaced.

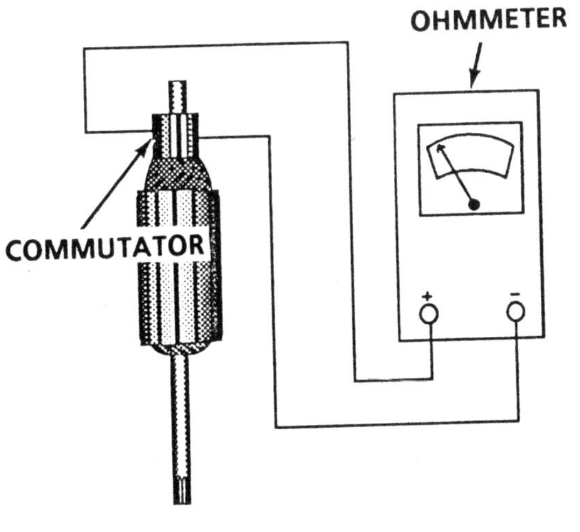

3

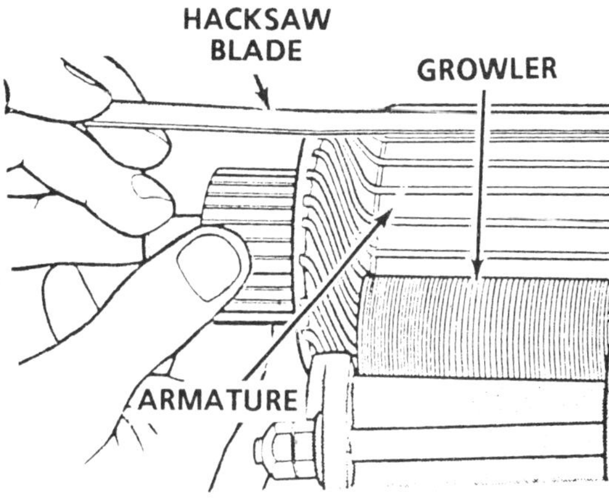

4

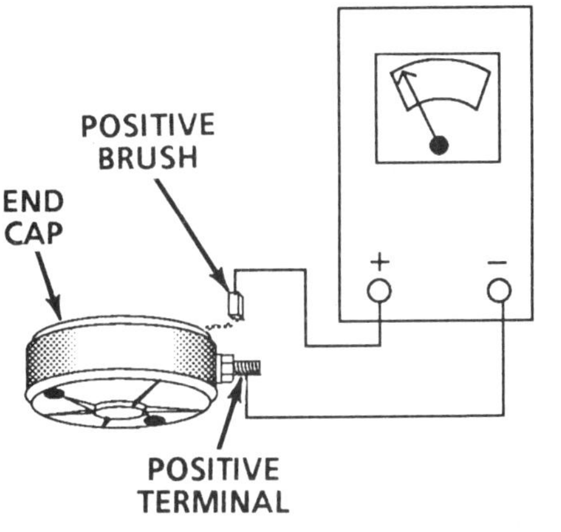

5

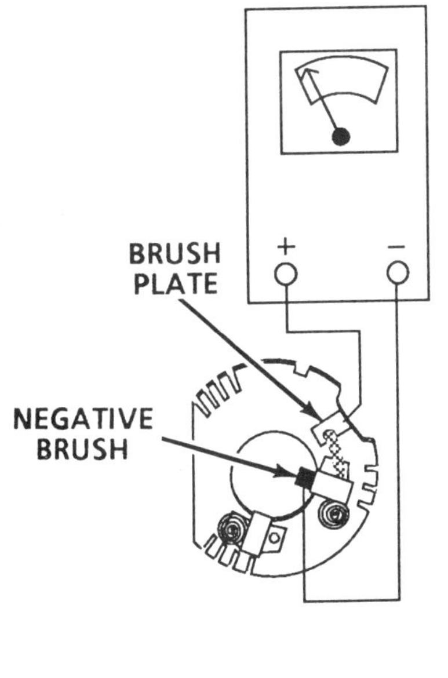

6

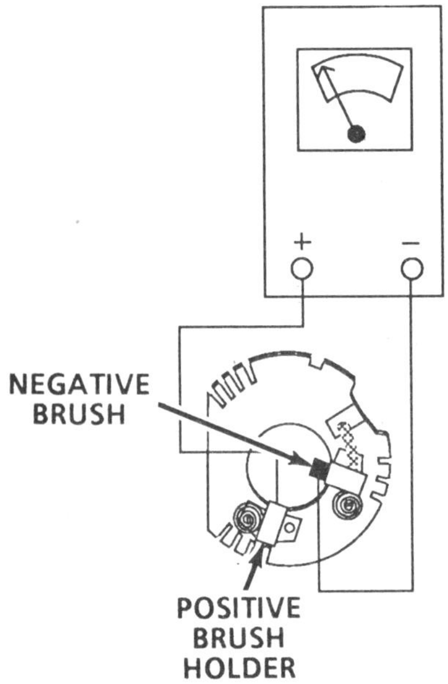

7

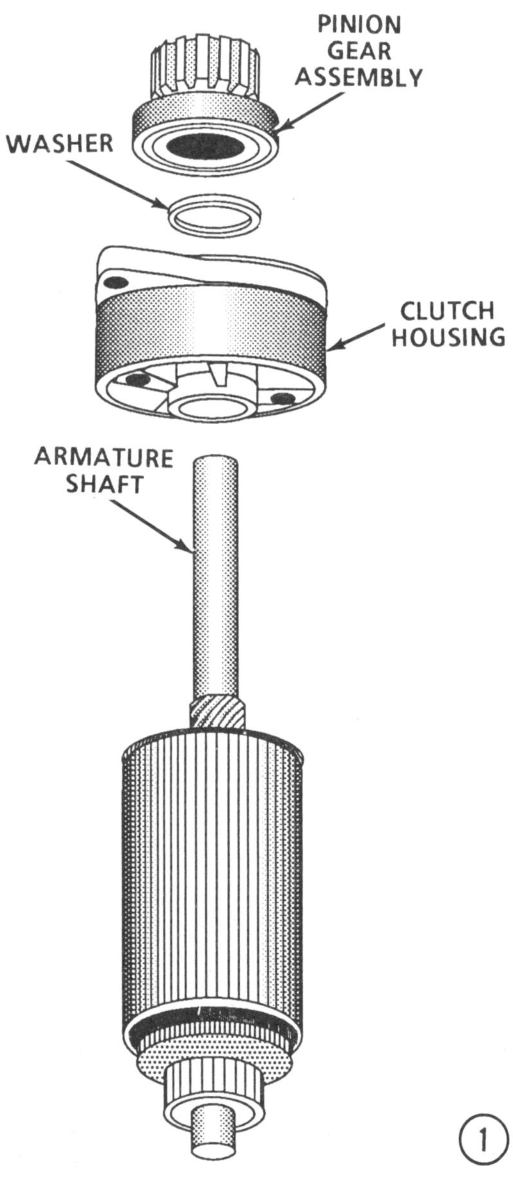

1

CRANKING MOTOR ASSEMBLING

1- Slide the clutch housing onto the armature shaft, followed by a washer, and the pinion gear assembly.

2- Install the following components onto the armature shaft in the sequence given: first the spring, followed by the collar, and finally the snap ring.

SAFETY WORDS

Snap rings and C-lockrings are made of spring steel and may pop out of the groove with considerable force. Therefore, warn others in the area and **WEAR** eye protection glasses or a shield while removing this type ring.

3- Hold the collar down against the spring while the snap ring is installed. The ring must be crimped around the groove in the shaft with a pair of pliers. When properly installed the collar will completely hide the spring. Place the end of the armature shaft on the workbench and use a small punch to tap down the collar, as shown.

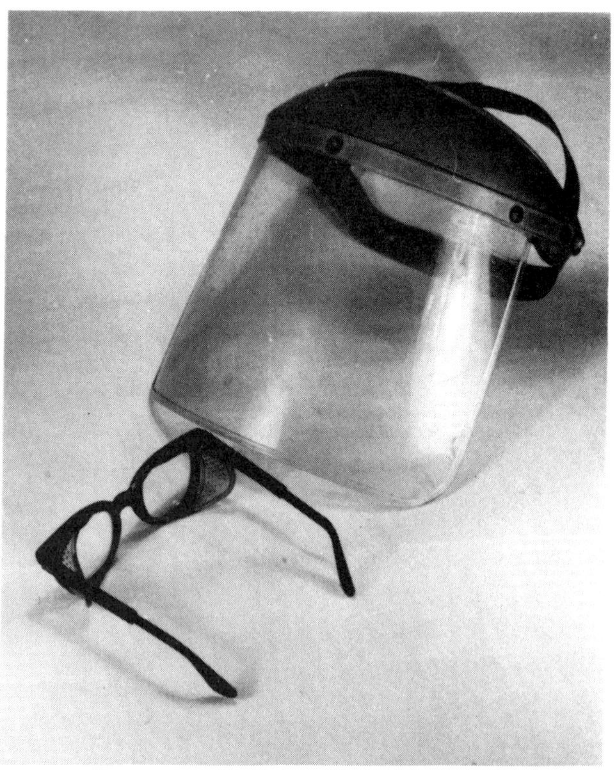

C-lockrings are under tremendous tension during removal and installation, presenting a potential hazard. As an eye protection measure, safety glasses or a shield should ***ALWAYS*** *be worn during work with C-lockrings or snap rings. Warn others in the area such work is in progress.*

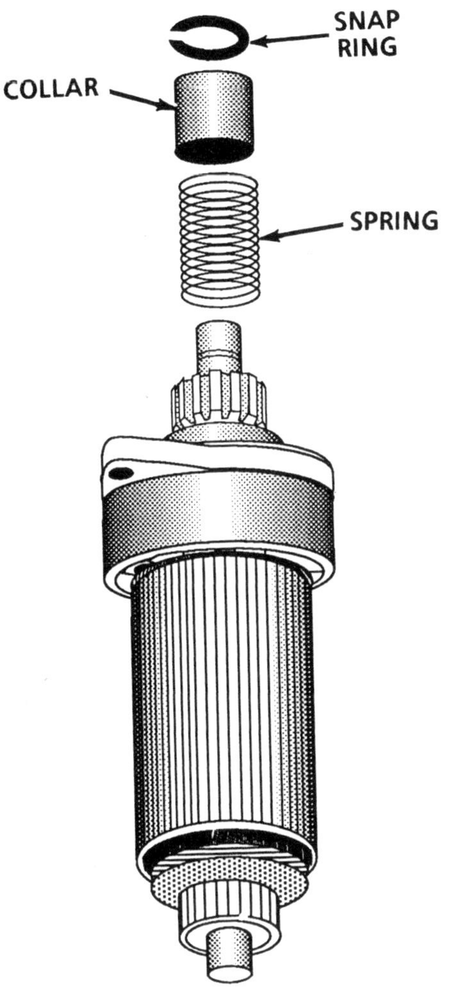

2

4- Slide a new O-ring over the armature. Install the motor frame over the armature with the protrusions inside the frame facing the commutator on the shaft when installed. Align the mark made on the motor frame with the mark on the clutch housing.

5- If the positive brush was removed or replaced, apply Loctite Superflex to the bolt hole in the end cap. Insert the terminal, with the positive brush attached, through the bolt hole from the inside. Install a washer and an insulator over the terminal and then secure it in place with a locknut.

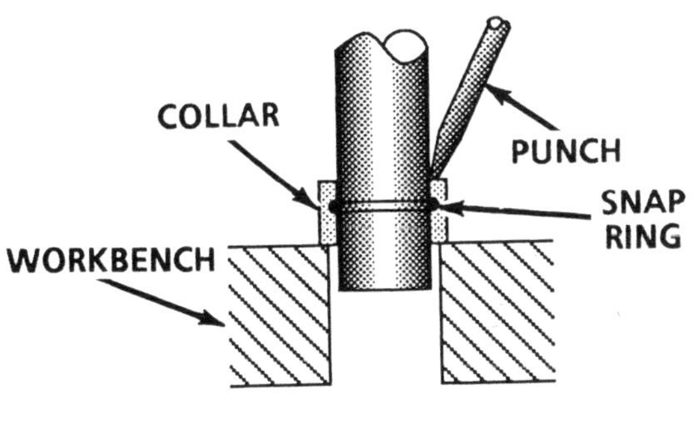

3

Place the washer over the armature shaft. Position a new O-ring around the end cap. Hold back the spring on the positive brush holder and insert the positive brush.

Push both brushes into their holders and slide the brush plate, with end cap attached, over the commutator.

Check the two brush springs for tension. Rotate the armature shaft. The brushes must stay in contact with the commutator surface without binding.

6- Align the highest protrusion inside the motor frame with the locating notch in the brush holder.

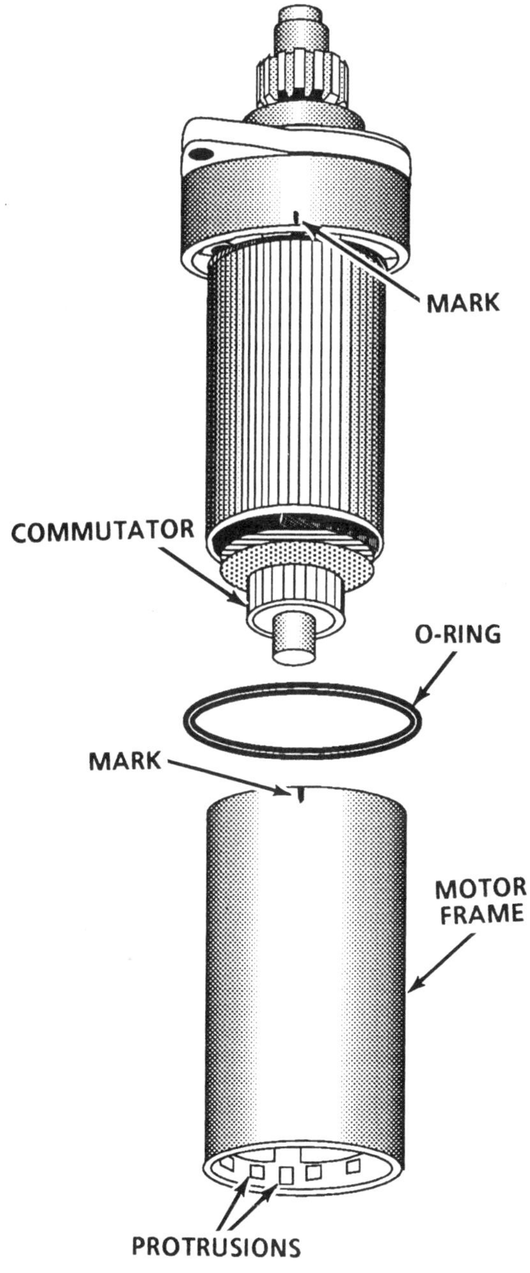

4

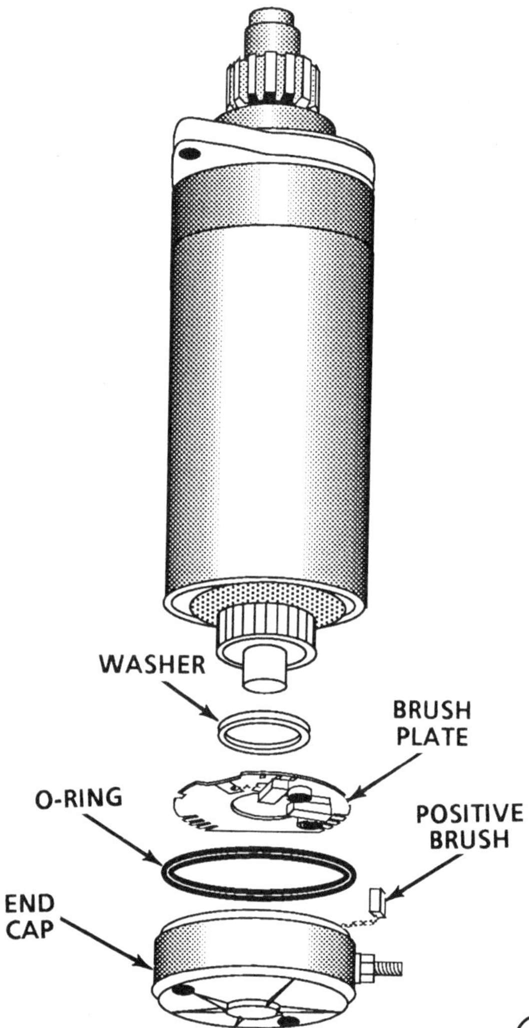

5

7- Align the mark on the end cap with the mark on the motor frame. Slide a lockwasher, plain washer, and O-ring onto each thru bolt. Hold all the pieces of the motor together with index marks aligned, while the thru bolts are installed.

WORDS OF ENCOURAGEMENT

Installing the two thru-bolts is not as simple a task as it sounds. The reason: as the bolts pass the magnets inside the motor frame, the magnets will attract the bolts, "pulling" them out of line. However, with a bit of patience, the bolts can be made to align with the holes in the upper end cap and the threads started.

After the bolts are successfully threaded into the end cap, tighten them securely.

8- Slide a plain washer over each thru-bolt end, and then install the support bracket with the mounting hole tab on the opposite side of the terminal. Next, install another washer and nut onto each thru-bolt. Tighten the nuts securely.

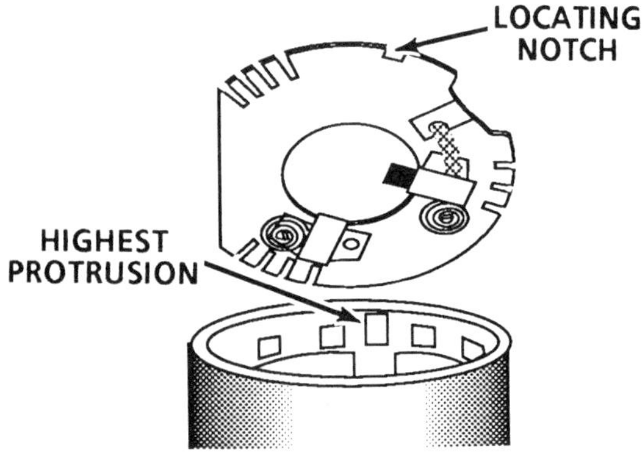

6

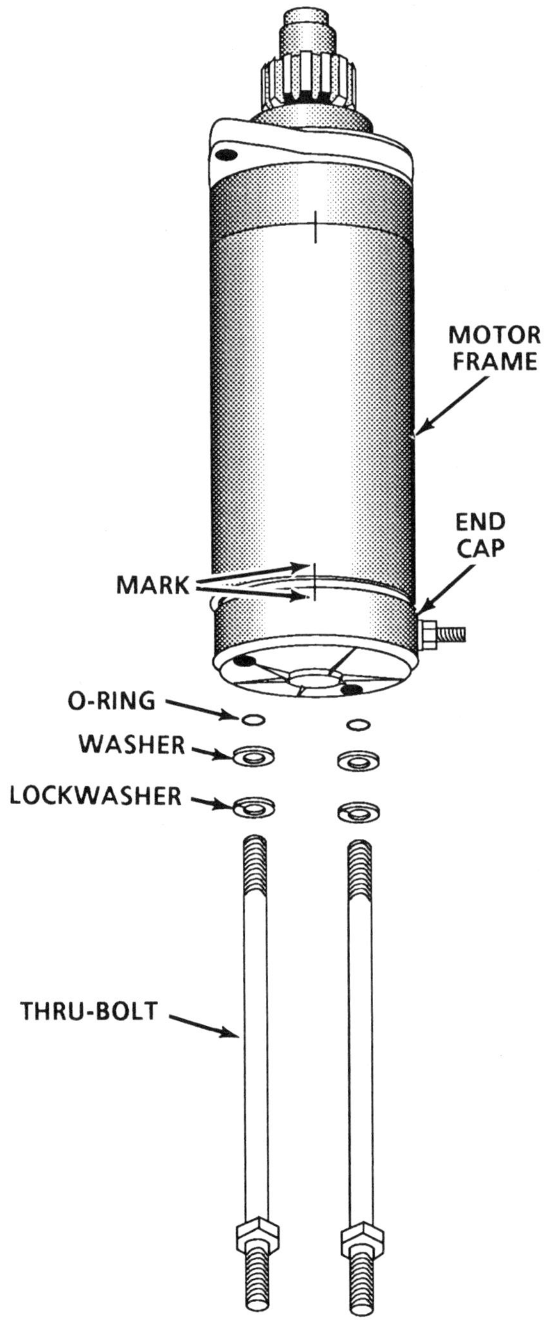

7

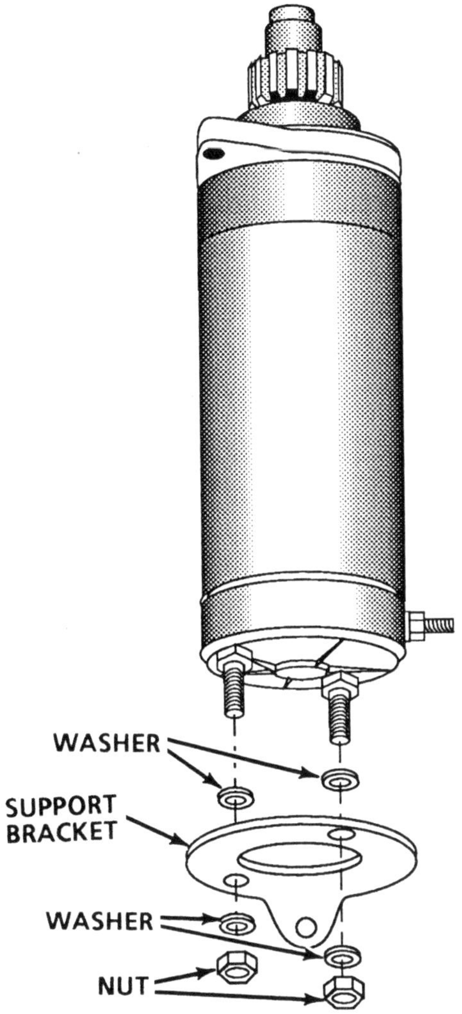

8

Cranking Motor Installation

1- Position a new gasket over the mounting flange and insert the cranking motor into the aft end of the magneto housing.

2- Apply Loctite 242 onto the threads of the three mounting bolts. Position the Black ground lead over the upper mounting bolt. Install the other two bolts, and then tighten all three to a torque value of 16 ft lb (22Nm).

Apply Dielectric lubricant onto the cranking motor terminal threads.

3- Connect the large Red lead onto the cranking motor terminal and tighten the securing nut to a torque value of 62 in lb (7Nm).

Check Completed Work

Secure the craft in a test tank or body of water. If a body of water or tank is not possible, connect a flush attachment and hose to the engine.

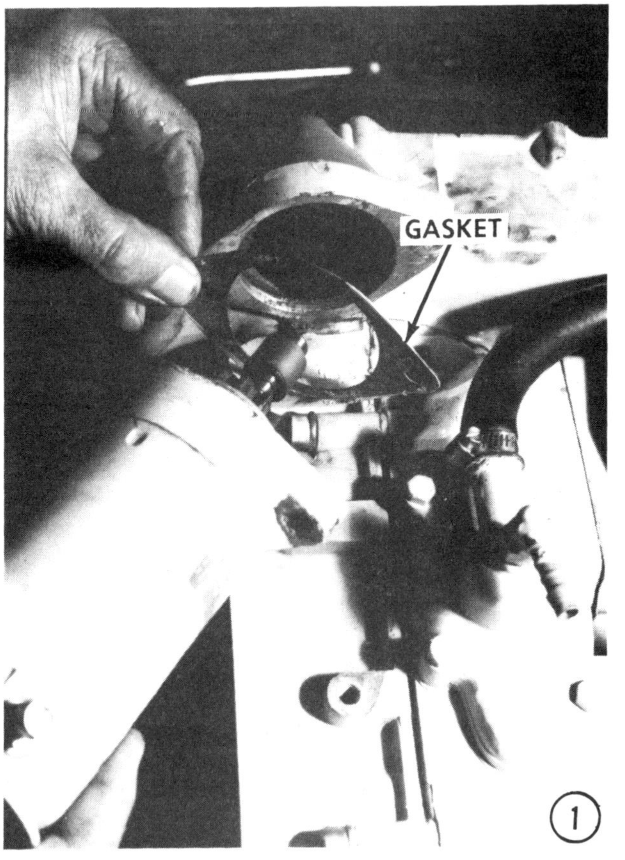

1

CAUTION

The following procedure MUST be followed whenever the engine is started with a hose attached for cooling the engine. Start the engine, and then IMMEDIATELY turn the water on. The engine MUST be started first to prevent water from backing up into

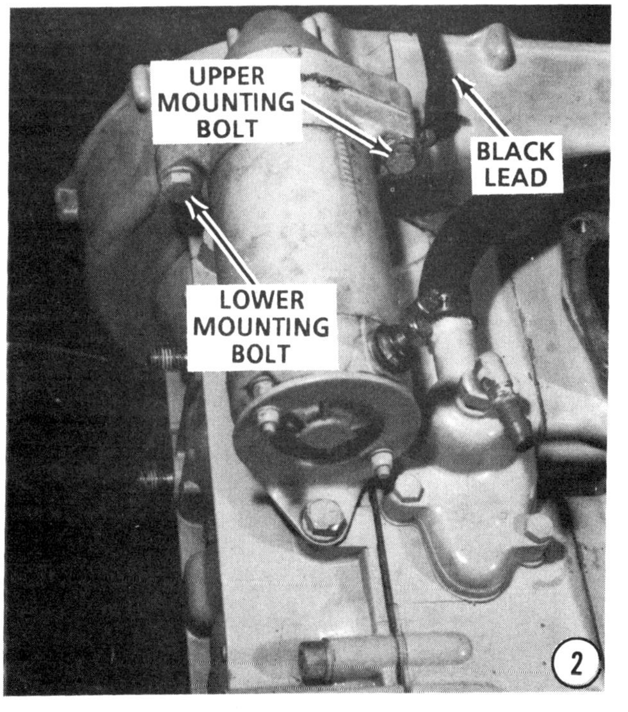

2

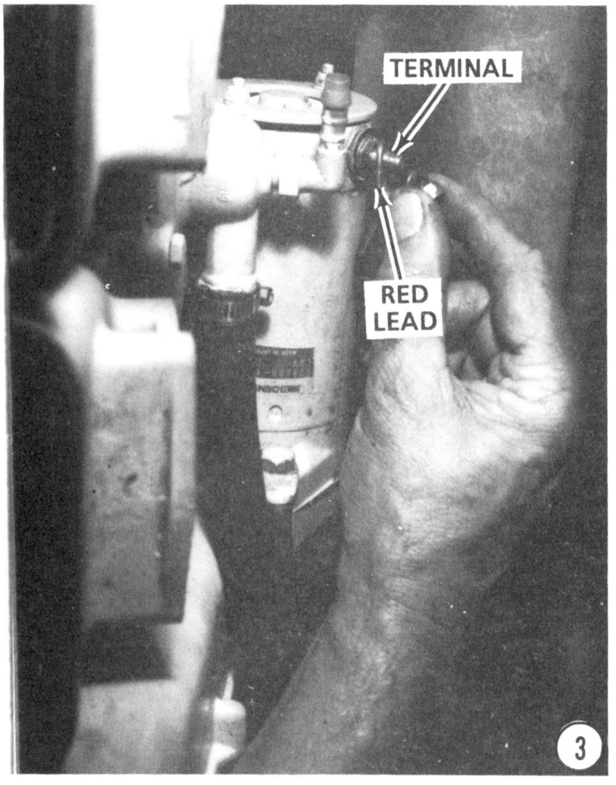

the tuned exhaust pipe and into the engine. Water must circulate anytime the engine is operating to prevent overheating -- which could cause damage to moving parts and possibly cause engine seizure. NEVER allow the engine to operate without cooling water for more than 15 seconds. NEVER run the engine with a flushing attachment for more than 5 minutes. The driveline seal is not provided with cooling water if it is not immersed in water. To shut the engine down, first shut off the water supply, and then shut down the engine. If this sequence of tasks is not followed as stated, the engine could be severely damaged.

REMEMBER

Start the engine.
Turn the water on.
Turn off the water.
Shut down the engine.

NEVER, again **NEVER** operate the engine at high speed with a flush device attached. The engine, operating at high speed with such a device attached, would **RUNAWAY** from lack of a load on the impeller, causing extensive damage.

Crank the engine with the cranking motor and start the unit. Shut the engine down and start it several times to check operation of the cranking motor and circuit.

9-9 TESTING OTHER ELECTRICAL COMPONENTS

This short section provides testing procedures for other electrical parts installed on or around the engine. If a unit fails the testing, the faulty part must be replaced. In most cases, removal and installation is through attaching hardware.

Colored Lead Identification

Test leads are identified by their color to aid in locating the correct lead. However, the manufacturer has been known to change lead colors. Therefore, identify the component and consult the wiring diagram for the correct component connections if colors vary.

NOW, THESE WORDS

The overheat buzzer, engine "kill" switch, tether cord switch -- safety lanyard -- and engine start switch are all mounted on the control panel. On some models, the start switch may also be used as the stop switch.

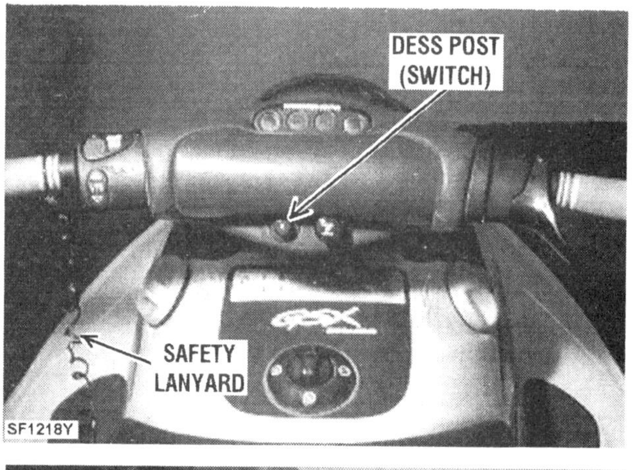

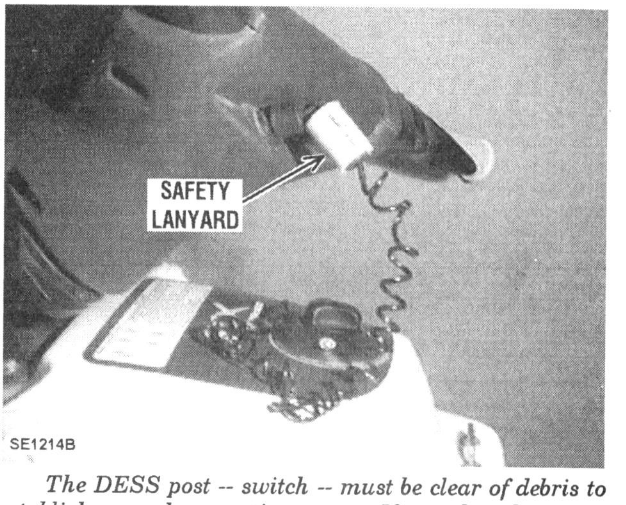

The DESS post -- switch -- must be clear of debris to establish a good connection -- top. If two short beeps are heard when connecting the safety lanyard, the craft is ready to go.

Resistance tests on these components are performed at the ends of their harness leads. These quick-disconnect fittings are typically inside the electrical box. Therefore, remove the electrical box cover. It may be possible to identify the leads without tracing the leads from the component.

START/STOP SWITCH

This start/stop switch, depending on the model being serviced, may be two unique switches, or combined as one unit. Therefore, a procedure for testing each follows.

Stop Switch/Tether Cord

These two switches may be tested together because they are wired in series. Identify and disconnect the stop switch Black lead in the electrical box. Disconnect the MPEM Black/Yellow lead from the MPEM. Obtain an ohmmeter and connect probes to the Black and Black/Yellow leads. Install the tether cord cap over the switch. If the measured reading is infinity, verify both switches separately, as described below.

Stop Switch

Disconnect the Black and Black/Yellow lead in the electrical box. Connect the multitester probes to the stop switch Black and Black/Yellow leads. The measured resistance must be close to zero ohm when in the closed -- operating -- condition and an open circuit -- infinity -- when depressed.

Tether Cord

Identify and disconnect the Black/Yellow leads in the electrical box. Connect a multitester probe to each of the Black/Yellow leads. The measured reading must be close to zero ohm when the cap is placed over the switch and an open circuit when removed.

Start/Stop Switch

Identify and disconnect the start/stop Yellow/Red lead. Connect one multitester probe to the start/stop Yellow/Red lead, connect the other tester probe to ground. The measured reading should indicate an open circuit -- infinity. The reading should be close to zero ohm when the switch is depressed.

Overheat Buzzer Test

Disconnect the White lead from the thermosensor switch on the cylinder head cover. Ground the lead -- a small Allen wrench may be used, as shown in the accompanying illustration. The buzzer should sound when the lead is grounded and be silent when the lead is not grounded. If the buzzer fails this test, check the continuity of the lead from the terminal on the end of the disconnected White lead to the quick disconnect fitting at the rectifier/regulator. The lead changes color -- White to Brown. Check the continuity of the Brown lead from the rectifier/regulator to the small terminal on the cranking motor solenoid. If all the leads pass the test, obtain a 12-volt battery. Connect the negative battery ter-

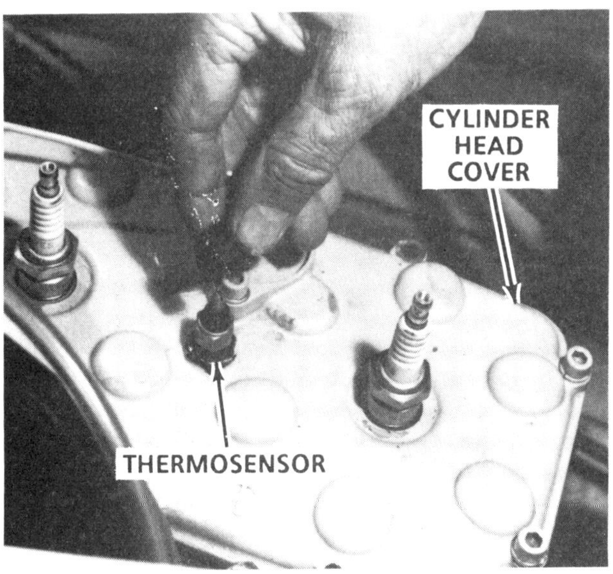

Disconnecting the White lead from the thermosensor, in preparation to testing the lead.

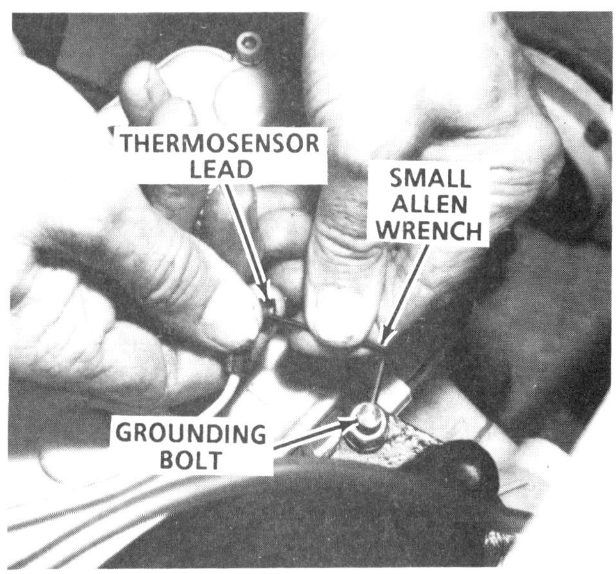

Testing the thermosensor lead, as explained in the text.

minal to the Black or White lead on the buzzer. Connect the positive battery terminal to the Red (changing color to Gray) lead between the buzzer and the "kill" switch. The buzzer should sound. If it fails to sound, the buzzer is defective and **MUST** be replaced.

Thermosensor Switch Test

The thermosensor switch is located in the cylinder head cover. The sensor measures the temperature of the cooling water circulating through the engine. If the cooling water temperature rises above 203°F (95°C), the sensor sends a signal to the warning buzzer -- the buzzer sounds and the craft operator is warned of the heating condition. If the warning buzzer sounds, the operator should immediately shut the engine down to prevent extensive and expensive damage to the engine from overheating.

To test operation of the temperature sensor, disconnect the White lead at the quick disconnect fitting. Remove the sensor from the cylinder head cover. Obtain an ohmmeter and select the Rx1000 (Rx1K) ohm scale. Connect the meter leads across the sensor terminal and the sensor threads. Heat a suitable container of water on the stove and have a thermometer handy to measure the water temperature.

Submerge only the sensor end in the water, as shown, without the sensor touching the sides or the bottom of the container. Observe the meter reading as the water temperature rises. Stir the water with the thermometer to evenly distribute the heat.

At temperatures above approximately 190°F (88°C), the meter should register more than 1,000,000 ohms (1,000K ohms). At temperatures less than 190°F (88°C), the meter should register almost zero ohms.

If the sensor does not perform or pass the tests, the sensor **MUST** be replaced.

VTS ELECTRIC TRIM

The SPX, GSI, GSX and XP Models feature an electronic variable nozzle trim system. This function serves to adjust the running height of the nose of the watercraft, adjusting for dynamic loading and speed of the craft for optimum performance.

VTS Switch

To verify the switch is functioning properly, first check the condition of the fuse. Then, disconnect the switch Black lead, Blue/White lead and Green/White leads. Obtain a digital multitester and connect the probes to switch Black and Blue/White leads to measure the resistance. Observe the meter reading. Next, connect the tester probes to the switch Black and Green/White leads. Observe the meter reading.

The resistance in both tests should be close to zero when the switch is activated. The reading should be high when the button is released.

VTS Motor

Remove the VTS cover from the trim motor housing. Check to be sure the fuse is in satisfactory condition before proceeding with testing the motor system. Obtain a digital multitester and connect the tester probes to the motor Blue/Orange lead and Green/Orange leads. The measured resistance should be approximately 1.5 ohm.

If the motor tests satisfactory but seems to jam prior to reaching the full "up" or "down" setting, refer to the VTS Actuator Mechanism removal and inspection procedures, outlined at the end of Chapter 11 -- **ADJUSTMENTS**.

10
JET PUMP

10-1 INTRODUCTION

Bombardier Sea-Doo watercraft use only axial type jet pumps. The axial pump used has a two-piece pump shaft -- a long driveshaft and a short impeller shaft. The driveshaft has splines at the forward end and is indexed with the aft flywheel. A rubber coupling is not used.

A single needle bearing in the seal carrier supports the drive shaft. The aft end of the driveshaft is indexed into the impeller. The impeller shaft is support by a needle bearing at each end. The forward end of the shaft indexes into the impeller.

10-2 PRINCIPLES OF OPERATION

One of the first lessons to be learned in any elementary physics class is Newton's basic law: "for every force, there is an opposite and equal force". This statement is the basic principle of the jet pump. Water is "sucked" and "scooped" in from under the craft by a powerful pump rotating at incredible speed, and then discharged, "blown" out, sternward in the opposite direction. In this manner the watercraft is propelled forward.

The personal watercraft covered in this manual are all equipped with a twin cylinder water-cooled two-stroke engine, matched with a single stage (one impeller) jet pump. The Sea-Doo GT is equipped with a "gate" which can be lowered down over the pump outlet nozzle forcing the exhausted water back in a forward direction, thus moving the craft sternward. Personal watercraft jet pumps may be classified as "axial flow" or "mixed flow".

All Sea-Doo Personal Watercraft are equipped with an axial flow jet pump.

Axial Flow

Water in an axial pump moves on a single axis, as depicted in the illustration on the following page, thus the word "axial". In simple terms, water is ingested and discharged parallel to the axis of impeller rotation. The inside of an axial pump case is parallel to the water flow -- unlike the inside of a mixed flow pump

EDUCATIONAL INFORMATION

Even though Bombardier Sea-Doo personal watercraft utilize only an axial flow pump, the following paragraphs are included in this chapter to acquaint the reader with a mixed flow pump -- how it differs from the axial flow pump and why.

Mixed Flow

Many experts consider the mixed flow system to the most efficient for personal watercraft application. The water flow is discharged at an inclined angle to the axis of impeller rotation, as indicated by the arrows in the accompanying illustration. The impeller blades are inclined in such a man-

A Bombardier Sea-Doo watercraft, driven by an axial flow pump and powered by a Rotax 587 water-cooled engine sits in the California sunshine waiting for a new owner.

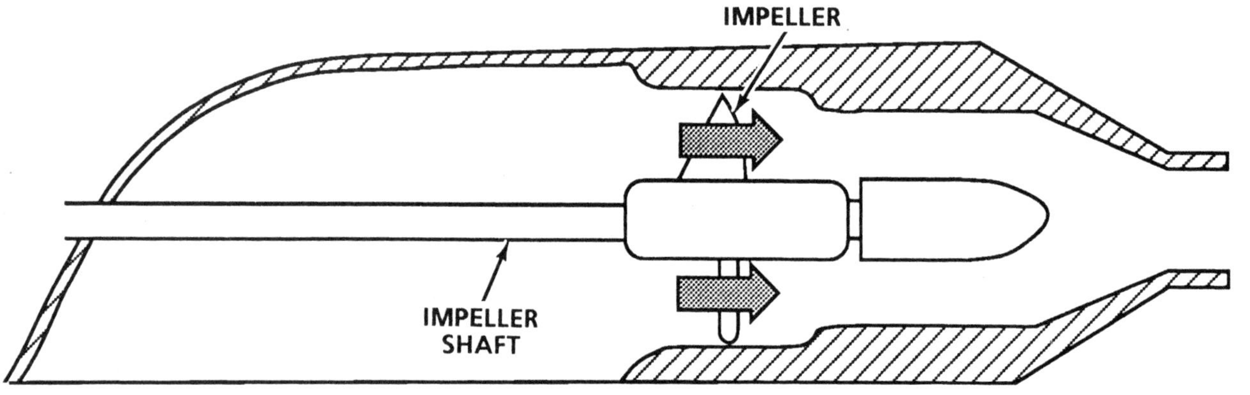

Simple cross-section line drawing to depict water flow through an axial flow jet pump. The water passes through parallel to the axis of impeller rotation.

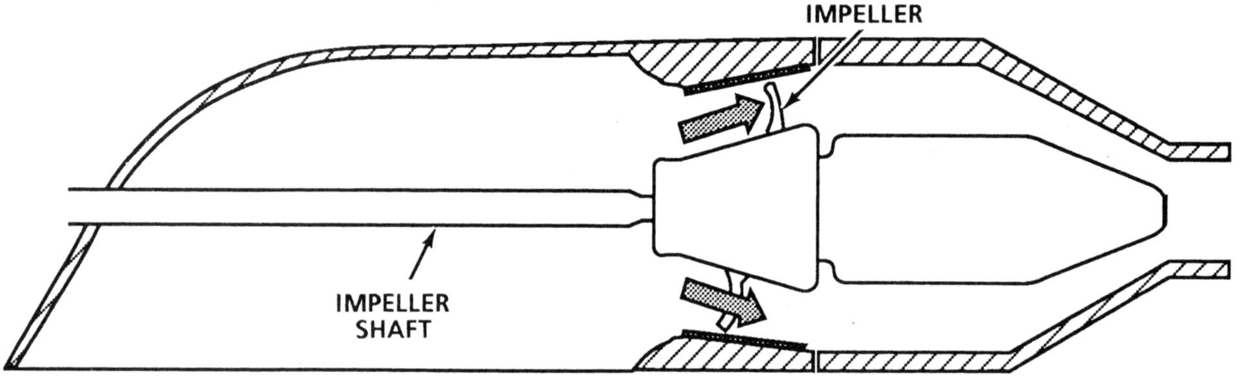

Line drawing to illustrate water flow through a mixed flow jet pump. Notice how the water is directed upward at an angle to the axis of impeller rotation.

ner to cause the water to be thrown off at a 45 degree angle to the impeller shaft within the pump case. From the pump case, the water is then forced out through the outlet nozzle and steering nozzle. The vanes in the pump case help straighten the water flow through the nozzle.

The inside of the pump case is inclined at an angle to the water flow. Therefore, this design pump incorporates shim material to space the impeller blades at a specified distance from the pump case. Mixed flow pump efficiency decreases by 30% as blade-to-pump case clearance increases by 0.04" (1mm).

Axial or Mixed?
Axial and mixed flow pumps are intended for different applications. One is not necessarily better than the other. The axial pump can be operated at higher engine rpm and is more suitable for racing. A mixed flow pump adds thrust at low engine rpm due to the pump design. Therefore, the mixed flow pump performs more efficiently at low engine rpm. This type pump is more suitable for a craft equipped to carry extra weight.

A jet pump seldom requires service except for the following:

Impeller-to-wear ring clearance becomes excessive.
Impeller is damaged.
The oil seal around the driveshaft fails -- allowing water to enter the hull.

10-3 ENGINE AND JET DRIVE

The basic principle and arrangement of engine and pump are almost identical for all manufacturers.

The engine crankshaft is coupled either directly to the pump shaft or through a short driveshaft without any gear reduction. An impeller mounted on the pump shaft draws water into an opening in the hull through a suction and intake casting. Once the craft has attained forward motion the intake also serves as a "scoop" adding to the volume of water moved through the pump.

Volume and Velocity

The outlet nozzle is slightly funnel shaped. This design coupled with the capacity of the pump (impeller), causes the volume of water entering the pump to exit the nozzle with increased velocity. This principle is similar to air passing through the venturi of a carburetor.

The amount of water ejected sternward from the nozzle at high velocity is the force propelling the craft through the water. This force can actually be measured and calculated in foot pounds or Newton meters. The greater the velocity of water mass moving through the nozzle, the greater the thrust to move the craft.

Cooling System

A fitting on the jet pump, aft of the impeller, syphons off cooling water. This water is delivered first to the exhaust manifold -- the hottest part of the engine. The water is then channeled around the cylinder walls, cylinder head, and exhaust pipe. Some water is mixed with exhaust gases in the tuned exhaust pipe and finally discharged from the exhaust outlet. This mixture of water and exhaust gases has the effect of cooling and somewhat quieting the exhaust emission.

10-4 IMPELLERS

Impellers provide thrust using a combination of water flow and water pressure in a closed environment.

The blades of an impeller overlap, thus water is trapped and forced through the impeller into the pump case and nozzle. The water moving past the impeller resists cavitation because it is under pressure. Blades in the pump case are angled exactly opposite to those of the impeller. These blades redirect the water to establish a concentrated flow through the nozzle.

The manufacturer recommends a radius of about 1/64" (0.3-0.5mm) for the leading edge of the impeller blade. A sharper radius will result in cavitation and a greater radius will reduce pump efficiency.

Cavitation Burns

Cavitation burns are the worst enemy of an impeller. These burns literally "eat away" material from the impeller, leaving holes and weakening the impeller structure. In extreme cases entire blades have been known to "depart" from the impeller hub because the cavitation burns at the base of the blade were so severe.

Cavitation burns are the result of imperfections (damage) on the impeller blades or air mixed with the water flow.

Cavitation burns may be caused by:

Wave jumping -- air is sucked into the pump as the craft leaves the surface of the water.

Bad seal between the pump case and outlet nozzle -- allowing air to enter.

Leading edge of the impeller becoming damaged.

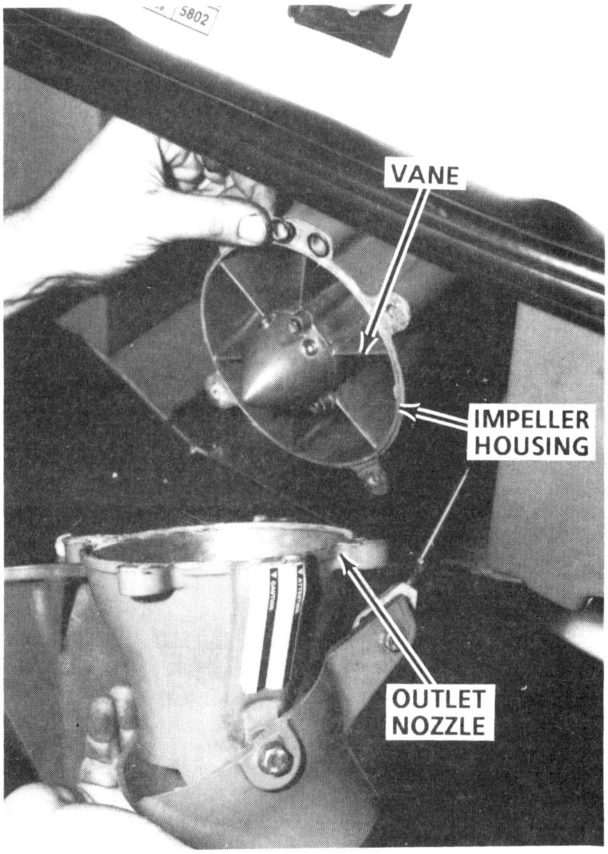

The vanes in the impeller housing straighten water flow and the conical shape of the outlet nozzle increases flow velocity.

To explain exactly what causes cavitation burns, first, let us examine the last cause. If the impeller leading edge is damaged and "mushrooms" over the blade, a low pressure area will form under the mushroomed lip.

At atmospheric pressure, water boils at 212 degrees F. Anytime there is an imperfection on the surface of an impeller blade and water passes over that imperfection, the water pressure is lowered. If the pressure is lowered, water will boil at a much lower temperature. Therefore, air bubbles will form in the water boiling under the mushroomed lip.

The bubbles will creep down the surface of the blade and accumulate at the impeller hub. A high pressure area is formed at the base of the blade. Here, the air bubbles will collapse and reform back into water with a release of energy. This energy is absorbed by the impeller and results in material being eaten away.

The impeller should be dressed to a slightly rounded edge. It is impossible to achieve a perfectly straight sharp edge, because one side will always be convex and the other side concave. The concave side will form an area of low pressure and encourage cavitation, as described above.

The other two causes of cavitation -- defective seals and wave jumping -- occurs when air is sucked in with the water flow. The same principles apply, as just described, in the previous paragraphs. Air bubbles will creep down the surface of the blade and accumulate at the impeller hub. A high pressure area is formed at the base of the blade. Here, the air bubbles will collapse and reform back into water with a release of energy. This energy is absorbed by the impeller and results in material being "eaten away".

Cooling Water and Bilge Hoses

On most pumps, three hoses are attached to the pump case and outlet nozzle. One hose channels some of the water flowing through the pump to the exhaust manifold -- the hottest part of the engine -- to cool the block during engine operation. The other two hoses siphon water out of the bottom of the hull. These hoses are referred to as the bilge system.

The cooling hose has water flowing **FROM** the pump. The bilge hoses carry

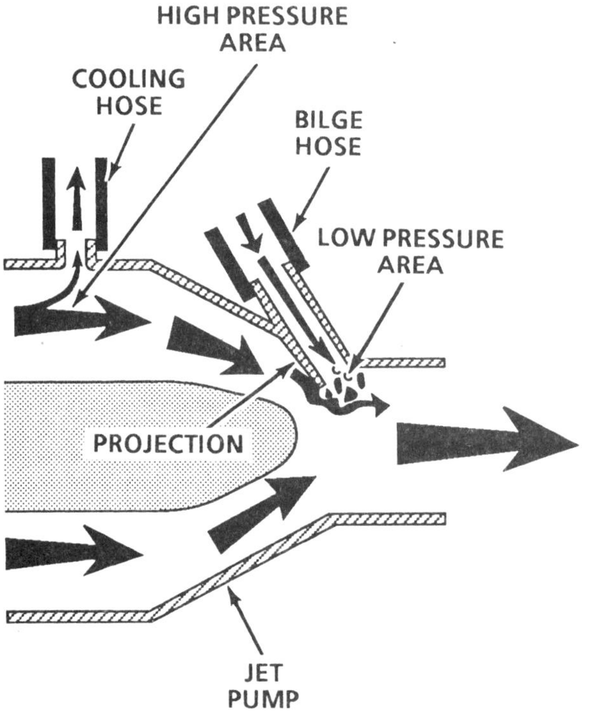

The direction of water flow through the cooling hose and the bilge hose is opposite. The cooling hose routes water from the pump to the engine. Water is "vacuumed" from the bilge through the bilge hose to the pump where it is forced out through the nozzle.

water **TO** the pump. All three hoses attach to the pump in a similar manner.

What determines the direction of water flow?

The answer to this question is in a simple explanation of high and low pressure areas along the inside surface of the pump.

Close inspection of the area inside the pump at the cooling water hose fitting reveals a smooth rounded shoulder with no obstruction or obstacle to impede the flow of water. A high pressure area develops here and draws the water down the hose attached to the fitting.

Further inspection of the area inside the housing at the bilge hose fitting reveals a small protrusion around the opening. This protrusion causes disturbance to the water flow and creates an area of low pressure. This low pressure area will have the effect of emptying air and water from the hose aft with the impeller water flow -- similar to the action of a vacuum cleaner, as depicted in the accompanying illustration. If the other end of the hose is submerged in water inside the hull, the water will be "vacuumed" out. In this manner, the bilge system keeps the bilge relatively "dry".

Many an owner -- with good intentions -- thinking his/her action would smooth water flow and increase pump performance -- has filed the protrusion described in the previous paragraph smooth with the surrounding area of the pump. During operation of the watercraft, the bilge system worked in **REVERSE.** Water was pumped from the impeller housing into the bilge area, quickly filling the bilge and possibly sinking the craft before the rider had time to realize what was happening.

10-5 CHECKING IMPELLER TO WEAR RING CLEARANCE

The following procedure may be performed with the watercraft elevated on a set of saw horses enabling a person to work underneath, or with the watercraft "laid over" on its **PORT** side, allowing access to the rock grate and the impeller.

SPECIAL WORDS

A decal on the transom includes a warning regarding which direction the craft must be "laid over" -- **ALWAYS** to port.

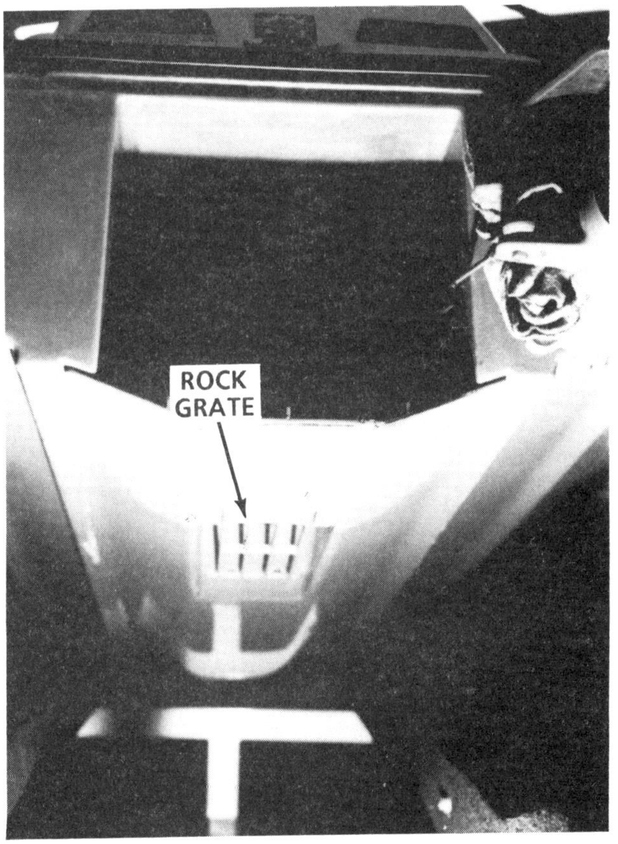

With the watercraft elevated on a set of saw horses and the bow down, the rock grate can be easily removed to measure clearance between the impeller and the wear ring, as explained in the text.

The accompanying illustration shows the craft elevated with the bow down, giving clearance to the rock grate.

If the impeller-to-wear ring clearance is to be determined with the watercraft raised on saw horses, the battery may be left secured in the craft. **HOWEVER**, if the watercraft is to be positioned on its side, first disconnect the electrical leads at both battery terminals, and then remove the battery.

Once the craft is secured in a favorable position, remove the four bolts securing the rock grate to the jet pump. If the seal between the hull and the pump housing is broken, or the foam packing is damaged, reseal the housing and intake area with silicone sealant to prevent impeller cavitation.

Insert a feeler gauge in the center between each impeller blade and the wear ring. Determine the clearance.

If the clearance is less than 0.040" (1.00mm) for each blade, the condition is considered satisfactory. No action is necessary.

If the clearance is more than the service limit, inspect the condition of the wear

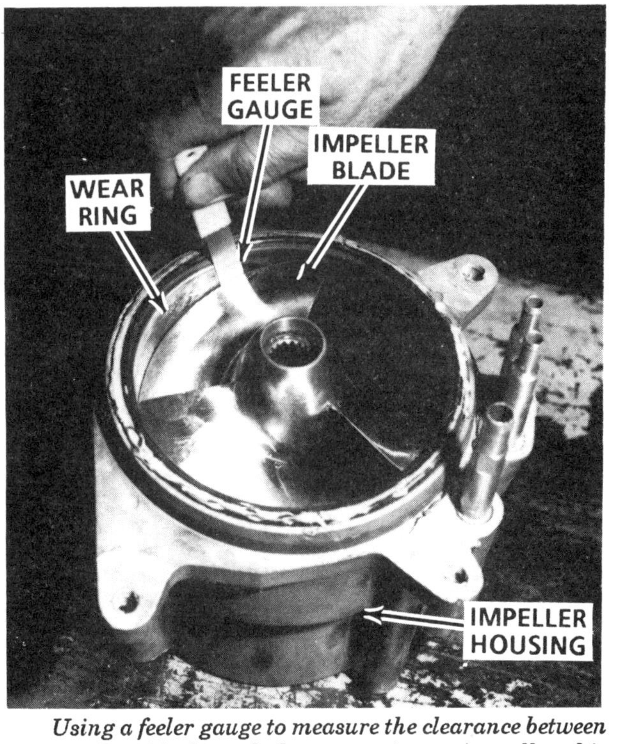

Using a feeler gauge to measure the clearance between an impeller blade and the wear ring. Actually, this measurement can be made after just the rock grate has been removed. Here, the pump and impeller housing have been removed for photographic clarity. Note: The hose fittings shown are plastic on late model units.

ring. If the wear ring has deep scratches, replace the wear ring. If the wear ring appears in good condition, the problem must be with the impeller. Visually inspect the impeller for nicks, scratches, pitting or a "mushroomed" edge.

If the clearance is not kept to a minimum, the jet pump will lose efficiency by allowing excess water to pass between the blades and the housing. This water has no thrust and will actually slow the stream of water under pressure from the impeller.

The jet pump must be removed from the craft to replace the wear ring or to service the impeller.

Install the rock grate and tighten the attaching bolts securely.

10-6 JET PUMP SERVICE

The procedures in this section include all the necessary steps to disassemble, inspect, and assemble virtually all parts of the jet pump. **THEREFORE**, if a part is found to be in satisfactory condition for further service, simply skip the steps involved for such part and continue with the necessary work to restore the pump to first class condition.

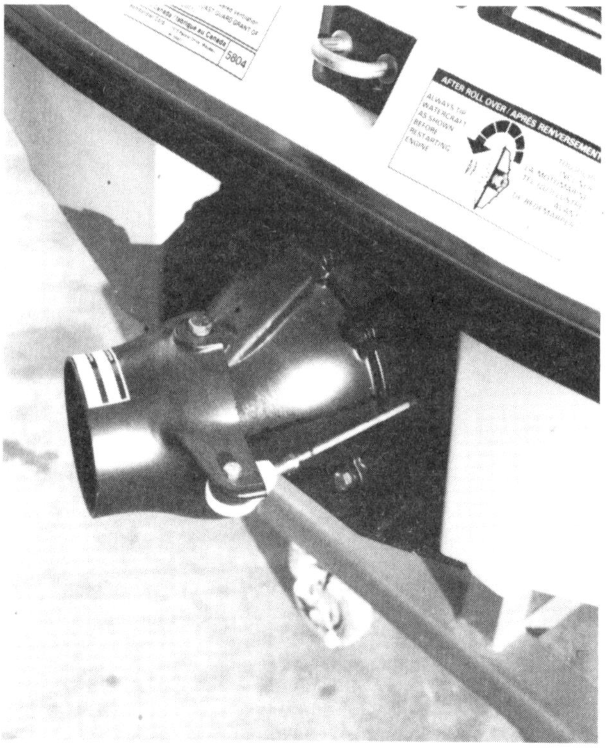

Proper service -- disassembly and assembly -- of the jet pump is almost impossible without the use of some special tools, "called out" in the text. In rare instances, a homemade tool may be fabricated.

Preliminary Tasks

Disconnect both battery cables from the battery. Remove the battery from the craft. If the craft is to be raised on a set of saw horses or similar blocking or fixture, with the stern higher than the bow, the battery may remain in place.

CAUTION

The battery **MUST** be removed to avoid spilling electrolyte into the bilge through the battery vent holes when the craft is moved over on its port side.

Rotate the fuel control valve to the **OFF** position. If the craft is to be rolled over on its port side, drain and remove the fuel tank and the oil tank.

JET PUMP REMOVAL

1- Loosen the clamps and remove the two bilge hoses and the engine cooling water hose from inside the engine compartment. Work the hoses free of the fittings on the jet pump.

2- Remove the four bolts and washers securing the nozzle to the impeller housing.

3- Separate the nozzle from the impeller housing, with the steering nozzle attached. Remove and discard the two small O-rings from the water passages at the top of the impeller housing. Leave the steering cable attached to the nozzle.

4- Obtain a "bungee" cord. Hook one end of the cord into one of the mounting holes in the nozzle, and then hook the other end of the cord into the upper tie ring. Swing the nozzle out of the way, as shown.

1

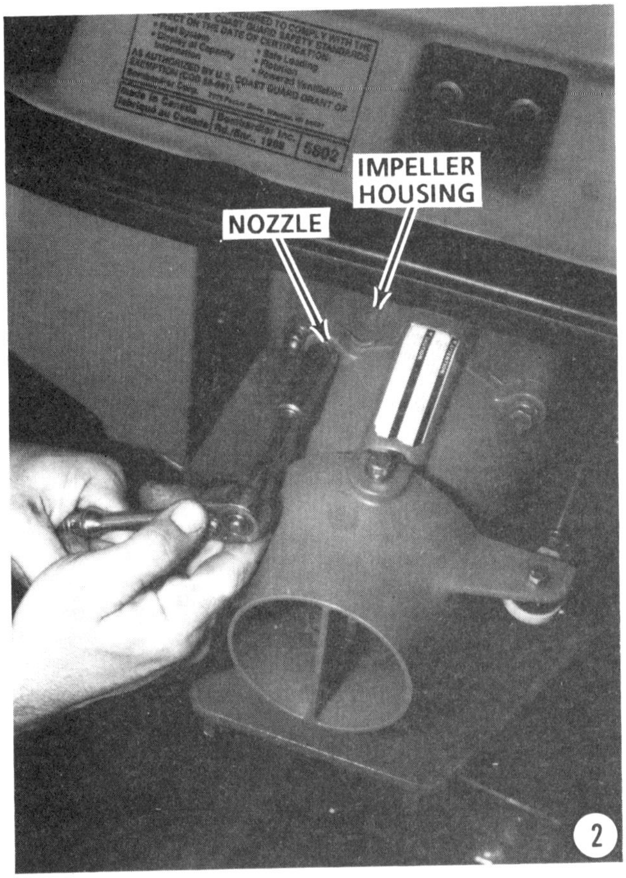

2

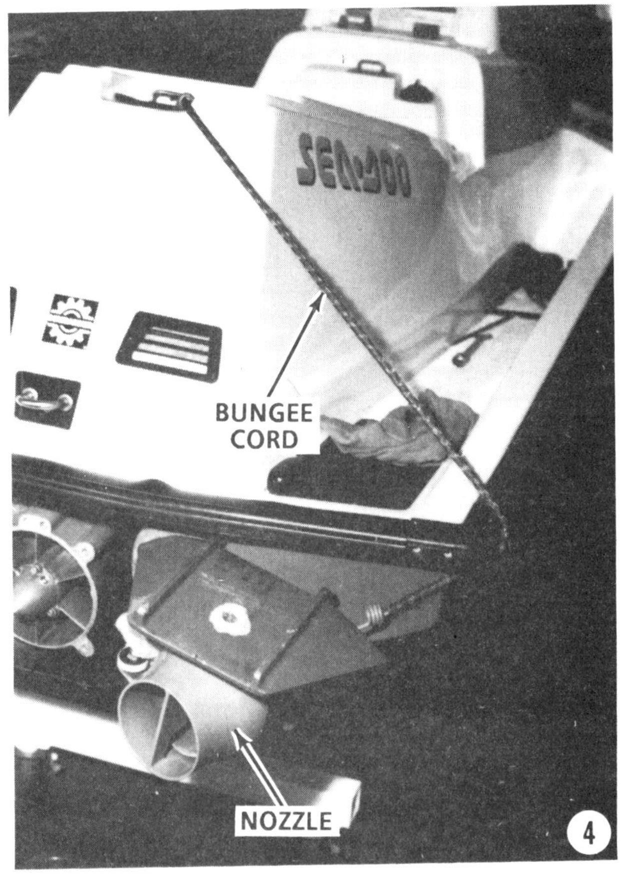

4

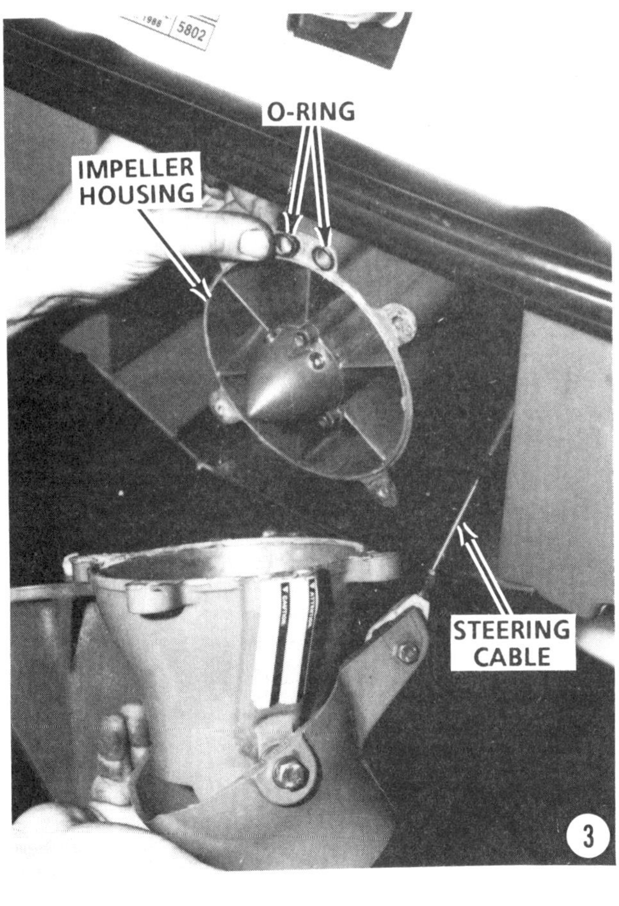

3

5- Position a suitable container under the jet pump to catch the oil from the reservoir in the end cap. Remove the three bolts securing the end cap to the impeller housing.

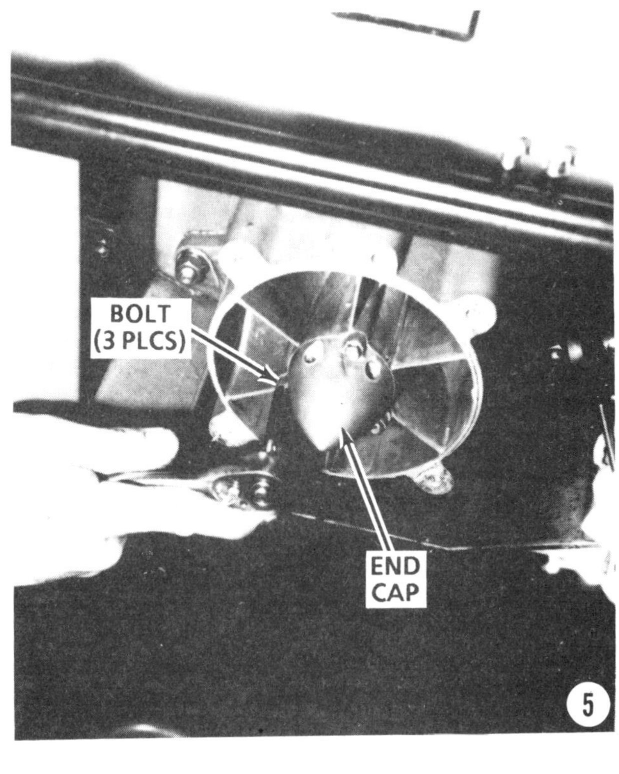

5

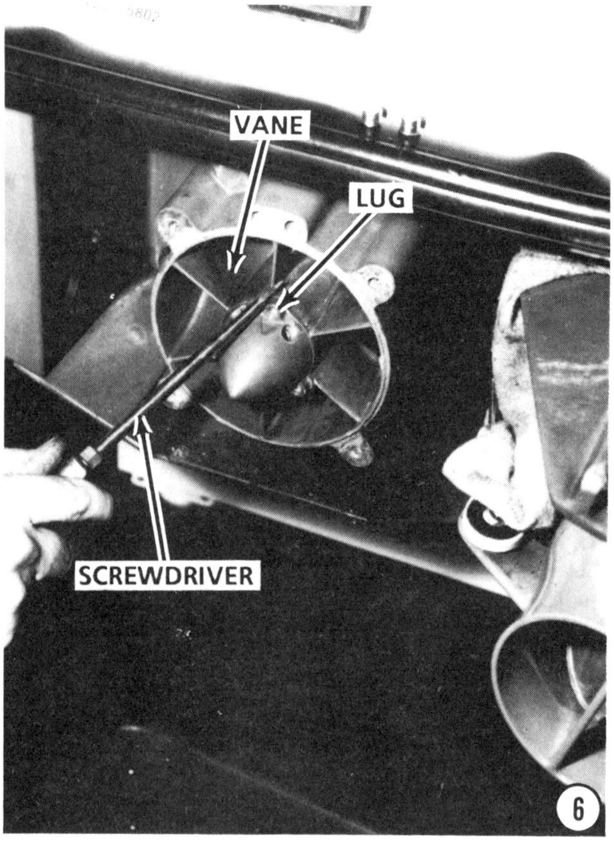

6- Insert a slotted screwdriver between one of the lugs on the end cap and adjacent vane on the impeller housing. Use the the screwdriver to rotate the end cap and break the seal between the cap and the housing. Continue to rotate the end cap until the hub is clear of the vane.

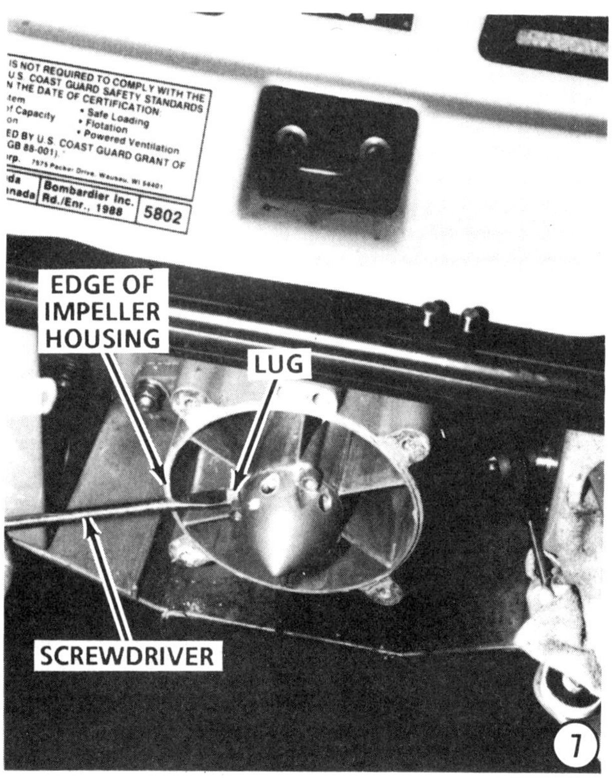

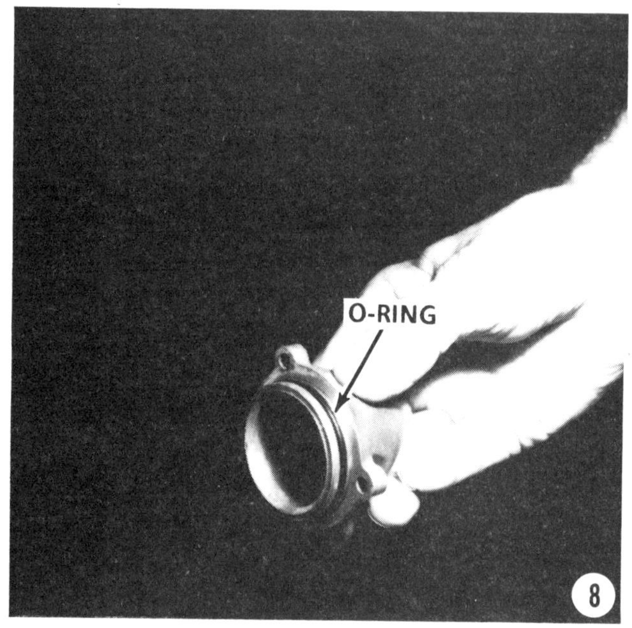

7- With the screwdriver behind the end cap lug, use the edge of the impeller housing as a pivot point and pry the end cap free of impeller housing.

8- Remove the end cap, and then remove and discard the O-ring.

9- Obtain special clamp pliers tool P/N 295 000 069. Squeeze the hooks of the aft flywheel boot large clamp together to release the band. Remove the large clamp from the boot. Obtain special clamp pliers tool P/N 295 000 054. Remove the small metal clamp from the boot by twisting the clamp to release it.

Slide the boot aft on the driveshaft. The driveshaft is now able to move aft and slide free of the boot when the jet pump is removed.

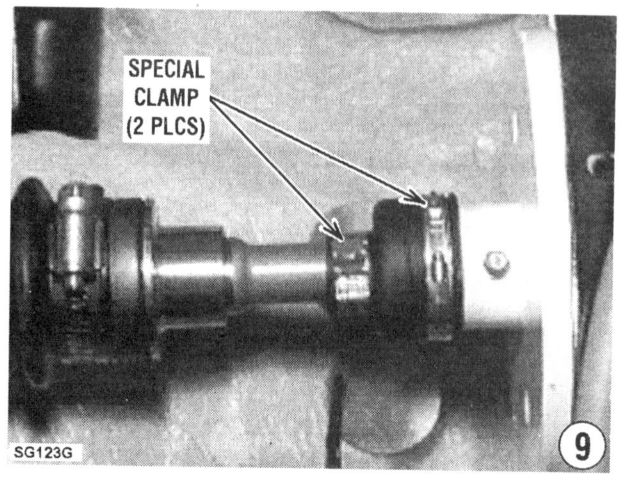

10- Remove the four nuts and washers securing the impeller housing to the mounting studs on the craft hull.

Jet Pump Puller
Tool Fabrication

The impeller housing can be pulled from the hull with the use of two homemade tools. Each is fabricated from a metal bar strap approximately 16" long by one inch wide. For adequate strength, the bar should be 1/2" thick -- (400mm x25mmx 12mm). Drill two 3/8" (10mm) holes through each bar in the position shown in the accompanying illustration. Wrap tape around the ends of the bars to prevent scratching the hull surface.

11- Use the bolts removed in the previous step to secure the two fabricated bars across the impeller housing against the hull. Tighten the bolts alternately and evenly to separate the impeller housing from the hull.

12- Withdraw the impeller housing from the hull. The driveshaft may or may not come away with the housing.

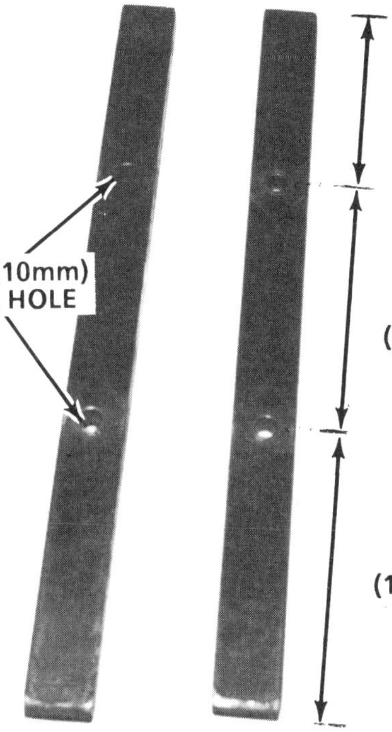

If special manufacturer's tools are not available, a set of homemade tools may be easily fabricated using the dimensions given. The holes drilled are used to pull the jet pump from the hull, as explained in the text.

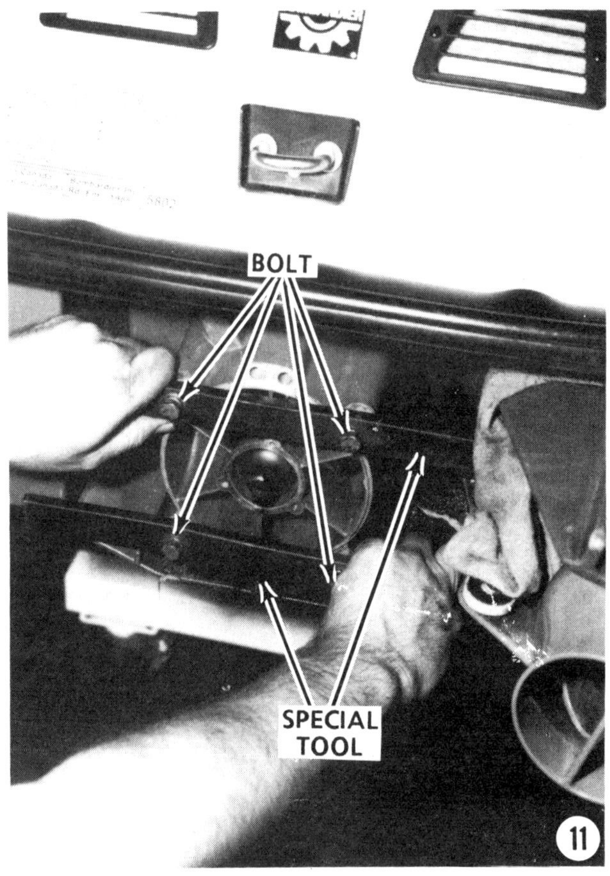

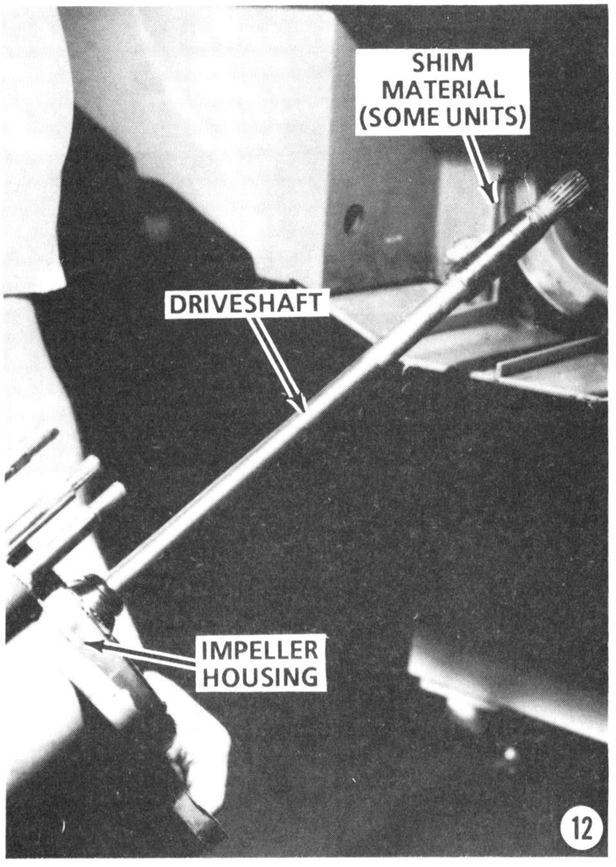

SPECIAL SHIM MATERIAL WORDS

Some jet pump may have a piece of shim material installed at the factory, between the impeller housing and the hull.

Look for and **SAVE** any shim material found at this location.

If the driveshaft came out with the impeller housing, skip the following step and proceed directly to Step 14. If the driveshaft remained in the hull, perform the next step.

13- Pull the driveshaft from the seal carrier.

SPECIAL DRIVESHAFT WORDS

Do not attempt to pull the driveshaft free of the impeller housing without first prying the boot free of the housing. Such action will surely damage the boot. If the boot is stuck, apply liquid soap at the sealing surfaces around the driveshaft and impeller housing. Allow the liquid soap to "seep" in for a few minutes, and then try to pull the driveshaft free again.

14- Attempt to pull the driveshaft free of the impeller housing. If the attempt fails, hold the impeller housing by the driveshaft just a few inches above the workbench. Strike the impeller housing lugs alternately and evenly with a soft head mallet to free the shaft.

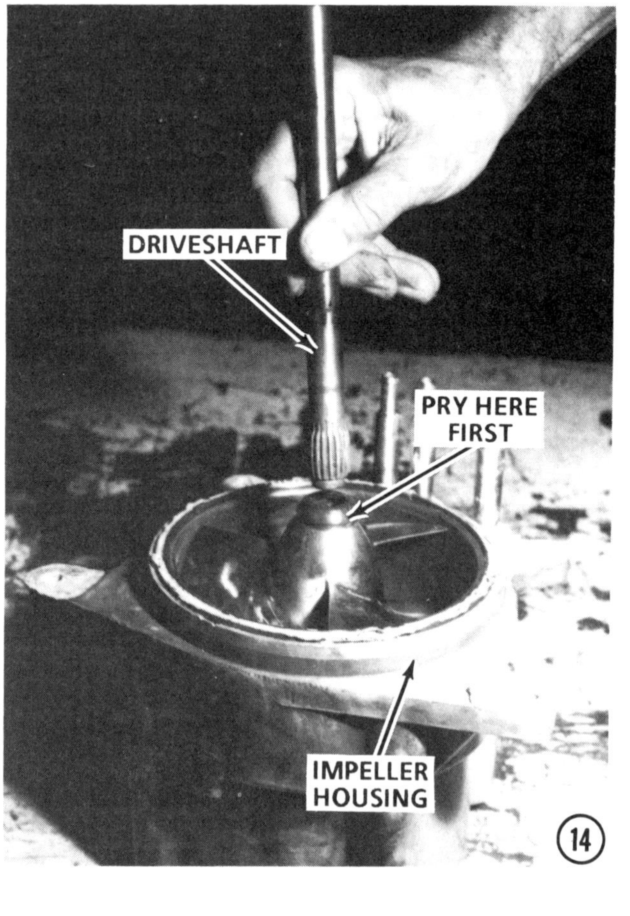

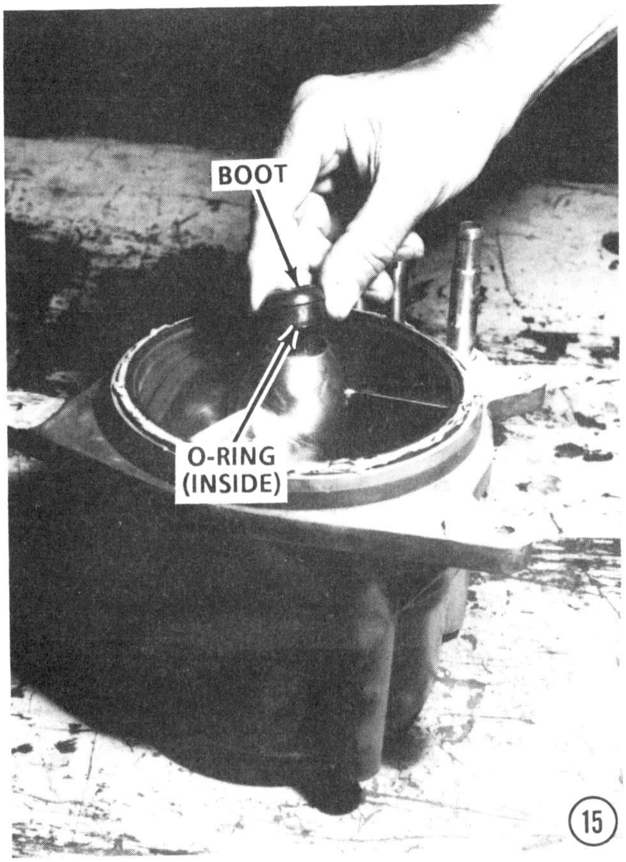

15- Remove the boot from the impeller housing. Remove and discard the **O**-ring inside the boot.

16- Remove the threaded dampener from the forward end of the driveshaft.

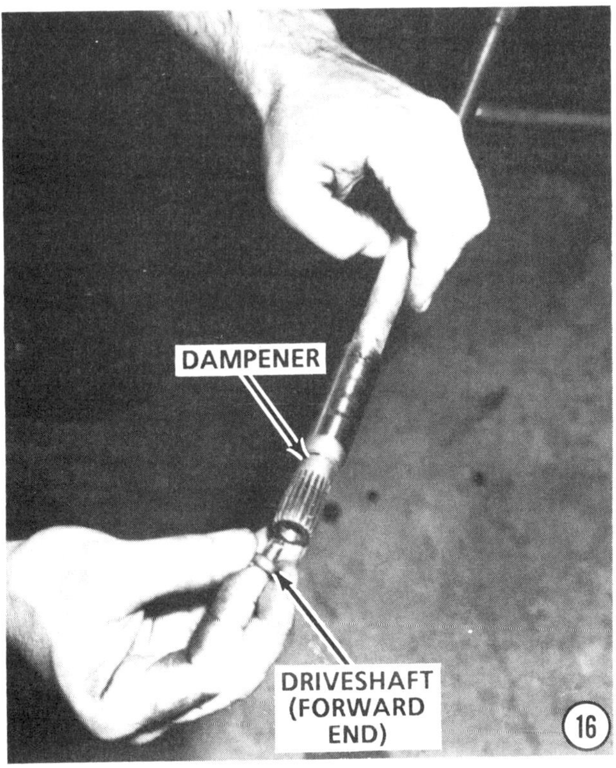

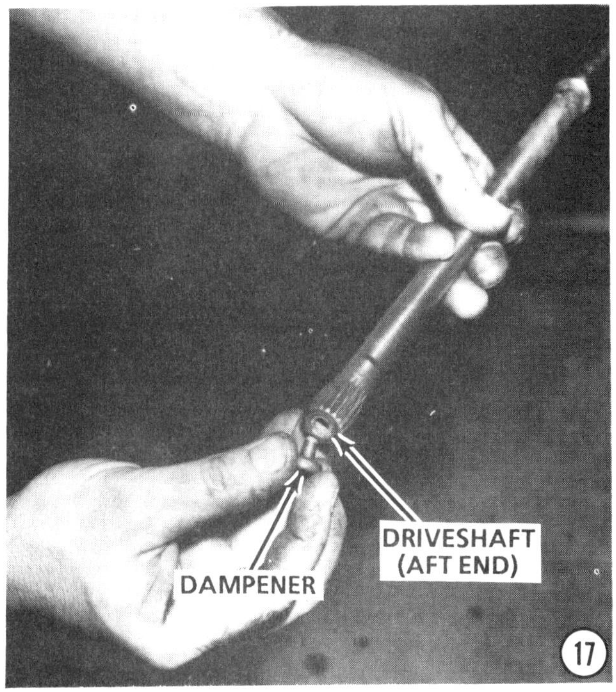

17- Remove another threaded dampener from the aft end of the driveshaft.

Impeller Removal

18- Clamp the impeller shaft in a vise equipped with soft jaws, with the housing upward over the vise and the vise jaws making contact with the two flats on the end of the impeller shaft.

Obtain Impeller Remover P/N 295-000-001. Insert the tool into the impeller splines.

When the impeller is secured in a vise, to disassemble the impeller housing, soft jaws, even a couple pieces of wood will do, ***MUST*** *be used to clamp on the shaft flats.* ***NEVER*** *attempt to clamp the shaft without a cushion between the shaft and the vise.*

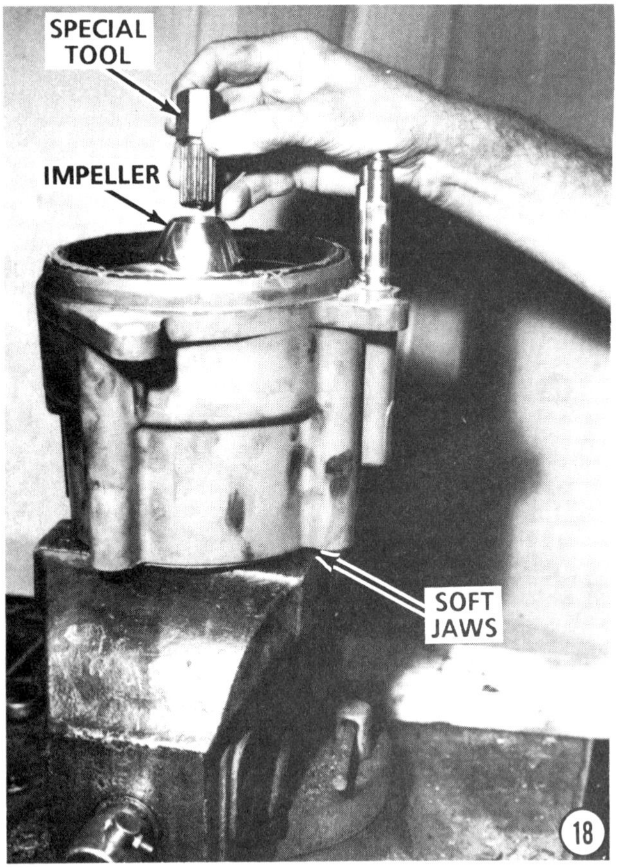

19- Place a suitable socket and breaker bar over the special tool and "break" the impeller free of the shaft.

20- Continue to unthread the impeller from the shaft. Release the impeller shaft from the vise.

21- Remove the impeller shaft and thrust bearing from the housing.

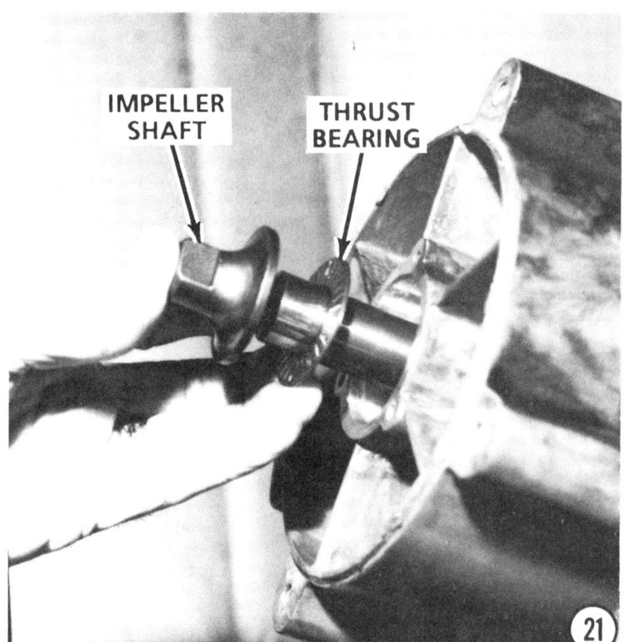

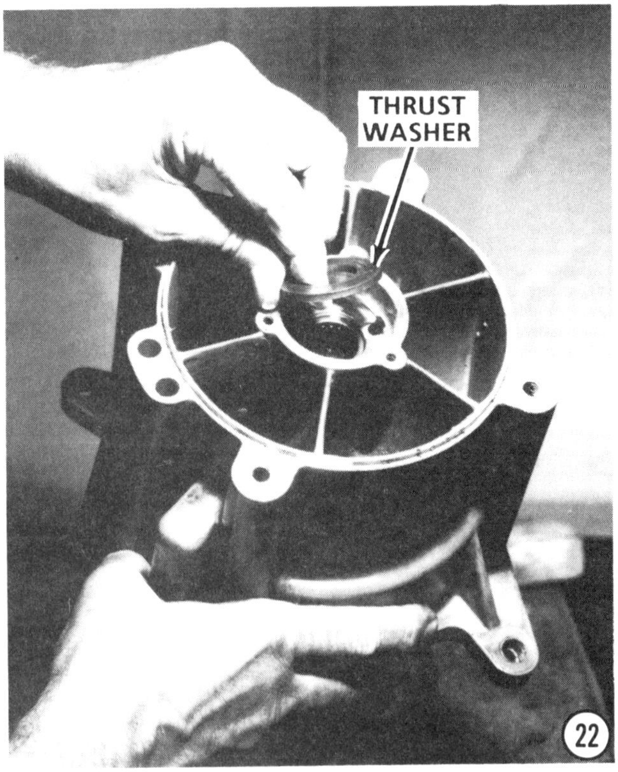

22- Lift out the thrust washer from the forward end of the impeller housing.

23- Pry out the oil seal from the forward end of the impeller housing. If the needle bearing behind the seal is no longer fit for service, obtain a slide hammer with a puller jaw attachment. Hook the end of the jaws behind the caged needle bearing and work the bearing free of the housing.

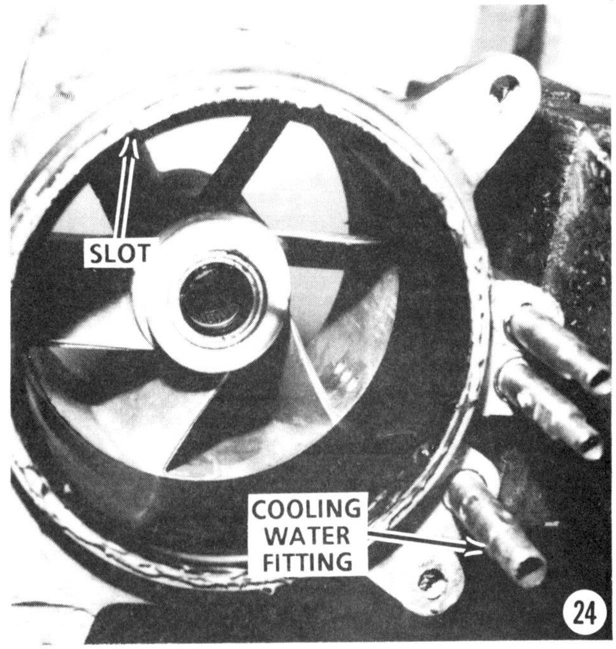

Wear Ring Removal

24- Secure the impeller housing in a vise equipped with soft jaws. Observe the slot in the housing directly across from the cooling water fitting. This slot is machined into the housing to avoid damage to the housing when the wear ring is cut through. Cut the wear ring in line with the slot provided using a small jig saw or low clearance hack saw. After the ring is cut, insert a screwdriver blade under a cut end. Work the ring up and out of the housing.

25- If the needle bearing inside the impeller housing is no longer fit for service, use the same slide hammer and puller jaws and pull the bearing free of the housing.

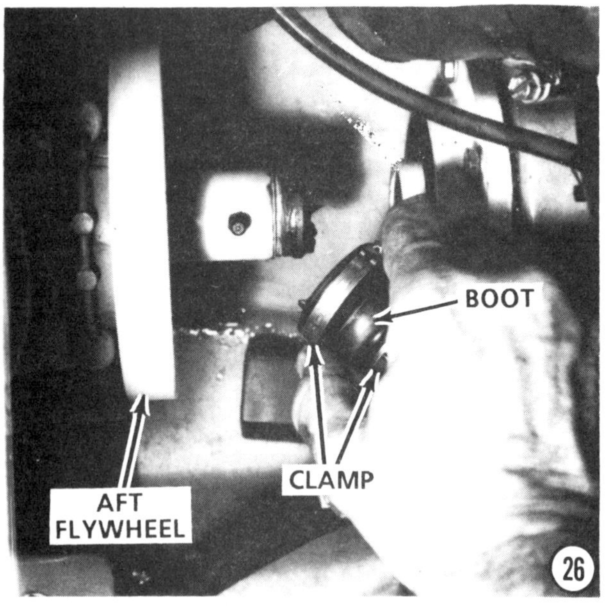

Seal Carrier Service

26- Loosen the two hose clamps on the boot between the seal carrier and the aft flywheel. Remove the protective boot. Note: If the boot is securely attached to the seal carrier and the seal carrier is to be removed for inspection, leave the boot in place.

GOOD CARRIER WORDS

Seldom is it necessary to remove a new seal carrier. Such action would only be required to replace one or both of the seals, as described in the next Step (27).

27- If the protective boot is attached to the seal carrier, grasp the rubber boot firmly and gently pull the boot/seal carrier assembly free of the hull housing. There is no attaching hardware.

28- Pry both oil seals out of the carrier. Use a suitable driver and drive the caged needle bearing free of the carrier.

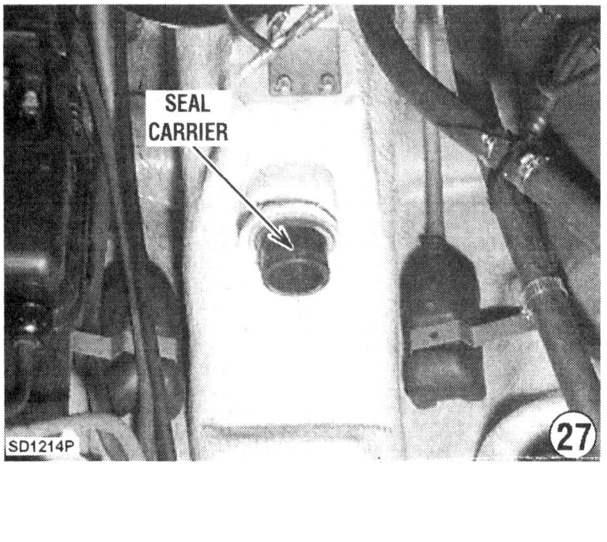

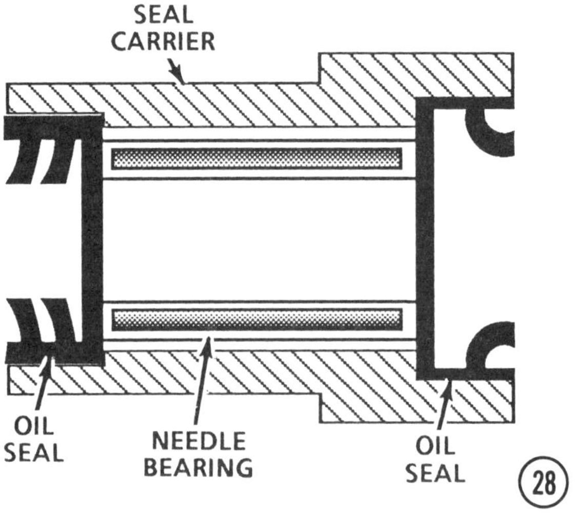

CLEANING AND INSPECTING

Inspect the leading edge of each impeller blade. Ideally, the edge should be rounded to a 1/32" radius, not a knife edge. The cross-section shape of an impeller blade is similar to the shape of an airplane wing.

If the craft is consistently operated in shallow water with a sandy or "pebbly" bottom, the leading edge of the impeller blades will become rounded or will become chipped with pieces missing.

Very little material may be removed from the impeller blades during reconditioning work. Therefore, if the edge is rounded off or corroded, a file can be carefully used to "dress" the edge.

If an impeller blade is chipped more than 1/4", the impeller should be replaced.

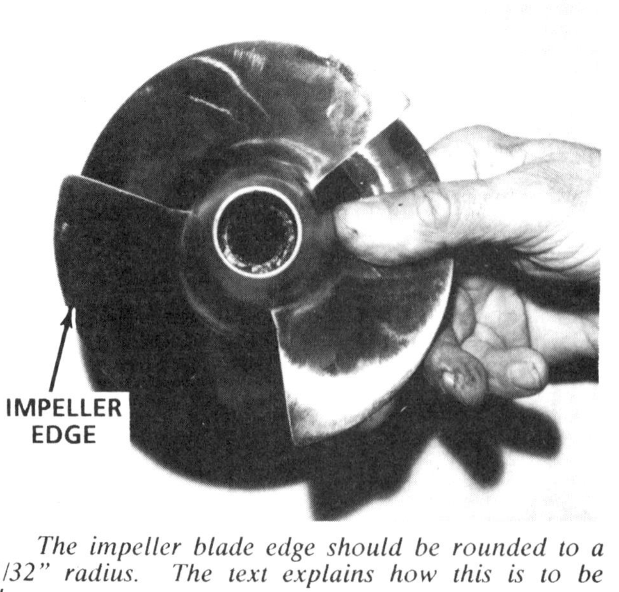

The impeller blade edge should be rounded to a 1/32" radius. The text explains how this is to be done.

Any chip less than 1/4" can be evenly filed away along the length of the blade. Whatever work is done to one blade, must be done to all blades in order to maintain impeller balance. If the impeller becomes excessively out of balance, a definite vibration condition will be quite noticeable, as the craft moves through the water. Such vibration will also cause premature bearing and shaft wear.

Measure the outside diameter of the impeller. If the blades of the impeller are worn, reducing the impeller diameter close to or beyond the service limit, 5.446" (138.33mm), the impeller must be replaced. If the wear ring was not removed, inspect the interior surface. If the surface of the wear ring has deep scratches, the ring should be replaced.

Driveshaft Runout

Set the long drive shaft on two **"V"** blocks -- one at each end. Position a dial indicator gauge over the center portion of the shaft, and measure the runout. The maximum allowable runout is 0.020" (0.5mm). Excessive driveshaft runout causes damage to driveshaft splines, impeller, flywheel, and the seal carrier.

Check to be sure the rock grate fins are not bent or missing. Such a condition might allow a large rock to pass through and cause serious damage to the impeller.

Inspect the vanes in the impeller housing. File the blades to a sharp edge on the forward end. Be sure to keep all the edges the same distance from the front of the housing.

Use a scraper to remove all traces of silicone sealer from both surfaces of the

In order to remove the wear ring, the ring must be cut through using a small jig saw or a low clearance hacksaw. Notice the deep scratches in the ring surface -- requiring the ring to be replaced.

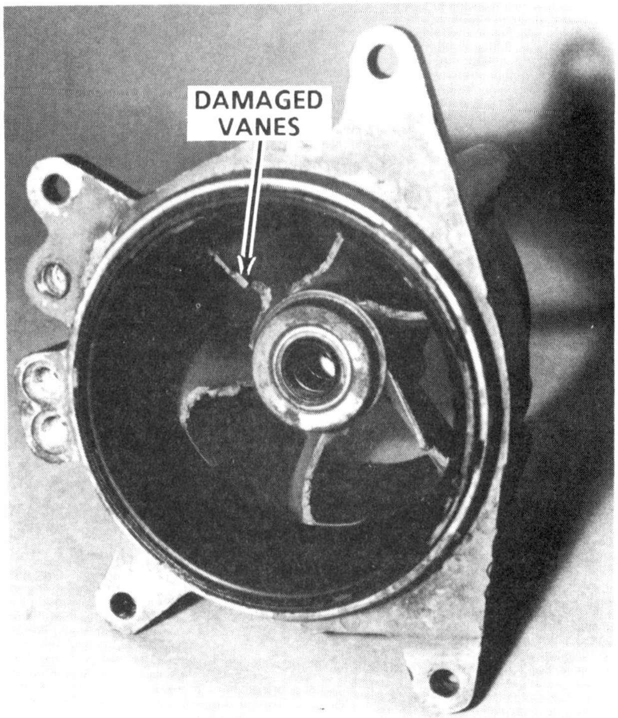

The vanes in this impeller housing were "chewed away" by the impeller, when the impeller worked loose on the shaft.

impeller housing, the hull/ride shoe, and the nozzle. Final traces can be removed with a steel brush mounted in an electric drill. A clean air tight seal is important when the parts are assembled, otherwise air could be drawn in and cause cavitation -- damage to the impeller and loss of power.

Check the water passages in the impeller housing -- three openings between two of the vanes and two on an outside lug. Check the two oil passages in the aft impeller housing hub. Blow low compressed air through each passage to verify the passage is not restricted or blocked.

Inspect the splines on both ends of the driveshaft, the impeller, and the engine flywheel. If the original impeller shaft is to be used again, clean the threads with Loctite Safety Solvent 755.

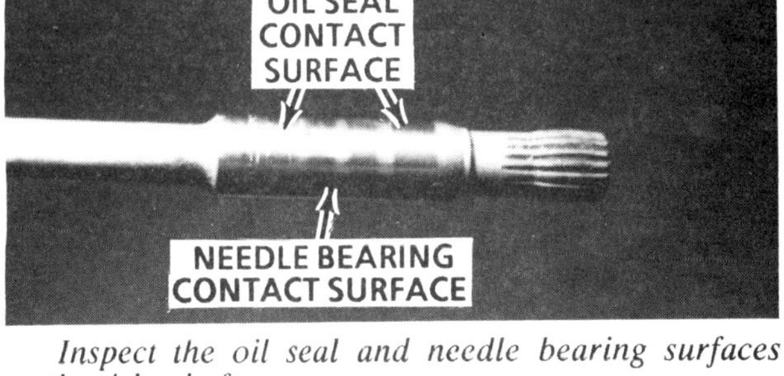

Inspect the oil seal and needle bearing surfaces on the driveshaft.

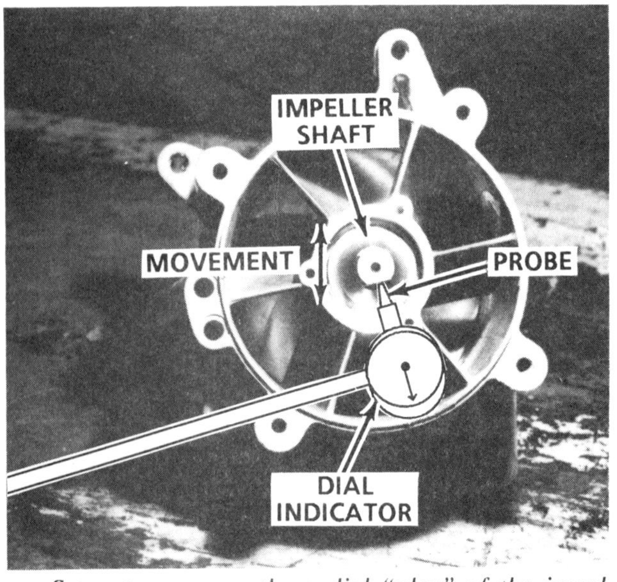

Setup to measure the radial "play" of the impeller shaft. The setup to measure the end "play" is similar, except the dial indicator probe rests against the end of the shaft.

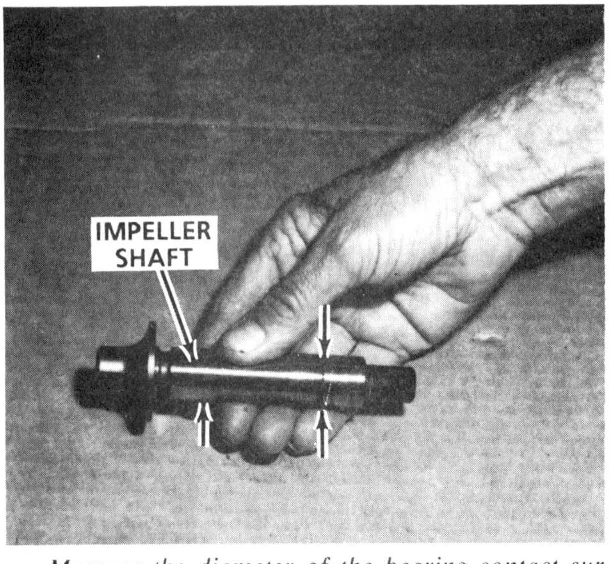

Measure the diameter of the bearing contact surface, as shown.

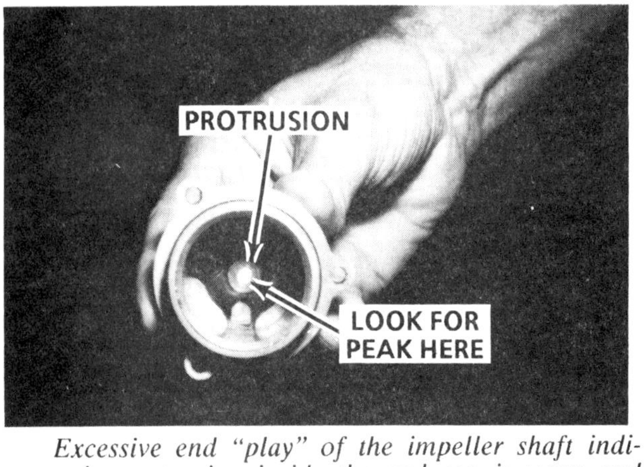

Excessive end "play" of the impeller shaft indicates the protrusion inside the end cap is worn and the cap MUST be replaced.

Impeller Shaft Radial "Play"

A minimum amount of radial "play" in the impeller shaft is necessary to ensure long life for the jet pump. To check for radial "play", insert the impeller shaft into the impeller housing, without the impeller installed, but with both needle bearings in place.

Clamp the impeller housing lug in a vise equipped with soft jaws to prevent housing movement. Make a dial indicator gauge setup with the gauge probe making contact with one of the flats on the end of the impeller shaft. Now, move the shaft up and down and make a note of the maximum deflection indicated by the indicator gauge.

If the deflection is more than 0.002" (0.05mm), the needle bearings may be worn or the impeller shaft may be worn. Use a micrometer and measure the diameter of the impeller shaft at the needle bearing surface. If the diameter is less than 0.8753" (22.23mm), the shaft is worn and **MUST** be replaced. A diameter greater than the figure specified, means the needle bearing is worn and **MUST** be replaced.

GOOD SHOP WORDS

Good shop practice dictates the shaft, the two needle bearings, the thrust washer, and the thrust bearing, be replaced, if excessive "play" is detected.

Impeller Shaft End "Play"

To check the end "play" of the impeller shaft, use the same setup as for checking the radial "play", except this time, the impeller is installed on the shaft and the end cap is installed on the housing.

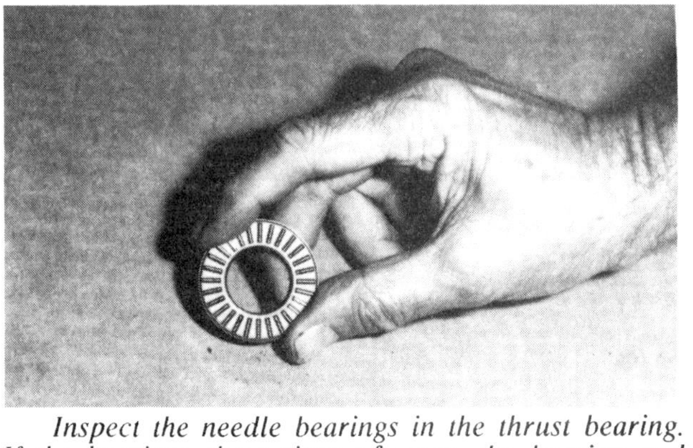

Inspect the needle bearings in the thrust bearing. If the bearings show signs of wear, the bearing and thrust washer should be replaced.

Position the probe of the dial indicator on the end of the impeller shaft -- on the same axis as the shaft. Move the shaft in and out of the housing. Make a note of the maximum deflection registered by the dial indicator. If the deflection is greater than 0.21" (0.54mm), the protrusion inside the end cap is worn and the cap **MUST** be replaced.

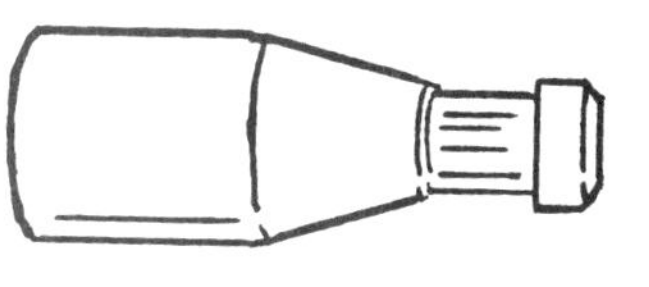

Simple line drawing of impeller shaft thread protector, P/N 295-000-002, used to protect the pump housing oil seals during assembling.

JET PUMP ASSEMBLING

The following procedures cover assembly of all parts to the jet pump. The procedures outline detailed instructions to assemble virtually all parts of the jet pump. As the work progresses, if a particular item was not removed or disassembled -- simply skip the step involved for the part and proceed directly to the work required.

Seal Carrier Assembling

1- Make a suitable setup in an arbor press and install a new needle bearing into the seal carrier. Use a suitable driver which presses on the needle bearing cage -- not the needle bearing. The bearing is correctly installed when each side is flush with shoulders inside the carrier.

Pack both seal lips with Lithium lubricant. Apply a coating of Loctite 515 around the circumference of each seal. Install the single lip seal on the engine side of the carrier with the seal lip facing **OUTWARD**. Install the double lip seal on the jet pump side of the carrier with the lip of the seal facing **OUTWARD.** The seals are correctly installed when the seal surface is flush with the housing.

2- Install the seal carrier by firmly pressing it into place in the hull. There is no attaching hardware.

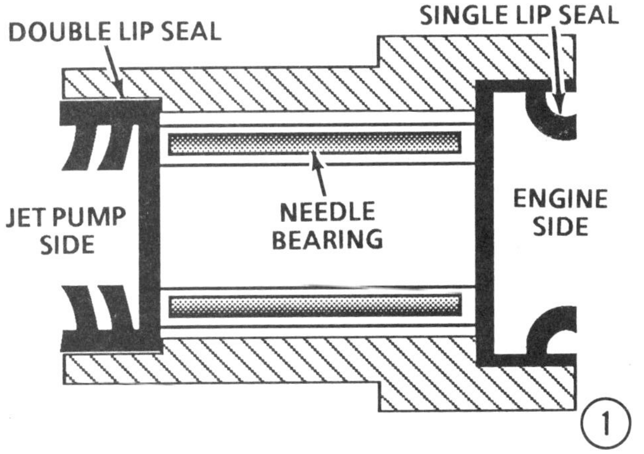

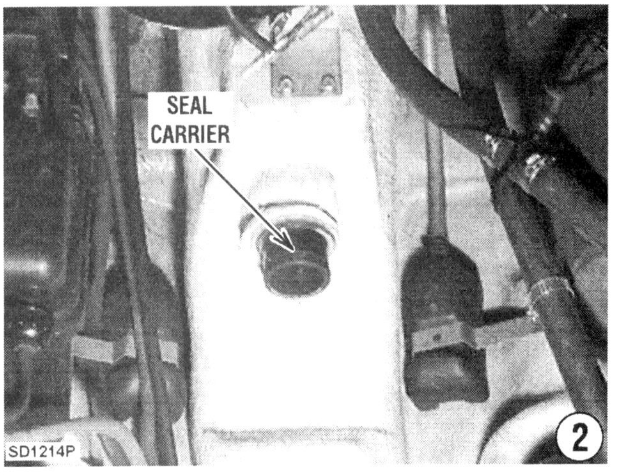

3- Slide a clamp onto each end of the boot, and then install the boot between the aft flywheel and the seal carrier. Tighten the clamps just **SNUG**. These clamps must not be permanently tightened until after the driveshaft has been installed, or the engine will be misaligned.

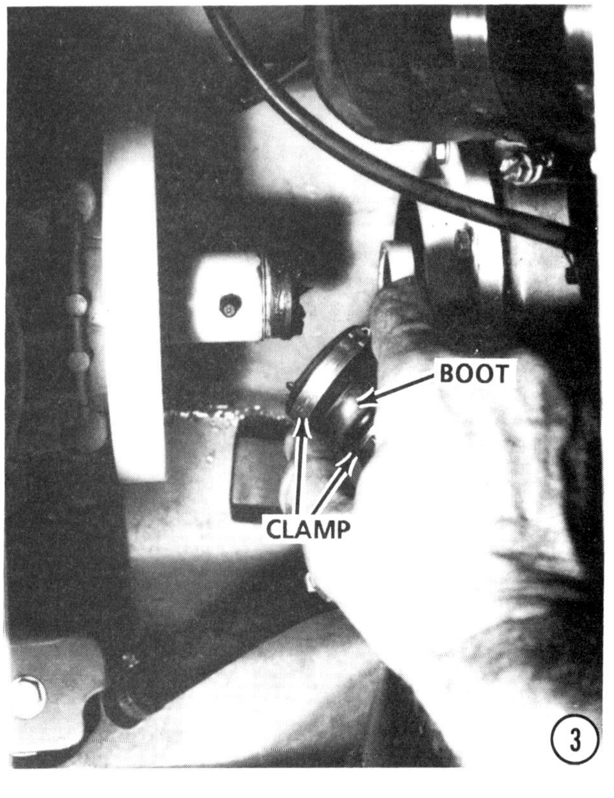

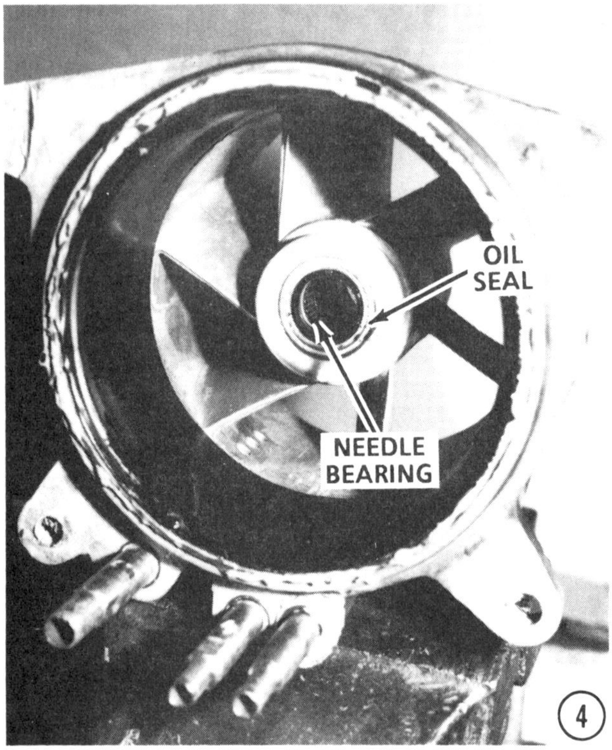

Impeller Housing Bearing and Seal Installation

4- Obtain Bearing Installer P/N 295-000-014. This tool will properly position the bearing within the impeller housing hub. If this tool is not available, a second but more difficult method of installation is presented at the end of this step.

Assuming the special tool is available: Position a new bearing, with the stamped surface facing **OUTWARD,** into the forward housing hub. Insert the tool into the bearing, using a rotating motion. Press the bearing in, using an arbor press, until a shoulder on the tool makes contact with the housing hub. The manufacturer strongly recommends the use of an arbor press -- and further states the bearing should **NEVER** be driven into place with a hammer.

If the special tool is not available: Obtain a suitable driver which contacts the bearing cage only -- not the needle bearing. Using an arbor press, work the bearing into place, with the stamped surface facing **OUTWARD**, until a clearance of 0.060-0.100" (1.5-2.5mm), exists between the stamped face of the bearing and the oil seal seat.

Pack the seal lip with Lithium lubricant. Coat the outside of the seal with Loctite 515. Install the seal using the special tool or suitable driver, with the seal lip facing **OUTWARD**, until the seal seats inside the housing.

5- Position a new needle bearing, with the stamped surface **OUTWARD**, into the aft housing hub. Insert the same tool used in the previous step, into the bearing, using a rotating motion. Using an arbor press, move the bearing, until the proper shoulder on the tool makes contact with housing hub surface.

The manufacturer strongly recommends the use of an arbor press -- and further states the bearing should **NEVER** be driven into place with a hammer.

If the special tool is not available, obtain a suitable driver which contacts only

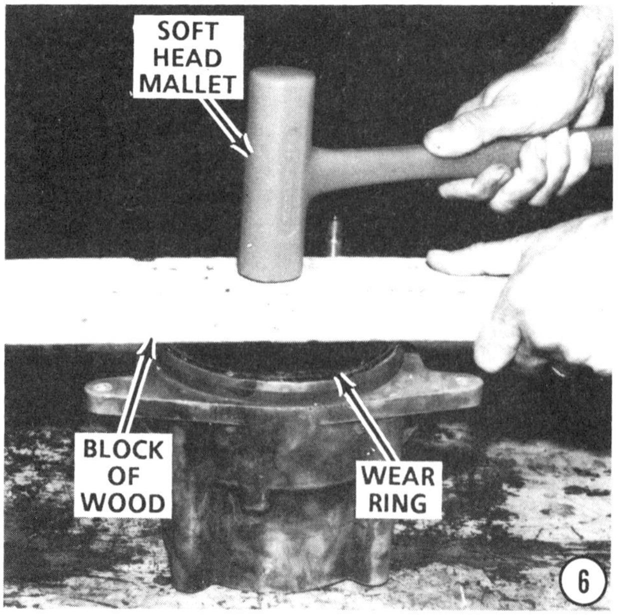

the bearing cage -- **NOT** the needle bearing. Make a suitable setup in the arbor press and press the bearing into the housing hub, with the stamped surface facing **OUTWARD**. The bearing is fully seated when the clearance between the stamped bearing surface and the thrust washer seat is 0.080-0.120" (2-3mm).

The manufacturer strongly recommends the use of an arbor press -- and further states the bearing should **NEVER** be driven into place with a hammer.

Wear Ring Installation

6- Cover the exterior surface of the wear ring with a coating of liquid soap. If the replacement ring has a lip on one face, position this lip to be on the outside of the housing.

If the ring is to be pressed in using an arbor press, obtain a steel plate approximately 7" x 7" x 1/4" thick (180mm x 180mm x 6mm). Fit the ring squarely in place over the housing. Place the plate over the ring and press the wear ring into the housing until it seats in the bottom of the housing.

If an arbor press is not available, use a 2" x 4" x 12" piece of wood. Fit the ring squarely into the housing, and then place the piece of wood over the ring. Now, work the ring into the housing by striking the wood with a hefty hammer, rotating the piece of wood 90°, striking again. Continue rotating the wood and striking

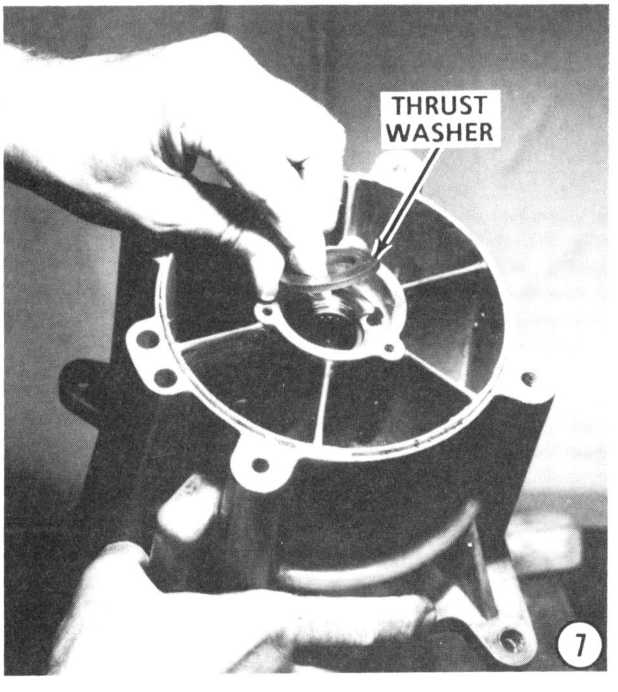

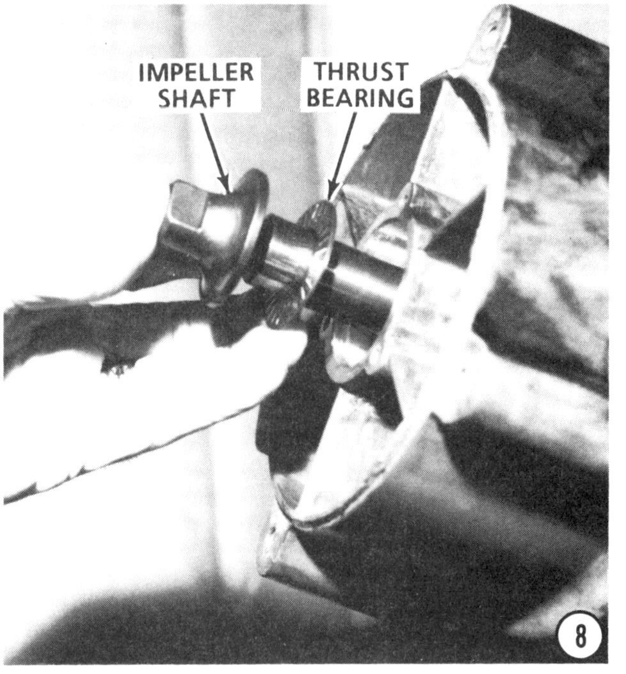

with the hammer, until the wear ring seats in the housing.

Impeller Shaft Installation

7- Coat both surfaces of the thrust washer with two-stroke engine oil, and then insert the washer into the aft impeller housing hub.

8- Coat both surfaces of the thrust bearing with two-stroke engine oil. Obtain impeller shaft thread protector P/N 295 000 002 and install it over the threads of the shaft. Slide the bearing over the impeller shaft, and then insert the shaft and bearing into the aft housing hub.

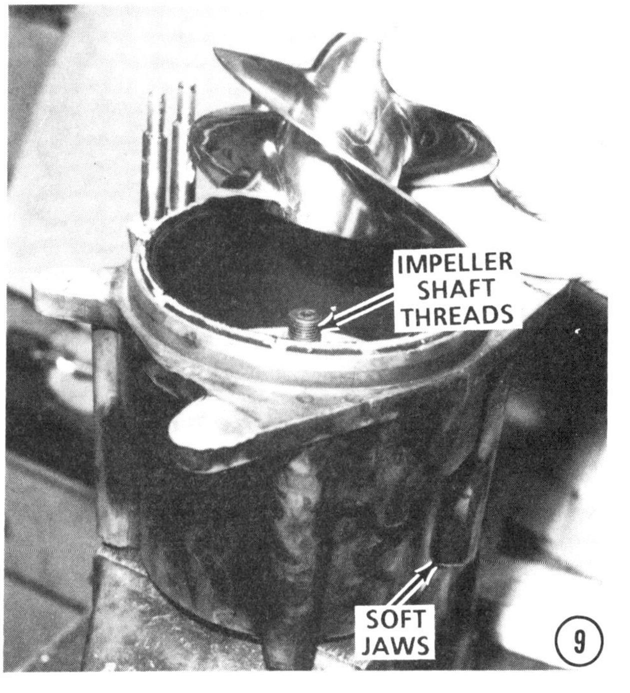

The protector ensures the oil seals are not damaged by the shaft threads during assembly.

Impeller Installation

9- Clamp the end of the impeller shaft with the housing installed on the shaft, in a

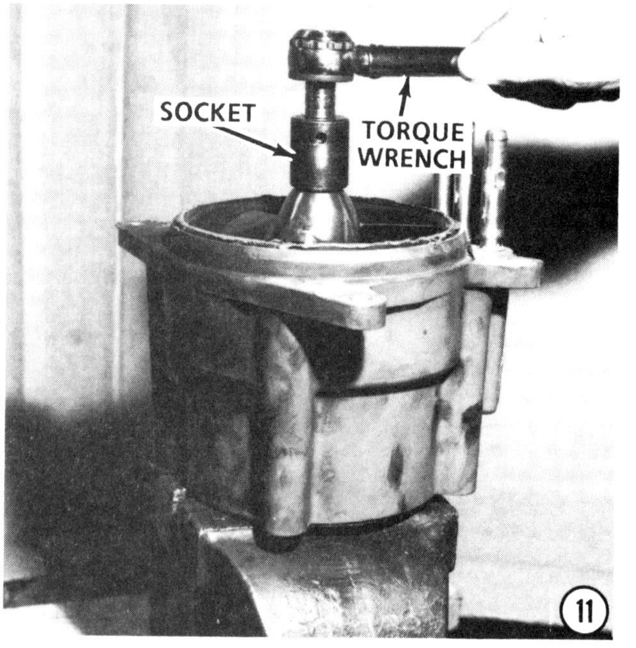

vise equipped with soft jaws. Check to be sure the impeller shaft splines are clean. Apply Loctite Primer N to the shaft threads. Allow the primer to dry for about twenty minutes. Apply Loctite 271 Red over the primer. The primer reduces the curing time of the Loctite 271 to six hours -- the time required for the 271 to reach maximum effectiveness. If the Loctite Primer N is not used, the 271 has a curing time of about 24 hours.

Make a final check to ensure the thrust washer and bearing are correctly seated in the hub, before the impeller is installed. Thread the impeller onto the impeller shaft.

The vanes in this impeller housing were damaged when the impeller worked loose on the shaft and moved forward.

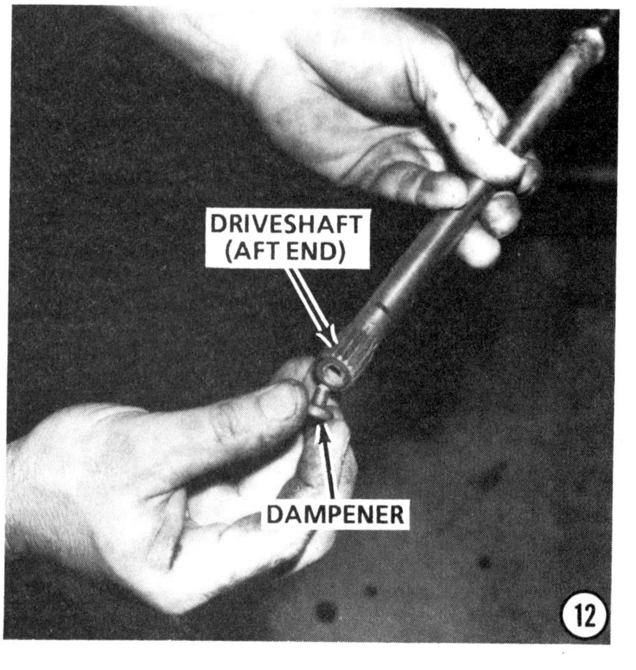

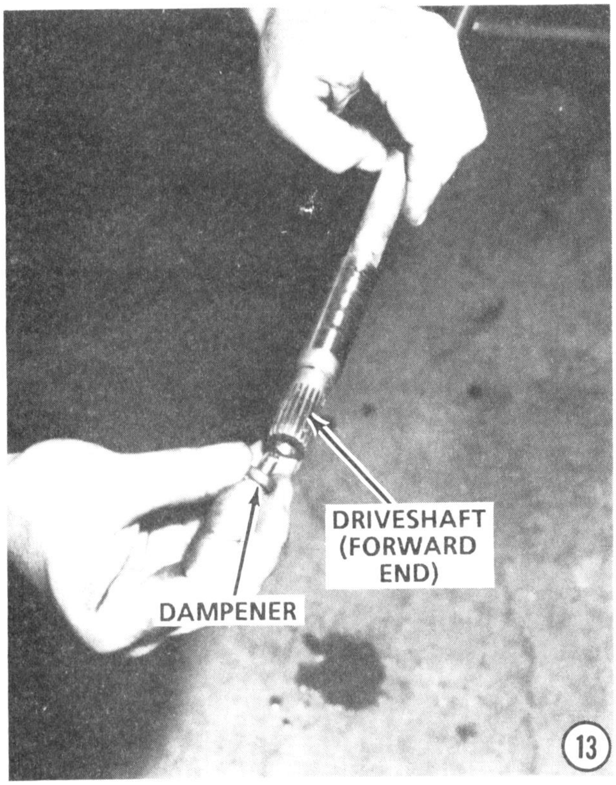

10- Slide Impeller Remover, P/N 295-000-001, into the impeller, with the splines of the tool and the splines of the impeller indexed.

11- Use the correct size socket over the tool, along with a torque wrench, and tighten the impeller to a torque value of 52 ft lb (70Nm).

Driveshaft Assembling

12- Install and tighten the threaded dampener into the aft end of the driveshaft.

13- Install and tighten another dampener into the forward end of the driveshaft.

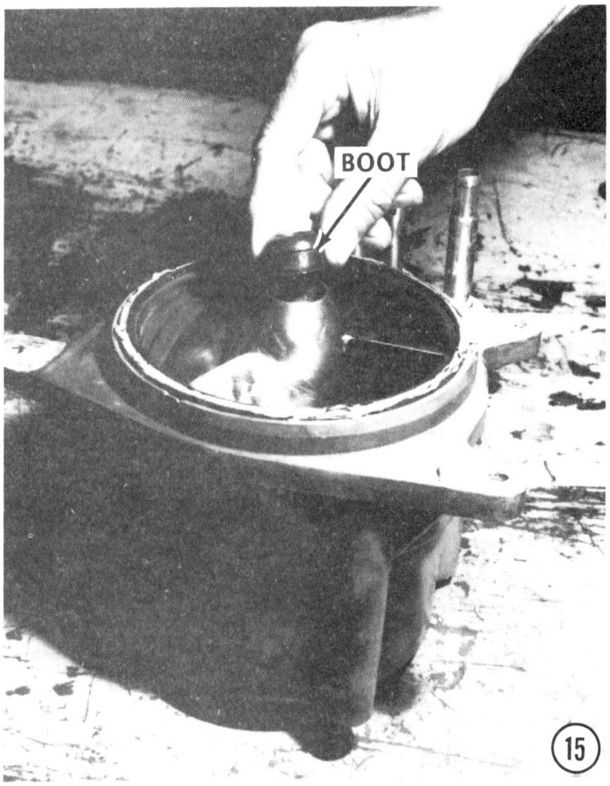

14- Lay down a bead of silicone sealer around the perimeter of the impeller housing, looping around the three water passages and all the bolt holes.

15- Dip the driveshaft boot in a container of liquid soap, and then work it into the impeller.

16- Slide a new O-ring onto the splines at the aft end of the driveshaft. Apply a

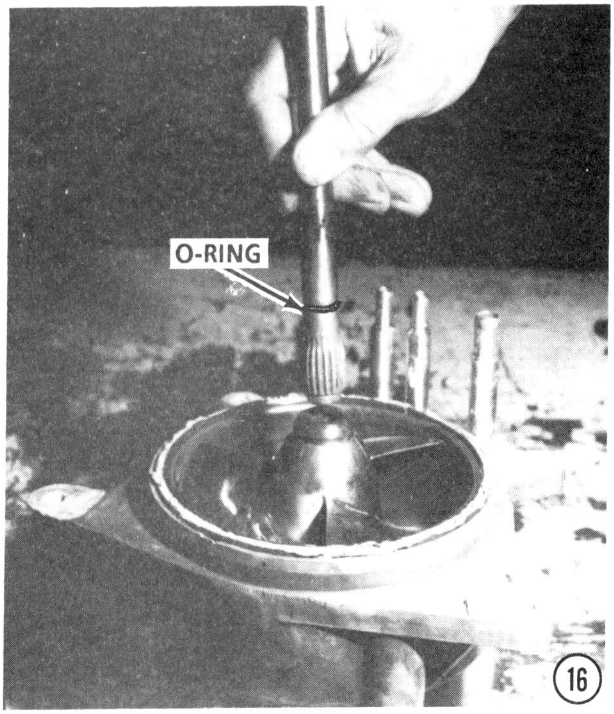

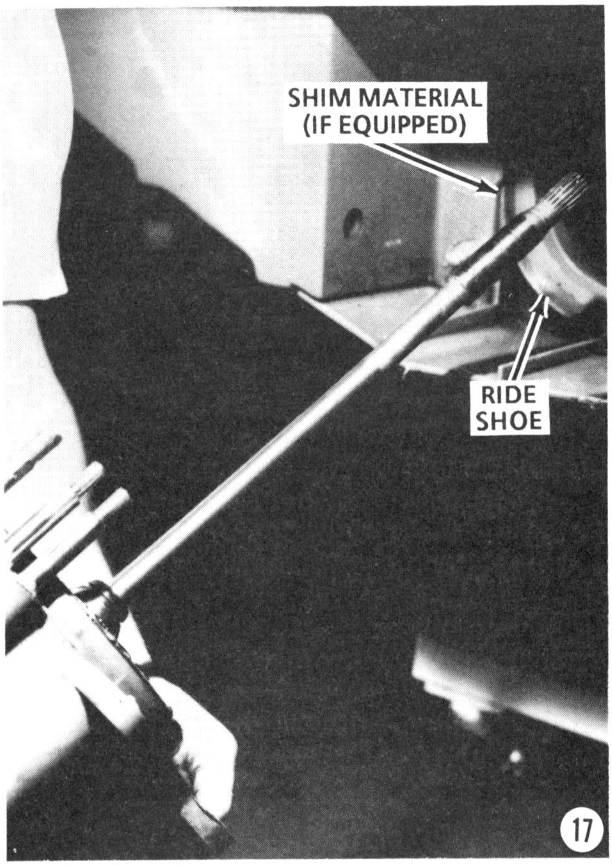

coating of synthetic lubricant onto the O-ring and splines. Work the splines through the boot and into the impeller, with the splines of the shaft indexed with the splines in the impeller. To prevent folding the lip of the seal backward, while the driveshaft is being installed, move the shaft back-and-

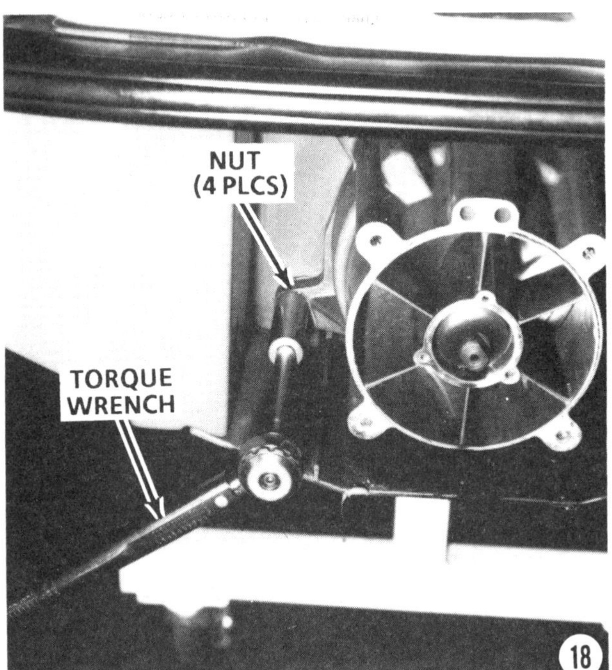

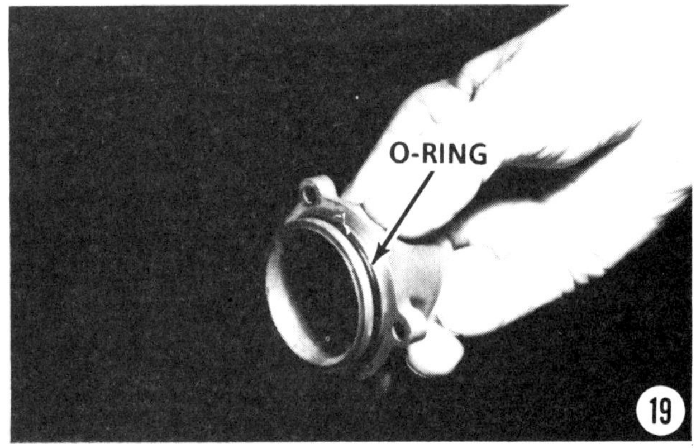

forth as the splines pass through the seal. Push the O-ring down the driveshaft and work it into the boot.

17- If shim material was found between the hull and the impeller housing during removal, install the same amount of shim material against the hull surface. Some models do not have shim material at this location. Insert the driveshaft through the ride shoe, seal carrier, and into the aft flywheel. Rotate the driveshaft until the splines of the shaft index with the splines in the flywheel. Install the impeller housing against the hull/shim material.

18- Apply a coating of Loctite 242 to the hull studs. Install the washers, lockwashers and nuts. Tighten the nuts alternately and evenly to a torque value of 26 ft lb (35Nm).

19- Coat a new end cap **O**-ring with Loctite PST 567, and then install the ring into the cap groove.

20- Install the end cap over the impeller housing, with the three bolt holes aligned and the plug hole facing **UPWARD**, as shown.

21- Coat the threads of the three attaching bolts with Loctite 242. Install the bolts and tighten them alternately and evenly to a torque value of 62 in lb (7Nm).

22- Remove the filler plug from the end cap reservoir. Fill the reservoir through the filler hole with Sea-Doo synthetic oil, until the oil is close to the top of the hole. Now, allow the oil to seep into the reservoir cavities -- i.e. bearings -- for at least five minutes, and then fill the reservoir again with oil. Approximately 2.2 ounces (65ml) of oil should be used.

*Bombardier Sea-Doo Jet Pump Synthetic Oil is the **ONLY** lubricant recommended by the manufacturer for use in the jet pump.*

LUBRICANT CAUTION

The manufacturer strongly states: Sea-Doo synthetic oil must **NEVER** be mixed with any other type of oil, especially mineral based oil.

Apply Loctite 242 to the plug threads. Install the plugs and tighten them securely.

23- Release the bungee cord holding the nozzle and nozzle housing to one side.

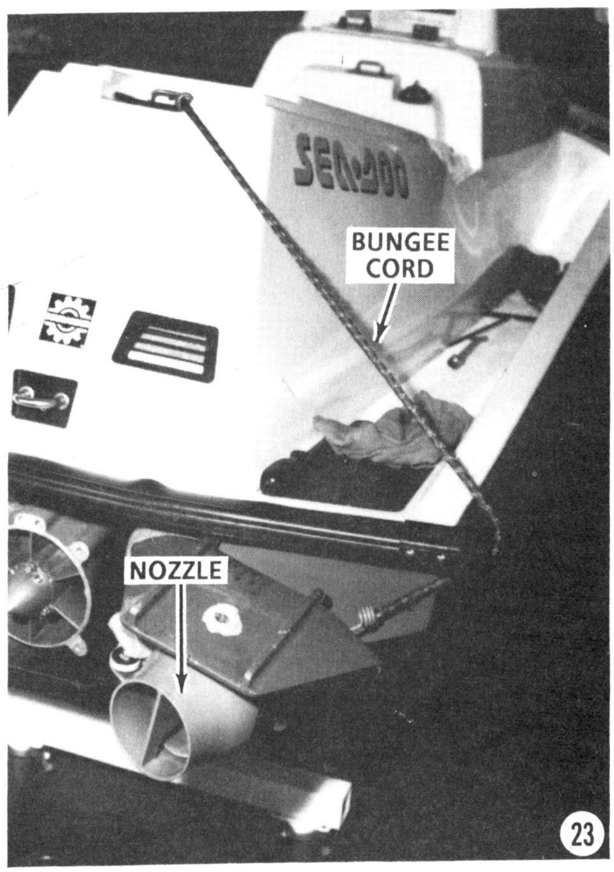

24- Install two new O-rings in the water passages in the top of the impeller housing.

25- Install the nozzle, with the steering nozzle and steering cable attached, onto the impeller housing. Coat the threads of the four attaching bolts with Loctite 242.

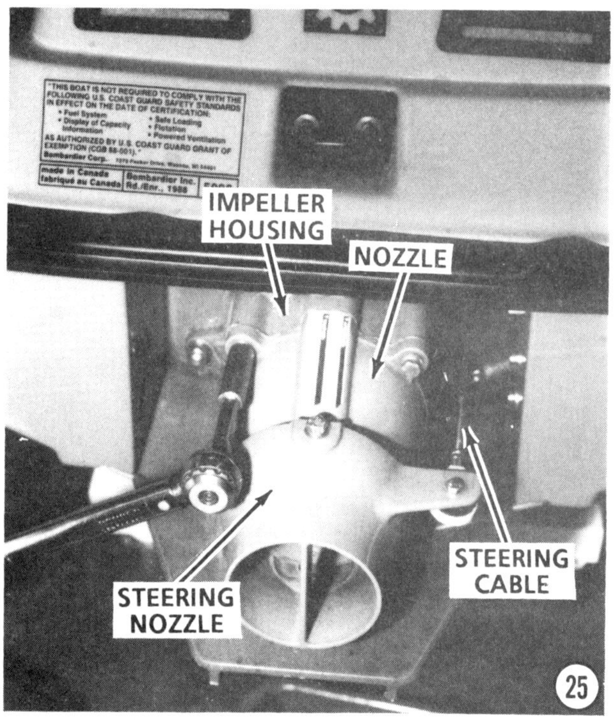

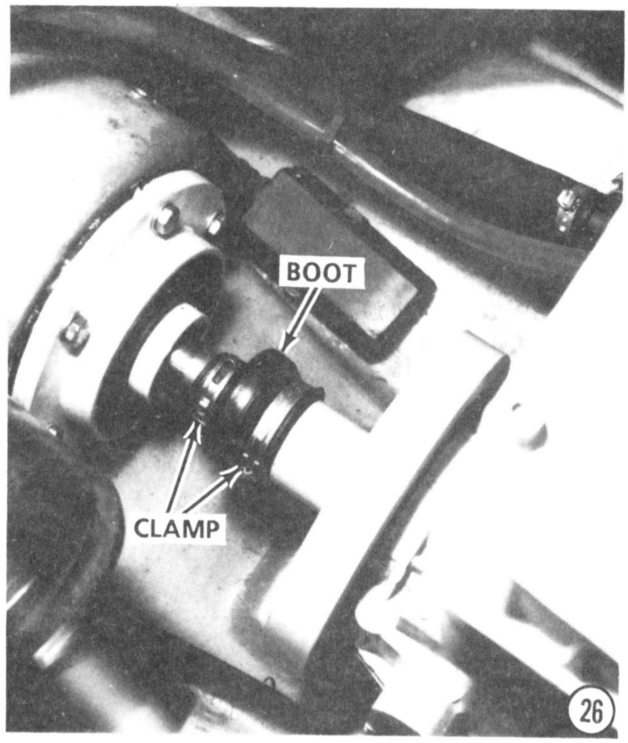

Install and tighten the bolts alternately and evenly to a torque value of 18 ft lb (25Nm). If the steering cable needs adjustment, see Chapter 11 -- **CONTROL ADJUSTMENTS**.

26- Tighten the two clamps on the boot between the aft flywheel and the seal carrier.

27- Slide clamps over the three hoses and then work the hoses onto the jet pump fittings. Tighten the clamps securely.

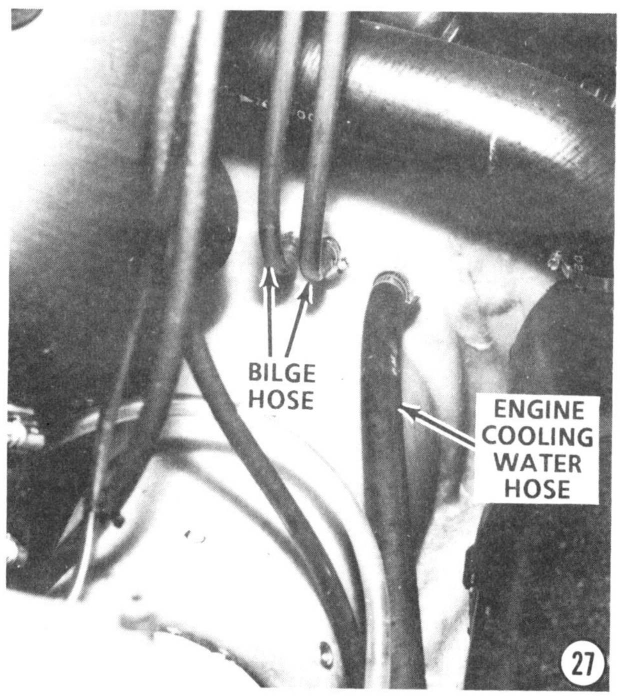

11
CONTROL ADJUSTMENTS

11-1 INTRODUCTION

Control adjustments on Bombardier Sea-Doo personal watercraft are limited to changing the length of the steering cable and the reverse shift cable. Normally, adjustments are not necessary for extended periods of time, unless the craft has experienced an accident or the control cable has been disconnected to service other components.

Detailed procedures to adjust the steering cable and the reverse shift cable are presented in this short chapter. Adjustment of the throttle cable and the choke cable is thoroughly covered -- in detail -- within the pages of Chapter 6 -- **FUEL.**

Inspection of the mechanical portion of the VTS nozzle trim system follows at the end of this chapter for procedures to test the trim switch and motor, refer to the end of Chapter 9 -- **ELECTRICAL.**

The "business" end of a Bombardier Sea-Doo personal watercraft, with the steering cable and trim pushrod linkage clearly visible.

11-2 STEERING CABLE ADJUSTMENT

The steering nozzle axis must be positioned exactly parallel to the axis of the craft to prevent the craft from veering to port or to starboard when the handlebar is positioned for "dead ahead".

Adjustment

1- Measure the distance from the bottom of the handlebar grip to the foot rest, as shown. Position the handlebar with equal vertical distances on both sides -- the "dead ahead" position. Perform steps 2, 3, and 4 **WITHOUT** disturbing the handlebar position.

2- Place a straight edge -- a piece of wood will do nicely -- against the nozzle -- parallel to the transom. Hold the piece of wood in place, as shown, and measure the

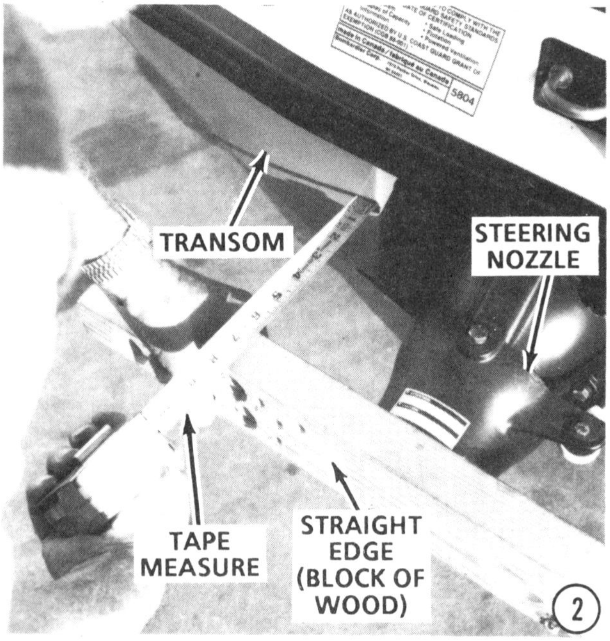

distance, on one side, from the straight edge piece of wood to the transom. Record or remember the dimension.

3- Keep the straight edge in place against the nozzle and measure the distance between the straight edge and the transom on the other side of the nozzle.

4- If the distances measured in Steps 2 and 3 are not identical, adjust the two nuts on the cable at the steering bracket to bring the straight edge an equal distance from the transom on both sides of the nozzle.

5- Turn the handlebar and swing the steering nozzle hard over to port. If the clearance between the outlet nozzle and the steering nozzle is less than 0.005" (0.13mm), adjust the nuts on the steering cable bracket until this minimum clearance is obtained. **DO NOT** exceed this clearance on one side, because the added clearance on one side of the nozzle will take away clearance on the other side. Repeat this procedure with the handlebars turned in the opposite direction -- hard over to starboard -- swinging the nozzle hard over to starboard.

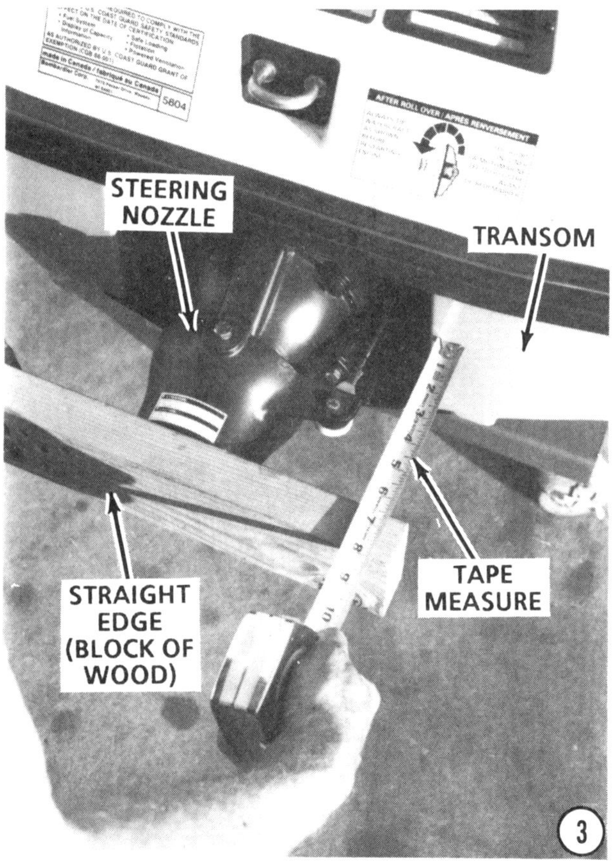

SPECIAL WORDS

A final check **MUST** be made with the craft in a body of water or in a test tank to **ENSURE** proper cable adjustment. Start the engine and allow it to warm to operating temperature.

CAUTION

Water must circulate through the jet pump -- to and from the engine, anytime the engine is operating. Circulating water will

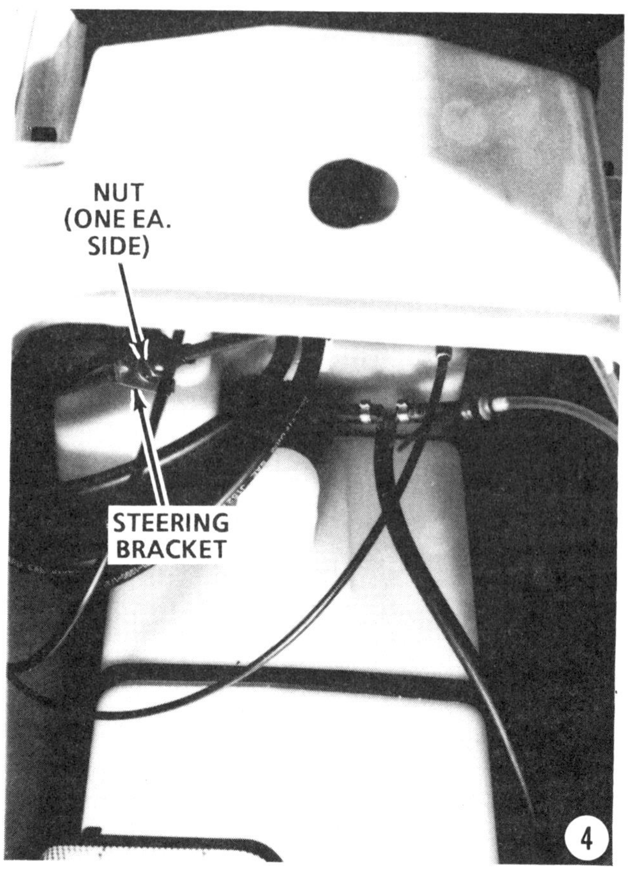

prevent overheating — which could cause damage to moving engine parts and possible engine seizure.

With the craft in the water; with the engine operating; and holding the handlebar in the "dead ahead" position; advance the throttle and observe if the bow tends to veer to port or to starboard. If the bow appears to remain relatively on a true course, the adjustment is satisfactory -- adjustment task completed.

Plastic Handlebar Cover

The plastic handlebar cover fits very snugly over the two foam inserts. If the old cover has become tattered and torn, cut it clear with a pair of scissors. To ease installation of a new cover, first soak it in hot water for, say 1/4 hour or until it seems to be a bit pliable. The cover could also be heated with a hair dryer. If a dryer is used, keep the dryer moving slowly over the entire surface of the cover. Once the cover is ready, work quickly before it has a chance to cool. Close the zipper.

Handlebar Grips

A grip which has worked loose on the handlebar is a nuisance and presents a definite danger to the operator. The old grip, if still in good condition may be reinstalled using the proper adhesion techniques, or a new set may be installed. Do not plan on using the craft following installation of new grips for at least **TWO** full days. This 48-hour period is absolutely necessary for the sealant to cure for maximum effectiveness.

Remove the old grip -- cut it away -- if necessary. Clean all traces of old sealant from the knurled section of the handlebar end. If the old grip is to be used, clean all traces of old sealant from inside the bore of the grip.

Apply Sikaflex primer 260 P/N 293-530-013 to the knurled portion of the handlebar and onto the inside bore of the grip. Allow the primer to cure for **AT LEAST** one full hour. Next, apply Sikaflex sealant 221 over the primer on the bar and the grip. Slide the grip over the end of the handlebar as far as possible. Now, wait at least **TWO** full days before putting the watercraft back into action on the water.

11-3 REVERSE CAPABILITY

The Bombardier Sea-Doo Model GT Series are equipped with reverse capability. As mentioned in Chapter 1 -- **DESCRIPTION AND OPERATION** -- "reverse" is accomplished by a gate installed in such a manner to swing down over the steering nozzle. The reverse gate is

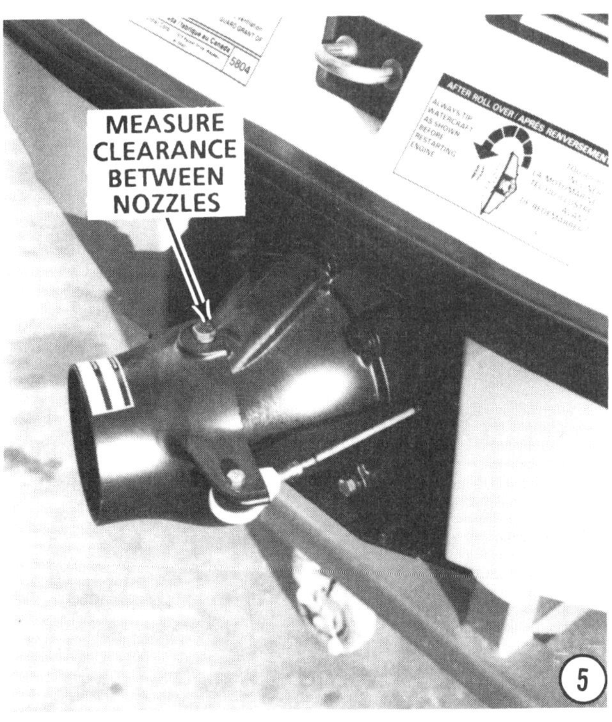

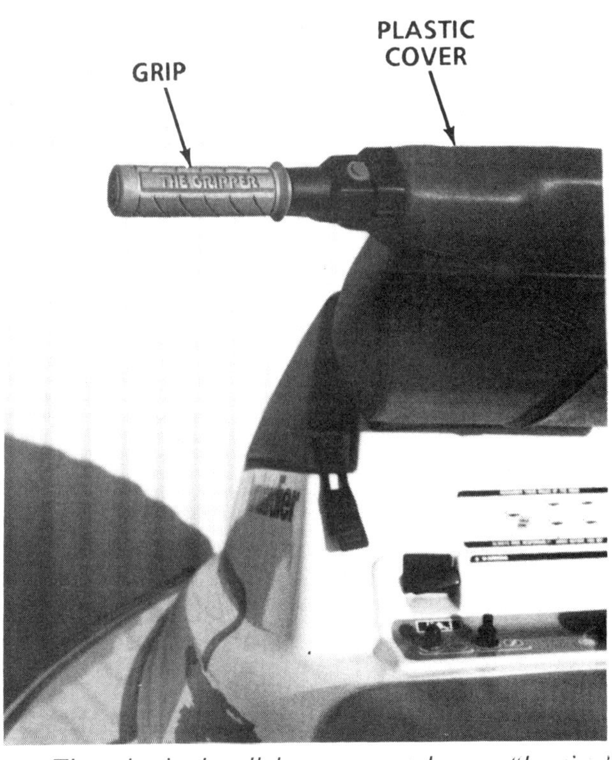

The plastic handlebar cover takes a "beating" from the sun, water, and operator abuse. Specific, detailed, and critical instructions to replace the cover are given in the text.

controlled with a cable. When the gate is in the fully raised position, allowing all water from the nozzle to be discharged sternward -- the craft moves forward.

When the gate is swung down just part way, some of the discharged water moves sternward and the other portion of water is deflected downward and forward under the craft. Positioning the gate in such a manner places the jet pump in **NEUTRAL**, because the water moving aft equals the water moving forward and the craft is static ("dead in the water").

When the gate is moved to the full down position, virtually all discharged water from the nozzle is deflected down and forward -- the craft moves sternward.

The position of the reverse bucket determines the direction of thrust from water expelled from the nozzle. If the bucket is not adjusted properly, forward motion of the craft will be seriously restricted because some of the expelled water will be forced back under the craft --as for going astern.

More importantly, improper adjustment of the reverse bucket can cause the bucket to be unintentionally lowered -- throwing the craft into reverse -- suddenly raising the bow out of the water -- causing possible injury to the operator and/or passengers.

A reverse shift cable is attached indirectly between the shift handle and a ball joint on the portside of the reverse bucket. When the shift handle is pulled back -- toward the operator -- the aft end of the cable is pulled forward -- toward the craft, and the reverse bucket is raised. Depending on the position of the shift handle, the bucket will be partially lowered -- the neutral position, or fully lowered -- the reverse position.

A reverse support plate, mounted directly behind the control panel, supports the cable and the shift handle. A triangular lever pivots at the base of the support bracket and transfers the motion of the shift handle to the shift cable. A locking lever, secured to the shift handle shaft, indents with two rectangular notches cut in the plate to hold the forward and reverse positions of the shift handle. Reverse shift cable adjustment is made at the point where the cable is supported -- at the rear of the plate.

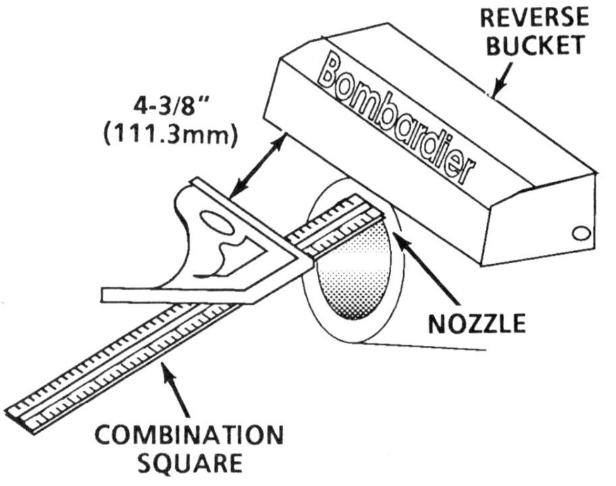

Simple line drawing to indicate placement of the square against the end face of the nozzle when making the NEUTRAL position adjustment, as shown in the drawing below.

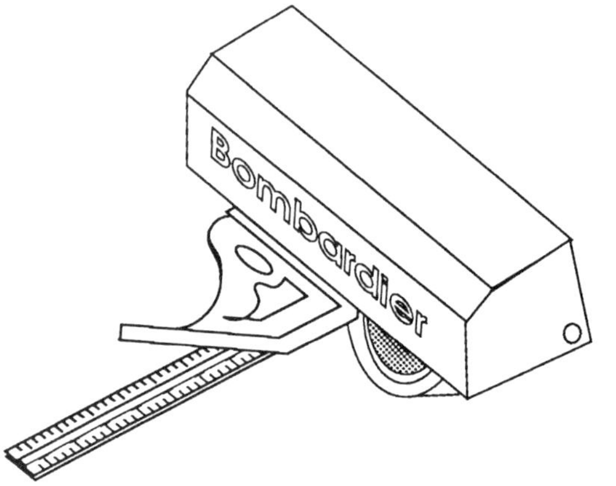

When making the reverse bucket adjustment for the NEUTRAL position, a combination square is used to set the outer lower face edge of the bucket the required distance from the nozzle. The return springs must be unhooked for this adjustment.

Two return springs, one on each side of the reverse bucket, provide tension on the bucket to keep it in the forward -- up -- position.

REVERSE CABLE ADJUSTMENT

Obtain a combination square (set square), and set it to 4-3/8" (111.3mm). Unhook both return springs from the reverse bucket. Set the shift handle in the **NEUTRAL** position. Place the handlebar in the "dead ahead" position, as described in the steering cable adjustment -- the back face of the nozzle parallel with the transom. Hold the square up to the top edge of the nozzle, under the reverse bucket, as shown. The lower edge of the bucket must

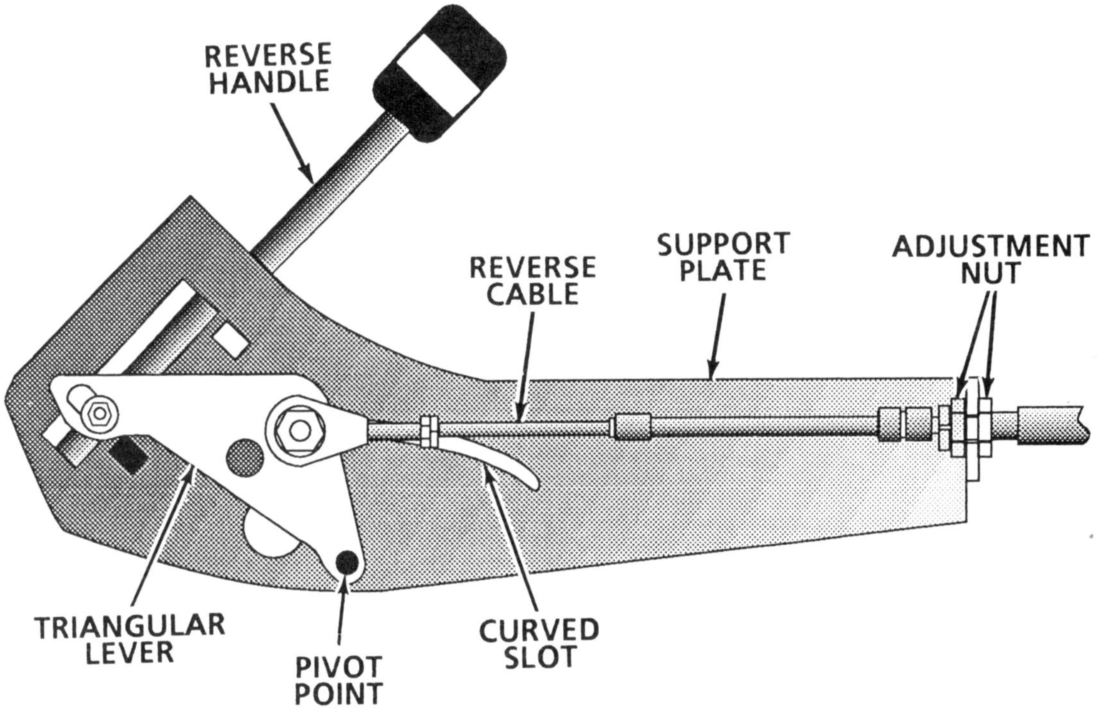

*Line drawing to illustrate the portside of the reverse bracket, with the handle in the **NEUTRAL** position.*

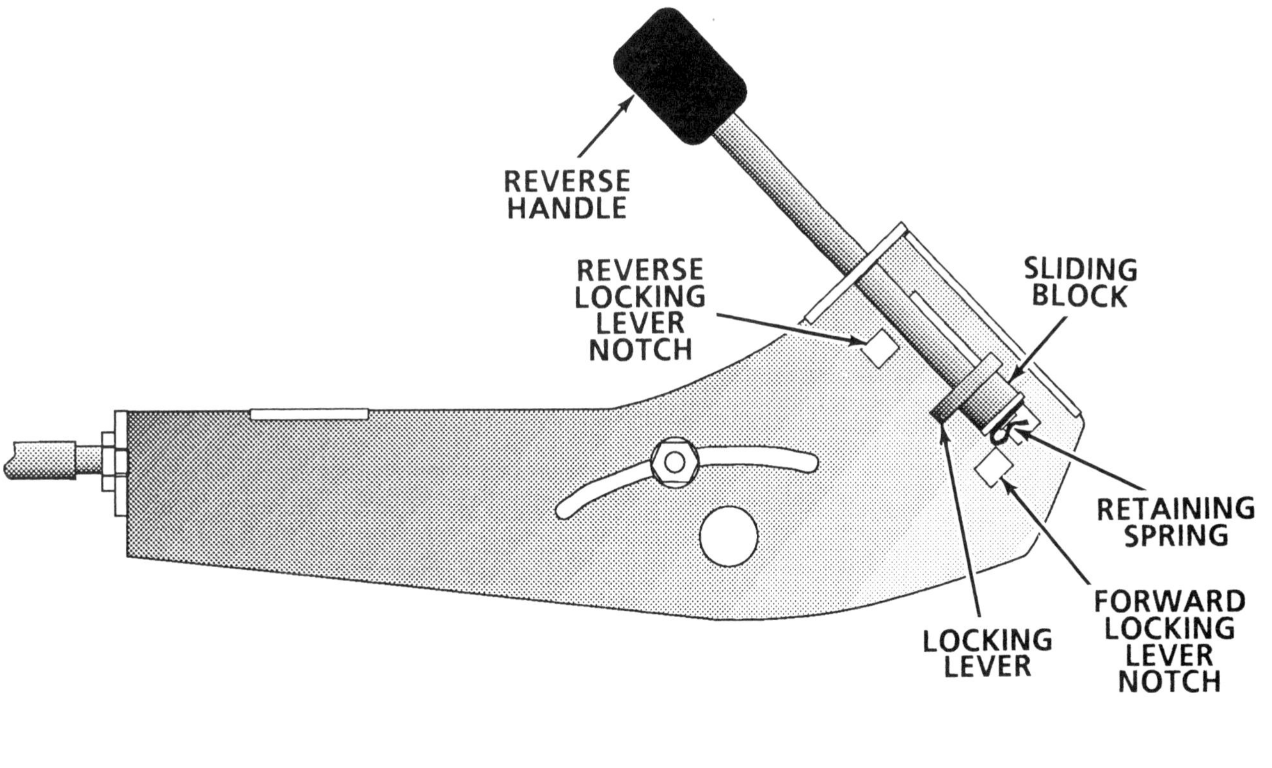

*Line drawing to illustrate the starboard side of the reverse bracket, with the handle in the **NEUTRAL** position.*

be between 4-3/8" to 4-7/8" (111.3-111.7mm) from the upper edge of the nozzle. If the bucket is positioned incorrectly, make the necessary adjustment at the support cable plate adjustment nuts, to lengthen or shorten the shift cable.

SPECIAL WORDS

A final check **MUST** be made with the craft in the water or in a test tank. Start the engine and allow it to warm to operating temperature.

CAUTION

Water must circulate through the jet pump -- to and from the engine, anytime the engine is operating. Circulating water will prevent overheating — which could cause damage to moving engine parts and possible engine seizure.

With the craft in the water; with the engine operating; and the shift handle in the **NEUTRAL** position; the craft **MUST** not move forward or sternward at any engine rpm.

VTS Actuator Mechanism Removal and Inspection

Remove the VTS cover from the trim motor housing. Before disconnecting the trim push rod ball joint connector, observe the position of the ball joint in relation to the sliding shaft, left or right. It should be connected on the same side during assembly. Disconnect the trim push rod ball joint attached to the nozzle. Loosen the hose clamps securing the boot to the actuator sliding shaft. To loosen the large retaining nut located at the aft end of the unit, a long socket extension must be used. Remove the nut and inspect the seal for wear. Loosen and remove the four retaining nuts securing the motor to the housing. Disconnect the electrical lead quick disconnect fittings and pull the motor free of the housing. Pull the worm gear and sliding shaft assembly from the housing. Inspect the threads and splines of the worm gear and sliding shaft. If either is worn, they should be replaced.

VTS Actuator Assembly and Installation

Apply an even coating of synthetic grease to the worm gear. If the worm gear was removed from the sliding shaft assembly for inspection, screw the gear into position under the sliding shaft.

Install the sliding shaft assembly on the trim motor housing, check to be sure the worm gear splines index with the motor.

Install the motor and sliding shaft assembly into the motor housing and check to be sure the motor seats firmly around the housing studs. Install and tighten the retaining nuts to a torque value of 62 in lbs (7Nm). Connect the motor electrical leads.

Install the seal previously removed from the large nut, check to be sure the lips face the hull of the craft. Install the nut and tighten it to a torque value of 10 ft lbs (13Nm) using the long socket extension tool.

Install the protective boot over the sliding shaft and fasten the hose clamps securely.

Connect the push rod ball joint, check to be sure it is installed on the same side of the sliding shaft from which it was removed. Install the trim motor cover in position over the motor housing.

APPENDIX

METRIC CONVERSION CHART

LINEAR

inches	X 25.4	= millimetres (mm)
feet	X 0.3048	= metres (m)
yards	X 0.9144	= metres (m)
miles	X 1.6093	= kilometres (km)
inches	X 2.54	= centimetres (cm)

AREA

inches^2	X 645.16	= millimetres^2 (mm^2)
inches^2	X 6.452	= centimetres^2 (cm^2)
feet^2	X 0.0929	= metres^2 (m^2)
yards^2	X 0.8361	= metres^2 (m^2)
acres	X 0.4047	= hectares (10^4 m^2) (ha)
miles^2	X 2.590	= kilometres^2 (km^2)

VOLUME

inches^3	X 16387	= millimetres^3 (mm^3)
inches^3	X 16.387	= centimetres^3 (cm^3)
inches^3	X 0.01639	= litres (l)
quarts	X 0.94635	= litres (l)
gallons	X 3.7854	= litres (l)
feet^3	X 28.317	= litres (l)
feet^3	X 0.02832	= metres^3 (m^3)
fluid oz	X 29.60	= millilitres (ml)
yards^3	X 0.7646	= metres^3 (m^3)

MASS

ounces (av)	X 28.35	= grams (g)
pounds (av)	X 0.4536	= kilograms (kg)
tons (2000 lb)	X 907.18	= kilograms (kg)
tons (2000 lb)	X 0.90718	= metric tons (t)

FORCE

ounces - f (av)	X 0.278	= newtons (N)
pounds - f (av)	X 4.448	= newtons (N)
kilograms - f	X 9.807	= newtons (N)

ACCELERATION

feet/sec^2	X 0.3048	= metres/sec^2 (m/S^2)
inches/sec^2	X 0.0254	= metres/sec^2 (m/s^2)

ENERGY OR WORK (watt-second - joule - newton-metre)

foot-pounds	X 1.3558	= joules (j)
calories	X 4.187	= joules (j)
Btu	X 1055	= joules (j)
watt-hours	X 3500	= joules (j)
kilowatt - hrs	X 3.600	= megajoules (MJ)

FUEL ECONOMY AND FUEL CONSUMPTION

miles/gal	X 0.42514	= kilometres/litre (km/l)

Note:
235.2/(mi/gal) = litres/100km
235.2/(litres/100 km) = mi/gal

LIGHT

footcandles	X 10.76	= lumens/metre^2 (lm/m^2)

PRESSURE OR STRESS (newton/sq metre - pascal)

inches HG (60° F)	X 3.377	= kilopascals (kPa)
pounds/sq in	X 6.895	= kilopascals (kPa)
inches H_2O (60° F)	X 0.2488	= kilopascals (kPa)
bars	X 100	= kilopascals (kPa)
pounds/sq ft	X 47.88	= pascals (Pa)

POWER

horsepower	X 0.746	= kilowatts (kW)
ft-lbf/min	X 0.0226	= watts (W)

TORQUE

pound-inches	X 0.11299	= newton-metres (N·m)
pound-feet	X 1.3558	= newton-metres (N·m)

VELOCITY

miles/hour	X 1.6093	= kilometres/hour (km/h)
feet/sec	X 0.3048	= metres/sec (m/s)
kilometres/hr	X 0.27778	= metres/sec (m/s)
miles/hour	X 0.4470	= metres/sec (m/s)

TEMPERATURE

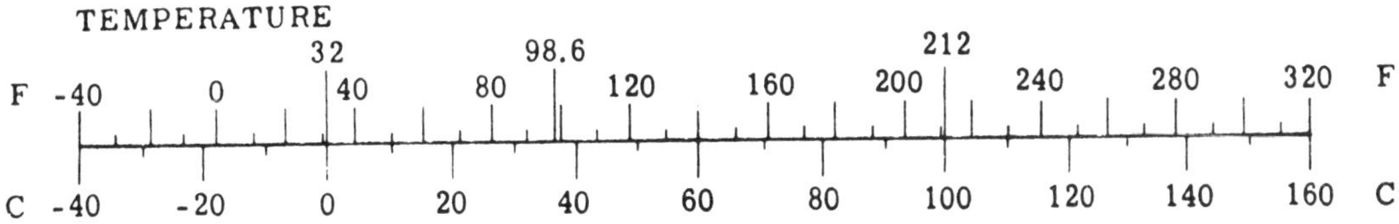

Celsius = 0.556 (°F -32)   F = (1.8°C) +32

CARBURETOR ADJUSTMENT SETTINGS

ENGINE	MODEL	YEAR	IDLE SPEED SCREW INITIAL TURNS OUT	IDLE SPEED	
				IN WATER	OUT OF WATER
587					
	GTS, SP, SPI	1992-94	3/4 - 1 1/4	1400	2500 GTS / 2400 SP,SPI
		1995-96	1 - 1 1/2	1500	2500 GTS,SPI / 2400 SP
	GTX	1992-94	1 - 1 1/2	1400	2800
	SPX	1993-94	1 - 1 1/2	1400	3000
	XP	1992	1 - 1 1/2	1400	3000
657					
	GTX	1995	1 - 1 1/2	1500	3000
	SPX	1995	7/8 - 1 3/8	1500	3000
	XP	1993	1 - 1 1/2	1500	3300
		1994	1 - 1 1/2	1500	3100
	XPI	1994	1 - 1 1/2	1500	3000
717					
	HX, XP	1995	1 1/2 - 2	1500	3000
	HX	1996-97	1 1/4 - 1 3/4	1500	3000
	GTI, SPX	1996	1 - 1 1/2	1500	3000
	GS, GSI, GTI, GTS, SP	1997	1 1/2 - 2	1500	3000
787					
	GSX, GTX, SPX	1997	3/4 - 1 1/4	1500	3000
	XP800	1995	5/8	1500	3000
	XP	1996-97	3/4 - 1 1/4	1500	3000

ENGINE TORQUE VALUES

Year/Model	Exhaust Manifold Bolts	Flywheel Nut Magneto -- FWD -- End	PTO -- Aft -- Flywheel	Crankcase Bolts -- M8	Crankcase Bolts -- M10	Crankcase/Engine Support Nuts	Engine Support Bolts	Cylinder Head Bolts	Cylinder Head Cover Bolts	Crankcase/Cylinder Bolts	Tuned Pipe Bolts/Nuts	Flame Arrestor Bolts	Magneto Housing Cover Bolts	Cranking Motor Bolts
	ft lbs (Nm)	ft lbs (Nm)	ft lbs (Nm)	ft lbs (Nm)	ft lbs (Nm)	ft lbs (Nm)	ft lbs (Nm)	ft lbs (Nm)	ft lbs (Nm)	ft lbs (Nm)	ft lbs (Nm)	ft lbs (Nm)	ft lbs (Nm)	ft lbs (Nm)
1992														
SP	15 (20)	77 (105)	81 (110)	16 (22)	28 (38)	26 (35)	18 (25)	15 (20)	7 (9)	15 (20)	18 (25)	7 (10)	3 (4)	16 (22)
XP	15 (20)	77 (105)	81 (110)	16 (22)	28 (38)	26 (35)	18 (25)	15 (20)	7 (9)	15 (20)	18 (25)	7 (10)	3 (4)	16 (22)
GTS	15 (20)	77 (105)	81 (110)	16 (22)	28 (38)	26 (35)	18 (25)	15 (20)	7 (9)	15 (20)	18 (25)	7 (10)	3 (4)	16 (22)
GTX	15 (20)	77 (105)	81 (110)	16 (22)	28 (38)	26 (35)	18 (25)	15 (20)	7 (9)	15 (20)	18 (25)	7 (10)	3 (4)	16 (22)
1993														
SP	15 (20)	77 (105)	81 (110)	16 (22)	28 (38)	26 (35)	18 (25)	15 (20)	7 (9)	15 (20)	18 (25)	7 (10)	4 (5)	16 (22)
SPX	15 (20)	77 (105)	81 (110)	16 (22)	28 (38)	26 (35)	18 (25)	15 (20)	7 (9)	15 (20)	18 (25)	7 (10)	4 (5)	16 (22)
SPI	15 (20)	77 (105)	81 (110)	16 (22)	28 (38)	26 (35)	18 (25)	15 (20)	7 (9)	15 (20)	18 (25)	7 (10)	4 (5)	16 (22)
XP	15 (20)	77 (105)	81 (110)	16 (22)	30 (40)	26 (35)	18 (25)	16 (22)	N/A	16 (22)	18 (25)	7 (10)	4 (5)	16 (22)
GTS	15 (20)	77 (105)	81 (110)	16 (22)	28 (38)	26 (35)	18 (25)	15 (20)	7 (9)	15 (20)	18 (25)	7 (10)	4 (5)	16 (22)
GTX	15 (20)	77 (105)	81 (110)	16 (22)	28 (38)	26 (35)	18 (25)	15 (20)	7 (9)	15 (20)	18 (25)	7 (10)	4 (5)	16 (22)

ENGINE TORQUE VALUES

Year/Model	Exhaust Manifold Bolts	Flywheel Nut Magneto -- FWD -- End	PTO -- Aft -- Flywheel	Crankcase Bolts -- M8	Crankcase Bolts -- M10	Crankcase/Engine Support Nuts	Engine Support Bolts	Cylinder Head Bolts	Cylinder Head Cover Bolts	Crankcase/Cylinder Bolts	Tuned Pipe Bolts/Nuts	Flame Arrestor Bolts	Magneto Housing Cover Bolts	Cranking Motor Bolts
	ft lbs (Nm)	ft lbs (Nm)	ft lbs (Nm)	ft lbs (Nm)	ft lbs (Nm)	ft lbs (Nm)	ft lbs (Nm)	ft lbs (Nm)	ft lbs (Nm)	ft lbs (Nm)	ft lbs (Nm)	ft lbs (Nm)	ft lbs (Nm)	ft lbs (Nm)
1994														
SP	17 (24)	107 (145)	81 (110)	16 (22)	28 (38)	26 (35)	18 (25)	16 (22)	7 (9)	16 (22)	18 (25)	7 (10)	4 (5)	16 (22)
SPX	17 (24)	107 (145)	81 (110)	16 (22)	28 (38)	26 (35)	18 (25)	16 (22)	N/A	16 (22)	18 (25)	7 (10)	4 (5)	16 (22)
SPI	17 (24)	107 (145)	81 (110)	16 (22)	28 (38)	26 (35)	18 (25)	16 (22)	7 (9)	16 (22)	18 (25)	7 (10)	4 (5)	16 (22)
XP 5854	17 (24)	107 (145)	81 (110)	16 (22)	30 (40)	26 (35)	18 (25)	16 (22)	N/A	16 (22)	18 (25)	7 (10)	4 (5)	16 (22)
XP 5855	17 (24)	107 (145)	81 (110)	16 (22)	28 (38)	26 (35)	18 (25)	16 (22)	N/A	16 (22)	18 (25)	7 (10)	4 (5)	16 (22)
GTS	17 (24)	107 (145)	81 (110)	16 (22)	28 (38)	26 (35)	18 (25)	16 (22)	7 (9)	16 (22)	18 (25)	7 (10)	4 (5)	16 (22)
GTX	17 (24)	107 (145)	81 (110)	16 (22)	28 (38)	26 (35)	18 (25)	16 (22)	N/A	16 (22)	18 (25)	7 (10)	4 (5)	16 (22)

ENGINE TORQUE VALUES

Year/Model	Exhaust Manifold Bolts	Flywheel Nut Magneto -- FWD -- End	PTO -- Aft -- Flywheel	Crankcase Bolts -- M8	Crankcase Bolts -- M10	Crankcase/Engine Support Nuts	Engine Support Bolts	Cylinder Head Bolts	Cylinder Head Cover Bolts	Crankcase/Cylinder Bolts	Tuned Pipe Bolts/Nuts	Flame Arrestor Bolts	Magneto Housing Cover Bolts	Cranking Motor Bolts
	ft lbs (Nm)	ft lbs (Nm)	ft lbs (Nm)	ft lbs (Nm)	ft lbs (Nm)	ft lbs (Nm)	ft lbs (Nm)	ft lbs (Nm)	ft lbs (Nm)	ft lbs (Nm)	ft lbs (Nm)	ft lbs (Nm)	ft lbs (Nm)	ft lbs (Nm)
1995														
SP	17 (24)	107 (145)	81 (110)	17 (24)	30 (40)	26 (35)	18 (25)	17 (24)	7 (9)	17 (24)	18 (25)	7 (10)	7 (9)	16 (22)
SPX	17 (24)	107 (145)	81 (110)	17 (24)	30 (40)	26 (35)	18 (25)	17 (24)	N/A	17 (24)	18 (25)	7 (10)	7 (9)	16 (22)
SPI	17 (24)	107 (145)	81 (110)	17 (24)	30 (40)	26 (35)	18 (25)	17 (24)	7 (9)	17 (24)	18 (25)	7 (10)	7 (9)	16 (22)
XP 800	30 (40)	77 (105)	81 (110)	17 (24)	30 (40)	N/A	18 (25)	17 (24)	N/A	30 (40)	18 (25)	7 (10)	7 (9)	16 (22)
XP 5857	17 (24)	107 (145)	81 (110)	17 (24)	30 (40)	26 (35)	18 (25)	17 (24)	N/A	17 (24)	18 (25)	7 (10)	7 (9)	16 (22)
GTS	17 (24)	107 (145)	81 (110)	17 (24)	30 (40)	26 (35)	18 (25)	17 (24)	7 (9)	17 (24)	18 (25)	7 (10)	7 (9)	16 (22)
GTX	17 (24)	107 (145)	81 (110)	17 (24)	30 (40)	26 (35)	18 (25)	17 (24)	N/A	17 (24)	18 (25)	7 (10)	7 (9)	16 (22)
HX	17 (24)	107 (145)	81 (110)	17 (24)	30 (40)	26 (35)	18 (25)	17 (24)	N/A	17 (24)	18 (25)	7 (10)	7 (9)	16 (22)

ENGINE TORQUE VALUES

Year/Model	Exhaust Manifold Bolts	Flywheel Nut Magneto -- FWD -- End	PTO -- Aft -- Flywheel	Crankcase Bolts -- M8	Crankcase Bolts -- M10	Crankcase/Engine Support Nuts	Engine Support Bolts	Cylinder Head Bolts	Cylinder Head Cover Bolts	Crankcase/Cylinder Bolts	Tuned Pipe Bolts/Nuts	Flame Arrestor Bolts	Magneto Housing Cover Bolts	Cranking Motor Bolts
	ft lbs (Nm)	ft lbs (Nm)	ft lbs (Nm)	ft lbs (Nm)	ft lbs (Nm)	ft lbs (Nm)	ft lbs (Nm)	ft lbs (Nm)	ft lbs (Nm)	ft lbs (Nm)	ft lbs (Nm)	ft lbs (Nm)	ft lbs (Nm)	ft lbs (Nm)
1996														
SP	17 (24)	107 (145)	81 (110)	17 (24)	30 (40)	26 (35)	18 (25)	17 (24)	7 (9)	17 (24)	18 (25)	7 (10)	7 (9)	16 (22)
SPX	17 (24)	107 (145)	81 (110)	17 (24)	30 (40)	26 (35)	18 (25)	17 (24)	N/A	17 (24)	18 (25)	7 (10)	7 (9)	16 (22)
SPI	17 (24)	107 (145)	81 (110)	17 (24)	30 (40)	26 (35)	18 (25)	17 (24)	7 (9)	17 (24)	18 (25)	7 (10)	7 (9)	16 (22)
XP	30 (40)	77 (105)	81 (110)	17 (24)	30 (40)	N/A	18 (25)	17 (24)	N/A	30 (40)	18 (25)	7 (10)	7 (9)	16 (22)
GSX	30 (40)	77 (105)	81 (110)	17 (24)	30 (40)	N/A	18 (25)	17 (24)	N/A	30 (40)	18 (25)	7 (10)	7 (9)	16 (22)
GTS	17 (24)	107 (145)	81 (110)	17 (24)	30 (40)	26 (35)	18 (25)	17 (24)	7 (9)	17 (24)	18 (25)	7 (10)	7 (9)	16 (22)
GTI	17 (24)	107 (145)	81 (110)	17 (24)	30 (40)	26 (35)	18 (25)	17 (24)	7 (9)	17 (24)	18 (25)	7 (10)	7 (9)	16 (22)
GTX	30 (40)	77 (105)	81 (110)	17 (24)	30 (40)	N/A	18 (25)	17 (24)	N/A	30 (40)	18 (25)	7 (10)	7 (9)	16 (22)
HX	17 (24)	107 (145)	81 (110)	17 (24)	30 (40)	26 (35)	18 (25)	17 (24)	N/A	17 (24)	18 (25)	7 (10)	7 (9)	16 (22)

ENGINE TORQUE VALUES

Year/Model	Exhaust Manifold Bolts	Flywheel Nut Magneto -- FWD -- End	PTO -- Aft -- Flywheel	Crankcase Bolts -- M8	Crankcase Bolts -- M10	Crankcase/Engine Support Nuts	Engine Support Bolts	Cylinder Head Bolts	Cylinder Head Cover Bolts	Crankcase/Cylinder Bolts	Tuned Pipe Bolts/Nuts	Flame Arrestor Bolts	Magneto Housing Cover Bolts	Cranking Motor Bolts
	ft lbs (Nm)	ft lbs (Nm)	ft lbs (Nm)	ft lbs (Nm)	ft lbs (Nm)	ft lbs (Nm)	ft lbs (Nm)	ft lbs (Nm)	ft lbs (Nm)	ft lbs (Nm)	ft lbs (Nm)	ft lbs (Nm)	ft lbs (Nm)	ft lbs (Nm)
1997														
SP	17 (24)	107 (145)	81 (110)	17 (24)	30 (40)	26 (35)	18 (25)	17 (24)	7 (9)	17 (24)	18 (25)	7 (10)	7 (9)	16 (22)
SPX	17 (24)	107 (145)	81 (110)	17 (24)	30 (40)	26 (35)	18 (25)	17 (24)	N/A	17 (24)	18 (25)	7 (10)	7 (9)	16 (22)
SPI	17 (24)	107 (145)	81 (110)	17 (24)	30 (40)	26 (35)	18 (25)	17 (24)	7 (9)	17 (24)	18 (25)	7 (10)	7 (9)	16 (22)
XP	30 (40)	77 (105)	81 (110)	17 (24)	30 (40)	N/A	18 (25)	17 (24)	N/A	30 (40)	18 30 (25) (40)	7 (10)	7 (9)	16 (22)
GSX	30 (40)	77 (105)	81 (110)	17 (24)	30 (40)	N/A	18 (25)	17 (24)	N/A	30 (40)	18 30 (25) (40)	7 (10)	7 (9)	16 (22)
GTS	17 (24)	107 (145)	81 (110)	17 (24)	30 (40)	26 (35)	18 (25)	17 (24)	7 (9)	17 (24)	18 (25)	7 (10)	7 (9)	16 (22)
GTI	17 (24)	107 (145)	81 (110)	17 (24)	30 (40)	26 (35)	18 (25)	17 (24)	7 (9)	17 (24)	18 (25)	7 (10)	7 (9)	16 (22)
GTX	30 (40)	77 (105)	81 (110)	17 (24)	30 (40)	N/A	18 (25)	17 (24)	N/A	30 (40)	18 30 (25) (40)	7 (10)	7 (9)	16 (22)
HX	17 (24)	107 (145)	81 (110)	17 (24)	30 (40)	26 (35)	18 (25)	17 (24)	N/A	17 (24)	18 (25)	7 (10)	7 (9)	16 (22)

ENGINE SPECIFICATIONS

MODEL	ENGINE TYPE	DISP. cu. in. (cm3)	BORE STANDARD	BORE 1st OVERSIZE	STROKE	MFG. COMPRESSION
1992					**1992**	
SP 5805	587	35.4 (580.7)	2.99" (76.0mm)	3.002" (76.25mm)	2.52" (64.0mm)	140 - 150
XP 5851	587	35.4 (580.7)	2.99" (76.0mm)	3.002" (76.25mm)	2.52" (64.0mm)	140 - 150
GTS 5812	587	35.4 (580.7)	2.99" (76.0mm)	3.002" (76.25mm)	2.52" (64.0mm)	140 - 150
GTX 5860	587	35.4 (580.7)	2.99" (76.0mm)	3.002" (76.25mm)	2.52" (64.0mm)	140 - 150
1993					**1993**	
SP 5806	587	35.4 (580.7)	2.99" (76.0mm)	3.002" (76.25mm)	2.52" (64.0mm)	140 - 150
SPX 5807	587	35.4 (580.7)	2.99" (76.0mm)	3.002" (76.25mm)	2.52" (64.0mm)	140 - 150
SPI 5808	587	35.4 (580.7)	2.99" (76.0mm)	3.002" (76.25mm)	2.52" (64.0mm)	140 - 150
XP 5852	657	39.7 (650.0mm)	3.071" (78.0mm)	3.081" (78.25mm)	2.68" (68.0mm)	140 - 150
GTS 5813	587	35.4 (580.7)	2.99" (76.0mm)	3.002" (76.25mm)	2.52" (64.0mm)	140 - 150
GTX 5861	587	35.4 (580.7)	2.99" (76.0mm)	3.002" (76.25mm)	2.52" (64.0mm)	140 - 150
1994					**1994**	
SP 5870	587	35.4 (580.7)	2.99" (76.0mm)	3.002" (76.25mm)	2.52" (64.0mm)	140 - 150
SPX 5871	587	39.7 (650.0mm)	3.071" (78.0mm)	3.081" (78.25mm)	2.68" (68.0mm)	140 - 150
SPI 5872	587	35.4 (580.7)	2.99" (76.0mm)	3.002" (76.25mm)	2.52" (64.0mm)	140 - 150
XP 5854	657	39.7 (650.0mm)	3.071" (78.0mm)	3.081" (78.25mm)	2.68" (68.0mm)	140 - 150
GTS 5814	587	35.4 (580.7)	2.99" (76.0mm)	3.002" (76.25mm)	2.52" (64.0mm)	140 - 150
GTX 5862	587	39.7 (650.0mm)	3.071" (78.0mm)	3.081" (78.25mm)	2.68" (68mm)	140 - 150

ENGINE SPECIFICATIONS

MODEL	ENGINE TYPE	DISP. cu. in. (cm3)	BORE STANDARD	BORE 1st OVERSIZE	STROKE	MFG. COMPRESSION
	1995				1995	
SP 5873	587	35.4 (580.7)	2.99" (76.0mm)	3.002" (76.25mm)	2.52" (64.0mm)	140 - 150
SPI 5875	587	35.4 (580.7)	2.99" (76.0mm)	3.002" (76.25mm)	2.52" (64.0mm)	140 - 150
SPX 5874	657x	39.67 (650.0)	3.071" (78.0mm)	3.081" (78.25mm)	2.68" (68.0mm)	140 - 150
XP	717	43.81 (718.0)	3.228" (82.0mm)	3.238" (82.25mm)	2.68" (68.0mm)	140 - 150
GTS	587	35.4 (580.7)	2.99" (76.0mm)	3.002" (76.25mm)	2.52" (64.0mm)	140 - 150
GTX	657x	39.67 (650.0)	3.071" (78.0mm)	3.081" (78.25mm)	2.68" (68.0mm)	140 - 150
	1996				1996	
SP 5876	587	35.4 (580.7)	2.99" (76.0mm)	3.002" (76.25mm)	2.52" (64.0mm)	140 - 150
SPI 5875	587	35.4 (580.7)	2.99" (76.0mm)	3.002" (76.25mm)	2.52" (64.0mm)	140 - 150
SPX 5877	717	43.81 (718.0)	3.228" (82.0mm)	3.238" (82.25mm)	2.68" (68.0mm)	140 - 150
GTS 5817	587	35.4 (580.7)	2.99" (76.0mm)	3.002" (76.25mm)	2.52" (64.0mm)	140 - 150
GTI 5865/5866	717	43.81 (718.0)	3.228" (82.0mm)	3.238" (82.25mm)	2.68" (68.0mm)	140 - 150
XP 5858/5859	787	47.7 (781.6)	3.228" (82.0mm)	3.238" (82.25mm)	2.99" (74.0mm)	140 - 150
HX 5881	717	43.81 (718.0)	3.228" (82.0mm)	3.238" (82.25mm)	2.68" (68.0mm)	140 - 150
	1997				1997	
SP 5879	717	43.81 (718.0)	3.228" (82.0mm)	3.238" (82.25mm)	2.68" (68.0mm)	140 - 150
SPX 5877	787	47.7 (781.6)	3.228" (82.0mm)	3.238" (82.25mm)	2.99" (74.0mm)	140 - 150
GS 5621	717	43.81 (718.0)	3.228" (82.0mm)	3.238" (82.25mm)	2.68" (68.0mm)	140 - 150
GSI 5622	717	43.81 (718.0)	3.228" (82.0mm)	3.238" (82.25mm)	2.68" (68.0mm)	140 - 150
GSX 5624	787	47.7 (781.6)	3.228" (82.0mm)	3.238" (82.25mm)	2.99" (74.0mm)	140 - 150
GTS 5818	717	43.81 (718.0)	3.228" (82.0mm)	3.238" (82.25mm)	2.68" (68.0mm)	140 - 150
GTI 5641	717	43.81 (718.0)	3.228" (82.0mm)	3.238" (82.25mm)	2.68" (68.0mm)	140 - 150
GTX 5642	787	47.7 (781.6)	3.228" (82.0mm)	3.238" (82.25mm)	2.99" (74.0mm)	140 - 150
XP 5662	787	47.7 (781.6)	3.228" (82.0mm)	3.238" (82.25mm)	2.99" (74.0mm)	140 - 150
HX 5882	717	43.81 (718.0)	3.228" (82.0mm)	3.238" (82.25mm)	2.68" (68.0mm)	140 - 150

ENGINE SPECIFICATIONS

MODEL	RING END GAP LIMIT	PISTON/ CYLINDER CLRNC. LIMIT	CYLINDER TAPER (MAX)	CYLINDER OUT OF ROUND (Max)	CONNECTING ROD BIG END LIMIT
	1992			**1992**	
SP 5805	0.039" (1.00mm)	0.008" (0.2mm)	0.003" (0.08mm)	0.002" (0.050mm)	0.047" (1.2mm)
XP 5851	0.039" (1.00mm)	0.008" (0.2mm)	0.003" (0.08mm)	0.002" (0.050mm)	0.047" (1.2mm)
GTS 5812	0.039" (1.00mm)	0.008" (0.2mm)	0.003" (0.08mm)	0.002" (0.050mm)	0.047" (1.2mm)
GTX 5860	0.039" (1.00mm)	0.008" (0.2mm)	0.003" (0.08mm)	0.002" (0.050mm)	0.047" (1.2mm)
	1993			**1993**	
SP 5806	0.039" (1.00mm)	0.008" (0.2mm)	0.003" (0.08mm)	0.002" (0.050mm)	0.047" (1.2mm)
SPX 5807	0.039" (1.00mm)	0.008" (0.2mm)	0.003" (0.08mm)	0.002" (0.050mm)	0.047" (1.2mm)
SPI 5808	0.039" (1.00mm)	0.008" (0.2mm)	0.003" (0.08mm)	0.002" (0.050mm)	0.047" (1.2mm)
XP 5852	0.039" (1.00mm)	0.006" (0.15mm)	0.003" (0.08mm)	0.002" (0.050mm)	0.047" (1.2mm)
GTS 5813	0.039" (1.00mm)	0.008" (0.2mm)	0.003" (0.08mm)	0.002" (0.050mm)	0.047" (1.2mm)
GTX 5861	0.039" (1.00mm)	0.008" (0.2mm)	0.003" (0.08mm)	0.002" (0.050mm)	0.047" (1.2mm)
	1994			**1994**	
SP 5870	0.039" (1.00mm)	0.008" (0.2mm)	0.003" (0.08mm)	0.002" (0.050mm)	0.047" (1.2mm)
SPX 5871	0.039" (1.00mm)	0.006" (0.15mm)	0.003" (0.08mm)	0.002" (0.050mm)	0.047" (1.2mm)
SPI 5872	0.039" (1.00mm)	0.008" (0.2mm)	0.003" (0.08mm)	0.002" (0.050mm)	0.047" (1.2mm)
XP 5854	0.039" (1.00mm)	0.006" (0.15mm)	0.003" (0.08mm)	0.002" (0.050mm)	0.047" (1.2mm)
GTS 5814	0.039" (1.00mm)	0.008" (0.2mm)	0.003" (0.08mm)	0.002" (0.050mm)	0.047" (1.2mm)
GTX 5862	0.039" (1.00mm)	0.006" (0.15mm)	0.003" (0.08mm)	0.002" (0.050mm)	0.047" (1.2mm)

ENGINE SPECIFICATIONS

MODEL	RING END GAP LIMIT	PISTON/ CYLINDER CLRNC. LIMIT	CYLINDER TAPER (MAX)	CYLINDER OUT OF ROUND (Max)	CONNECTING ROD BIG END LIMIT
	1995			1995	
SP 5873	0.039" (1.00mm)	0.008" (0.2mm)	0.004" (0.100mm)	0.003" (0.080mm)	0.047" (1.2mm)
SPI 5875	0.039" (1.00mm)	0.008" (0.2mm)	0.004" (0.100mm)	0.003" (0.080mm)	0.047" (1.2mm)
SPX 5874	0.039" (1.00mm)	0.008" (0.2mm)	0.004" (0.100mm)	0.003" (0.080mm)	0.047" (1.2mm)
XP	0.039" (1.00mm)	0.008" (0.2mm)	0.004" (0.100mm)	0.003" (0.080mm)	0.047" (1.2mm)
GTS	0.039" (1.00mm)	0.008" (0.2mm)	0.004" (0.100mm)	0.003" (0.080mm)	0.047" (1.2mm)
GTX	0.039" (1.00mm)	0.008" (0.2mm)	0.004" (0.100mm)	0.003" (0.080mm)	0.047" (1.2mm)
	1996			1996	
SP 5876	0.039" (1.00mm)	0.008" (0.2mm)	0.004" (0.100mm)	0.003" (0.080mm)	0.047" (1.2mm)
SPI 5875	0.039" (1.00mm)	0.008" (0.2mm)	0.004" (0.100mm)	0.003" (0.080mm)	0.047" (1.2mm)
SPX 5877	0.039" (1.00mm)	0.008" (0.2mm)	0.004" (0.100mm)	0.003" (0.080mm)	0.047" (1.2mm)
GTS 5817	0.039" (1.00mm)	0.008" (0.2mm)	0.004" (0.100mm)	0.003" (0.080mm)	0.047" (1.2mm)
GTI 5865/5866	0.039" (1.00mm)	0.008" (0.2mm)	0.004" (0.100mm)	0.003" (0.080mm)	0.047" (1.2mm)
XP 5858/5859	0.039" (1.00mm)	0.008" (0.2mm)	0.004" (0.100mm)	0.003" (0.080mm)	0.047" (1.2mm)
HX 5881	0.039" (1.00mm)	0.008" (0.2mm)	0.004" (0.100mm)	0.003" (0.080mm)	0.047" (1.2mm)
	1997			1997	
SP 5879	0.039" (1.00mm)	0.008" (0.2mm)	0.004" (0.100mm)	0.003" (0.080mm)	0.047" (1.2mm)
SPX 5877	0.039" (1.00mm)	0.008" (0.2mm)	0.004" (0.100mm)	0.003" (0.080mm)	0.047" (1.2mm)
GS 5621	0.039" (1.00mm)	0.008" (0.2mm)	0.004" (0.100mm)	0.003" (0.080mm)	0.047" (1.2mm)
GSI 5622	0.039" (1.00mm)	0.008" (0.2mm)	0.004" (0.100mm)	0.003" (0.080mm)	0.047" (1.2mm)
GSX 5624	0.039" (1.00mm)	0.008" (0.2mm)	0.004" (0.100mm)	0.003" (0.080mm)	0.047" (1.2mm)
GTS 5818	0.039" (1.00mm)	0.008" (0.2mm)	0.004" (0.100mm)	0.003" (0.080mm)	0.047" (1.2mm)
GTI 5641	0.039" (1.00mm)	0.008" (0.2mm)	0.004" (0.100mm)	0.003" (0.080mm)	0.047" (1.2mm)
GTX 5642	0.039" (1.00mm)	0.008" (0.2mm)	0.004" (0.100mm)	0.003" (0.080mm)	0.047" (1.2mm)
XP 5662	0.039" (1.00mm)	0.008" (0.2mm)	0.004" (0.100mm)	0.003" (0.080mm)	0.047" (1.2mm)
HX 5882	0.039" (1.00mm)	0.008" (0.2mm)	0.004" (0.100mm)	0.003" (0.080mm)	0.047" (1.2mm)

TUNE-UP ADJUSTMENTS

SERIES	SPARK PLUG TYPE	Ignition Timing Degrees BTDC @ rpm	Piston Location BTDC (Ignition Spark)
1992			
SP	NGK BR7ES	21° @ 6,000; engine cold	0.104" (2.65mm)
XP	NGK BR7ES	21° @ 6,000; engine cold	0.104" (2.65mm)
GTS	NGK BR7ES	21° @ 6,000; engine cold	0.104" (2.65mm)
GTX	NGK BR7ES	21° @ 6,000; engine cold	0.104" (2.65mm)
1993			
SP	NGK BR7ES	21° @ 6,000; engine cold	0.104" (2.65mm)
SPX	NGK BR7ES	21° @ 6,000; engine cold	0.104" (2.65mm)
SPI	NGK BR7ES	21° @ 6,000; engine cold	0.104" (2.65mm)
XP	NGK BR8ES	20° @ 6,000; engine cold	0.102" (2.59mm)
GTS	NGK BR7ES	21° @ 6,000; engine cold	0.104" (2.65mm)
GTX	NGK BR7ES	21° @ 6,000; engine cold	0.104" (2.65mm)
1994			
SP	NGK BR7ES	21° @ 6,000; engine cold	0.104" (2.65mm)
SPX	NGK BR8ES	21° @ 6,000; engine cold	0.112" (2.85mm)
SPI	NGK BR7ES	21° @ 6,000; engine cold	0.104" (2.65mm)
XP (5854)	NGK BR8ES	20° @ 6,000; engine cold	0.102" (2.59mm)
XP (5855)	NGK BR8ES	21° @ 6,000; engine cold	0.112" (2.85mm)
GTS	NGK BR7ES	21° @ 6,000; engine cold	0.104" (2.65mm)
GTX	NGK BR8ES	21° @ 6,000; engine cold	0.112" (2.85mm)
1995			
SP	NGK BR7ES	20° @ 6,000; engine cold	0.095" (2.41mm)
SPX	NGK BR8ES	20° @ 6,000; engine cold	0.102" (2.59mm)
SPI	NGK BR7ES	20° @ 6,000; engine cold	0.095" (2.41mm)
XP 800	NGK BR8ES	22° @ 3,500	0.133" (3.38mm)
XP (5857)	NGK BR8ES	20° @ 6,000; engine cold	0.102" (2.59mm)
GTS	NGK BR7ES	20° @ 6,000; engine cold	0.095" (2.41mm)
GTX	NGK BR8ES	20° @ 6,000; engine cold	0.102" (2.59mm)
HX	NGK BR8ES	20° @ 6,000; engine cold	0.102" (2.59mm)

TUNE-UP ADJUSTMENTS

SERIES	SPARK PLUG TYPE	Ignition Timing Degrees BTDC @ rpm	Piston Location BTDC (Ignition Spark)
1992			
SPX	NGK BR8ES	20° @ 6,000; engine cold	0.102" (2.59mm)
SPI	NGK BR7ES	20° @ 6,000; engine cold	0.095" (2.41mm)
XP 800	NGK BR8ES	22° @ 3,500	0.133" (3.38mm)
GSX	NGK BR8ES	22° @ 3,500	0.133" (3.38mm)
GTS	NGK BR7ES	20° @ 6,000; engine cold	0.095" (2.41mm)
GTI	NGK BR8ES	20° @ 6,000; engine cold	0.102" (2.59mm)
GTX	NGK BR8ES	22° @ 3,500	0.133" (3.38mm)
HX	NGK BR8ES	20° @ 6,000; engine cold	0.102" (2.59mm)
1997			
SP	NGK BR7ES	20° @ 6,000; engine cold	0.095" (2.41mm)
SPX	NGK BR8ES	22° @ 3,500	0.133" (3.38mm)
GS	NGK BR8ES	20° @ 6,000; engine cold	0.102" (2.59mm)
GSI	NGK BR8ES	20° @ 6,000; engine cold	0.102" (2.59mm)
GSX	NGK BR8ES	22° @ 3,500	0.133" (3.38mm)
GTS	NGK BR8ES	20° @ 6,000; engine cold	0.102" (2.59mm)
GTI	NGK BR8ES	20° @ 6,000; engine cold	0.102" (2.59mm)
GTX	NGK BR8ES	22° @ 3,500	0.133" (3.38mm)
HX	NGK BR8ES	20° @ 6,000; engine cold	0.102" (2.59mm)
XP	NGK BR8ES	22° @ 3,500	0.133" (3.38mm)

TIMING CORRECTION CHART 787 Engines w/DC CDI	
Programmer Number (MPEM)	**Ignition Timing Correction**
2	3°
3	2°
4	1°
1	0°
5	-1°
6	-2°
7	-3°
8	-4°

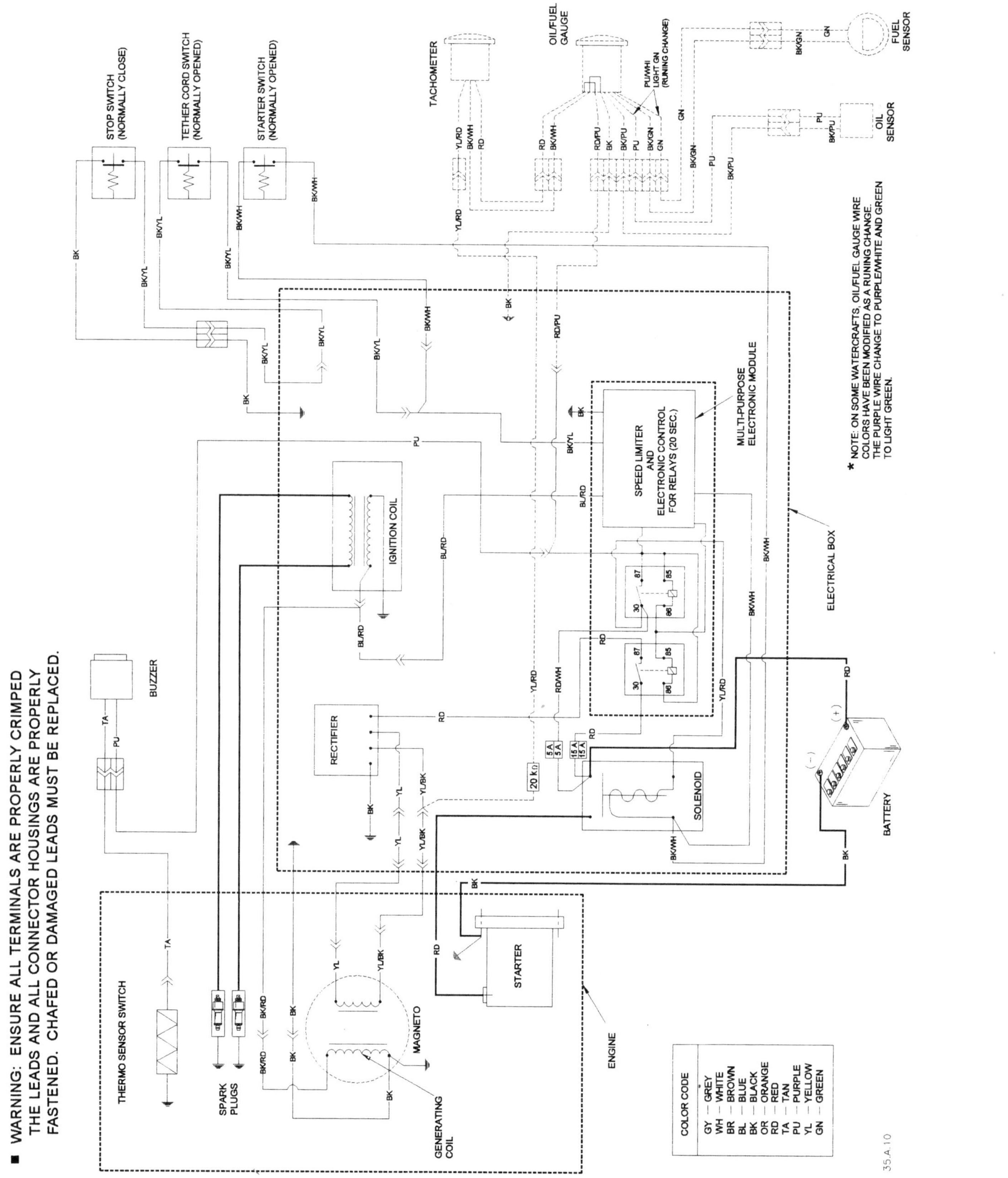

***1992 SP, GTS, XP and GTX Models** -- Electrical wiring diagram and color code identification. Major components of the electrical and ignition systems are identified.*

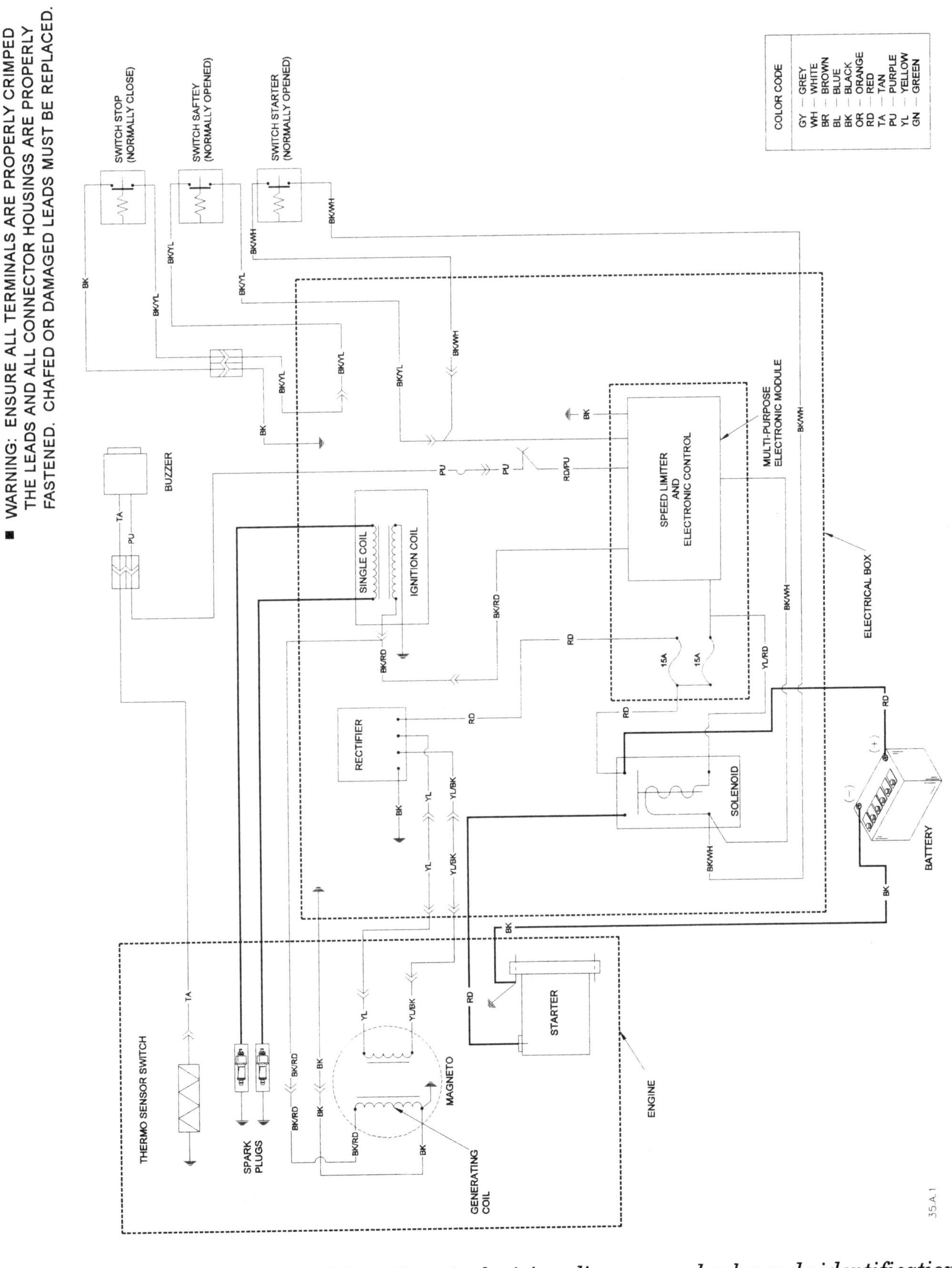

***1993 SP, SPI and GTS Models** -- Electrical wiring diagram and color code identification. Major components of the electrical and ignition systems are identified.*

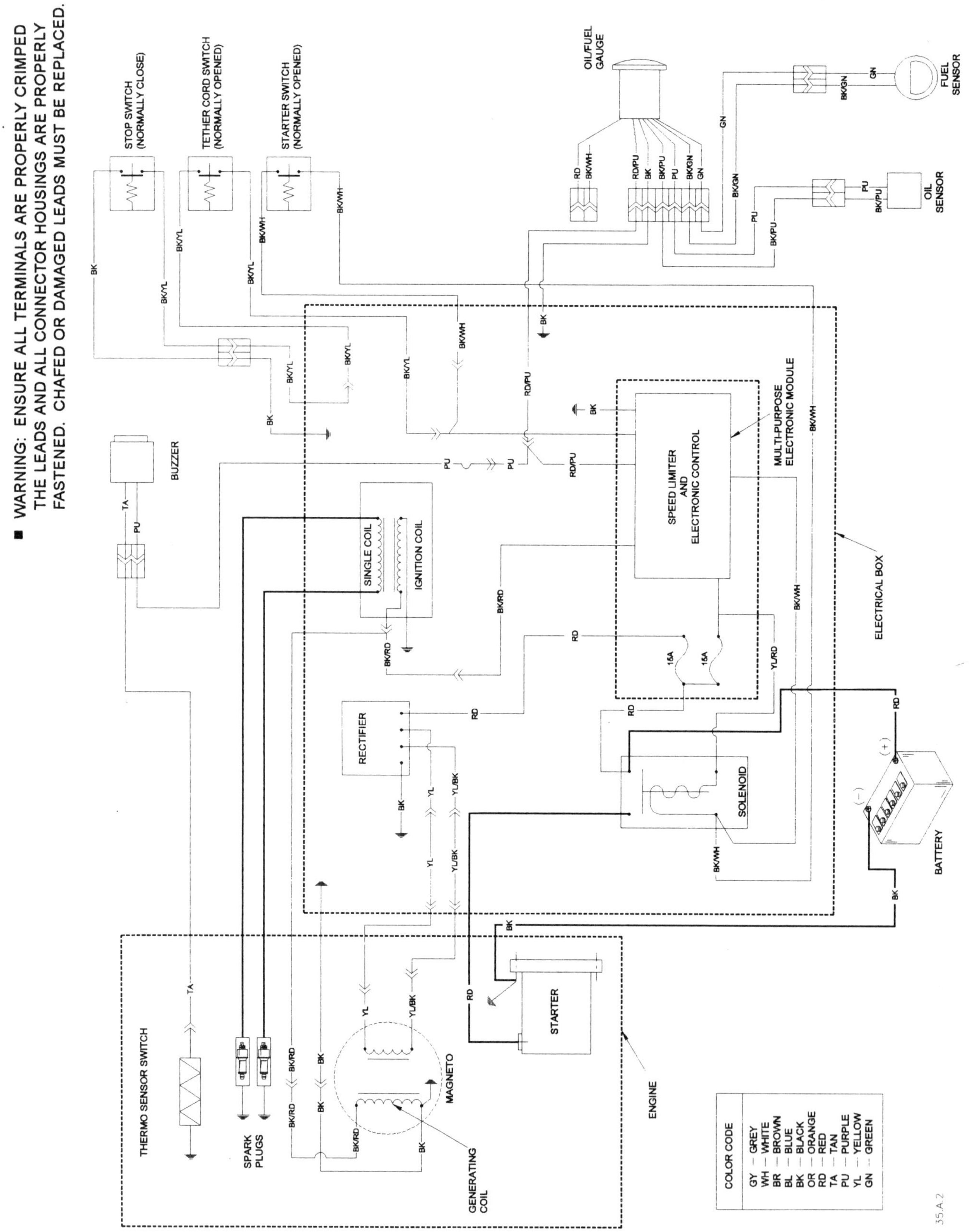

1993 SPX -- Electrical wiring diagram and color code identification. Major components of the electrical and ignition systems are identified.

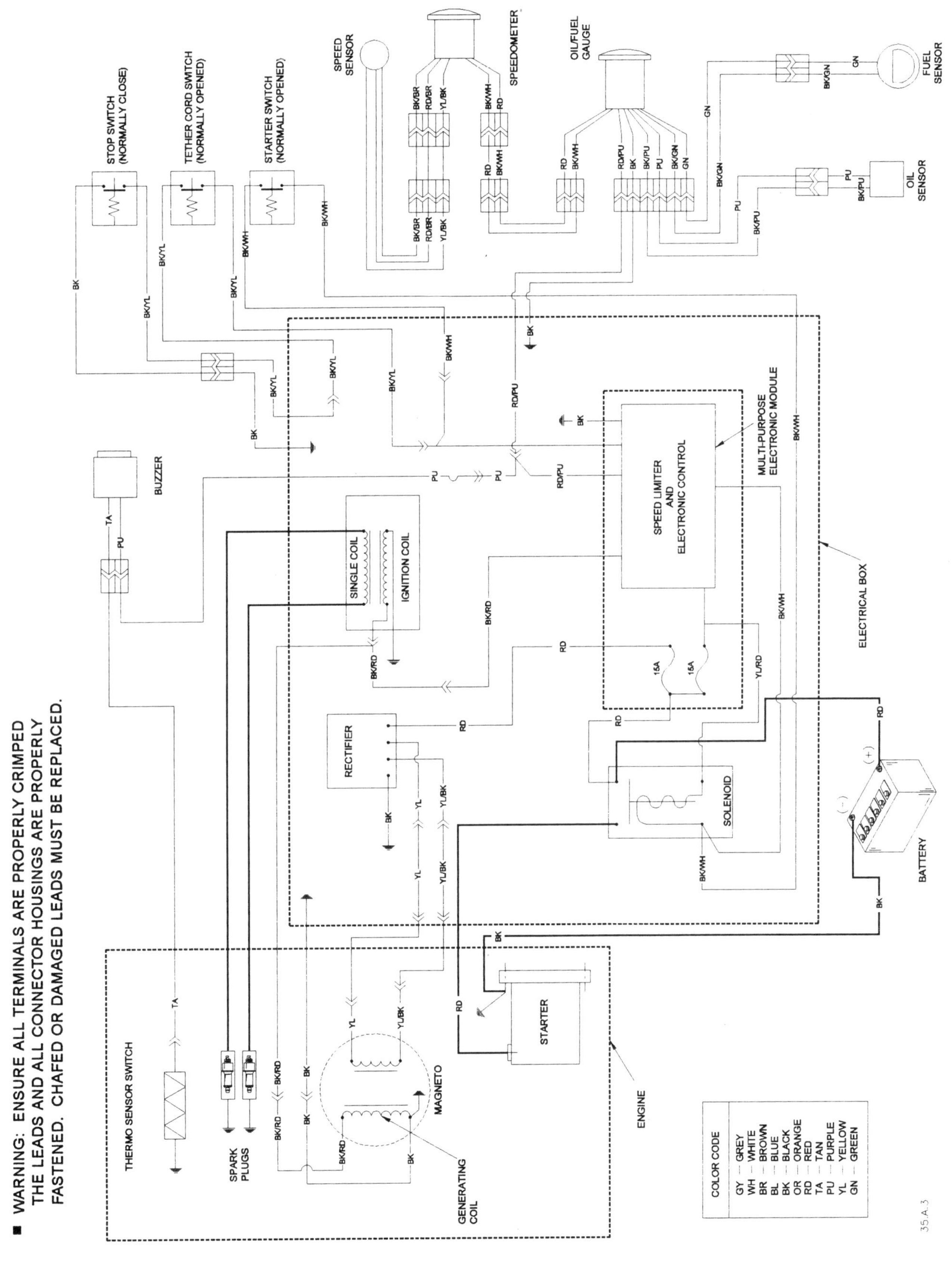

***1993 GTX** -- Electrical wiring diagram and color code identification. Major components of the electrical and ignition systems are identified.*

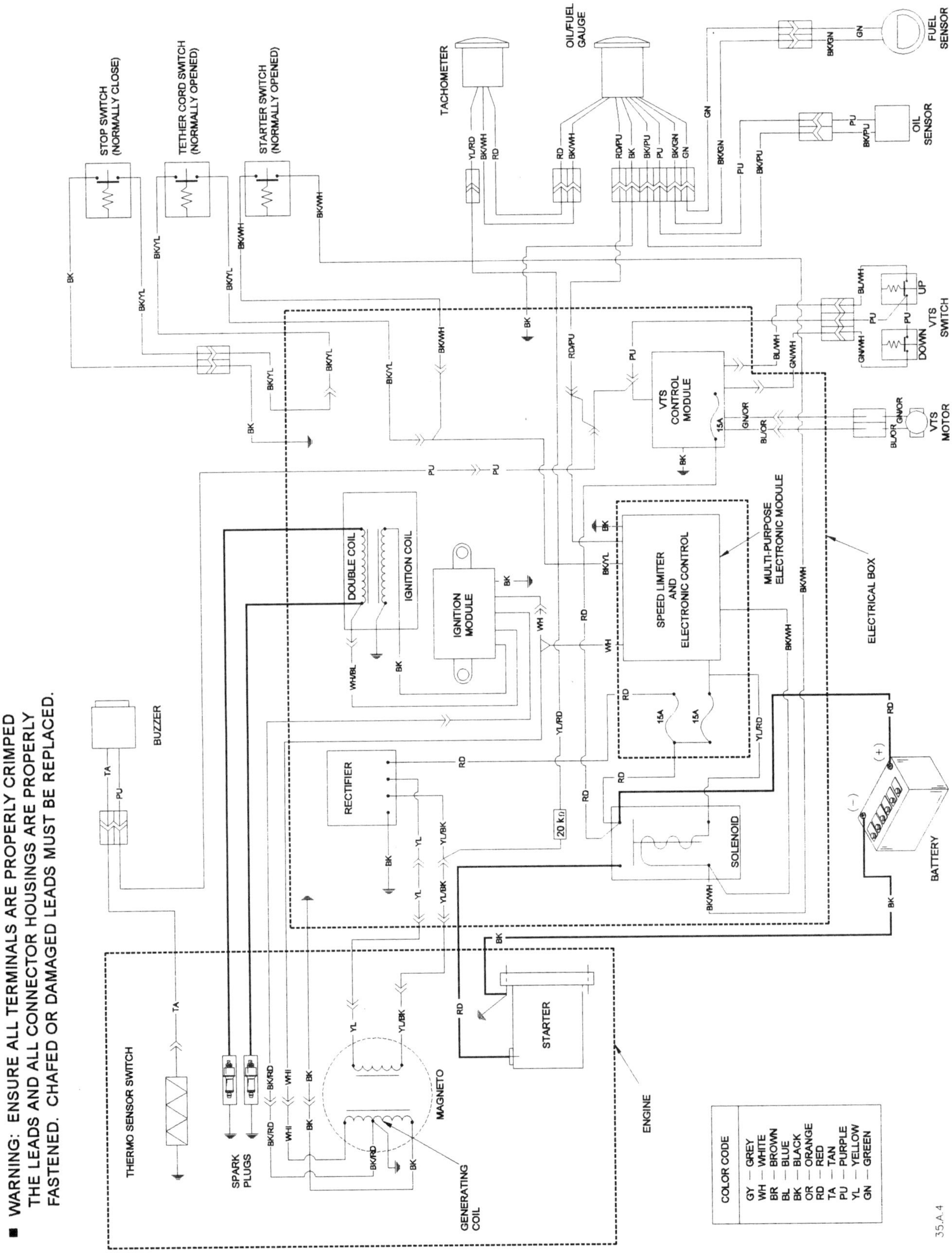

***1993 XP** -- Electrical wiring diagram and color code identification. Major components of the electrical and ignition systems are identified.*

■ WARNING: ENSURE ALL TERMINALS ARE PROPERLY CRIMPED THE LEADS AND ALL CONNECTOR HOUSINGS ARE PROPERLY FASTENED. CHAFED OR DAMAGED LEADS MUST BE REPLACED.

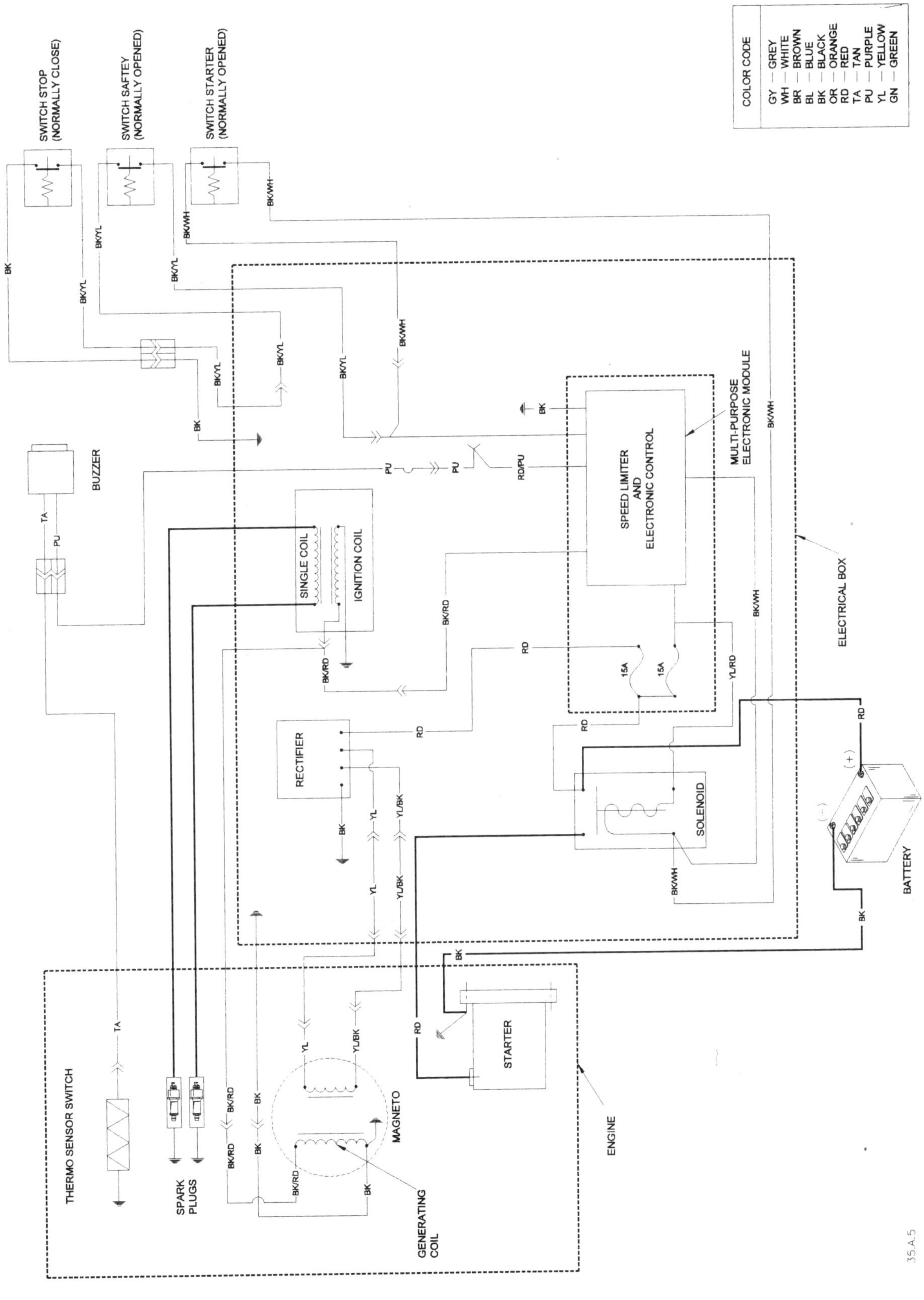

1994 SP and GTS Models -- Electrical wiring diagram and color code identification. Major components of the electrical and ignition systems are identified.

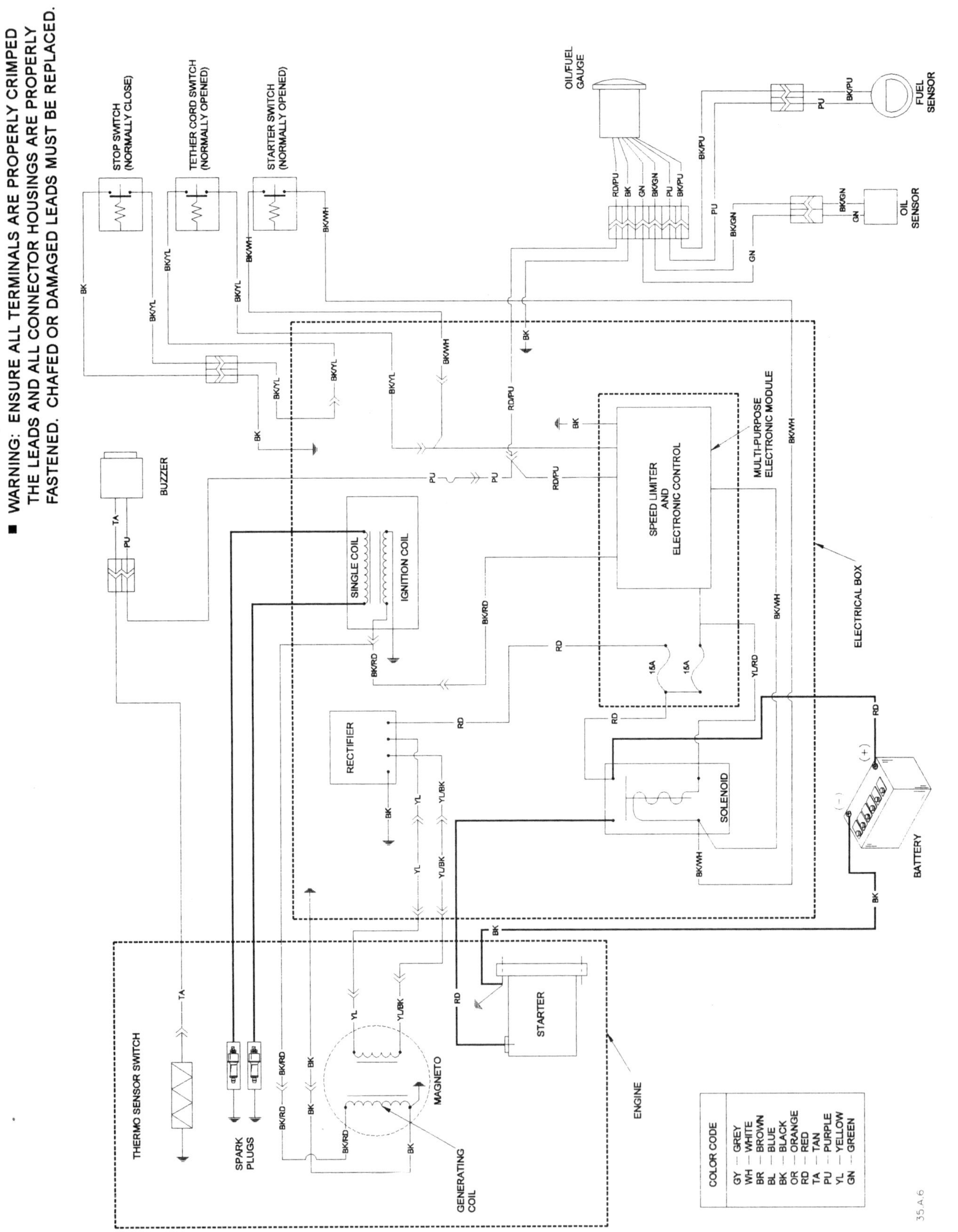

1994 SPI and SPX Models -- *Electrical wiring diagram and color code identification. Major components of the electrical and ignition systems are identified.*

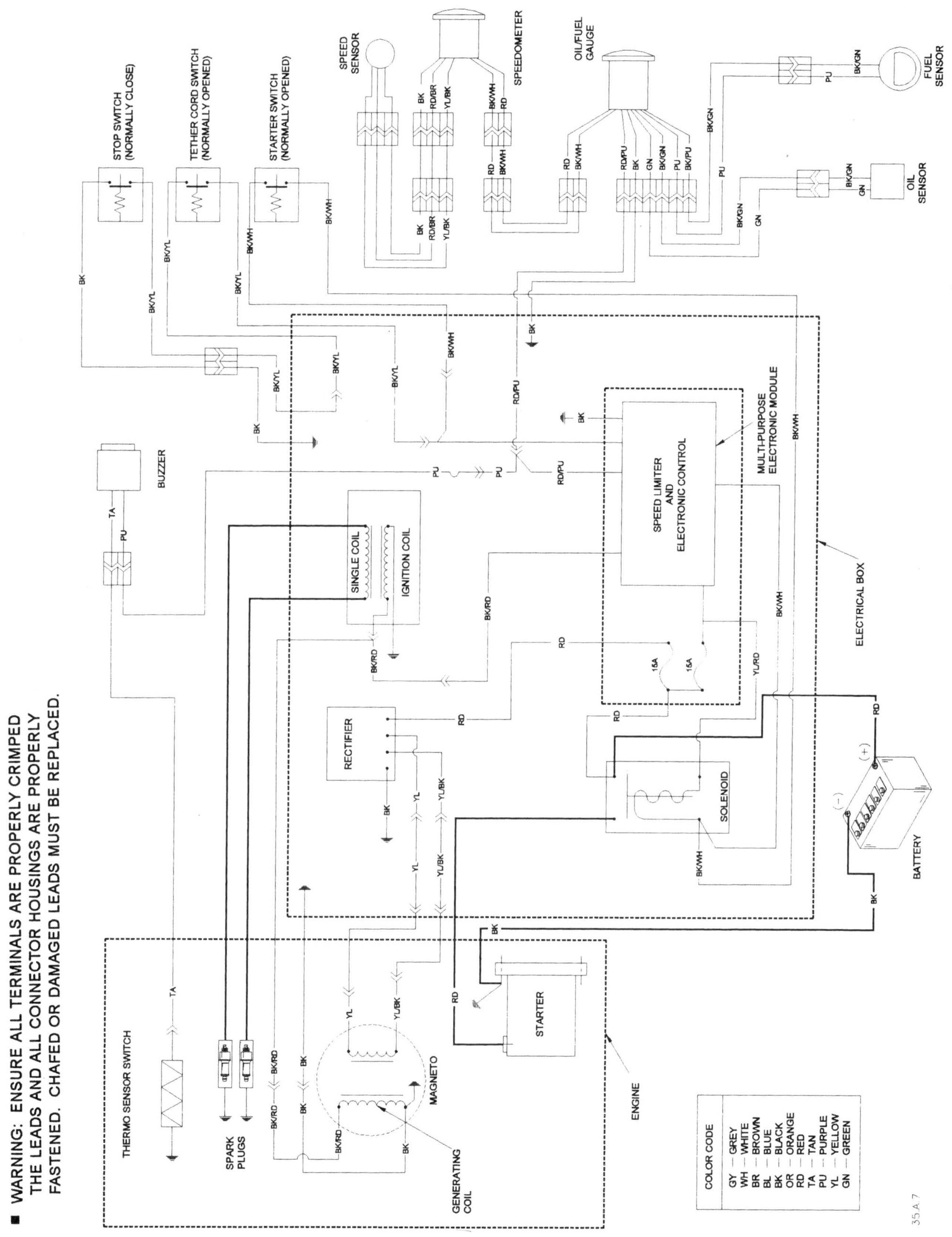

1994 GTX Model -- Electrical wiring diagram and color code identification. Major components of the electrical and ignition systems are identified.

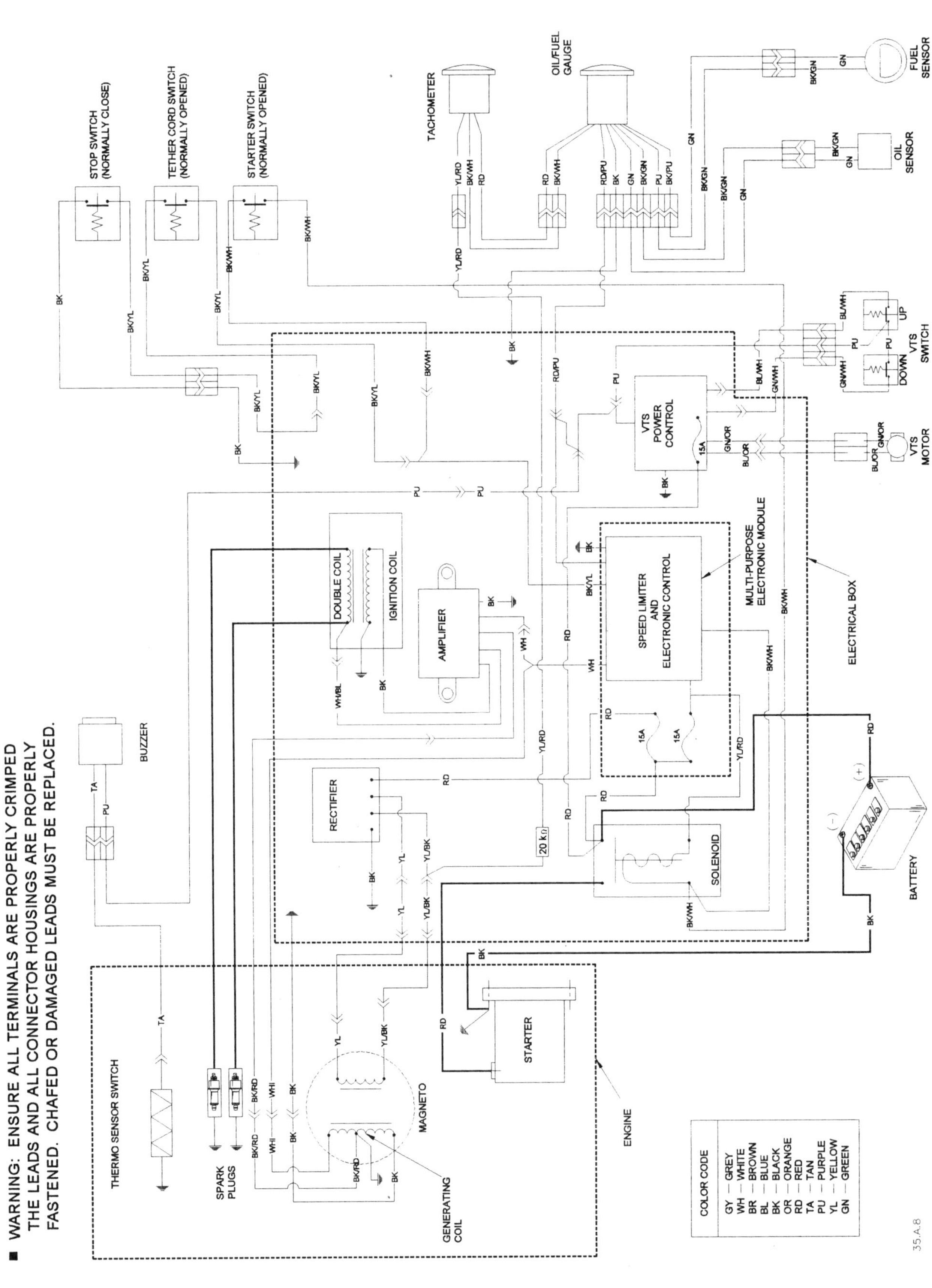

__1994 XP Model__ -- Electrical wiring diagram and color code identification. Major components of the electrical and ignition systems are identified.

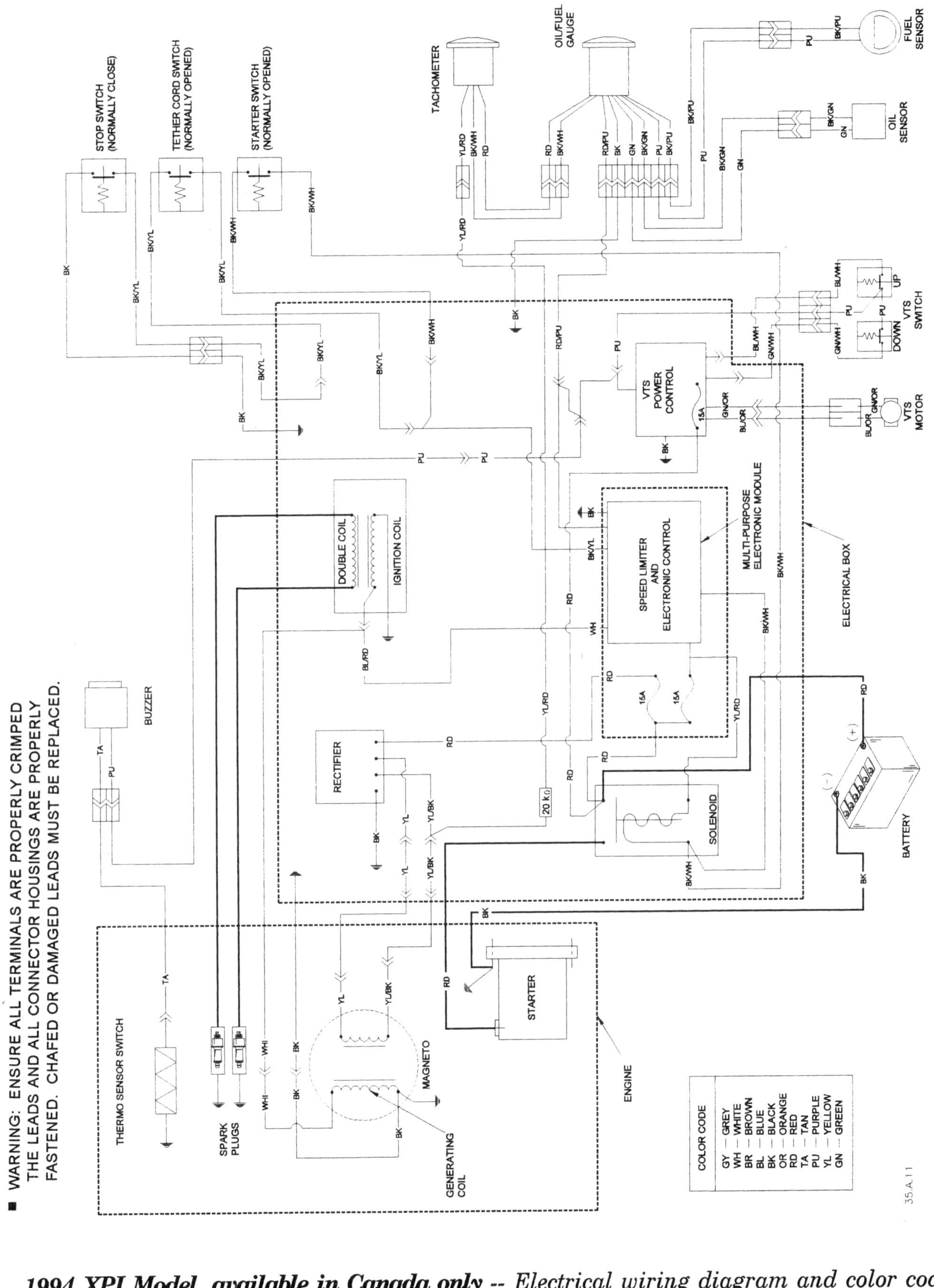

1994 XPI Model, available in Canada only -- *Electrical wiring diagram and color code identification. Major components of the electrical and ignition systems are identified.*

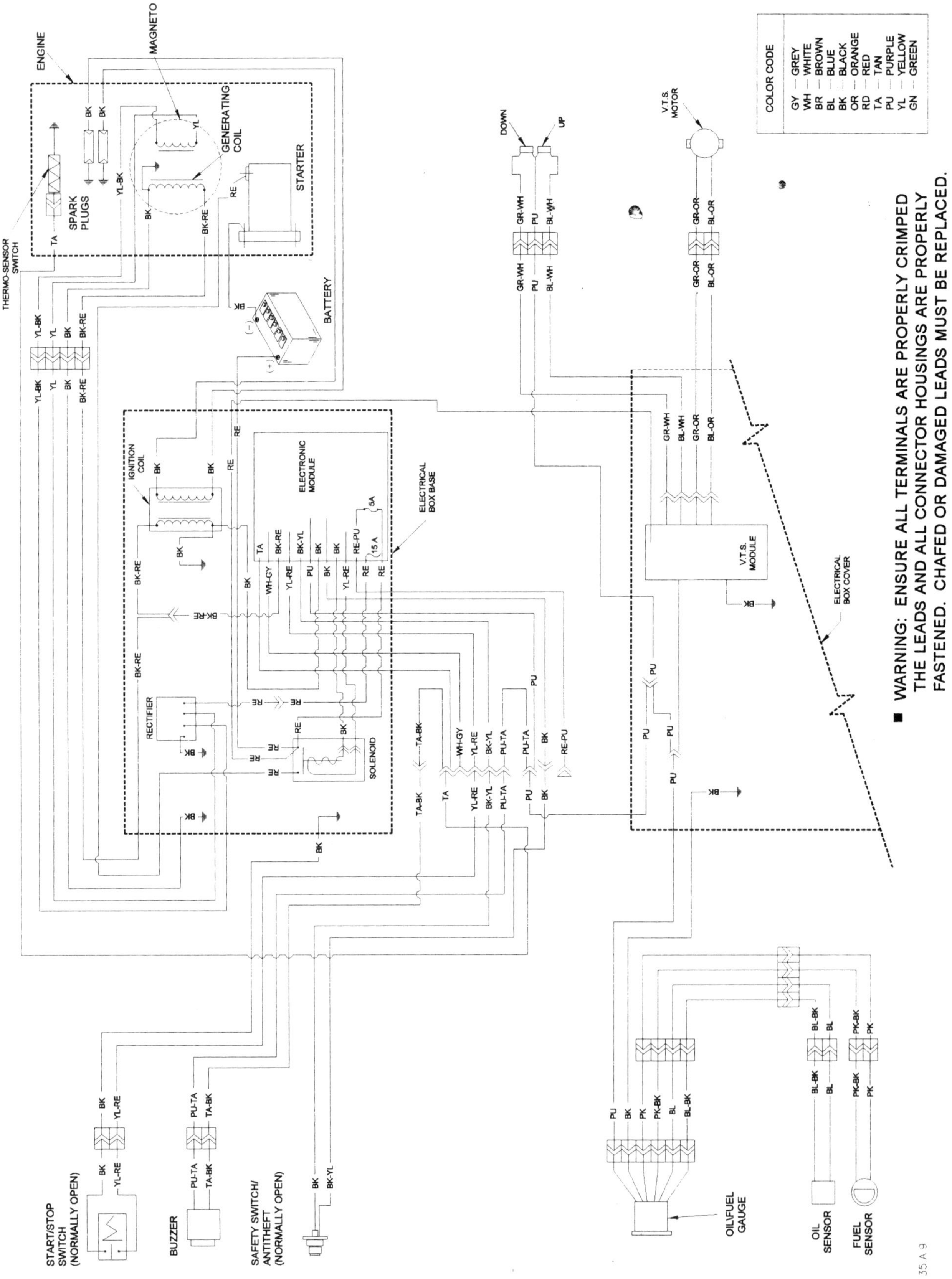

***1995 SPX Model** -- Electrical wiring diagram and color code identification. Major components of the electrical and ignition systems are identified.*

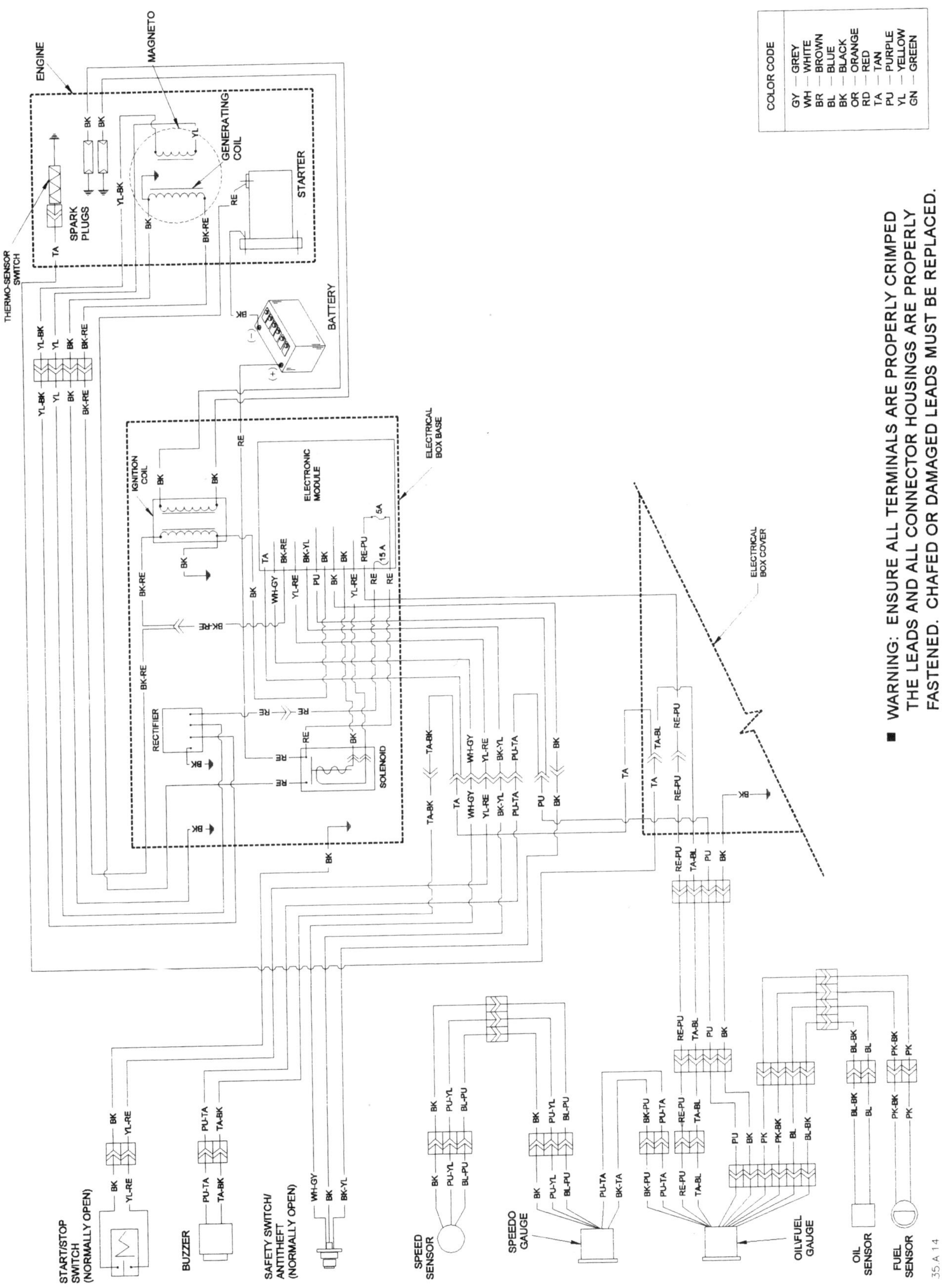

1995 GTX Model *-- Electrical wiring diagram and color code identification. Major components of the electrical and ignition systems are identified.*

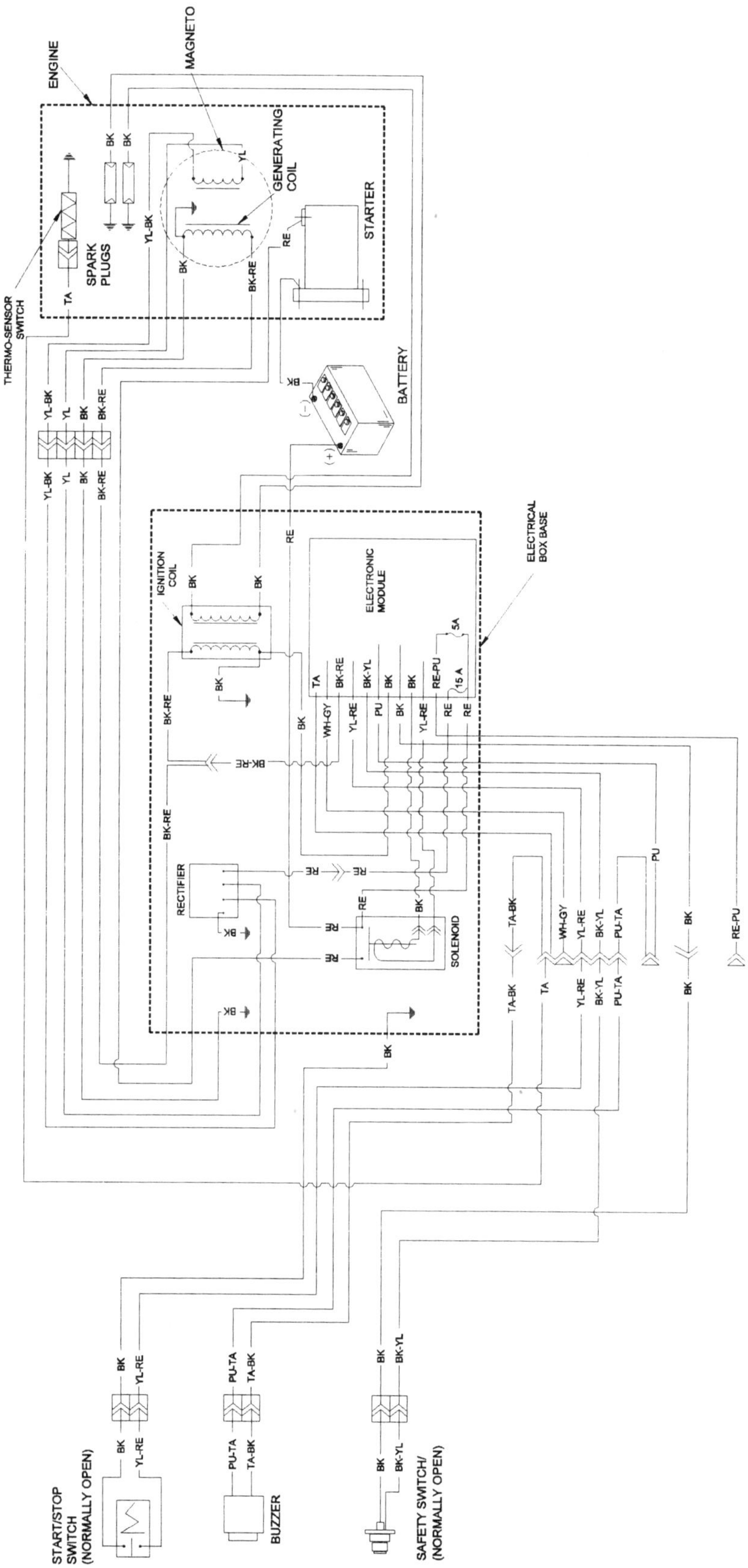

COLOR CODE
GY — GREY
WH — WHITE
BR — BROWN
BL — BLUE
BK — BLACK
OR — ORANGE
RD — RED
TA — TAN
PU — PURPLE
YL — YELLOW
GN — GREEN

■ WARNING: ENSURE ALL TERMINALS ARE PROPERLY CRIMPED THE LEADS AND ALL CONNECTOR HOUSINGS ARE PROPERLY FASTENED. CHAFED OR DAMAGED LEADS MUST BE REPLACED.

***1995 and 1996 SX and HX Models** -- Electrical wiring diagram and color code identification. Major components of the electrical and ignition systems are identified.*

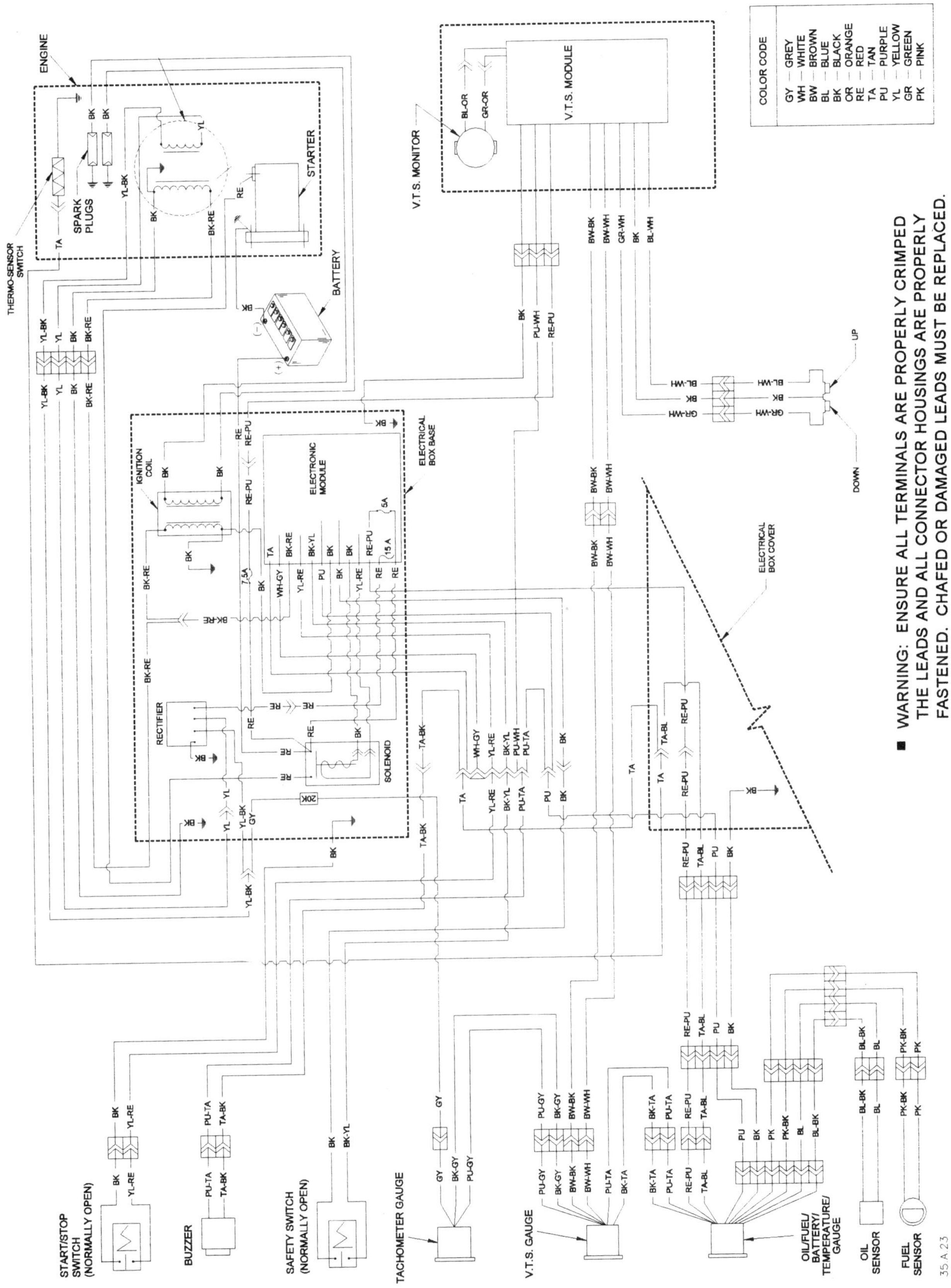

***1995 and 1996 SPX and XP Models** -- Electrical wiring diagram and color code identification. Major components of the electrical and ignition systems are identified.*

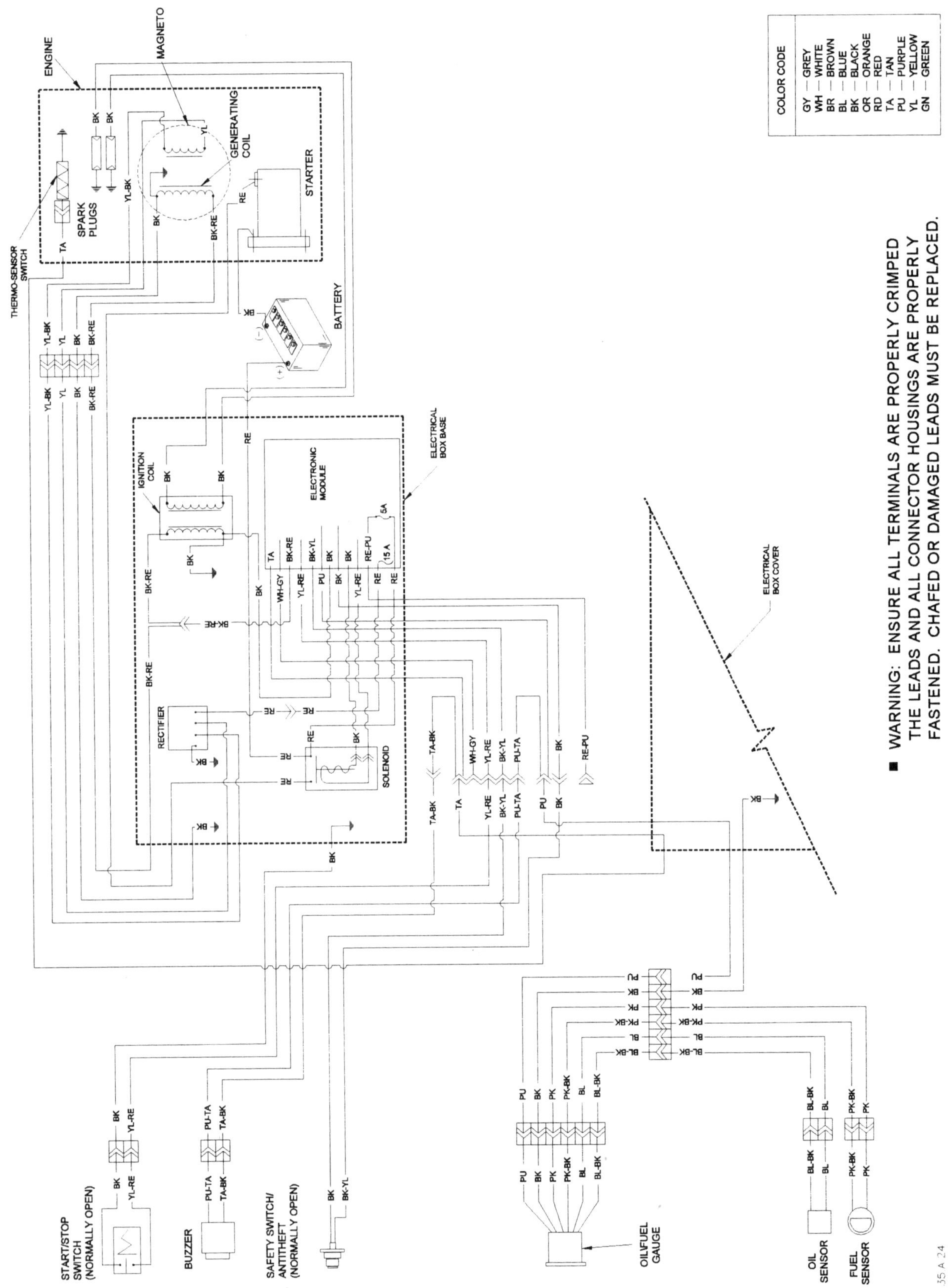

***1995 and 1996 SPI and GTS Models, 1997 GTS and HX Models** -- Electrical wiring diagram and color code identification. Major components of the electrical and ignition systems are identified.*

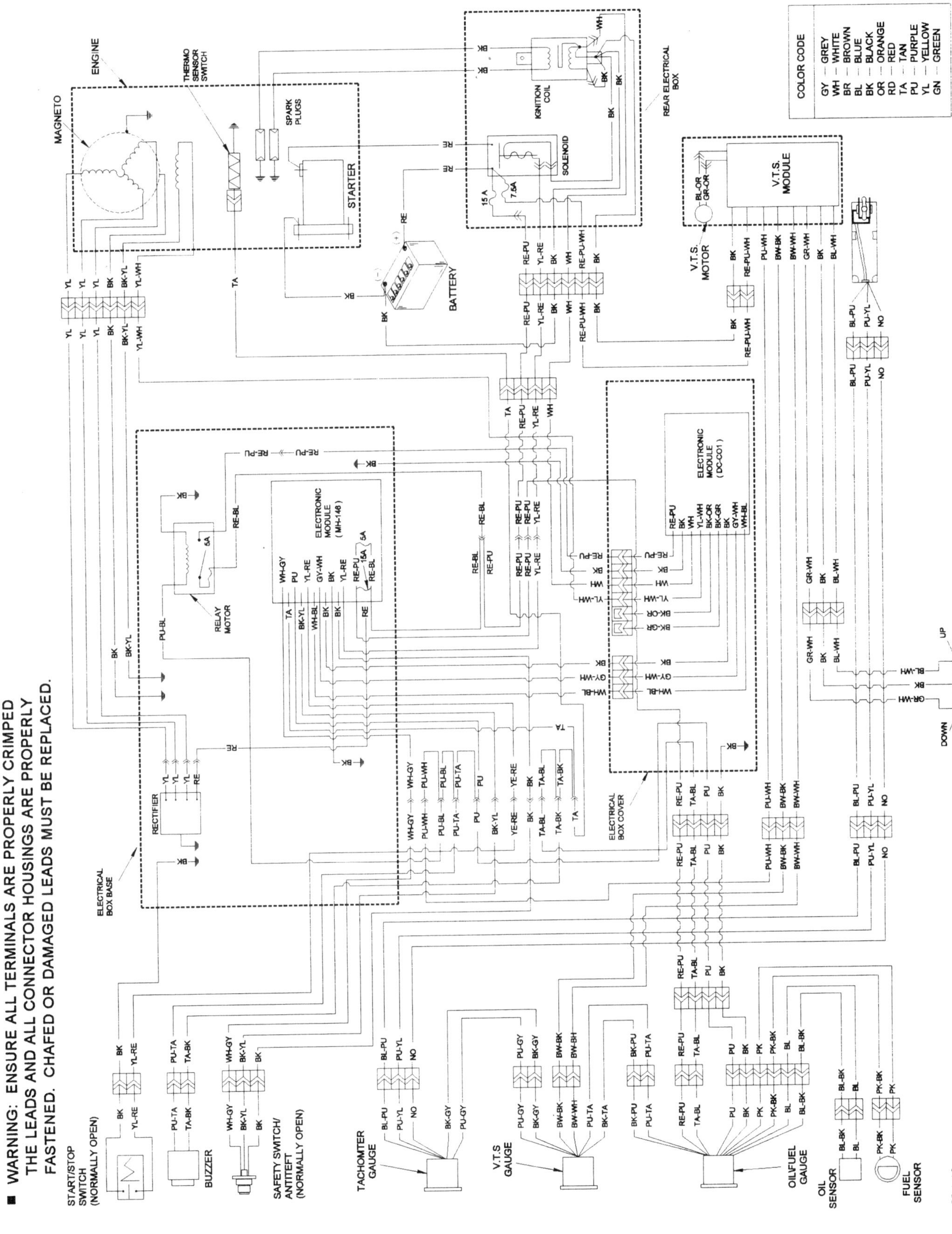

***1995 XP Model** -- Electrical wiring diagram and color code identification. Major components of the electrical and ignition systems are identified.*

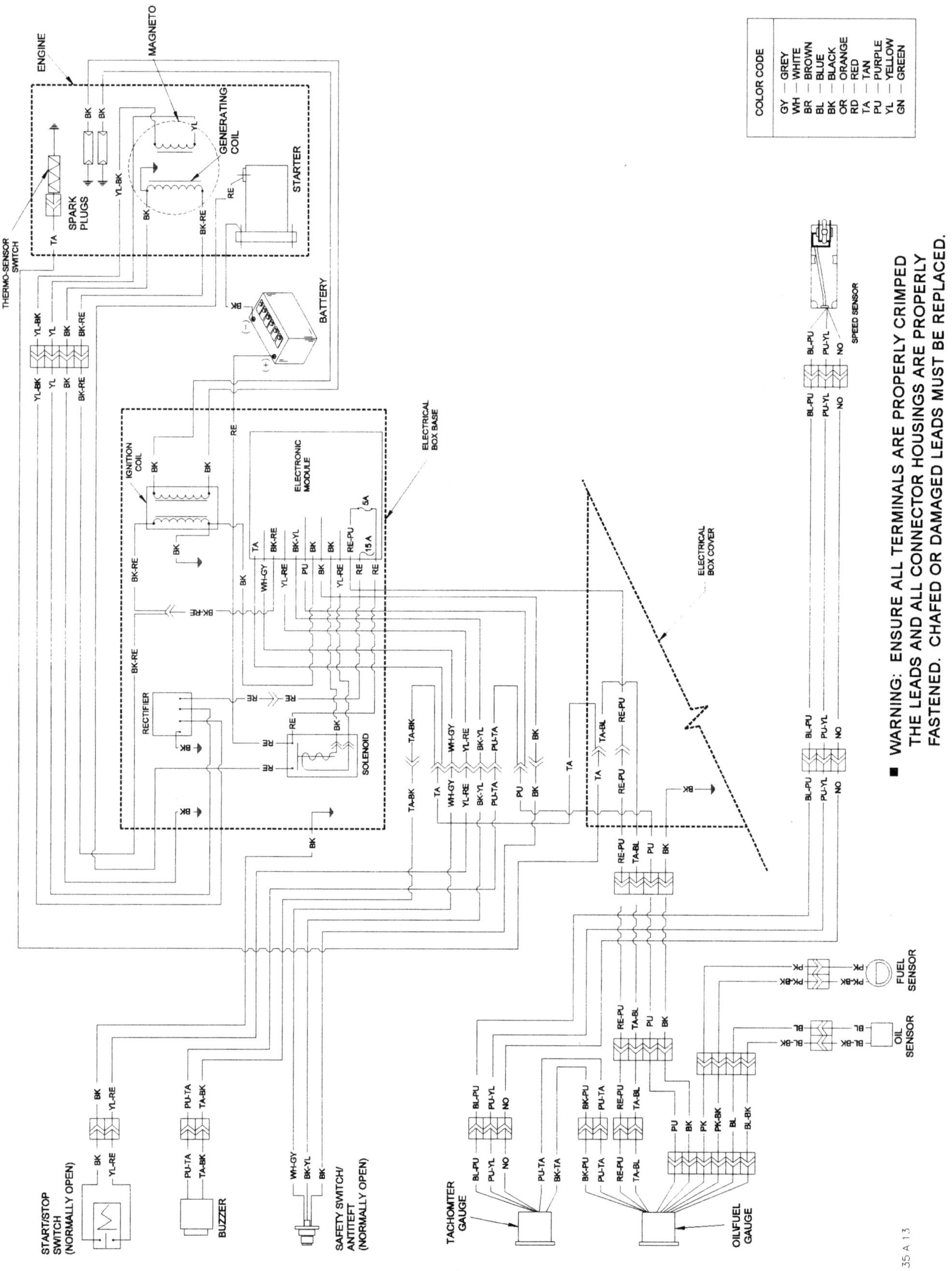

__1996 GTI Model__ -- Electrical wiring diagram and color code identification. Major components of the electrical and ignition systems are identified.

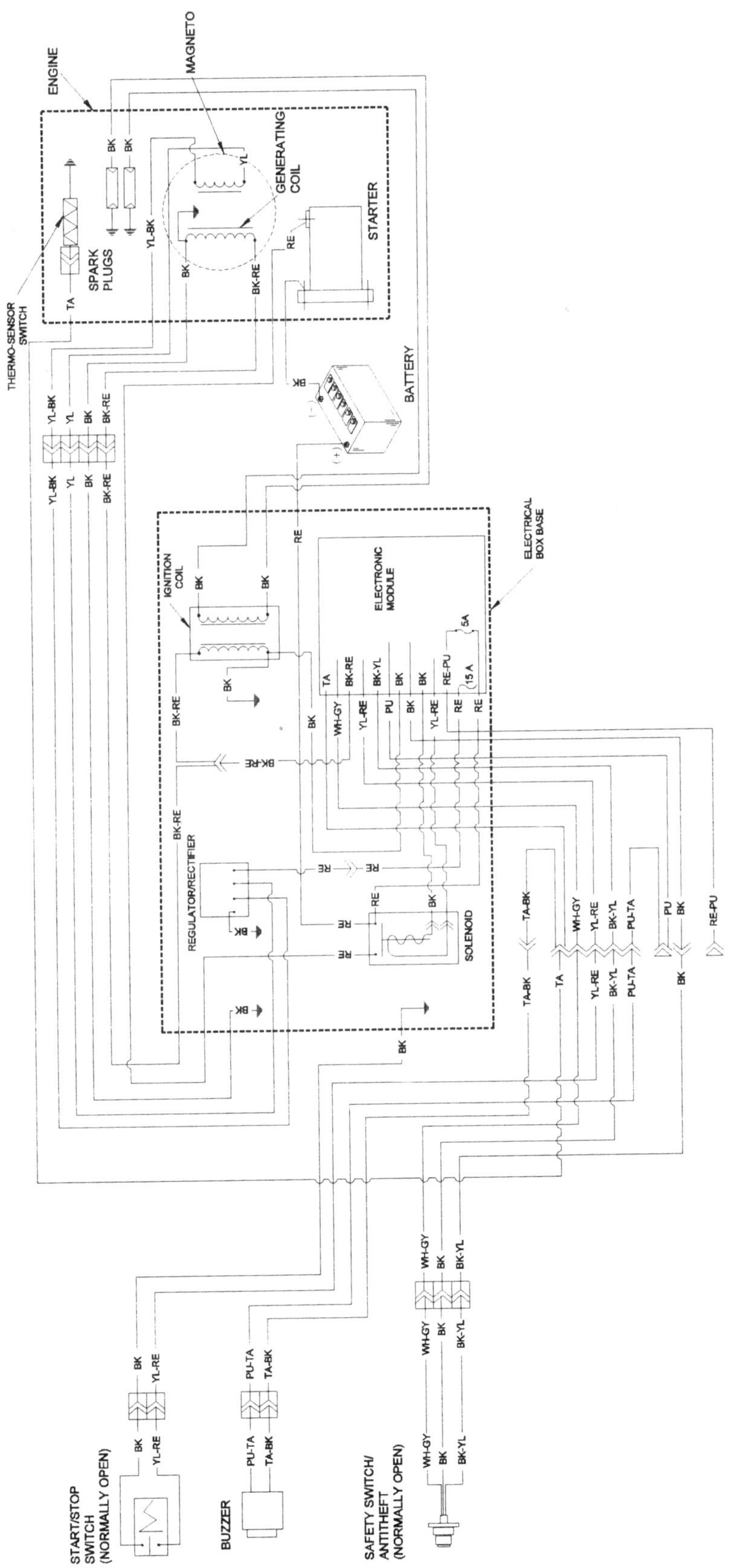

COLOR CODE

GY — GREY
WH — WHITE
BR — BROWN
BL — BLUE
BK — BLACK
OR — ORANGE
RD — RED
TA — TAN
PU — PURPLE
YL — YELLOW
GN — GREEN

■ WARNING: ENSURE ALL TERMINALS ARE PROPERLY CRIMPED THE LEADS AND ALL CONNECTOR HOUSINGS ARE PROPERLY FASTENED. CHAFED OR DAMAGED LEADS MUST BE REPLACED.

***1997 SP Model** -- Electrical wiring diagram and color code identification. Major components of the electrical and ignition systems are identified.*

■ WARNING: ENSURE ALL TERMINALS ARE PROPERLY CRIMPED THE LEADS AND ALL CONNECTOR HOUSINGS ARE PROPERLY FASTENED. CHAFED OR DAMAGED LEADS MUST BE REPLACED.

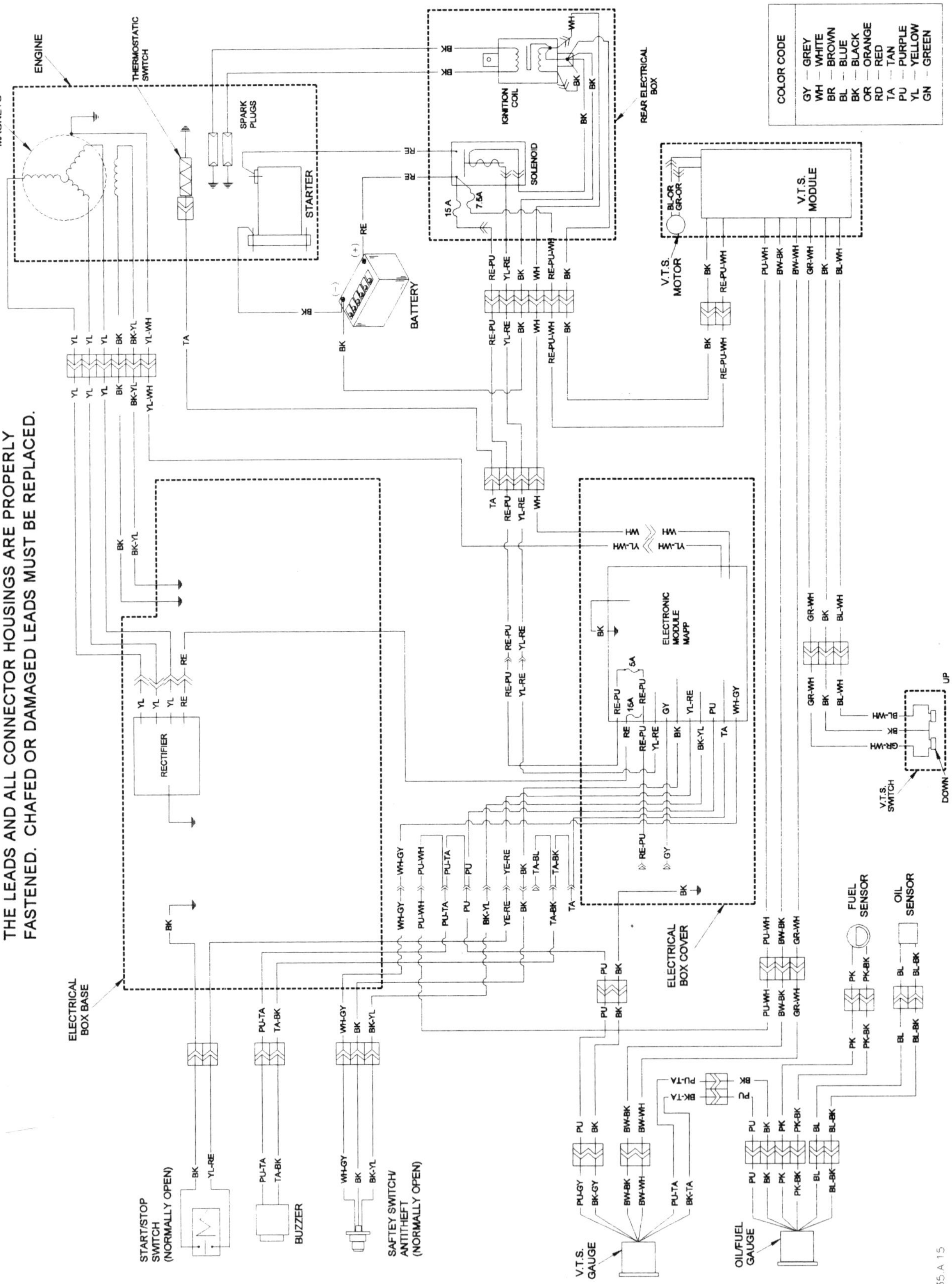

1997 SPX Model -- Electrical wiring diagram and color code identification. Major components of the electrical and ignition systems are identified.

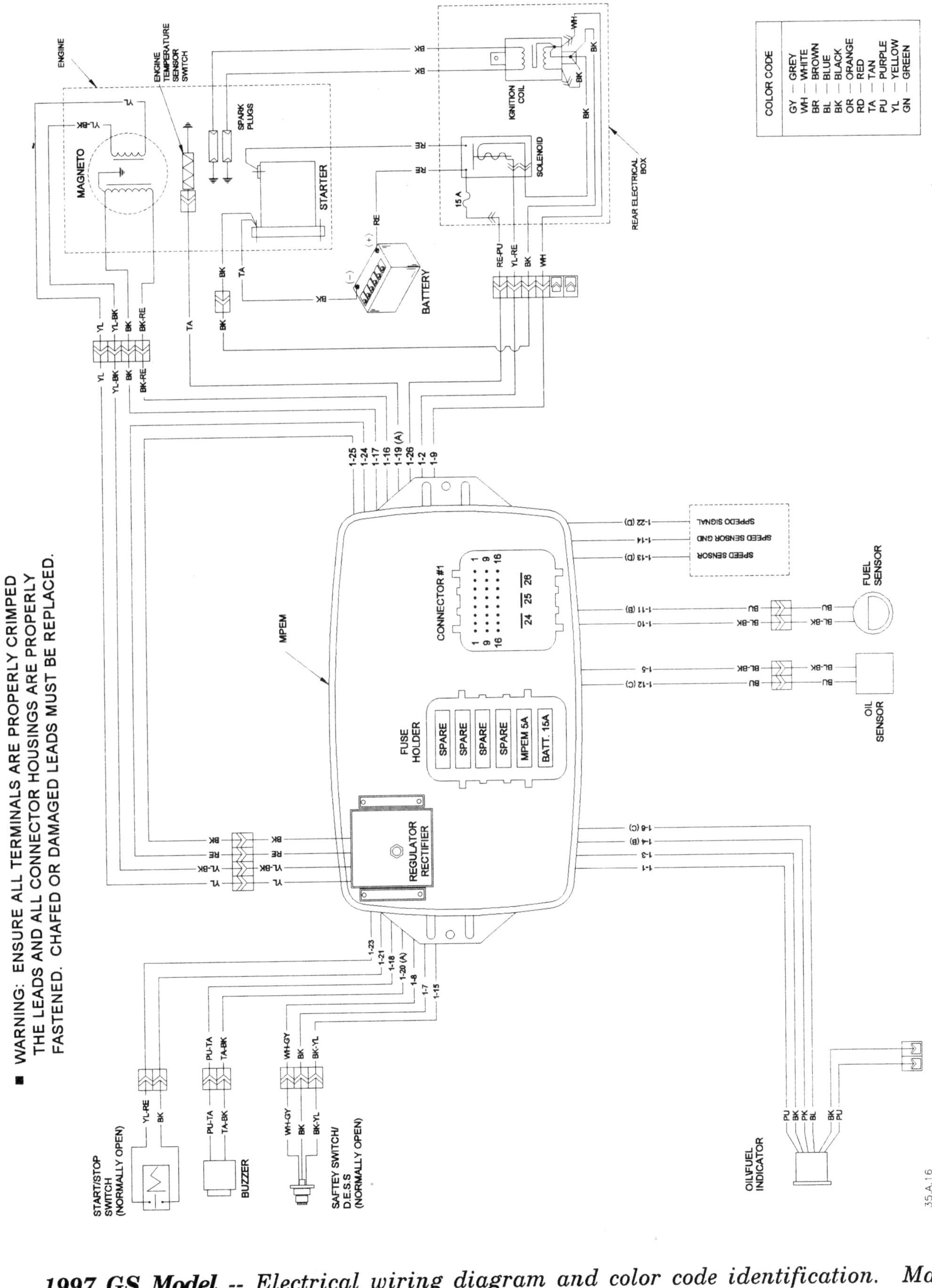

1997 GS Model -- Electrical wiring diagram and color code identification. Major components of the electrical and ignition systems are identified.

■ WARNING: ENSURE ALL TERMINALS ARE PROPERLY CRIMPED THE LEADS AND ALL CONNECTOR HOUSINGS ARE PROPERLY FASTENED. CHAFED OR DAMAGED LEADS MUST BE REPLACED.

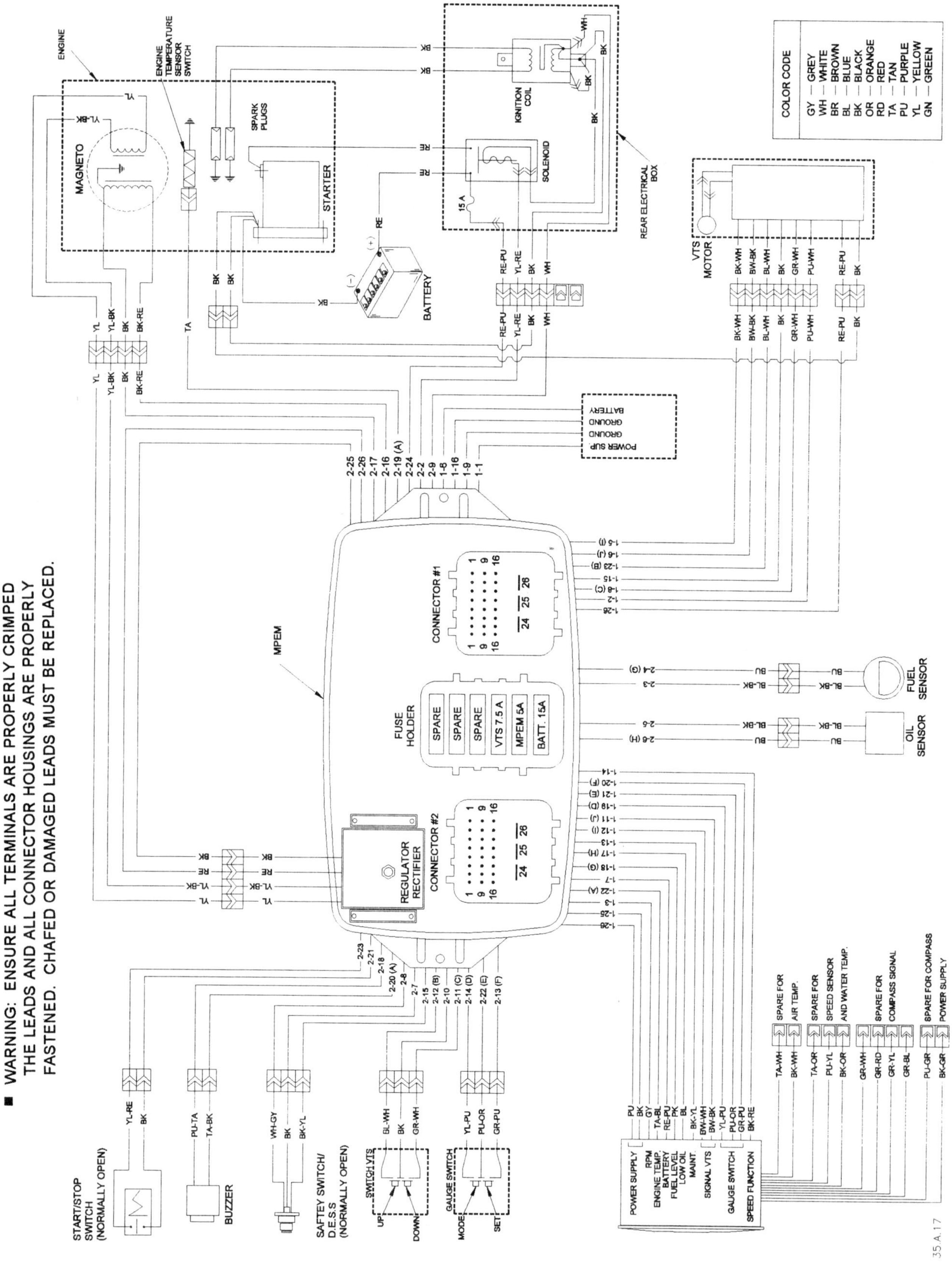

1997 GSI Model *-- Electrical wiring diagram and color code identification. Major components of the electrical and ignition systems are identified.*

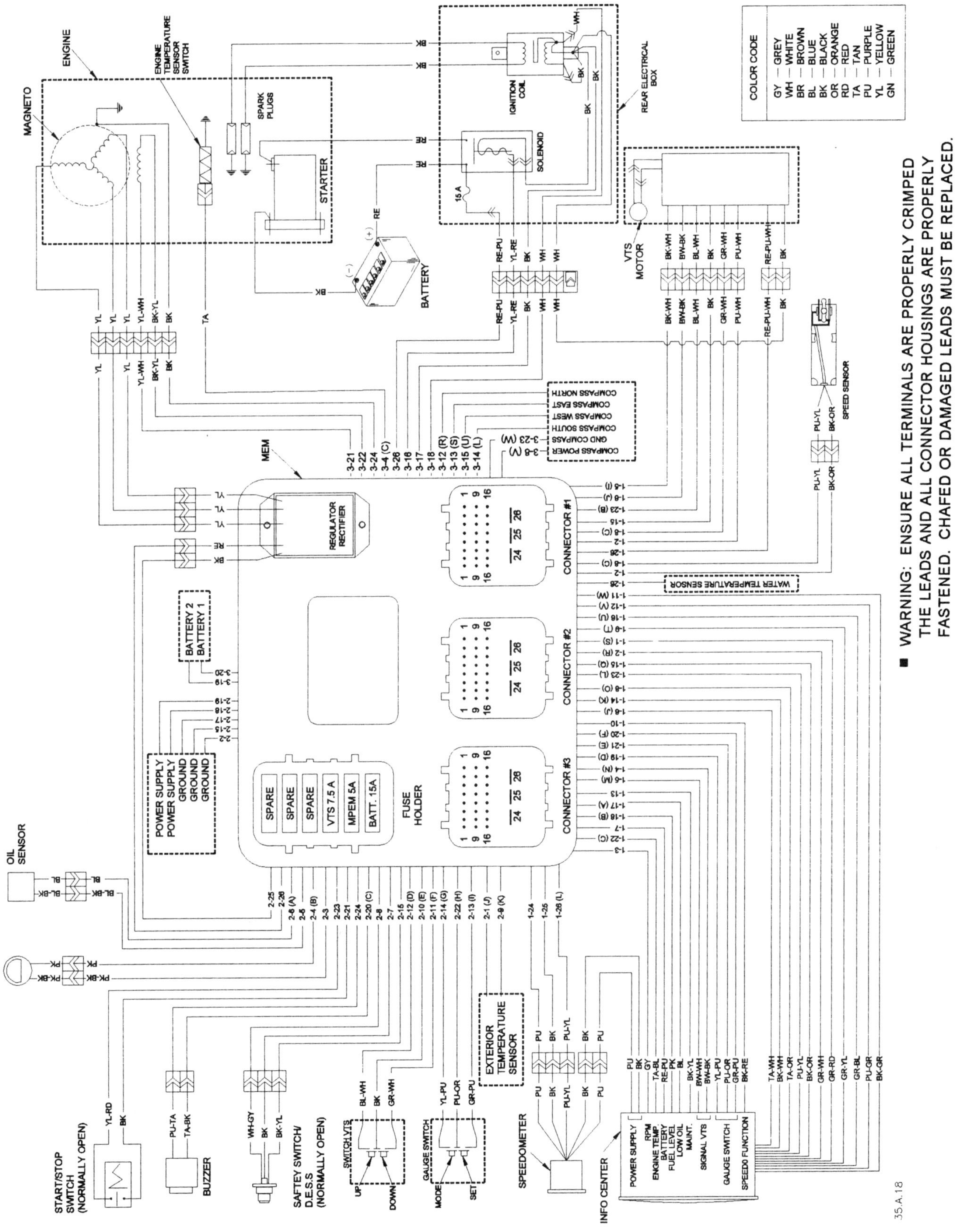

***1997 GSX Model** -- Electrical wiring diagram and color code identification. Major components of the electrical and ignition systems are identified.*

■ WARNING: ENSURE ALL TERMINALS ARE PROPERLY CRIMPED THE LEADS AND ALL CONNECTOR HOUSINGS ARE PROPERLY FASTENED. CHAFED OR DAMAGED LEADS MUST BE REPLACED.

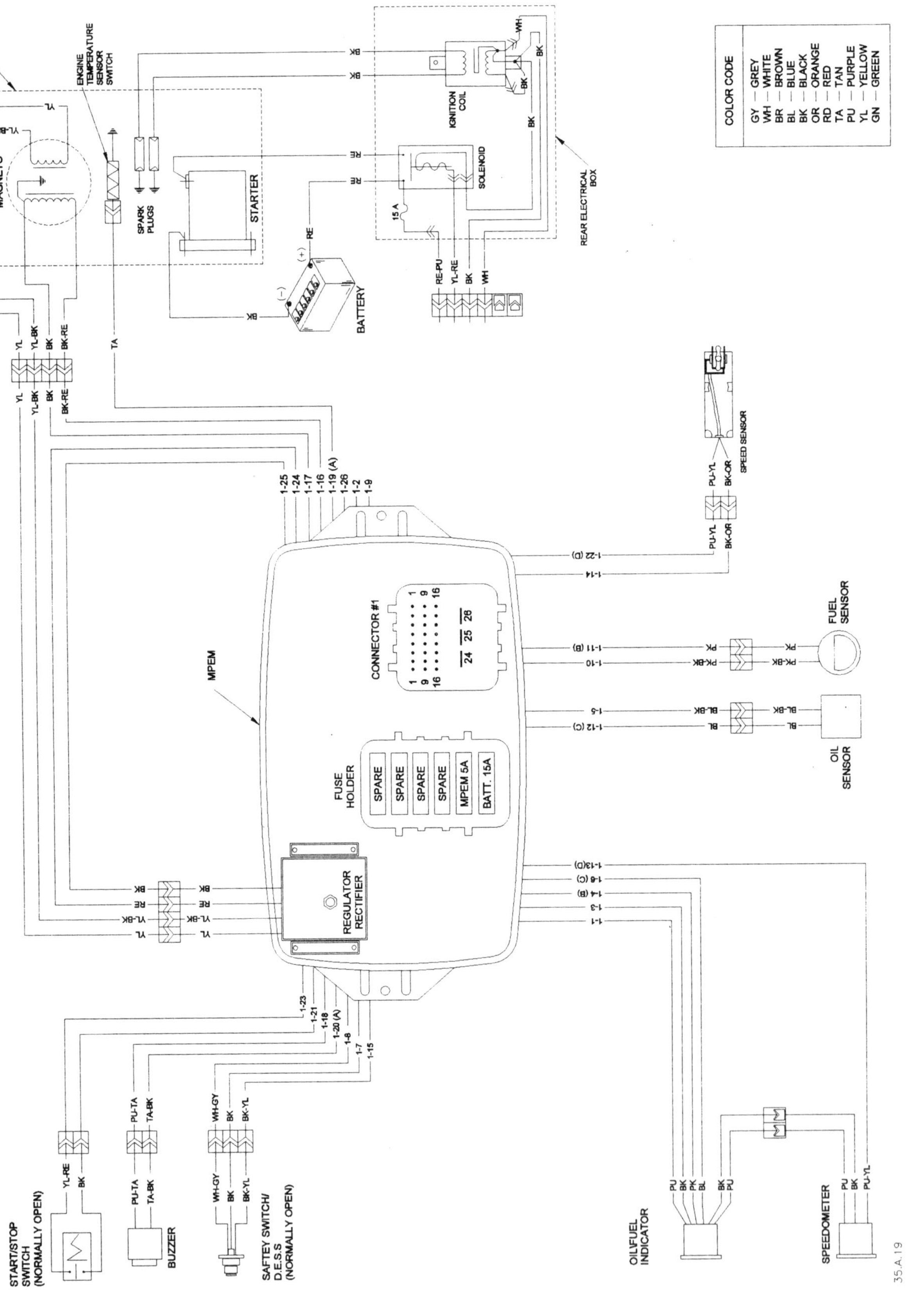

***1997 GTI Model** -- Electrical wiring diagram and color code identification. Major components of the electrical and ignition systems are identified.*

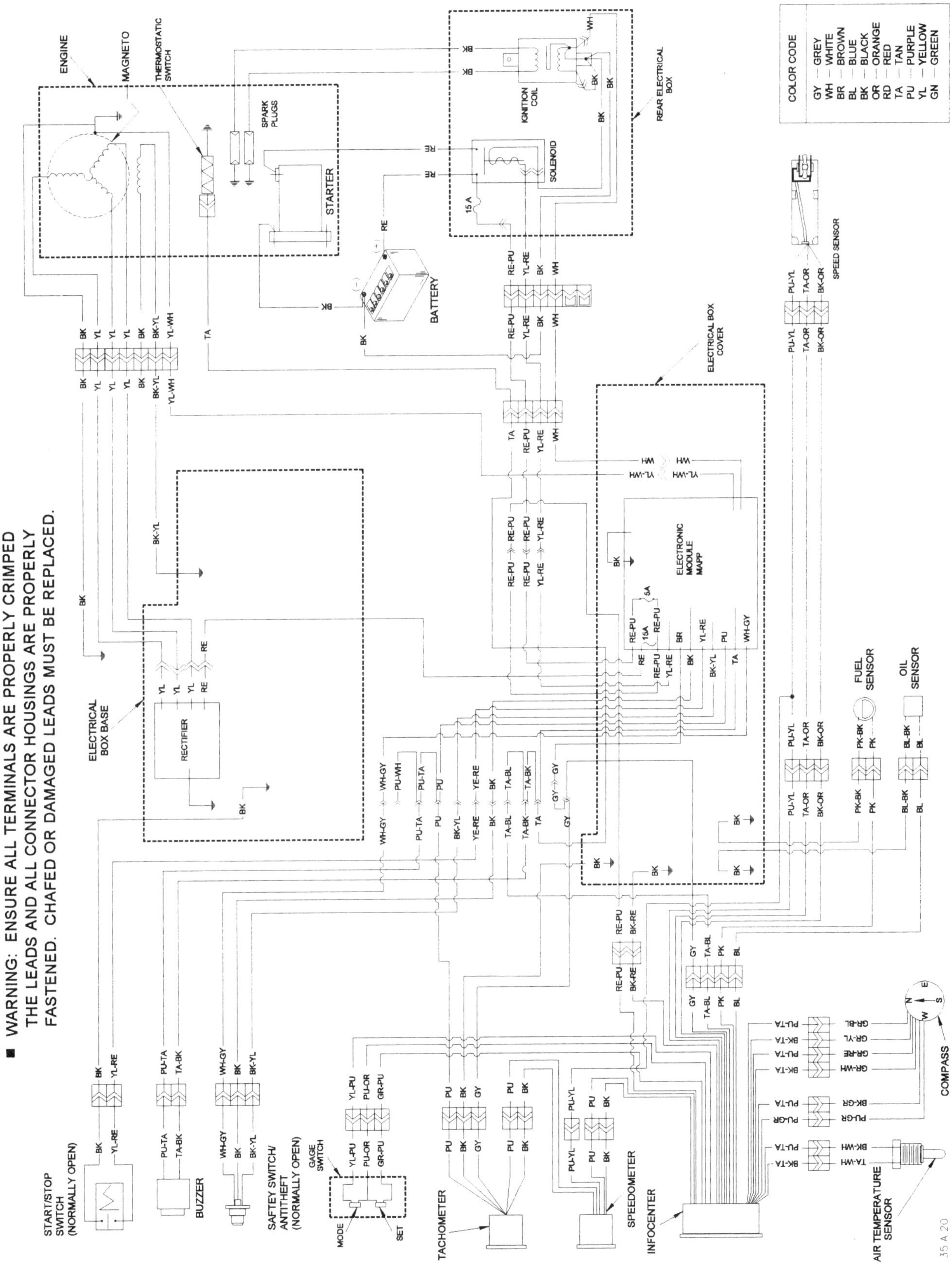

***1997 GTX Model** -- Electrical wiring diagram and color code identification. Major components of the electrical and ignition systems are identified.*

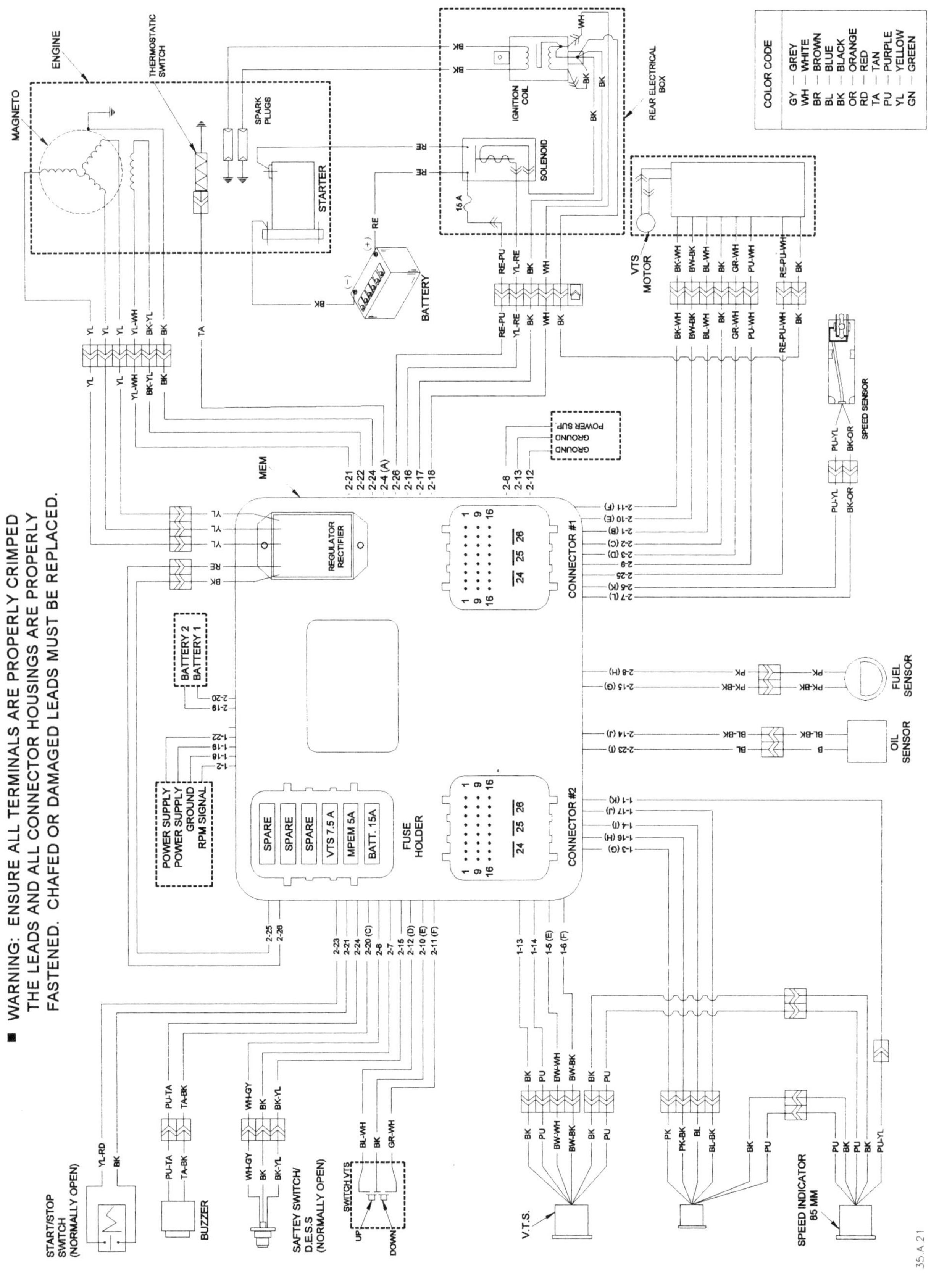

***1997 XP Model** -- Electrical wiring diagram and color code identification. Major components of the electrical and ignition systems are identified.*

DEGREE WHEEL FOR ROTARY VALVE TIMING

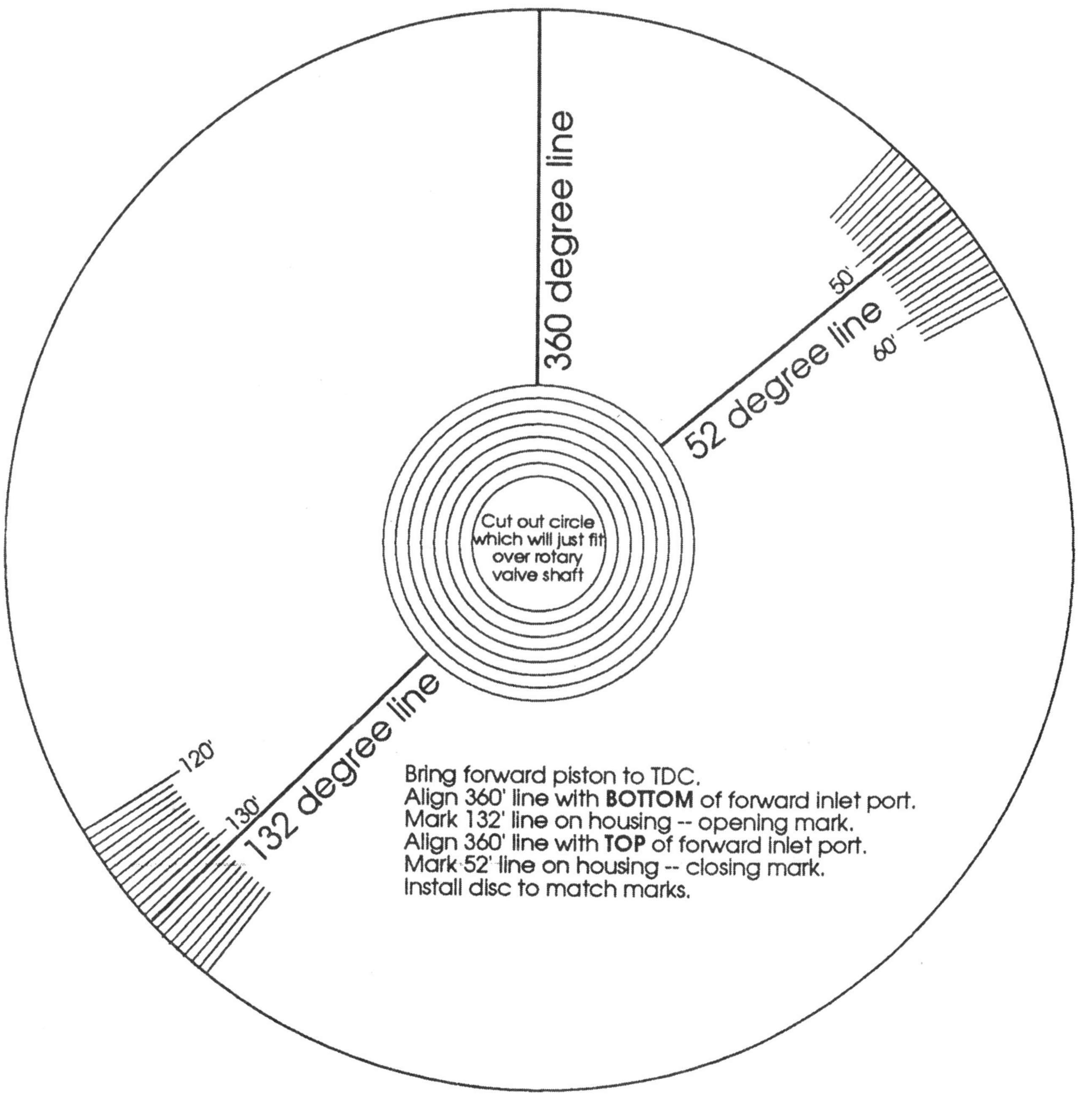

NOTES & NUMBERS

NOTES & NUMBERS

OMC COBRA STERN DRIVE
OMC STERN DR
MERCRUISER STERN DRIVE
Includes Alpha and Bravo
1984-1992
VOLVO PENTA STERN DRIVE
YAMAHA OUTBOARD
MARINER OUTBOARD
JOHNSON/EVINRUDE OUTBOARD
MARINE JET DRIVE
MERCURY OUTBOARD
PERSONAL WATERCRAFT
Yamaha Wetjet and Funjet
6-In-Line and V6
Joan and Clarence Coles

SELOC PUBLISHING'S FULL-LINE MASTER LIST

	ISBN	PART NO	TITLE/DESCRIPTION	YEARS
OUTBOARDS				
089330018-7	018-7	1000	Chrysler Outboards, All Engines	1962-84
089330055-1	055-1	1100	Force Outboards, All Engines	1984-99
089330048-9	048-9	1200	Honda Outboards, All Engines	1978-99
089330007-1	007-1	1300	Johnson/Evinrude Outboards, 1-2 Cyl	1956-70
089330008-X	008-X	1302	Johnson/Evinrude Outboards, 1-2 Cyl	1971-89
089330026-8	026-8	1304	Johnson/Evinrude Outboards, 1-2 Cyl	1990-95
089330009-8	009-8	1306	Johnson/Evinrude Outboards, 3-4 Cyl	1958-72
089330010-1	010-1	1308	Johnson/Evinrude Outboards, 3, 4 & 6 Cyl	1973-91
089330040-3	040-3	1310	Johnson/Evinrude Outboards - All V Engines	1992-96
089330063-2		1310	Johnson/Evinrude Outboards - All V Engines	1992-01
089330052-7		1312	Johnson/Evinrude Outboards, All In-line engines/2 & 4 Stroke	1996-01
089330015-2	015-2	1400	Mariner Outboards, 1-2 Cyl	1977-89
089330016-0	016-0	1402	Mariner Outboards, 3, 4 & 6 Cyl	1977-89
089330012-8	012-8	1404	Mercury Outboards, 1-2 Cyl	1965-91
089330013-6	013-6	1406	Mercury Outboards, 3-4 Cyl	1965-89
089330014-4	014-4	1408	Mercury Outboards, 6 Cyl	1965-89
089330051-9	051-9	1416	Mercury/Mariner Outboards, All Engines	1990-00
089330050-0	050-0	1600	Suzuki Outboards, All Engines	1988-99
089330021-7	021-7	1700	Yamaha Outboards, 1-2 Cyl	1984-91
089330022-5	022-5	1702	Yamaha Outboards, 3 Cyl	1984-91
089330023-3	023-3	1704	Yamaha Outboards, 4 & 6 Cyl	1984-91
089330047-0	047-0	1706	Yamaha Outboards, All Engines	1992-98
STERN DRIVES				
089330029-2	029-2	3000	Marine Jet Drive	1961-96
089330005-5	005-5	3200	Mercruiser Stern Drive	1964-91
089330053-5		3206	Mercruiser Stern Drive - All	1992-01
089330004-7	004-8	3400	OMC Stern Drive	1964-86
089330025-X	025-X	3402	OMC Cobra Stern Drive	1985-95
089330056-X	025-X	3402	OMC Cobra Stern Drive	1985-99
089330011-X	011-X	3600	Volvo/Penta Stern Drives	1968-91
089330038-1	038-1	3602	Volvo/Penta Stern Drives	1992-93
089330041-1	041-1	3604	Volvo/Penta Stern Drives	1992-95
089330057-8	041-1	3604	Volvo/Penta Stern Drives	1992-02
INBOARDS				
089330049-7	049-7	7400	Yanmar Inboards	1975-98
PERSONAL WATERCRAFT				
089330032-2	032-2	9200	Kawasaki	1973-91
089330042-X	042-X	9202	Kawasaki	1992-97
089330045-4	045-4	9400	Polaris	1992-97
089330033-0	033-0	9000	Sea-Doo/Bombardier	1988-91
089330043-8	043-8	9002	Sea-Doo/Bombardier	1992-97
089330034-9	034-9	9600	Yamaha	1987-91
089330044-6	044-6	9602	Yamaha	1992-97
Seloc-On-Line (Internet Access)				
089330075-6	5000		One Mfg/Model - Subscription per user 3 years	1990-01
	5002		Master Pack -Case of 6 CD's including POP Display	
PROFESSIONAL TECHNICIANS MANUALS				
089330060-8		4500	Labor Guide - Johnson and Evinrude	1980-00
089330061-6		4550	Labor Guide - Yamaha	1980-00
089330062-4		4600	Labor Guide - Mercury	1980-00
BRIGGS AND STRATTON				
096376610-4	610-4	610	Briggs & Stratton 4 Stroke Horizontal Crankshaft	1950+
096376611-2	611-2	611	Briggs & Stratton 4 Stroke Vertical Crankshaft	1950+
096376612-0	612-0	612	Briggs & Stratton 4 Stroke Overhead Crankshaft	1950+

ENGINE FINDER

The following listings contain all engines covered in this manual

Engine	Model	Year
581cc, 2cyl	GTS	1992-96
581cc, 2cyl	GTS Touring	1995
581cc, 2cyl	GTX	1992-93
581cc, 2cyl	SP, SPI	1992-96
581cc, 2cyl	XP	1992
581cc, 2cyl	SPX	1993
581cc, 2cyl	SLTX	1996-97
650cc, 2cyl	GTX	1994-95
650cc, 2cyl	GTX Touring	1995
650cc, 2cyl	SPX	1994-95
650cc, 2cyl	XP	1993-94
718cc, 2cyl	GS, GSI, SP	1997
718cc, 2cyl	GSX	1996-97
718cc, 2cyl	GTI, Sportster	1996-97
718cc, 2cyl	GTS	1997
718cc, 2cyl	HX	1995-97
718cc, 2cyl	SPX	1996
718cc, 2cyl	XP	1995
782cc, 2cyl	GTX, XP	1996-97
782cc, 2cyl	SPX	1997

SelocOnLine

SelocOnLine is a maintenance and repair database accessed via the Internet.
Always up-to-date, skill level and special tool icons, quick access buttons to wiring diagrams, specification charts, maintenance charts, and a parts database.
Contact your local marine dealer or see our demo at www.seloconline.com

SelocOnLine

Step 1

Select your manufacturer
Year / Model
Engine

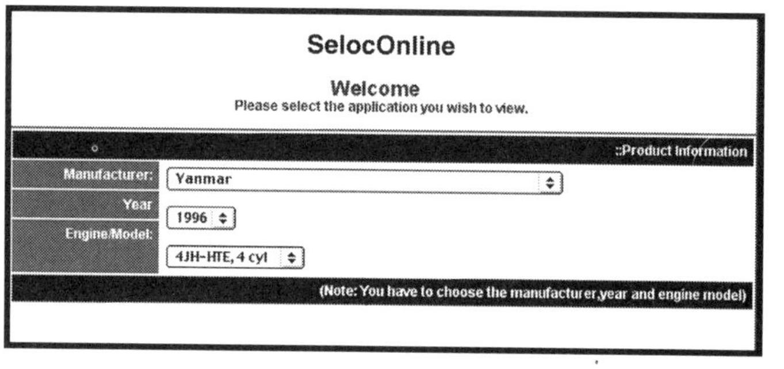

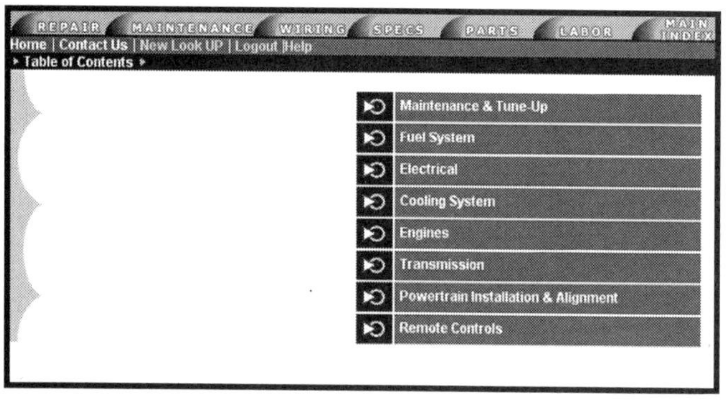

Step 2

Select an Engine System

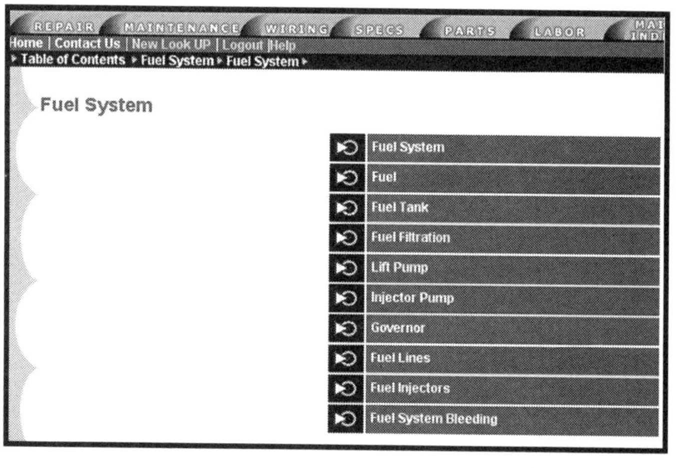

Step 3

Select the repair

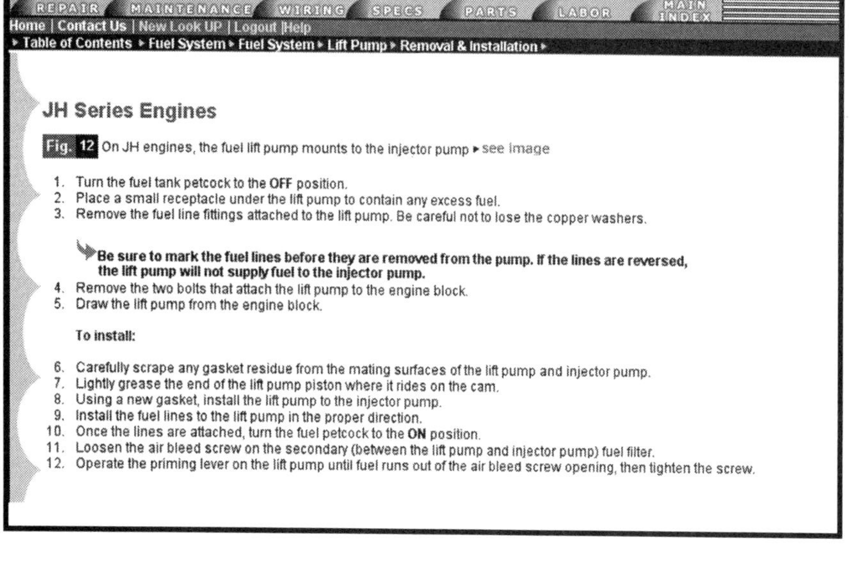

Step 4

Print the repair procedure